Lecture Notes in Computer Science   9050

Commenced Publication in 1973
Founding and Former Series Editors:
Gerhard Goos, Juris Hartmanis, and Jan van Leeuwen

More information about this series at http://www.springer.com/series/7409

Matthias Renz · Cyrus Shahabi
Xiaofang Zhou · Muhammad Aamir Cheema (Eds.)

Database Systems for Advanced Applications

20th International Conference, DASFAA 2015
Hanoi, Vietnam, April 20–23, 2015
Proceedings, Part II

 Springer

Editors

Matthias Renz
Universität München
München
Germany

Cyrus Shahabi
University of Southern California
Los Angeles
USA

Xiaofang Zhou
University of Queensland
Brisbane
Australia

Muhammad Aamir Cheema
Monash University
Clayton
Australia

ISSN 0302-9743 ISSN 1611-3349 (electronic)
Lecture Notes in Computer Science
ISBN 978-3-319-18122-6 ISBN 978-3-319-18123-3 (eBook)
DOI 10.1007/978-3-319-18123-3

Library of Congress Control Number: 2015936691

LNCS Sublibrary: SL3 – Information Systems and Applications, incl. Internet/Web, and HCI

Springer Cham Heidelberg New York Dordrecht London

Springer International Publishing AG Switzerland is part of Springer Science+Business Media
(www.springer.com)

Preface

It is our great pleasure to welcome you to DASFAA 2015, the 20th edition of the International Conference on Database Systems for Advanced Applications (DASFAA 2015), which was held in Hanoi, Vietnam during April 20–23, 2015. Hanoi (Vietnamese: *Hà Nội*), the capital of Vietnam, is the second largest city in Vietnam and has collected all the essence, unique features, and diversification of Vietnamese culture. The city is preserving more than 4000 historical and cultural relics, architecture and beauty spots, in which nearly 900 relics have been nationally ranked with hundreds of pagodas, temples, architectural works, and sceneries. Handcraft streets and traditional handcraft villages remain prominent and attractive to tourists when visiting Hanoi, many of which centered around the Hoan Kiem Lake in the Old Quarter, close to the conference venue. Hanoi has recently been included on TripAdvisor's list of best destinations in the world, ranked 8th among the world's top 25 destinations.

We are delighted to offer an exciting technical program, including two keynote talks by Amr El Abbadi (University of California, Santa Barbara) and Luc Vincent (Google Inc.); one 10-year best paper award presentation; a panel session on "Big Data Search and Analysis;" a poster session with 18 papers; a demo session with 6 demonstrations; an industry session with 3 full paper presentations; 3 tutorial sessions; and of course a superb set of research papers. This year, we received 287 submissions, each of which went through a rigorous review process. That is, each paper was reviewed by at least three Program Committee members, followed by a discussion led by the meta-reviewers, and a final meta-review prepared for each paper. At the end, DASFA 2015 accepted 63 full papers (the acceptance ratio is 22%).

Two workshops were selected by the Workshop Co-chairs to be held in conjunction with DASFAA 2015. They are the Second International Workshop on Big Data Management and Service (BDMS 2015), and the Second International Workshop on Semantic Computing and Personalization (SeCoP 2015). The workshop papers are included in a separate volume of proceedings also published by Springer in its Lecture Notes in Computer Science series.

The conference received generous financial support from the Hanoi University of Science and Technology (HUST). We, the conference organizers, also received extensive help and logistic support from the DASFAA Steering Committee and the Conference Management Toolkit Support Team at Microsoft.

We are grateful to all conference organizers, Han Su (University of Queensland) and many other people, for their great effort in supporting conference organization. Special thanks also go to the DASFAA 2015 Local Organizing Committee: Tuyet-Trinh Vu, Hong-Phuong Nguyen, and Van Thu Truong, all from the Hanoi University of Science and Technology, Vietnam. Finally, we would like to take this opportunity to thank all the meta-reviewers, Program Committee members, and external reviewers for their expertise and help in evaluating papers, and all the authors who submitted their papers to this conference.

February 2015

Quyet-Thang Huynh
Qing Li
Matthias Renz
Cyrus Shahabi
Xiaofang Zhou

Organization

General Co-chairs

Qing Li   City University of Hong Kong, HKSAR,
                                         Hong Kong
Quyet-Thang Huynh   Hanoi University of Science and Technology,
                                           Vietnam

Program Committee Co-chairs

Cyrus Shahabi   University of Southern California, USA
Matthias Renz   Ludwig-Maximilians-Universität München,
                                         Germany
Xiaofang Zhou   University of Queensland, Australia

Tutorial Co-chairs

Arbee L.P. Chen   NCCU, Taiwan
Pierre Senellart   Télécom ParisTech, France

Workshops Co-chairs

An Liu   Soochow University, China
Yoshiharu Ishikawa   Nagoya University, Japan

Demo Co-chairs

Haiwei Pan   Harbin Engineering University, China
Binh Minh Nguyen   Hanoi University of Science and Technology,
                                         Vietnam

Panel Co-chairs

Bin Cui   Peking University, China
Katsumi Tanaka   Kyoto University, Japan

Poster Co-chairs

Sarana Nutanong   City University of Hong Kong, China
Tieyun Qian   Wuhan University, China

Industrial/Practitioners Track Co-chairs

Mukesh Mohania	IBM, India
Khai Tran	Oracle, USA

PhD Colloquium

Khoat Than	Hanoi University of Science and Technology, Vietnam
Ge Yu	Northeastern University, China
Tok Wang Ling	National University of Singapore, Singapore
Duong Nguyen Vu	John Von Neumann Institute - VNU-HCMUS, Vietnam

Publication Chair

Muhammad Aamir Cheema	Monash University, Australia

Publicity Co-chairs

Yunjun Gao	Zhejiang University, China
Bao-Quoc Ho	VNU-HCMUS, Vietnam
Jianfeng Si	Institute for Infocomm Research, Singapore
Wen-Chih Peng	National Chiao Tung University, Taiwan

Local Organizing Committee

Tuyet-Trinh Vu	Hanoi University of Science and Technology, Vietnam
Hong-Phuong Nguyen	Hanoi University of Science and Technology, Vietnam
Van Thu Truong	Hanoi University of Science and Technology, Vietnam

Steering Committee Liaison

Stephane Bressan	National University of Singapore, Singapore

Webmaster

Viet-Trung Tran	Hanoi University of Science and Technology, Vietnam

Program Committees

Senior PC members

Ira Assent	Aarhus University, Denmark
Lei Chen	Hong Kong University of Science and Technology (HKUST), China
Reynold Cheng	University of Hong Kong, China
Gabriel Ghinita	University of Massachusetts Boston, USA
Panos Kalnis	King Abdullah University of Science and Technology, Saudi Arabia
Nikos Mamoulis	University of Hong Kong, China
Kyriakos Mouratidis	Singapore Management University, Singapore
Mario Nascimento	University of Alberta, Canada
Dimitris Papadias	Hong Kong University of Science and Technology (HKUST), China
Stavros Papadoupoulos	MIT, USA
Torben Bach Pedersen	Aalborg University, Denmark
Jian Pei	Simon Fraser University, Canada
Thomas Seidl	RWTH Aachen University, Germany
Timos Sellis	RMIT University, Australia
Raymond Wong	Hong Kong University of Science and Technology (HKUST), China

PC Members

Nikolaus Augsten	University of Salzburg, Austria
Spiridon Bakiras	City University of New York, USA
Zhifeng Bao	University of Tasmania, Australia
Srikanta Bedathur	IBM Research, Delhi, India
Ladjel Bellatreche	University of Poitiers, France
Boualem Benatallah	University of New South Wales, Australia
Bin Cui	Peking University, China
Athman Bouguettaya	Commonwealth Scientific and Industrial Research Organisation (CSIRO), Australia
Panagiotis Bouros	Humboldt-Universität zu Berlin, Germany
Selcuk Candan	Arizona State University, USA
Jianneng Cao	A*STAR, Singapore
Marco Casanova	Pontifical Catholic University of Rio de Janeiro, Brazil
Sharma Chakravarthy	University of Texas at Arlington, USA
Jae Chang	Chonbuk National University, Korea
Rui Chen	Hong Kong Baptist University, China
Yi Chen	New Jersey Institute of Technology, USA
James Cheng	The Chinese University of Hong Kong (CUHK), China
Gao Cong	Nanyang Technological University (NTU), Singapore

Ugur Demiryurek	University of Southern California (USC), USA
Prasad Deshpande	IBM Research, India
Gill Dobbie	University of Auckland, New Zealand
Eduard Dragut	Temple University, USA
Cristina Dutra de Aguiar Ciferri	Universidade de São Paulo, Brazil
Sameh Elnikety	Microsoft Research Redmond, USA
Tobias Emrich	Ludwig-Maximilians-Universität München, Germany
Johann Gamper	Free University of Bozen-Bolzano, Italy
Xin Gao	King Abdullah University of Science and Technology (KAUST), Saudi Arabia
Chenjuan Guo	Aarhus University, Denmark
Ralf Hartmut Güting	University of Hagen, Germany
Takahiro Hara	Osaka University, Japan
Haibo Hu	Hong Kong Baptist University, China
Yoshiharu Ishikawa	Nagoya University, Japan
Mizuho Iwaihara	Waseda University, Japan
Adam Jatowt	Kyoto University, Japan
Vana Kalogeraki	Athens University of Economy and Business, Greece
Panos Karras	Skoltech, Russia
Norio Katayama	National Institute of Informatics, Japan
Sang-Wook Kim	Hanyang University, Korea
Seon Ho Kim	University of Southern California (USC), USA
Hiroyuki Kitagawa	University of Tsukuba, Japan
Peer Kröger	Ludwig-Maximilians-Universität München, Germany
Jae-Gil Lee	Korea Advanced Institute of Science and Technology (KAIST), Korea
Wang-Chien Lee	Portland State University (PSU), USA
Sang-Goo Lee	Seoul National University, Korea
Hou Leong	University of Macau, China
Guoliang Li	Tsinghua University, China
Hui Li	Xidian University, China
Xiang Lian	University of Texas–Pan American (UTPA), USA
Lipyeow Lim	University of Hawaii, USA
Sebastian Link	University of Auckland, New Zealand
Bin Liu	NEC Laboratories, USA
Changbin Liu	AT & T, USA
Eric Lo	Hong Kong Polytechnic University, China
Jiaheng Lu	Renmin University of China, China
Qiong Luo	Hong Kong University of Science and Technology (HKUST), China
Matteo Magnani	Uppsala University, Sweden
Silviu Maniu	University of Hong Kong (HKU), China
Essam Mansour	Qatar Computing Research Institute, Qatar

Marco Mesiti	University of Milan, Italy
Yasuhiko Morimoto	Hiroshima University, Japan
Wilfred Ng	Hong Kong University of Science and Technology (HKUST), China
Makoto Onizuka	Osaka University, Japan
Balaji Palanisamy	University of Pittsburgh, USA
Stefano Paraboschi	Università degli Studi di Bergamo, Italy
Sanghyun Park	Yonsei University, Korea
Dhaval Patel	IIT Roorkee, India
Evaggelia Pitoura	University of Ioannina, Greece
Pascal Poncelet	Université Montpellier 2, France
Maya Ramanath	Indian Institute of Technology, New Delhi, India
Shazia Sadiq	University of Queensland, Australia
Sherif Sakr	University of New South Wales, Australia
Kai-Uwe Sattler	Ilmenau University of Technology, Germany
Peter Scheuermann	Northwestern University, USA
Markus Schneider	University of Florida, USA
Matthias Schubert	Ludwig-Maximilians-Universität München, Germany
Shuo Shang	China University of Petroleum, Beijing, China
Kyuseok Shim	Seoul National University, Korea
Junho Shim	Sookmyung Women's University, Korea
Shaoxu Song	Tsinghua University, China
Atsuhiro Takasu	National Institute of Informatics, Japan
Kian-Lee Tan	National University of Singapore (NUS), Singapore
Nan Tang	Qatar Computing Research Institute, Qatar
Martin Theobald	University of Antwerp, Belgium
Dimitri Theodoratos	New Jersey Institute of Technology, USA
James Thom	RMIT University, Australia
Wolf Tilo-Balke	University of Hannover, Germany
Hanghang Tong	City University of New York (CUNY), USA
Yongxin Tong	Hong Kong University of Science and Technology (HKUST), China
Kristian Torp	Aalborg University, Denmark
Goce Trajcevski	Northwestern University, USA
Vincent S. Tseng	National Cheng Kung University, Taiwan
Stratis Viglas	University of Edinburgh, UK
Wei Wang	University of New South Wales, Australia
Huayu Wu	Institute for Infocomm Research (I^2R), Singapore
Yinghui Wu	University of California, Santa Barbara (UCSB), USA
Xiaokui Xiao	Nanyang Technological University (NTU), Singapore
Xike Xie	Aalborg University, Denmark
Jianliang Xu	Hong Kong Baptist University, China
Bin Yang	Aalborg University, Denmark

Yin Yang	Advanced Digital Sciences Center, Singapore
Man-Lung Yiu	Hong Kong Polytechnic University, China
Haruo Yokota	Tokyo Institute of Technology, Japan
Jeffrey Yu	The Chinese University of Hong Kong (CUHK), China
Zhenjie Zhang	Advanced Digital Sciences Center (ADSC), Singapore
Xiuzhen Zhang	RMIT University, Australia
Kevin Zheng	University of Queensland, Australia
Wenchao Zhou	Georgetown University, USA
Bin Zhou	University of Maryland, Baltimore County, USA
Roger Zimmermann	National University of Singapore (NUS), Singapore
Lei Zou	Beijing University, China
Andreas Züfle	Ludwig-Maximilians-Universität München, Germany

External Reviewers

Yeonchan Ahn	Seoul National University, Korea
Cem Aksoy	New Jersey Institute of Technology, USA
Ibrahim Alabdulmohsin	King Abdullah University of Science and Technology, Saudi Arabia
Yoshitaka Arahori	Tokyo Institute of Technology, Japan
Zhuowei Bao	Facebook, USA
Thomas Behr	University of Hagen, Germany
Jianneng Cao	A*STAR, Singapore
Brice Chardin	LIAS/ISAE-ENSMA, France
Lei Chen	Hong Kong Baptist University, China
Jian Dai	The Chinese Academy of Sciences, China
Ananya Dass	New Jersey Institute of Technology, USA
Aggeliki Dimitriou	National Technical University of Athens, Greece
Zhaoan Dong	Renmin University of China, China
Hai Dong	RMIT University, Australia
Zoé Faget	LIAS/ISAE-ENSMA, France
Qiong Fang	Hong Kong University of Science and Technology (HKUST), China
ZiQiang Feng	Hong Kong Polytechnic University, China
Ming Gao	East China Normal University, China
Azadeh Ghari-Neiat	RMIT University, Australia
Zhian He	Hong Kong Polytechnic University, China
Yuzhen Huang	The Chinese University of Hong Kong, China
Stéphane Jean	LIAS/ISAE-ENSMA, France
Selma Khouri	LIAS/ISAE-ENSMA, France
Hanbit Lee	Seoul National University, Korea
Sang-Chul Lee	Carnegie Mellon University, USA

Feng Li	Microsoft Research, Redmond, USA
Yafei Li	Hong Kong Baptist University, China
Jinfeng Li	The Chinese University of Hong Kong, China
Xin Lin	East China Normal University, China
Yu Liu	Renmin University of China, China
Yi Lu	The Chinese University of Hong Kong, China
Nguyen Minh Luan	A*STAR, Singapore
Gerasimos Marketos	University of Piraeus, Greece
Jun Miyazaki	Tokyo Institute of Technology, Japan
Bin Mu	City University of New York, USA
Johannes Niedermayer	Ludwig-Maximilians-Universität München, Germany
Konstantinos Nikolopoulos	City University of New York, USA
Sungchan Park	Seoul National University, Korea
Youngki Park	Samsung Advanced Institute of Technology, Korea
Jianbin Qin	University of New South Wales, Australia
Kai Qin	RMIT University, Australia
Youhyun Shin	Seoul National University, Korea
Hiroaki Shiokawa	NTT Software Innovation Center, Japan
Masumi Shirakawa	Osaka University, Japan
Md. Anisuzzaman Siddique	Hiroshima University, Japan
Reza Soltanpoor	RMIT University, Australia
Yifang Sun	University of New South Wales, Australia
Erald Troja	City University of New York, USA
Fabio Valdés	University of Hagen, Germany
Jan Vosecky	Hong Kong University of Science and Technology (HKUST), China
Jim Jing-Yan Wang	King Abdullah University of Science and Technology, Saudi Arabia
Huanhuan Wu	The Chinese University of Hong Kong, China
Xiaoying Wu	Wuhan University, China
Chen Xu	Technische Universität Berlin, Germany
Jianqiu Xu	Nanjing University of Aeronautics and Astronautics, China
Takeshi Yamamuro	NTT Software Innovation Center, Japan
Da Yan	The Chinese University of Hong Kong, China
Fan Yang	The Chinese University of Hong Kong, China
Jongheum Yeon	Seoul National University, Korea
Seongwook Youn	University of Southern California, USA
Zhou Zhao	Hong Kong University of Science and Technology (HKUST), China
Xiaoling Zhou	University of New South Wales, Australia

Tutorials

Scalable Learning Technologies
for Big Data Mining

Gerard de Melo[1] and Aparna S. Varde[2]

[1] Institute for Interdisciplinary Information Sciences, Tsinghua University, Beijing, China
gdm@demelo.org
[2] Department of Computer Science, Montclair State University, Montclair, NJ, USA
vardea@montclair.edu

Abstract. As data expands into big data, enhanced or entirely novel data mining algorithms often become necessary. The real value of big data is often only exposed when we can adequately mine and learn from it. We provide an overview of new scalable techniques for knowledge discovery. Our focus is on the areas of cloud data mining and machine learning, semi-supervised processing, and deep learning. We also give practical advice for choosing among different methods and discuss open research problems and concerns.

Keywords: Big Data · Cloud Data Mining · Deep Learning · Semi-Supervised Learning

1 Cloud Data Mining and Machine Learning

When moving from data to big data, we often need fundamentally different technologies for mining and learning. For data mining, one can rely on SQL over Big Data systems such as Hive, Cloudera's Impala, and Hortonworks' Stinger for efficient querying. One can also opt for more general processing frameworks like Scalding and Apache Spark, or stream-based alternatives like Apache Storm and Flink.

For cloud-based machine learning, Apache Mahout provides open source implementations of recommendation, clustering and classification, with MLBase and its MLLib component as a Spark-based alternative. Applications of this include news gathering, email classification, and recommender systems. Additionally, custom algorithms can be used for specific applications such as scalable rule mining and frequent item-set mining.

2 Semi-Supervised Processing and Deep Learning

Since training data is hard to obtain, semi-supervised learning and distant supervision can be used to exploit large amounts of unlabeled data, e.g. for sentiment analysis on Twitter. Additional signals for certain tasks can come from specific sources such as Wikipedia. Web-scale statistics, e.g. from Google's Web N-Grams, improve performance in co-reference analysis and information extraction.

Deep Learning and representation learning aim at learning multiple layers of abstraction from original features to capture more complex interactions. One can exploit unlabeled textual data to improve over bag-of-words feature representations or to create multimodal embeddings that combine words and images in the same space.

Overall, we see numerous new possibilities arising from the availability of Big Data in conjunction with such novel methods for exploiting it.

Acknowledgments. Gerard de Melo's research is in part by the National Basic Research Program of China Grants 2011CBA00300, 2011CBA00301, and NSFC Grants 61033001, 61361136003.

Large Scale Video Management Using Spatial Metadata and Their Applications

Seon Ho Kim[1] and Roger Zimmermann[2]

[1] Integrated Media Systems Center, Univ. of Southern California, Los Angeles, USA
seonkim@usc.edu
[2] Interactive & Digital Media Institute, National University of Singapore, Singapore
rogerz@comp.nus.edu.sg

1 Introduction

Recently, there has been research to utilize spatial metadata of videos for the management of large numbers of videos. For example, the location of a camera from GPS and the viewing direction from an embedded digital compass are captured at the recording time so that they form the field of view (FOV) model for the coverage of a viewable scene. These spatial metadata effectively convert the challenging video search problem to well known spatial database problems, which can greatly enhance the utility of collected videos in many emerging applications. However, there are still many open fundamental research questions in utilizing spatial metadata for data collection, indexing, searching, etc., especially in the presence of a large video dataset, as well as more sophisticated questions such as whether a video database search can accommodate human friendly views.

This tutorial covers and describes existing technologies and methods in harnessing spatial metadata of videos in spatial databases for video applications, especially for mobile videos from smartphones. The discussion includes types of metadata, their acquisition, indexing, searching, and potential applications of geo-tagged videos in scale. Furthermore, a spatial crowdsourcing of mobile media contents is discussed with regard to the use of geographical information in crowdsourcing. Some use cases of the implemented techniques are presented too. This tutorial can be a good survey and summary of the techniques which bridge the spatial database and videos. The target audience can be any researchers and practitioners who are interested in designing and managing a large video databases and Big Data in social media.

2 Tutorial Outline

Motivation: why are spatial metadata of videos important?
Types of Geospatial Metadata: types and methods of acquisition.
Modeling the Coverage Area of Scene: Field of View model and its use.
Extracting Textual Keywords Using Metadata: methods to extract.
Spatial Crowdsourcing: crowdsourcing media content.
Indexing and Searching: how to handle a large number of FOVs.
Applications: use cases and futuristics applications.
Open Discussion and Q&A

Querying Web Data

Andrea Calì

[1] Dept. of Computer Science
University of London, Birkbeck College
[2] Oxford-Man Institute of Quantitative Finance
University of Oxford
andrea@dcs.bbk.ac.uk

Abstract. A vast amount of information is available on the Web, not only in HTML static pages (documents), but also and especially in more structured form in databases accessible on the Web in different forms. In the Semantic Web [1], ontologies provide shared conceptual specifications of domains, so that the data can be enhanced with intensional semantic information to provide answers to queries that are consistent with an ontology. Ontological query answering [4] amounts to returning the answers to a query, that are logically entailed by the union of a data set (a set of membership assertions) and an ontology, where the latter is a set of logical assertions. We give an overview of the most important ontology formalisms for the Semantic Web, and we illustrate the most relevant query answering techniques, with particular emphasis on their efficiency [3]. Then we illustrate techniques specifically developed to access Hidden Web data, which can be accessed only according to specific patterns, therefore making query answering a complex and generally inefficient task [2]; we show techniques to improve efficiency by determining whether an access to a certain source is relevant to a given query. Finally we show techniques to query Linked Data sets, and to source relevant data to a certain query in such sets. We show how Linked Data can be employed in semantic search and recommender systems [5], by extracing ontological information from Linked Data sources.

Acknowledgments. The author acknowledges support by the EPSRC project "Logic-based Integration and Querying of Unindexed Data" (EP/E010865/1).

References

1. A. Artale, D. Calvanese, R. Kontchakov, and M. Zakharyaschev. The DL-Lite family and relations. *J. Artif. Intell. Res.*, 36:1–69, 2009.
2. M. Benedikt, G. Gottlob, and P. Senellart. Determining relevance of accesses at runtime. In *Proc. of PODS 2011*, pages 211–222.
3. A. Calì, G. Gottlob, and A. Pieris. Towards more expressive ontology languages: The query answering problem. *Artif. Intell.*, 193:87–128, 2012.
4. R. Kontchakov, C. Lutz, D. Toman, F. Wolter, and M. Zakharyaschev. The combined approach to ontology-based data access. In *Proc. of IJCAI*, pages 2656–2661, 2011.
5. V. C. Ostuni, T. D. Noia, E. D. Sciascio, and R. Mirizzi. Top-N recommendations from implicit feedback leveraging linked open data. In *Proc. of RecSys 2013*, pages 85–92.

Contents – Part II

Spatio-temporal Data II

Query Processing

Database Storage and Index II

Social Networks II

Industrial Papers

Demo

Contents – Part I

Spatio-Temporal Data I

Modern Computing Platform

Social Networks I

Information Integration and Data Quality

Information Retrieval and Summarization

Security and Privacy

Outlier and Imbalanced Data Analysis

A Synthetic Minority Oversampling Method Based on Local Densities in Low-Dimensional Space for Imbalanced Learning

Zhipeng Xie[1,2(✉)], Liyang Jiang[1], Tengju Ye[1], and Xiao-Li Li[3]

[1] School of Computer Science, Fudan University, Shanghai, China
{xiezp,13210240017,13210240039}@fudan.edu.cn
[2] Shanghai Key Laboratory of Data Science, Fudan University, Shanghai, China
[3] Institute of InfoComm Research, Fusionopolis Way, Singapore, Singapore
xlli@i2r.a-star.edu.sg

Abstract. Imbalanced class distribution is a challenging problem in many real-life classification problems. Existing synthetic oversampling do suffer from the curse of dimensionality because they rely heavily on Euclidean distance. This paper proposed a new method, called Minority Oversampling Technique based on Local Densities in Low-Dimensional Space (or MOT2LD in short). MOT2LD first maps each training sample into a low-dimensional space, and makes clustering of their low-dimensional representations. It then assigns weight to each minority sample as the product of two quantities: local minority density and local majority count, indicating its importance of sampling. The synthetic minority class samples are generated inside some minority cluster. MOT2LD has been evaluated on 15 real-world data sets. The experimental results have shown that our method outperforms some other existing methods including SMOTE, Borderline-SMOTE, ADASYN, and MWMOTE, in terms of G-mean and F-measure.

Keywords: Imbalanced learning · Oversampling method · Local densities · Dimensionality reduction

1 Introduction

Imbalanced distribution of data samples among different classes is a common phenomenon in many real-world classification problems, such as fraud detection [1] and text classification [2]. In this paper, we focus on two-class classification problems for imbalanced data sets, where the class that contains few samples is called the minority class, and the other that dominates the instance space is called the majority class. The imbalanced data sets have degraded the learning performance of existing learning algorithms and posed a challenge to them for the hardness to learn the minority class samples.

Confronted with the problem of imbalanced learning, some simple yet effective methods have been proposed to generate extra synthetic minority samples in order to

© Springer International Publishing Switzerland 2015
M. Renz et al. (Eds.): DASFAA 2015, Part II, LNCS 9050, pp. 3–18, 2015.
DOI: 10.1007/978-3-319-18123-3_1

balance the distribution between the majority class and the minority class [3][4][5][6], which are called synthetic oversampling methods. SMOTE[3], Borderline-SMOTE[4], ADASYN[5], and MWMOTE[6] are typical examples of this kind of algorithms. All these algorithms generate synthetic minority samples in two main phases. The first phase is to identify those informative minority class samples, and the second phase is to interpolate a synthetic minority class sample between those informative minority class samples and their nearby ones. The difference among them exists in the way of how the synthetic samples are generated. SMOTE algorithm [3] is the first and simplest synthetic oversampling method, which treats all the minority class samples equally. To generate a synthetic minority sample, it first draws seed samples randomly from the whole set of minority class samples in the seed drawing phase, and then calculates the k nearest neighbors in the minority class for each seed sample and generates new synthetic samples along the line between the seed sample and its nearest minority neighbors. As an improvement over SMOTE, Borderline-SMOTE [4] only draw seed samples from those dangerous minority samples at borderline. A minority class sample is at borderline if there are more majority class samples than minority ones in its m nearest neighbors. Borderline-SMOTE first identifies the borderline minority class samples, and then uses them as seed samples for generating the synthetic samples because they are most likely to be misclassified by a classifier. However, all the borderline samples are treated equally. To adaptively draw seed samples, ADASYN algorithm [5] adaptively assigns weights to the minority class samples. A large weight enhances the chance for the minority class sample serving as a seed sample in the synthetic sample generation process. Both Borderline-SMOTE and ADASYN share a synthetic sample generation process that is similar to the one used by SMOTE: the synthetic minority class samples are generated by interpolation randomly between the seed samples and their K-nearest neighbors of the minority class. Recently, a new algorithm MWMOTE [6] is proposed to identify the hard-to-learn informative minority class samples, to assign them weights according to their Euclidean distance from the nearest majority class samples, and then to generate synthetic samples inside minority class clusters. It has been illustrated in [6] that MWMOTE can avoid some situations that the other methods will generate wrong and unnecessary synthetic samples.

Although these synthetic oversampling methods have achieved some satisfactory results for imbalanced learning, they still have their deficiencies. Firstly, all of them for these methods rely heavily on the Euclidean distance in the calculation of K-nearest neighbors, which may suffer from the curse of dimensionality, especially when the dimensionality of the sample space is high. Secondly, the synthetic generation process used by SMOTE, Borderline-SMOTE, and ADASYN does not take the cluster structure into consideration. Last but not least, we think the local minority density should have its position in determining the importance of a minority class sample for the generation of synthetic minority samples.

To solve these problems, we propose a new algorithm, which consists of three main steps. It first applies t-SNE algorithm to reduce the dimensionality of the training samples into a two-dimensional space, where each sample is represented as a two-dimensional vector. Then, a density-peak algorithm is used to learn the cluster structure of the training samples in the low-dimensional space. The importance of a minority class sample is measured by taking two factors into consideration: local majority count and local minority density. Local majority count indicates how many majority class samples appear in the K-nearest neighbors of the minority class sample. The higher the local majority count is, the harder is to make the correct decision for the sample. The local minority density of a given minority sample indicates the density of minority class samples around it. The lower the local minority density is, the more likely is to generate a synthetic sample from the sample. Finally, based on the importance measurement of the minority samples, synthetic minority samples are generated, which are located inside some minority cluster.

The whole paper is organized as follows. Section 2 describes our proposed method MOT2LD (Minority Oversampling Technique based on Local Densities in Low-Dimensional Space) in detail. Section 3 presents the experimental results. Finally, we summarize the whole paper and point out possible directions for future work.

2 The Proposed Method

The objective in this paper is to exploit and integrate modern dimensionality reduction and clustering techniques in order to solve the problems that existing synthetic over-sampling methods are facing. The proposed algorithm, called Minority Oversampling Technique based on Local Densities in Low-Dimensional Space (or **MOT2LD** in short), consists of five major steps as listed in Table 1.

- The first step is to reduce the dimensionality of the representation of training samples. By dimensionality reduction, each sample in high-dimensional space can be mapped into a point in a low-dimensional space. Dimensionality reduction can be thought of a kind of metric learning technique, leading to a better distance metric between samples.
- The second step is to discover the cluster structure of the minority class samples in the low-dimensional space. A new density-based clustering algorithm called DPCluster is exploited, which is capable of determining the cluster number automatically. It is desirable that the generated synthetic minority samples are within some cluster, instead of "between clusters".
- The third step is to detect and filter out outliers and noises in the set of minority samples, for the existence of outliers and noises may do harm to the quality of the generated synthetic minority samples.
- The fourth step is to assign weights to the minority samples, indicating their importance for oversampling. The weight is measured as the product of the local majority count and the inverse of local minority density.
- The final step is to generate the synthetic minority samples according to the importance weights of the training minority samples. We also restrict that the synthetic minority samples should be inside some minority cluster, to avoid generating synthetic samples between different clusters.

The details of MOT2LD are described below.

Table 1. The framework of MOT2LD algrorithm

Algorithm:	Minority Oversampling Technique based on Local Densities in Low-Dimensional Space

Input:

NSamples:	A set of majority class samples (Negative class)
PSamples:	A set of minority class samples (Positive class)
K:	The number of nearest neighbors observed when filtering noise samples
NumToGen:	The number of synthetic minority samples to be generated

Output:

Y:	The set of synthetic minority samples that are generated

Procedure Begin

Step 1: (Dimensionality Reduction)

Use t-SNE algorithm to reduce the dimensionality of the dataset, where each data sample x_i is represented as a low-dimensional image y_i in a low-dimensional space.

Step 2: (Clustering of Minority Class Samples)

Use Density Peak Clustering algorithm to partition the set of minority class samples into a number of clusters $Cl_1,...,Cl_s$, where s is the number of clusters. As byproduct, we can also get the local minority density ρ_i for each minority sample i.

Step 3: (Outlier Detection and Noise Filtering)

For each minority class sample, if its local minority density is zero, it will be treated as outlier and get removed. In addition, we also count the number of majority class samples in its K-nearest neighbors. If all the K neighbors are from majority class, then the minority sample is a noise to be filtered.

Step 4: (Weight Assignment)

Assign an importance weight $Importance(i)$ to each minority class sample i as a product of its local majority count $\gamma(i)$ and the inverse of its local minority density $\rho(i)$.

Step 5: (Synthetic Sample Generation)

For each minority sample i, set $prob(i) = \frac{Importance(i)}{Z}$ where $Z = \sum_{i \in minority\ class} Importance(i)$.

for $i := 1$ to NumToGen

1) Randomly draw a minority sample x_s as the seed sample, according to the probability distribution { $prob(i): i \in minority\ class$}
2) Choose another minority sample x_t from the minority cluster that x_s belongs to.
3) Generate one synthetic minority sample $x_{new} = \alpha \times x_s + (1 - \alpha) \times x_t$, where α is a random number between 0 and 1.
4) Add x_{new} into Y.

end for

Procedure End

2.1   Dimensionality Reduction via t-SNE

Due to the curse of dimensionality, the commonly-used distance metrics that work well in low-dimensional space may have significantly-degraded performance in high-dimensional space. To alleviate this problem, dimensionality reduction is an important preprocessing step for many machine learning tasks such as clustering and classification. A lot of methods have been proposed to embed objects, described by either high-dimensional vectors or pairwise dissimilarities, into a lower-dimensional space [7]. Principal component analysis (PCA) [8] seeks to capture as much variance as possible. Multidimensional scaling (MDS) [9] tries to preserve dissimilarities between items. Traditional dimensionality reduction methods such as Principal Component Analysis and Multidimensional Scaling usually focus on keeping the low-dimensional representations of dissimilar data points far apart. However, for high-dimensional data that lies on or near a low-dimensional non-linear manifold, it is usually more important to keep the low-dimensional representations of very similar data points close together. Locally linear embedding (LLE) [10] attempts to preserve local geometry. Stochastic Neighbor Embedding (SNE) [11] is an iterative technique that aims at retaining the pairwise distances between the data points in the low-dimension space, which is similar to MDS. However, SNE differs from MDS in that it makes use of a Gaussian kernel such that the similarities of nearby points contribute more to the cost function. As such, it preserves mainly local properties of the manifold.

In this paper, we adopt a recently developed dimensionality reduction algorithm, called t-Distributed Stochastic Neighbor Embedding (t-SNE) [12], which is an extension to the well-known original Stochastic Neighbor Embedding (SNE) algorithm [11]. The t-SNE algorithm was proposed originally for the visualization of high-dimensional data points, which can transform the high-dimensional data set into two or three-dimensional data. The reason why we choose to use t-SNE is that it is capable of capturing much of the local structure of high-dimensional data very well, while also revealing global structure such as the presence of clusters. The first capability provides the quality of K-nearest neighbors, while the second capability makes it easy to discover the cluster structure of the minority class samples. Both these two capabilities are fundamental to the proposed MOT2LD algorithm. A brief description of t-SNE goes as follows.

In SNE [11] or t-SNE [12] algorithm, the high-dimensional Euclidean distances between data points are transformed into conditional probabilities that one data point would pick another data point as its neighbor.

$$p_{j|i} = \frac{\exp\left(-\left\|x_i - x_j\right\|^2/(2\sigma_i^2)\right)}{\sum_{k \neq i} \exp(-\left\|x_i - x_k\right\|^2/(2\sigma_i^2))} \tag{1}$$

where σ_i is the variance of the Gaussian that is centered on data point x_i, and $\frac{\|x_i - x_k\|^2}{2\sigma_i^2}$ represents the dissimilarities between two data points that are measured as the scaled squared Euclidean distance. The value of σ_i is chosen by a binary search such that the Shannon entropy $H(P_i) = -\sum_j p_{j|i} \log p_{j|i}$ of the distribution over neighbors equals to $\log u$, where u is a user-specified perplexity with 15 as default value.

In the high-dimensional space, the joint probabilities p_{ij} is defined to be the symmetrized conditional probabilities, that is:

$$p_{ij} = \frac{p_{j|i} + p_{i|j}}{2n} \tag{2}$$

where n denotes the number of data points.

In t-SNE, a Student t-distribution with one degree of freedom is employed as the heavy-tailed distribution in the low-dimensional space. The joint distribution is defined as:

$$q_{ij} = \frac{\left(1 + \|y_i - y_j\|^2\right)^{-1}}{\sum_{k \neq l}(1 + \|y_k - y_l\|^2)^{-1}} \tag{3}$$

Based on the definitions (2) and (3), the goal of t-SNE is to minimize the difference between the two joint probability distributions P and Q. The Kullback-Leibler divergence between the two joint probability distributions P and Q, which measures their difference, is given by:

$$C = KL(P\|Q) = \sum_i \sum_j p_{ij} \log \frac{p_{ij}}{q_{ij}} = \sum_i \sum_j (p_{ij} \log p_{ij} + p_{ij} \log q_{ij}) \tag{4}$$

Therefore, we take the Kullback-Leibler divergence as the objective function to be minimized. Its gradient can be written as:

$$\frac{\partial C}{\partial y_i} = 4 \sum_j (p_{ij} - q_{ij})\left(1 + \|y_i - y_j\|^2\right)^{-1}(y_i - y_j) \tag{5}$$

For the detailed derivation procedure of the expression (5), please refer to the Appendix A in [12].

A gradient descent method can be implemented to find out the map points in the low-dimensional space that minimizes the Kullback-Leibler divergence, following the gradient (5). To initialize the gradient descent process, we sample map points randomly from an isotropic Gaussian with small variance (10^{-8} by default) that is centered at the origin.

Through applying the t-SNE algorithm described above to the training samples inclusive of minority class and majority class, each sample is mapped to a point in a low-dimensional space. The map points in the low-dimensional space can better reveal the implicit structure of the high-dimensional data, especially when the high-dimensional data are lying on several different low-dimensional manifolds.

2.2   Density Peak Clustering in Low-Dimensional Space

Dimensionality reduction can be thought of as unsupervised distance metric learning, in that every dimensionality reduction approach can essentially learn a distance metric in the low-dimensional space [13]. Equivalently speaking, after high-dimensional data get mapped to map points in a low-dimensional space, we can derive a new distance metric between data points in the low-dimensional space, which is normally of higher quality than the original distance metric in the high-dimensional space. A better distance metric usually leads to higher quality of calculated K-nearest neighbors or density, and in turn yields a better clustering result. The global cluster structure is helpful to synthetic oversampling methods, because each generated synthetic sample should be inside some minority cluster. MWMOTE [6] uses an average-linkage agglomerative clustering algorithm [14] to derive the cluster structure of minority class. In our method, we use a simple clustering algorithm called Density Peak Clustering (DPCluster in short) [15]. DPCluster assumes that the cluster centers are defined as local maxima in the density of data points, or in other words, the cluster centers are surrounded by neighbors with lower density. It also assumes that the cluster centers are at a relatively large distance from any points with a higher local density. According to these two assumptions, DPCluster calculates two quantities for each data point: one is its local density, and the other is its distance from points of higher density, which play important roles in the clustering solutions and are defined as follows [15].

Definition 1. (Local Density) The local density of a data point i is defined as:

$$\rho_i = \rho(i) = \sum_j \chi(d_{ij} - d_c) \tag{6}$$

where d_{ij} denotes the distance between two data points i and j, $\chi(x) = 1$ if $x < 0$ and $\chi(x) = 0$ otherwise, and d_c is a cutoff distance.

In this paper, the distance between two data points is calculated as the Euclidean distance in the low-dimensional space, the value of d_c is set such that the average local density over all points equal to 2% of the total number of points. Because the clustering is applied only on minority class samples, the local density is also called the **local minority density** in this paper.

Definition 2. (Distance from points of higher density) For any data point i, its distance δ_i from points of higher density is measured as the minimum distance between the point and any other point with higher density:

$$\delta_i = \delta(i) = \min_{j:\rho_j>\rho_i} d_{ij} \tag{7}$$

For the data point with the highest local density y_i, δ_i is defined to be its maximal distance from any other point, that is, $\delta_i = \max_j d_{ij}$.

Based on those quantities, the clustering process consists of two steps. The first step is to identify the cluster centers which are the points with anomalously large value of ρ and relatively large value of δ, because cluster centers normally has high densities. In our implementation, this paper, a point is thought of as a cluster center if its local density is larger than 80% of all the data points, and its distance from points of higher density is among the top 5% of all the data points. As such, the number of clusters is automatically determined as the number of cluster centers identified, where each cluster center represents a unique cluster. In this step, the points with a high δ value and a low ρ can be treated as outliers. The second step is to assign the remaining data points to the same cluster as its nearest neighbor of higher density. This assignment step is performed in a single pass, which is much faster than other clustering algorithms such as k-means [16].

2.3   Outlier Detection and Noise Filtering

In MOT2LD, we detect outliers and filter noises, in the low-dimensional space with two strategies (Step 3).

- Strategy 1 (Outlier Detection): During the clustering process described in section 2.2, we calculate the local minority density ρ_i for each minority class sample i. If the local minority density ρ_i equals to zero, then the sample i is likely to be an outlier, because there is no minority samples surrounding it. Therefore, it is deleted from the set of minority class samples, and gets removed from subsequent processing.

- Strategy 2 (Noise Filtering): We calculate the set $NN(i)$ of K-nearest neighbors for a minority class sample i in the low-dimensional space. If the K-nearest neighbors of a minority class sample in the low-dimensional space are all from the majority class, then the minority sample i is likely to be a noise sample because it is surrounded by only the majority class samples. It is then filtered out of the minority sample set.

2.4   Weight Assignment

As to measuring the importance of a minority class sample for synthetic minority sample generation, there are three facts that deserve our attention:

- The first fact is that the borderline points of a cluster normally have low local minority densities ρ, but the interior points usually have high local minority densities. For classification, the borderline points are more informative than the interior points. Therefore, the points with lower local densities should be given higher probabilities when chosen as the seed samples for generating synthetic minority samples. In other words, for a given minority sample i, if the number of minority class samples, whose distance to i is less than the cutoff distance d_c, is low, then its weight should be increased.

- The second fact is that for two minority clusters of different densities, the samples in the cluster of lower density should get more chances to serve as seed samples for generating synthetic minority samples than those in the cluster of higher density. This fact leads to the same conclusion as the first fact: minority samples of lower minority density should be given more weight.

- The third fact is that a minority sample is hard to make the correct decision if there are many majority class samples in its K-nearest neighbors. As such, we use the local majority count $\gamma(i)$ to indicate how many majority samples occur in the K-nearest neighbors of a given minority sample i. Minority samples with higher local majority count should be given higher probability of serving as seed samples for generating synthetic minority samples.

For a given minority class sample i, its importance $importance(i)$ is defined as the product of its local majority count $\gamma(i)$ and the inverse of its local minority density $\rho(i)$:

$$importance(i) = \frac{\gamma(i)}{\rho(i)} \tag{8}$$

The importance weight of a given minority class sample is an indicator of the importance for generating synthetic minority sample from it. A large weight implies that the sample needs to generate many synthetic minority samples nearby it.

To illustrate the rationale behind the weighting scheme, we construct a simple example explained as follows. In Fig. 1, there are totally 550 points. Among these points, 500 points represented as blue dots are drawn randomly from a bivariate Gaussian $\mathcal{N}(\mu_0, \Sigma_0)$, and 50 points as red plus-signs are drawn from another bivariate Gaussian $\mathcal{N}(\mu_1, \Sigma_1)$, where $\mu_0 = (1,2)$, $\mu_1 = (1,-1)$, and $\Sigma_0 = \Sigma_1 = \begin{bmatrix} 2 & 1 \\ 1 & 1 \end{bmatrix}$. Next, we shall examine three minority class samples labeled by 1, 2, and 3.

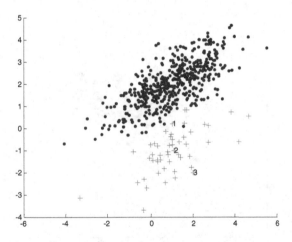

Fig. 1. An example of Gaussian-distributed Minority and Majority Samples

First, let us focus only on the minority class samples. The samples 1 and 3 are at the borderline of minority class, while the sample 2 is an interior point of minority class. As illustrated in Fig. 2, where each circle is centered at sample 1, sample 2, or sample 3, and the radius of each circle equals to the chosen cutoff value d_c. Clearly, the local minority density of sample 1, $\rho(1)$, equals to 5, because there are five minority class samples in its circle. Similarly, the local density of sample 2, $\rho(2)$, is 16, while the local density of sample 3, $\rho(3)$, is only 2.

Next, we examine the local majority count for the three minority samples. For the sample 1, there are two majority samples in its 5-nearest neighbors, so its majority count equals to 2, that is $\gamma(1) = 2$. For the sample 2 and the sample 3, there are no majority samples appearing in their 5-nearest neighbors, so we have $\gamma(2) = \gamma(3) = 0$.

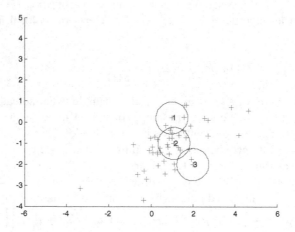

Fig. 2. Local minority densities for three minority class samples

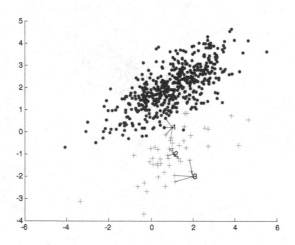

Fig. 3. Local majority counts for three minority class samples

Following the definition of importance weight in Equation (8), we can calculate the importance weights of the three minority samples:

$$Importance(1) = \frac{\gamma(1)}{\rho(1)} = \frac{2}{5};$$

$$Importance(2) = \frac{\gamma(2)}{\rho(2)} = 0;$$

$$Importance(3) = \frac{\gamma(3)}{\rho(3)} = 0.$$

Clearly, a minority class sample is given a high importance weight, if it has a high local majority count and a low local minority density.

2.5   Generation of Synthetic Minority Samples

Before we describe the generation of synthetic samples, we first transform the importance weights of minority samples into a probability distribution that indicates the probability that a minority sample is selected as the seed sample:

$$prob(i) = \frac{Importance(i)}{\sum_{j \in minority\ class} Importance(j)} \tag{9}$$

To generate a synthetic minority sample, a minority sample x_s is selected randomly as the seed sample according to the probability distribution. Let S denote the cluster that contains x_s. We then select a second minority sample x_t that belongs to the minority cluster S. A new synthetic minority sample is thus generated by random interpolation between the two minority samples x_s and x_t.

3   Experimental Results

To evaluate the effectiveness of the proposed MOT2LD method, we compare it with four other synthetic oversampling methods: SMOTE[3], Borderline-SMOTE[4], ADASYN[5], and MWMOTE[6], on 15 data sets from the UCI machine learning repository [17]. The data sets with more than two classes are transformed to two-class problems. Table 2 lists detailed information about the data sets and how the majority and minority classes.

On each data set, we randomly split it into two parts of (almost) the same size, one for training set and the other for testing set. Synthetic oversampling method is applied on the training set.

Accuracy is the most commonly-used evaluation metric for classification problems. However, the accuracy measure suffers greatly from the imbalanced class distribution, and thus is not suitable for imbalance classification [18]. To assess the classifier performance on imbalanced two-class classification problem, a confusion matrix is constructed as shown in Table 3, where TP denotes the number of true positive, FP denotes the number of false positive, FN denotes the number of false negative, and TN denotes the number of true negative. Two evaluation metrics derived from confusion matrix are used in this paper to assess learning from imbalanced data sets. They are G-mean and F-measure [18].

Table 2. Characteristics of the experimental data sets

Data Sets	Minority Class	Majority Class	Features	Minority	Majority	Imbalance Ratio
Statlandsat	4	other	37	415	4435	0.09:0.91
Yeast	ME3, ME2, EXC, VAC, POX, ERL	other	8	304	1180	0.21:0.79
Ecoli	im	other	7	77	259	0.23:0.77
PageBlocks	Graphic, Vert.line, Picture	other	10	231	5245	0.04:0.96
BreastCancer	Malignant	Benign	9	239	444	0.34:0.66
Glass	5, 6, 7	other	9	51	163	0.24:0.76
Vehicle	van	other	18	199	647	0.24:0.76
Libra	1, 2, 3	other	90	72	288	0.20:0.80
Abalone	18	other	7	42	689	0.06:0.94
Vowel	0	other	10	90	900	0.09:0.91
Pima	1	0	8	268	500	0.35:0.65
Ionosphere	bad radar	good radar	34	126	225	0.36:0.64
Segment	Grass	other	19	330	1980	0.14:0.86
BreastTissue	CAR, FAD	other	9	36	70	0.34:0.66
Wine	3	other	13	48	130	0.26:0.74

Table 3. Confusion Matrix

		True Class	
		Positive	Negative
Predicted Class	Positive	TP	FP
	Negative	FN	TN

Based on the confusion matrix in Table 3, the evaluation metrics, G-mean and F-measure, are defined as follows:

- G-mean is a good indicator for performance assessment of imbalanced learning by combining the accuracies on the positive class and negative class samples.

$$G_{Mean} = \sqrt{\frac{TP}{TP + FN} \times \frac{TN}{TN + FP}}$$

where $\frac{TP}{TP+FN}$ is the accuracy on the positive class and $\frac{TN}{TN+FP}$ is the accuracy on the negative class.

- F-measure make a combination of precision and recall of the positive samples:

$$F_{measure} = \frac{2 \times recall \times precision}{recall + precision}$$

where $recall = \frac{TP}{TP+FN}$ and $precision = \frac{TP}{TP+FP}$.

We use the CART [16] decision tree as the classification model in our experiments. Throughout the experiments, we do not fine-tune the parameters in our algorithm. All the parameters take default values as indicated in Section 2. Table 4 and Table 5 summarize the results of SMOTE, Borderline-SMOTE, ADASYN, and MOT2LD on the 15 experimental data sets. The reported performance results are all averaged over 20 independent runs. At each run, the data set is randomly divided into two parts of approximately equal size: one for training set and the other for testing set. The number of synthetic minority samples that generated by the compared oversampling methods is two times the number of minority samples in the training set. On each data set, the best result is highlighted with underlined bold-face type.

Table 4. Comparison of G-mean on experimental data sets

Data Sets	ADASYN	Borderline-SMOTE	SMOTE	MWMOTE	MOT2LD
Statlandsat	0.717	0.714	**0.723**	0.719	**0.723**
Yeast	**0.783**	0.781	**0.783**	0.779	0.778
Ecoli	0.830	0.824	0.844	0.836	**0.851**
PageBlocks	**0.869**	0.858	0.861	0.858	0.868
BreastCancer	0.898	0.903	0.909	**0.910**	0.907
Glass	0.884	**0.893**	0.887	0.880	0.892
Vehicle	0.891	0.900	0.900	0.894	**0.903**
Libra	0.744	0.742	0.764	0.761	**0.769**
Abalone	0.557	0.561	0.5998	0.580	**0.619**
Vowel	0.947	0.923	0.941	0.927	**0.947**
Pima	0.664	0.648	0.662	0.658	**0.669**
Ionosphere	0.835	**0.857**	0.829	0.849	0.824
Segment	**0.997**	0.996	0.996	**0.997**	0.996
BreastTissue	0.717	0.712	0.692	0.681	**0.725**
Wine	0.949	**0.949**	0.943	0.947	0.947

Table 5. Comparison of F-measures on experimental data sets

Data Sets	ADASYN	Borderline-SMOTE	SMOTE	MWMOTE	MOT2LD
Statlandsat	0.513	0.514	**0.519**	0.507	0.507
Yeast	0.646	**0.650**	0.647	0.643	0.642
Ecoli	0.741	0.735	0.758	0.749	**0.765**
PageBlocks	**0.728**	0.726	0.718	0.682	0.700
BreastCancer	0.890	0.896	0.902	**0.904**	0.900
Glass	0.835	**0.848**	0.832	0.828	**0.848**
Vehicle	0.836	0.845	0.846	0.835	**0.847**
Libra	0.630	0.641	**0.650**	0.643	0.636
Abalone	0.304	0.322	0.321	0.334	**0.359**
Vowel	**0.876**	0.863	0.867	0.864	0.858
Pima	0.575	0.554	0.573	0.567	**0.580**
Ionosphere	0.790	**0.818**	0.781	0.804	0.776
Segment	0.995	**0.996**	0.995	**0.996**	**0.996**
BreastTissue	0.635	0.630	0.605	0.594	**0.646**
Wine	**0.923**	**0.923**	0.917	0.916	**0.923**

From Table 4 and Table 5, it can be seen that MOT2LD has achieved 8 best results out of the 15 data sets among all the compared methods, in both G-mean and F-measure, which is much better than the others including SMOTE, Border-line-SMOTE, ADASYN, and MWMOTE, which have mostly achieved 2 or 3 best results out of the 15 data sets.

4   Conclusion and Future Work

In this paper, we propose a new synthetic oversampling method MOT2LD for imba-lanced learning. MOT2LD first maps samples into a low-dimensional space using t-SNE algorithm, and discovers the cluster structure of the minority class in the low-dimensional space by DPCluster. It then assigns importance weights to minority samples as the products of the local majority count and the inverse of local minority density.

To finalize this paper, we would like to list several directions for our future work:

Firstly, it may be interesting to study the effect of supervised dimensionality reduc-tion technique as a proprecessing step. If we could make use of supervised information in the dimensionality reduction algorithm to maximize the separation between minority class and majority class, it is expected that a better results would be achieved.

Secondly, although synthetic oversampling methods have achieved satisfactory results for imbalanced learning, a lot of other methods do exist. Recently, there are some model-based oversampling methods such as SPO [20][21] and MoGT [22]. SPO [20][21] assumes that the minority samples follow a multivariate Gaussian distribution. It estimates its mean vector and covariance matrix and then draws extra minority sample from the probability distribution. MoGT [22] assumes another probabilistic model called mixture of Gaussian Trees. It is similar to the Gaussian Mixture model, but differs in that Gaussian Tree can be thought of as a restricted kind of Gaussian distribution, which has much less parameters to be estimated. How to combine synthetic oversampling methods and model-based oversampling ones is a challenging problem.

Acknowledgements. This work is supported by National High-tech R&D Program of China (863 Program) (No. SS2015AA011809), Science and Technology Commission of Shanghai Municipality (No. 14511106802), and National Natural Science Fundation of China (No. 61170007). We are grateful to the anonymous reviewers for their valuable comments.

References

1. Fawcett, T.E., Provost, F.: Adaptive Fraud Detection. Data Min. Knowl. Disc. **3**(1), 291–316 (1997)
2. Mladenić, D., Grobelnik, M.: Feature selection for unbalanced class distribution and naive bayes. In: Proceedings of the 16th International Conference on Machine Learning, pp. 258–267 (1999)
3. Chawla, N.V., Bowyer, K.W., Hall, L.O., Kegelmeyer, W.P.: SMOTE: synthetic minority oversampling technique. J. Artif. Intell. Res. **16**, 321–357 (2002)
4. Han, H., Wang, W.Y., Mao, B.H.: Borderline-SMOTE: a new oversampling method in imbalanced data sets learning. In: Proceedings of International Conference on Intelligent Computing, pp. 878–887 (2005)
5. He, H., Bai, Y., Garcia, E.A., Li, S.: ADASYN: adaptive synthetic sampling approach for imbalanced learning. In: Proceedings of IEEE International Joint Conference on Neural Networks, pp. 1322–1328 (2008)
6. Barua, S., Islam, M.M., Yao, X., Murase, K.: MWMOTE - majority weighted minority oversampling technique for imbalanced data set learning. IEEE Trans. Knowl. Data Eng. **26**(2), 405–425 (2014)
7. van der Maaten, L.J.P., Postma, E.O., van den Herik, H.J.: Dimensionality reduction: a comparative review. Tilburg University Techical Report, TiCC-TR 2009–005 (2009)
8. Hotelling, H.: Analysis of a complex of statistical variables into principal components. J. Educ. Psychol. **24**, 417–441 (1933)
9. Torgerson, W.S.: Multidimensional scaling I: theory and method. Psychometrika **17**, 401–419 (1952)
10. Roweis, S.T., Saul, L.K.: Nonlinear dimensionality reduction by Locally Linear Embedding. Science **290**(5500), 2323–2326 (2000)
11. Hinton, G.E., Roweis, S.T.: Stochastic neighbor embedding. In: Advances in Neural Information Processing Systems, vol. 15, pp. 833–840 (2002)
12. Van der Maaten, L., Hinton, G.: Visualizing data using t-SNE. Journal of Machine Learning Research **9**, 2579–2605 (2008)

13. Liu, Y.: Distance metric learning: a comprehensive survey. Research Report, Michigan State University (2006)
14. Voorhees, E.M.: Implementing agglomerative hierarchic clustering algorithms for use in document retrieval. Inf. Process. Manage. **22**(6), 465–476 (1986)
15. Rodriguez, A., Laio, A.: Clustering by fast search and find of density peaks. Science **344**, 1492–1496 (2014)
16. MacQueen, J.: Some methods for classifications and analysis of multivariate observations. In: Proceedings of the Fifth Berkeley Symposium on Mathematics, Statistics and Probability, University of California Press, pp. 281–297 (1967)
17. Bache, K., Lichman, M.: UCI Machine Learning Repository. University of California, School of Information and Computer Science, 2013 [http://archive.ics.uci.edu/ml]
18. He, H., Garcia, E.A.: Learning from imbalanced data. IEEE Trans. Knowl. Data Eng. **21**(9), 1263–1284 (2009)
19. Breiman, L., Friedman, J., Stone, C.J., Olshen, R.A.: Classification and Regression Trees. CRC press (1984)
20. Cao, H., Li, X.L., Woon, Y.-K., Ng, S.K.: SPO: structure preserving oversampling for imbalanced time series classification. In: Proceedings of IEEE International Conference on Data Mining (2011)
21. Cao, H., Li, X.L., Woon, Y.K., Ng, S.K.: Integrated oversampling for imbalanced time series classification. IEEE Trans. Knowl. Data Eng. **25**(12), 2809–2822 (2013)
22. Pang, Z.F., Cao, H., Tan, Y.F.: MOGT: oversampling with a parsimonious mixture of Gaussian trees model for imbalanced time-series classification. In: Proceedings of IEEE International Workshop on Machine Learning for Signal Processing (MLSP), pp. 1–6 (2013)

Fast and Scalable Outlier Detection
with Approximate Nearest Neighbor Ensembles

Erich Schubert[(✉)], Arthur Zimek, and Hans-Peter Kriegel

Ludwig-Maximilians-Universität München, Oettingenstr. 67, 80538
München, Germany
{schube,zimek,kriegel}@dbs.ifi.lmu.de
http://www.dbs.ifi.lmu.de

Abstract. Popular outlier detection methods require the pairwise comparison of objects to compute the nearest neighbors. This inherently quadratic problem is not scalable to large data sets, making multidimensional outlier detection for big data still an open challenge. Existing approximate neighbor search methods are designed to preserve distances as well as possible. In this article, we present a highly scalable approach to compute the nearest neighbors of objects that instead focuses on preserving neighborhoods well using an ensemble of space-filling curves. We show that the method has near-linear complexity, can be distributed to clusters for computation, and preserves neighborhoods—but not distances—better than established methods such as locality sensitive hashing and projection indexed nearest neighbors. Furthermore, we demonstrate that, by preserving neighborhoods, the quality of outlier detection based on local density estimates is not only well retained but sometimes even improved, an effect that can be explained by relating our method to outlier detection ensembles. At the same time, the outlier detection process is accelerated by two orders of magnitude.

1 Introduction

Vast amounts of data require more and more refined data analysis techniques capable to process big data. While the volume of the data often decreases dramatically with selection, projection and aggregation, not all problems can be solved this way. The domain of outlier detection is a good example where individual records are of interest, not overall trends and frequent patterns. Summarization will lose the information of interest here and thus cannot be applied to outlier detection. In the data mining literature, a large variety of methods is based on object distances, assuming that outliers will essentially exhibit larger distances to their neighbors than inliers, i.e., the estimated local density is lower than the "usual" density level in the dataset. Without employing index structures, this requires the computation of all pairwise distances in the worst case.

Here we focus on methods for improving generically all such methods by fast approximations of the relevant neighbors. We demonstrate that the approximation error is negligible for the task of outlier detection. In the literature,

© Springer International Publishing Switzerland 2015
M. Renz et al. (Eds.): DASFAA 2015, Part II, LNCS 9050, pp. 19–36, 2015.
DOI: 10.1007/978-3-319-18123-3_2

approximations of neighborhoods and distances have been used as a filter step. Here, we show that an approximate identification of the neighborhood is good enough for the common outlier detection methods since these do not actually require the neighbors *as such* but only as a means to derive a local *density estimate*. Notably, the detection accuracy can even improve by using an approximate neighborhood. We explain this effect and argue that a tendency to improve the accuracy of outlier detection by approximate neighborhood identification is not accidental but follows a certain bias, inherent to our method.

In the remainder, we will discuss outlier detection methods and their efficiency variants (Section 2). Then (Section 3), we reason about the theoretical background of our method and consequences for its performance, and introduce the core of our method. We demonstrate the effectiveness and efficiency in an extensive experimental analysis (Section 4) and conclude the paper with rules of thumb on the expected usefulness of the different methods (Section 5).

2   Related Work

2.1   Outlier Detection

Existing outlier detection methods differ in the way they model and find the outliers and, thus, in the assumptions they, implicitly or explicitly, rely on. The fundamentals for modern, database-oriented outlier detection methods (i.e., methods that are motivated by the need of being scalable to large data sets, where the exact meaning of "large" has changed over the years) have been laid in the statistics literature. A broader overview for modern applications has been presented by Chandola et al. [10]. Here, we focus on techniques based on computing distances (and derived secondary characteristics) in Euclidean data spaces.

With the first database-oriented approach, Knorr and Ng [22] triggered the data mining community to develop many different methods, typically with a focus on scalability. A method in the same spirit [33] uses the distances to the k nearest neighbors (kNN) of each object to rank the objects. A partition-based algorithm is then used to efficiently mine top-n outliers. As a variant, the sum of distances to all points within the set of k nearest neighbors (called the "weight") has been used as an outlier degree [4]. The so-called "density-based" approaches consider ratios between the local density around an object and the local density around its neighboring objects, starting with the seminal LOF [7] algorithm. Many variants adapted the original LOF idea in different aspects [36].

2.2   Approximate Neighborhoods

For approximate nearest neighbor search, the Johnson-Lindenstrauss lemma [19] states the existence and bounds of a projection of n objects into a lower dimensional space of dimensionality $\mathcal{O}(\log n/\varepsilon^2)$, such that the distances are preserved within a factor of $1 + \varepsilon$. Matoušek [26] further improves these error bounds. The most interesting and surprising property is that the reduced dimensionality depends only logarithmically on the number of objects and on the error bound, but *not* on the original dimensionality d. Different ways of obtaining such a projection have been proposed for common norms such as Manhattan and Euclidean distance. A popular choice are the "database-friendly" random projections [1], where 2/3 of the terms are 0 and the others ±1 (along with a global scaling factor of $\sqrt{3}$), which can be computed more efficiently than the previously used matrices. Another popular choice are projections based on s-stable distributions [11], where the Cauchy distribution is known to be 1-stable and the Gaussian distribution to be 2-stable [44] (i.e., they preserve L_1 and L_2 norms well). An overview and empirical study on different variations of the Johnson-Lindenstrauss transform [38] indicates that a reduced dimensionality of $k = 2 \cdot \log n/\varepsilon^2$ will usually maintain the pairwise distances within the expected quality.

2.3   Outlier Detection with Approximate Neighborhoods

Wang et al. [39] propose outlier detection based on Locality Sensitive Hashing (LSH) [11,14,17]. However—in contrast to what the authors state—it cannot be used for "any distance-based outlier detection mechanism", but it will only be useful for global methods such as kNN-Outlier [4,33]: the key idea of this method is to use LSH to identify low-density regions, and refine the objects in these regions first, as they are more likely to be in the top-n global outliers. For local outlier detection methods there may be interesting outliers within a globally dense region, though. As a consequence, the pruning rules this method relies upon will not be applicable. Projection-indexed nearest-neighbours (PINN) [12] shares the idea of using a random projection to reduce dimensionality. On the reduced dimensionality, a spatial index is then employed to find neighbor candidates that are refined to k nearest neighbors in the original data space.

Much research aimed at improving efficiency by algorithmic techniques, for example based on approximations or pruning techniques for mining the top-n outliers only [6,29]. A broad and general analysis of efficiency techniques for outlier detection algorithms [30] identifies common principles or building blocks for efficient variants of the so-called "distance-based" models [4,22,33]. The most fundamental of these principles is "approximate nearest neighbor search" (ANNS). The use of this technique in the efficient variants studied by Orair et al. [30] is, however, different from the approach we are proposing here in a crucial point. Commonly, ANNS has been used as a filter step to discard objects from computing the *exact* outlier score. The exact kNN distance could only become smaller, not larger, in case some neighbor was missed by the approximation. Hence, if the upper bound of the kNN distance, coming along with the ANNS, is already too small to possibly qualify the considered point as a top-n outlier, the respective point will not be refined. For objects passing this filter step, the *exact neighborhood* is still required in order to compute the *exact outlier score*. All other efficiency techniques, as discussed by Orair et al. [30], are similarly based on this consideration and differ primarily in the pruning or ranking strategies. As opposed to using ANNS as a filter step, we argue to *directly* use approximate nearest neighbor search to compute outlier scores without this refinement i.e., we base the outlier score on the k *approximate* nearest neighbors directly.

2.4   Summary

The differentiation between "distance-based" and "density-based" approaches, commonly found in the literature, is somewhat arbitrary. For both families, the basic idea is to provide estimates of the density around some point. As opposed to statistical approaches, that fit specific distributions to the data, the density-estimates in the efficient database algorithms are, in a statistical sense, parameter-free, i.e., they do not assume a specific distribution but estimate the density-level, using typically some simple density model (such as the k nearest neighbor distance). This observation is crucial here. It justifies our technique also from the point of view of outlier detection models: the exact neighbors are usually not really important but just the estimate of the density-level around some point and the difference from estimated density-levels around other points. In most cases, this will derive just the same outliers as if the outlier detection model were based on the *exact* distances to the *exact* neighbors, since those outlier scores will always remain *just an estimate*, based on the data sample at hand, and not on the unknown true density-level. The same reasoning relates to ensemble methods for outlier detection [3,40], where a better overall judgment is yielded by diversified models. Models are diversified using approximations of different kinds: subsets of features [23], subsets of the dataset [42], even by adding *noise* components to the data points in order to yield diverse density-estimates, the results for outlier detection ensembles can be improved [41]. As opposed to outlier detection ensemble methods, here, we push the ensemble principle of diversity and combination to a deeper level: instead of creating an ensemble from different outlier models, we create an ensemble of different neighborhood approximations and use the combined, ensemble approximation of the neighborhood as the base for the outlier model.

3   Efficient Outlier Detection

Outlier detection methods based on density estimates, such as the kNN [33] or weighted kNN [4] outlier models, as well as the Local Outlier Factor (LOF) [7] and its many variants [36], rely on the retrieval of nearest neighbors for each data object. While the major part of database research on efficient outlier detection focused on retrieving the exact values of the top n outliers as fast as possible, using approximate neighborhoods as a filter, we maintain here that outlier detection methods do not heavily rely on the neighborhood sets to be *exact*: they use the distance to the kNN to estimate a density, local methods additionally use the kNN to compute an average neighbor density as reference. As long as the distances are not influenced heavily by the approximation, and the reference (approximate) neighbors still have a similar density, the results are expected to be similar. For approximations of neighborhoods by space filling curves the approximation error is not symmetric: it will never underestimate a k-distance, but by missing some true nearest neighbors it will instead return the $k + e$-distance for $e \geq 0$. The difference between the k and the $k + e$ distance is expected to be rather small in dense areas (i.e., for "inliers"), as there are many neighbors at similar distances. For an object in a sparse area (i.e., an "outlier"), the $k + 1$-distance can already be much larger than the k-distance. We can expect, on average, the scores of true outliers to further increase, which is well acceptable for the purpose of detecting outliers.

In computer science and data analysis, we rely on mathematics for correctness of the results. Yet, we also have to deal with the fact that neither our computation will be perfect—due to limited numerical precision—nor our data are exact: even unlimited data will still only yield an approximation of reality. With our data only being a finite sample, chances are that the exact computation will not be substantially closer to the truth than a good approximation. Of course we must not give up precision without having benefits from doing so. However, for large data sets we have an immediate problem to solve: finding the nearest neighbors by computing a distance matrix, or by repeated linear scans, will not scale to such data sets anymore. Trading some precision for reducing the runtime from quadratic to linear may be well worth the effort.

3.1   Approximate Indexing Techniques

There exist capable indexing methods for low-dimensional data such as the k-d tree and the R*-tree. In order to use such techniques for high dimensional data, the data dimensionality must be reduced. Approaches in the context of outlier detection are feature bagging [23] and PINN [12], using Achlioptas' database-friendly random projections [1]. Locality Sensitive Hashing (LSH, [14,17]) uses s-stable random projections [11] for indexing Minkowski distances and found use in outlier detection as well [39]. LSH (on dense vector data with L_p-norms) combines multiple established strategies:

1. dimensionality reduction by s-stable random projections to k dimensions;
2. grid-based data binning into $\mathbb{N}^d$ bins of width w;
3. reduction of grid size by hashing to a finite bin index;
4. similarity search in the bin of the query point only;
5. ensemble of ℓ such projections.

Individual parts of this approach can be substituted to accommodate different data types and similarity functions. For example, instead of computing the hash code on a regular grid, it can be based on the bit string of the raw data, or a bit string obtained by splitting the data space using random hyperplanes. LSH is an approximate nearest neighbor search algorithm both due to the use of random projections (which only approximately preserve distances) but also due to searching within the same bin as the query point only. The use of hash tables makes it easy to parallelize and distribute on a cluster.

Projection-indexed nearest-neighbours (PINN) [12] also uses random projection to reduce dimensionality. A spatial index is then employed to find neighbor candidates in the projected space:

1. dimensionality reduction using "database friendly" random projections;
2. build a spatial index (R*-tree, k-d tree) on the projected data;

3. retrieve the $c \cdot k$ nearest neighbors in the projection;
4. refine candidates to k nearest neighbors in original data space.

Due to the use of random projections, this method may also not return the true k nearest neighbors, but it has a high probability of retrieving the correct neighbors [12]. In contrast to LSH, it is also guaranteed to return the desired number of neighbors and thus to always provide enough data for density estimation and reference sets to be used in outlier detection. When a true nearest neighbor is not found, the false positives will still be spatially close to the query point, whereas with LSH they could be any data.

The type of random projections discussed here are not a general purpose technique: the Johnson-Lindenstrauss lemma only gives the existence of a random projection that preserves the distances, but we may need to choose different projections for different distance functions. The projections discussed here were for unweighted L_p-norm distances. Furthermore it should be noted, as pointed out by Kabán [20], that random projection methods are not suitable to defy the "concentration of distances"-aspect of the "curse of dimensionality" [43]: since, according to the Johnson-Lindenstrauss lemma, distances are preserved approximately, these projections will also preserve the distance concentration.

3.2 Space-Filling Curves

Space-filling curves are a classic mathematical method for dimensionality reduction [15,31]. In contrast to random projections, by space-filling curves the data are always reduced to a single dimension. In fact, the earliest proposed space-filling curves, such as the Peano curve [31] and the Hilbert curve [15], were defined originally for the two dimensional plane and have only later been generalized to higher dimensionality. A space-filling curve is a fractal line in a bounded d dimensional space (usually $[0; 1]^d$) with a Hausdorff dimensionality of d that will actually pass through every point of the space.

The first curve used for databases was the Z-order. Morton [27] used it for indexing multidimensional data for range searching, hence the Z-order is also known as Morton code and Lebesgue curve. This curve can be obtained by interleaving the bits of two bit strings x_i and y_i into a new bit string: $(x_1y_1x_2y_2x_3y_3x_4y_4 \ldots)$. The first mathematically analyzed space-filling curve was the Peano curve [31], closely followed by the Hilbert curve [15] which is considered to have the best mathematical properties. The Peano curve has not received much attention from the data indexing community because it splits the data space into thirds, which makes the encoding of coordinates complex. The Hilbert curve, while tricky in high dimensional data due to the different rotations of the primitives, can be implemented efficiently with bit operations [8], and has been used, e.g., for bulk-loading the R*-tree [21] and for image retrieval [28].

Indexing data with space-filling curves as suggested by Morton [27] is straightforward: the data are projected to the 1-dimensional coordinate, then indexed using a B-tree or similar data structure. However, *querying* such data is challenging: while this index can answer exact matches and rectangular window queries well, finding the exact k nearest neighbors is nontrivial. Thus, for outlier detection, we will need query windows of different size in order to find the k nearest neighbors. A basic version of this method for high dimensional similarity search [37] used a large number of "randomly rotated and shifted" curves for image retrieval. A variant of this approach [25] uses multiple systematically shifted – not rotated – copies of Hilbert curves and gives an error bound based on Chan's work [9]. To retrieve the k nearest neighbors, both methods look at the preceding and succeeding k objects in each curve, and refine this set of candidates.

Chan [9] gave approximation guarantees for grid-based indexes based on shifting the data diagonally by $1/(d+1)$ times the data extent on each axis. The proof shows that a data point must be at least $1/(2d+2) \cdot 2^{-\ell}$ away from the nearest border of the surrounding cell of size $2^{-\ell}$ (for any level $\ell \geq 0$) in at least one of these curves due to the pigeonhole principle. Within at least one grid cell, all neighborhoods within a radius of $1/(2d+2) \cdot 2^{-\ell}$ therefore are in the same grid cell (i.e., nearby on the same curve). By looking at the length of the shared bit string prefix, we can easily determine the ℓ which we have fully explored, and then stop as desired. An approximate k-nearest neighbor search on such curves – by looking at the k predecessors and successors on each of the $d+1$ curves only – returns approximate k nearest neighbors which are at most $\mathcal{O}(d^{1+1/p})$ farther than the exact k nearest neighbors, for any L_p-norm [25]. For the 1st-nearest neighbor, the error factor is at most $d^{1/p}(4d+4) + 1$ [25].

In our approach, we divert from using the systematic diagonal shifting for which these error bounds are proved. It can be expected that the errors obtained by randomized projections are on a similar scale on *average*, but we cannot guarantee such bounds for the worst case anymore. We do however achieve better scalability due to the lower dimensionality of our projections, we gain the ability to use other space filling curves, and are not restricted to using $d+1$ curves. Similar to how diversity improves outlier ensembles [35,40], we can expect diverse random subspaces to improve the detection result.

Space filling curves are easy to use in low dimensional space, but will not trivially scale up to high dimensionality due to the combinatorial explosion (just as any other grid based approach) [43]. They work on recursive subdivisioning of the data space, into 2^d (3^d for the Peano curve) cells, a number which grows exponentially with the dimensionality d. In most cases, the ordering of points will then be determined by binary splits on the first few dimensions only. HilOut [4] suffers both from this aspect of the curse of dimensionality, and from the distance concentration which reduces its capability to prune outlier candidates: since all distances are increasingly similar, the set of outlier candidates does not shrink much with each iteration of HilOut. For this top-n method to perform well, it must be able to shrink the set of candidates to a minimum fast, so that it can analyze a wider window of neighbors.

3.3   Fast Approximate kNN Search

Our proposed method to search for nearest neighbors is closely inspired by the methods discussed before, such as HilOut. However, it is designed with

---

Algorithm 1. Phase 1: Projection and Data Rearragement

---

distributed on *every node* **do**   // Project data locally
    foreach *block* **do**
        foreach *curve* **do**
            project data to curve
            store projected data
            send sample **to** coordination node
on *coordination node* **do**   // Estimate distribution for sorting
    foreach *curve* **do**
        Read sample
        Sort sample
        Estimate global data distribution
        send global quantiles **to** every node
distributed on *every node* **do**   // Rearrange data in cluster
    foreach *curve* **do**
        foreach *projected block* **do**
            split according to global quantiles
shuffle to new blocks

---

parallelism and distributed computation in mind: where HilOut uses a nested loops approach to refine the current top candidates for a single outlier detection model, we focus on a method to compute the k nearest neighbors of all objects (often called kNN-self-join), so that we can then use an arbitrary kNN-based outlier detection method. At the same time, our method becomes easy to parallelize.

The principle of the proposed method is:

1. Generate m space-filling curves, by varying
   (a) curve families (Hilbert, Peano, Z-curve),
   (b) random projections and/or subspaces,
   (c) shift offsets to decorrelate discontinuities.
2. Project the data to each space-filling-curve.
3. Sort data on each space-filling-curve.
4. Using a sliding window of width $w \times k$, generate candidates for each point.
5. Merge the neighbor candidates across all curves and remove duplicates.
6. Compute the distance to each candidate, and keep the k nearest neighbors.

The parameter m controls the number of curves, and w can be used to control the tradeoff between recall and runtime, with $w = 1$ being a reasonable default. The proposed algorithm can be broken into three phases. We assume that the data are organized in blocks in a distributed file system such as HDFS (which provides built-in functionality for data chunking) or Sparks sliced RDDs.

In the first phase (Algorithm 1), the data are projected to each space-filling curve. The resulting data are stored on the local node, and only a sample is sent to the central node for estimating the data distribution. The central node then reads the projected samples, and estimates the global data distribution.

Algorithm 2. Phase 2: Compute kNN and RkNN

---

  distributed on *every node* **do**   `// Process sliding windows`
     | **foreach** *curve* **do**
     | | **foreach** *projected, shuffled block* **do**
     | | | Sort block
     | | | **foreach** *object (using sliding windows)* **do**
     | | | | emit (object, neighbors)
  shuffle to (object, neighbor list)
  distributed on *every node* **do**   `// compute kNN and build RkNN`
     | **foreach** *(object, neighbor list)* **do**
     | | Remove duplicates from neighbor list
     | | Compute distances
     | | emit (object, neighbors, $\emptyset$)   `// Keep forward neighbors`
     | | **foreach** *neighbor* **do**
     | | | emit (neighbor, $\emptyset$, [object])   `// Build reverse neighbors`
  shuffle to (object, kNN, RkNN)

The resulting split points are then distributed to each node, and the data are read a second time and reorganized into the desired partitions via the shuffle process in map-reduce. This sorting strategy was shown to scale to 100 TB in TritonSort [34]. While this is not a formal map-reduce process (requiring the transmission and use of auxiliary data), an implementation of this sorting process can be found in the Hadoop "terasort" example. The first phase serves as preprocessing to avoid having to project the data twice, and partially sorts the data according to the spatial curves. The required storage and communication cost is obviously $O(n \cdot m)$, i.e., linear in the data size and number of curves.

Algorithm 2 reads the output of the first phase. Each block in this data is a contiguous part of a space filling curve. We first finish the distributed sorting procedure within this data block. Then we can use a sliding window over the sorted data set to obtain neighbor candidates of each point. By emitting *(object, neighbor)* pairs to the map-reduce shuffle, we can easily reorganize the data to a *(object, neighbor list)* data layout and remove duplicates. For many local outlier detection algorithms, we will also need the reverse k-nearest neighbors (RkNN) to orchestrate model redistribution. This can be achieved by emitting inverted triples *(neighbor, $\emptyset$, object)*. The shuffle process will then reorganize the data such that for each object we have a triple *(object, neighbors, reverse neighbors)*.

In the third phase (Algorithm 3), we then compute the outlier scores using the generalized model of Schubert et al. [36]. In the experiments, we will use the LOF [7] model in this phase. Obviously, one can run other kNN-, SNN- [16], and reverse-kNN-based [18, 32] algorithms on the precomputed neighborhoods as well in the same framework. The reverse-kNNs are computed by simple list inversion to optimize data communication: this makes it easy to transmit an object's density estimate to the neighbor objects for comparison.

Algorithm 3. Phase 3: Compute Outlier Scores

---

distributed on *every node* **do**   `// Compute models`
    foreach *(object, kNN, RkNN)* **do**
        Compute model for object   `// Build own model`
        emit (object, (object, model))   `// Retain own model`
        emit (reverse neighbor, (object, model))   `// Distribute model`
shuffle to (object, model list)   `// Collect models`
distributed on *every node* **do**   `// Compare models`
    foreach *(object, (neighbor, model))* **do**
        Compare model to neighbor models
        Store outlier score for model
        Collect outlier score statistics   `// (for normalization)`
    emit Send statistics to coordination node
on *coordination node* **do**   `// Normalize Outlier Scores`
    Merge outlier score statistics
    send statistics **to** every node
distributed on *every node* **do**
    foreach *(object, score)* **do**
        Normalize Outlier Score
        if *score above threshold* **then**
            emit (outlier, normalized score)

3.4 Favorable Bias of the Approximation

There exists an interesting bias in the approximation using space-filling curves (SFCs), which makes them particularly useful for outlier detection. The error introduced by SFCs scales with the density of the data: if the bit strings of two vectors agree on the first $d \cdot \ell$ bits, the vectors are approximately within a cube of edge length $2^{-\ell}$ times the original data space.

For query points in a dense region of the data, the explored neighbors will be closely nearby, whereas for objects in less dense areas (i.e., outliers) the error introduced this way will be much larger on average. Furthermore, for an object central to a cluster, "wrong" nearest neighbors tend to be still members of the same cluster, and will just be slightly farther away.

For an outlier however, missing one of the true nearest neighbors – which may be another outlier with low density – and instead taking an even farther object as neighbor actually increases the chance that we end up using a cluster member of a nearby cluster for comparison. So while the approximation will likely not affect inlier scores much, we can expect it to emphasize outliers.

This effect is related to an observation for subsampling ensembles for outlier detection [42]: when subsampling a relative share of s objects from a uniformly distributed ball, the kNN-distances are expected to increase by a relative factor of $(1 - s^{1/d})/s^{1/d}$. Since for outliers this distance is expected to be higher, the expected increase will also be larger, and thus the outlier will become more pronounced with respect to this measure.

For other use cases such as density based cluster analysis, the effects of this approximation may be much more problematic. Such methods may fail to discover connected components correctly when a cluster is cut into half by a discontinuity in the space-filling curve: in contrast to outlier detection which only requires representative neighbors, such methods may rely on complete neighbors.

3.5   Discussion

Though our approach and the related approaches, PINN [12] and LSH-based outlier detection [39], can all be used to find the approximate nearest neighbors efficiently, they are based on subtly different foundations of approximation.

Random projections are designed to approximately *preserve distances*, while reducing dimensionality. Using an exact index on the projected data, as done in PINN, will therefore find the k nearest neighbors with respect to the approximate distance. The index based on locality sensitive hashing (LSH) in contrary is lossy: it is designed to have a high chance of *preserving regions of a fixed size w*, where the size w is a critical input parameter: the smaller the size that needs to be preserved, the faster the index; when the parameter is chosen too high, all objects will be hashed into the same bin, and the index will degenerate to a linear scan.

Space-filling curves on the contrary neither aim at directly preserving distances, nor do they try to preserve regions of a given radius. Instead, space-filling curves try to *preserve closeness*: the nearest neighbors of an object will often be nearby on the curve, while far neighbors in the data space will often be far away on the curve as well. For the purpose of density-based outlier detection, this yields an important effect: the index based on space-filling curves is better at adapting to different densities in the data set than the other two indexes, which makes it more appropriate for local density-based outlier detection methods.

4   Experiments

For the experiments, all methods were implemented in ELKI [2]. We tested the behavior of our method as well as the related approaches PINN [12] and LSH-based outlier detection on several datasets. As a reference, we have also LOF results based on an exact index, using the R*-tree in different variants (i.e., different page sizes, different bulkload strategies).

In our experiments, we used a number of larger data sets. From the image database ALOI [13], containing 110,250 images, we extracted 27 and 64 dimensional color histogram vectors. In order to obtain an outlier data set, a random subset of the classes were downsampled, so that the final data set contains only 75,000 images, of which 717 are labeled as outliers [16]. We also extracted all geographic coordinates from DBpedia [24] (Wikipedia preprocessed as RDF triples). This 2-dimensional data set does not have labeled outliers, and thus we used the top-1% according to LOF as outliers. This allows to see how close methods based on approximate neighborhoods come to the exact LOF results. However, as it is a low dimensional dataset, runtime results demonstrate that R*-tree indexes can work well. The forest covertype data set from the UCI machine learning repository [5] is a well known classification data set. The type cottonwood/willow is used as outlier class. In the following, we analyse the results on the 27 dimensional ALOI dataset in detail, as this data set has labeled outliers and is only of medium dimensionality. For the other datasets, we can draw similar conclusions and show only some sample results.

Figure 1 visualizes the results for running the LOF algorithm on the ALOI data set with $k = 20$ on a single CPU core. In total, we evaluated over 4,000 different index variations for this data set. To make the results readable we visualize only the skyline results in Figure 1a. The skyline are all objects where no other result is both faster and has a higher score at the same time (upper skyline) or where no other result is both slower and scores less at the same time (lower skyline). The upper skyline is useful for judging the *potential* of a method, when all parameters were chosen optimal, whereas the lower skyline indicates the worst case. In Figure 1c, we give a sample of the full parameter space we explored. Obviously, all 4,000 runs will be an unreadable cloud of points, thus we filter the results to LSH and only one variant of SFC (random curve families and random subspaces), which are about 700 runs. To show the continuity of the explored parameter space, we connected similar parametrizations with a line, more specifically for SFC indexes we connected those that differ only in window width w and for LSH we connect those that vary the number of hash tables ℓ. For exact indexes, different results arise for different page sizes. Note, though, that the skylines used typically represent hundreds of experiments. The skylines for LSH in Figure 1a represent the same set of results as in Figure 1c whereas the results for SFC in Figure 1c are restricted to the random variant and the skylines for SFC in Figure 1a include the other variants. Based on the systematic exploration of the parameter space as sampled in Figure 1c, it is interesting to see that with SFC we repeatedly were able to get higher outlier detection quality at a more than 10-fold speedup over the exact indexes. Merely when using a *single* space-filling curve, the results were not substantially better. Figure 1a visualizes the skyline of outlier detection quality vs. runtime, whereas Figure 1b is an evaluation of the actual index by measuring the recall of the true 20 nearest neighbors. In both measures, the SFC based method has the best potential – it can even be better than the exact indexes – while at the same time it is usually an order of magnitude faster than PINN and two orders of magnitude faster than LSH (both at comparable quality). Even if the parameters are chosen badly (which for SFC usually means using too few curves), the results are still comparable to LSH and PINN. However, there is a surprising difference between these two charts. They are using the same indexes, but the LOF ROC AUC scores for the SFC index start improving quickly at a runtime of 1,000-2,000 ms. The recall however, starts rising much slower, in the range of 1,500-10,000 ms. When we choose an average performing combination for the SFC index, e.g., 8 curves of different families combined with a random subspace projection to 8 dimensions and a window width of $w = 1$, we get a runtime of 4,989 ms, a ROC AUC of 0.747, and an average recall of the true nearest neighbors of

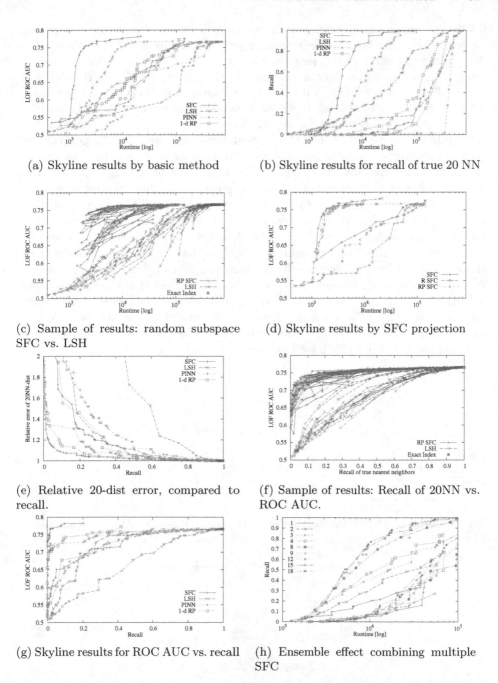

(a) Skyline results by basic method

(b) Skyline results for recall of true 20 NN

(c) Sample of results: random subspace SFC vs. LSH

(d) Skyline results by SFC projection

(e) Relative 20-dist error, compared to recall.

(f) Sample of results: Recall of 20NN vs. ROC AUC.

(g) Skyline results for ROC AUC vs. recall

(h) Ensemble effect combining multiple SFC

Fig. 1. Experiments on 27d ALOI data set. *SFC*: Space filling curves (proposed method), *R SFC*: Random features + SFC, *RP SFC*: Random projections + SFC, *LSH*: Locality sensitive hashing, *PINN*: Projection indexed nearest neighbors, *1-d RP*: one-dimensional random projections. *Exact index*: R*-tree

0.132. For an explanation for such a good performance *despite* the low recall, we refer the reader back to the reasoning provided in Section 3.4. In Figure 1d, we explore the effect of different random projections. The skyline, marked as "SFC", does not use random projections at all. The curves titled "R SFC" are space filling curves on randomly selected features only (i.e., feature bagging), while "RP SFC" uses full Achlioptas style random projections. As expected, the variant without projections is fastest due to the lower computational cost. Using randomly selected features has the highest potential gains and in general the largest variance. Achlioptas random projections offer a similar performance as the full-dimensional SFC, but come at the extra cost of having to project the data, which makes them usually slower. Figure 1e visualizes the relative error of the 20-nearest neighbor distance over the recall. The SFC curves, despite a very low recall of less than 0.2, often suffer much smaller relative error than the other approaches. While the method does make *more* errors, the errors are *less severe*, i.e., the incorrect nearest neighbors have a smaller distance than those retrieved by the other methods. This is again evidence for the outlier-friendly bias of space filling curves (Section 3.4). Figure 1f is the same sample as Figure 1c, but projected to LOF ROC AUC quality and recall. One would naïvely expect that a low recall implies that the method cannot work well. While algorithm performance and recall are correlated for locality sensitive hashing, the SFC approach violates this intuition: even with very low recall, it already works

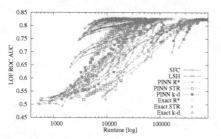

(a) Results for ROC AUC vs. Runtime on 64d ALOI

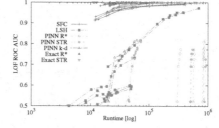

(b) Results for ROC AUC vs. Runtime on DBpedia

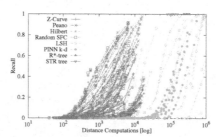

(c) Results for Recall vs. Dist. Calc. on Covertype

Fig. 2. Sample results on other datasets

surprisingly well; some of the best results have a recall of only around 0.25 – and outperform the exact solution. The bias (Section 3.4) again proves to be positive. When looking at the skylines of the complete data in Figure 1g, this even yields an upper skyline that ends at a recall of 0.2 – no result with a higher recall performed better than this. As particular 1-dimensional projections, space filling curves are by design more apt for low dimensional data rather than for high dimensional data. However, by combining multiple curves, i.e., building an ensemble of approximate neighborhood predictors, the performance gain is quite impressive also for high dimensional data. We show skylines for different numbers of curves combined in Figure 1h. While single curves are performing badly and remain unstable, combinations, here of up to 18 curves, improve considerably.

For other datasets, we show samples in Figure 2. For the 64 dimensional variant of ALOI (Figure 2a), including all variants of the exact indexes, backing PINN and backing LOF (i.e., "exact"), we can draw the same conclusions as for the 27d ALOI dataset. Again, our method is performing typically better at a shorter runtime. On the 2-dimensional DBpedia dataset (Figure 2b), as expected, we cannot beat the exact indices. However, in comparison with the other approximate methods, our method is performing excellent. For the large forest covertype dataset, let us study another aspect than those discussed before. We see in Figure 2c a reason for the faster runtime behavior of our method: we reach a good recall with far less distance computations than PINN or LSH.

5   Conclusion

We proposed a method, based on space filling curves (that can be used in combination with other approximation methods such as random projections or LSH), for highly scalable outlier detection based on ensembles for approximate nearest neighbor search. As opposed to competing methods, our method can be easily distributed for parallel computing. We are hereby not only filling the gap of possible approximation techniques for outlier detection between random projections and LSH. We also show that this particular technique is more apt for outlier detection. Its competitive or even improved effectiveness is explained by a bias of space-filling curves favourable for outlier detection. Furthermore, the principle of combining different approximations is related to ensemble approaches for outlier detection [40], where the diversity is created at a different level than usual.

We rely on the same motivation as outlier detection ensembles: diversity is more important than getting the single most accurate outlier score because the exact outlier score of *some* method is not more than just some variant of the *estimate* of the density-level around some point and the difference from *estimated* density-levels around other points. Therefore, an approximate density estimate, based on approximate neighborhoods, will be typically good enough to identify just the same outliers as when computing the *exact* distances to the *exact* neighbors which will still be *only an estimate*, based on the data sample at hand, of the *true* but *inevitably unknown* density-level. Often, the results based on approximate neighborhoods are even better.

Based on our reasoning and experimental findings, we conclude with the following rules of thumb for the practitioner:

1. If the data dimensionality is low, bulk-loaded R*-trees are excellent.
2. If the exact distances are of importance, PINN is expected to work best.
3. If neighborhoods for a known, small radius w are needed, LSH is expected to work best.
4. If k-nearest neighborhoods are needed, as it is the case for the most well-known outlier detection methods, SFC is the method of choice.

References

1. Achlioptas, D.: Database-friendly random projections: Johnson-Lindenstrauss with binary coins. JCSS **66**, 671–687 (2003)
2. Achtert, E., Kriegel, H.P., Schubert, E., Zimek, A.: Interactive data mining with 3D-parallel-coordinate-trees. In: Proc. SIGMOD, pp. 1009–1012 (2013)
3. Aggarwal, C.C.: Outlier ensembles. SIGKDD Explor. **14**(2), 49–58 (2012)
4. Angiulli, F., Pizzuti, C.: Outlier mining in large high-dimensional data sets. IEEE TKDE **17**(2), 203–215 (2005)
5. Bache, K., Lichman, M.: UCI machine learning repository (2013). http://www.archive.ics.uci.edu/ml
6. Bay, S.D., Schwabacher, M.: Mining distance-based outliers in near linear time with randomization and a simple pruning rule. In: Proc. KDD, pp. 29–38 (2003)
7. Breunig, M.M., Kriegel, H.P., Ng, R., Sander, J.: LOF: identifying density-based local outliers. In: Proc. SIGMOD, pp. 93–104 (2000)
8. Butz, A.R.: Alternative algorithm for Hilbert's space-filling curve. IEEE TC **100**(4), 424–426 (1971)
9. Chan, T.M.: Approximate nearest neighbor queries revisited. Disc. & Comp. Geom. **20**(3), 359–373 (1998)
10. Chandola, V., Banerjee, A., Kumar, V.: Anomaly detection: A survey. ACM CSUR **41**(3), Article 15, 1–58 (2009)
11. Datar, M., Immorlica, N., Indyk, P., Mirrokni, V.S.: Locality-sensitive hashing scheme based on p-stable distributions. In: Proc. ACM SoCG, pp. 253–262 (2004)
12. de Vries, T., Chawla, S., Houle, M.E.: Density-preserving projections for large-scale local anomaly detection. KAIS **32**(1), 25–52 (2012)
13. Geusebroek, J.M., Burghouts, G.J., Smeulders, A.W.M.: The amsterdam library of object images. Int. J. Computer Vision **61**(1), 103–112 (2005)
14. Gionis, A., Indyk, P., Motwani, R.: Similarity search in high dimensions via hashing. In: Proc. VLDB, pp. 518–529 (1999)
15. Hilbert, D.: Ueber die stetige Abbildung einer Linie auf ein Flächenstück. Math. Ann. **38**(3), 459–460 (1891)
16. Houle, M.E., Kriegel, H.-P., Kröger, P., Schubert, E., Zimek, A.: Can shared-neighbor distances defeat the curse of dimensionality? In: Gertz, M., Ludäscher, B. (eds.) SSDBM 2010. LNCS, vol. 6187, pp. 482–500. Springer, Heidelberg (2010)
17. Indyk, P., Motwani, R.: Approximate nearest neighbors: towards removing the curse of dimensionality. In: Proc. STOC, pp. 604–613 (1998)
18. Jin, W., Tung, A.K.H., Han, J., Wang, W.: Ranking outliers using symmetric neighborhood relationship. In: Ng, W.-K., Kitsuregawa, M., Li, J., Chang, K. (eds.) PAKDD 2006. LNCS (LNAI), vol. 3918, pp. 577–593. Springer, Heidelberg (2006)

19. Johnson, W.B., Lindenstrauss, J.: Extensions of Lipschitz mappings into a Hilbert space. In: Conference in Modern Analysis and Probability, Contemporary Mathematics, vol. 26, pp. 189–206. American Mathematical Society (1984)

20. Kabán, A.: On the distance concentration awareness of certain data reduction techniques. Pattern Recognition **44**(2), 265–277 (2011)

21. Kamel, I., Faloutsos, C.: Hilbert R-tree: an improved R-tree using fractals. In: Proc. VLDB, pp. 500–509 (1994)

22. Knorr, E.M., Ng, R.T.: Algorithms for mining distance-based outliers in large datasets. In: Proc. VLDB, pp. 392–403 (1998)

23. Lazarevic, A., Kumar, V.: Feature bagging for outlier detection. In: Proc. KDD, pp. 157–166 (2005)

24. Lehmann, J., Isele, R., Jakob, M., Jentzsch, A., Kontokostas, D., Mendes, P.N., Hellmann, S., Morsey, M., van Kleef, P., Auer, S., Bizer, C.: DBpedia - a large-scale, multilingual knowledge base extracted from wikipedia. Semantic Web J. (2014)

25. Liao, S., Lopez, M.A., Leutenegger, S.T.: High dimensional similarity search with space filling curves. In: Proc. ICDE, pp. 615–622 (2001)

26. Matoušek, J.: On variants of the Johnson-Lindenstrauss lemma. Random Structures & Algorithms **33**(2), 142–156 (2008)

27. Morton, G.M.: A computer oriented geodetic data base and a new technique in file sequencing. Tech. rep, International Business Machines Co. (1966)

28. Nguyen, G., Franco, P., Mullot, R., Ogier, J.M.: Mapping high dimensional features onto Hilbert curve: applying to fast image retrieval. In: ICPR12, pp. 425–428 (2012)

29. Nguyen, H.V., Gopalkrishnan, V.: Efficient pruning schemes for distance-based outlier detection. In: Buntine, W., Grobelnik, M., Mladenić, D., Shawe-Taylor, J. (eds.) ECML PKDD 2009, Part II. LNCS, vol. 5782, pp. 160–175. Springer, Heidelberg (2009)

30. Orair, G.H., Teixeira, C., Wang, Y., Meira Jr., W., Parthasarathy, S.: Distance-based outlier detection: Consolidation and renewed bearing. PVLDB **3**(2), 1469–1480 (2010)

31. Peano, G.: Sur une courbe, qui remplit toute une aire plane. Math. Ann. **36**(1), 157–160 (1890)

32. Radovanović, M., Nanopoulos, A., Ivanović, M.: Reverse nearest neighbors in unsupervised distance-based outlier detection. IEEE TKDE (2014)

33. Ramaswamy, S., Rastogi, R., Shim, K.: Efficient algorithms for mining outliers from large data sets. In: Proc. SIGMOD, pp. 427–438 (2000)

34. Rasmussen, A., Porter, G., Conley, M., Madhyastha, H., Mysore, R., Pucher, A., Vahdat, A.: TritonSort: a balanced large-scale sorting system. In: Proceedings of the 8th USENIX Conference on Networked Systems Design and Implementation (2011)

35. Schubert, E., Wojdanowski, R., Zimek, A., Kriegel, H.P.: On evaluation of outlier rankings and outlier scores. In: Proc. SDM, pp. 1047–1058 (2012)

36. Schubert, E., Zimek, A., Kriegel, H.P.: Local outlier detection reconsidered: a generalized view on locality with applications to spatial, video, and network outlier detection. Data Min. Knowl. Disc. **28**(1), 190–237 (2014)

37. Shepherd, J.A., Zhu, X., Megiddo, N.: Fast indexing method for multidimensional nearest-neighbor search. In: Proc. SPIE, pp. 350–355 (1998)

38. Venkatasubramanian, S., Wang, Q.: The Johnson-Lindenstrauss transform: an empirical study. In: Proc. ALENEX Workshop (SIAM), pp. 164–173 (2011)

39. Wang, Y., Parthasarathy, S., Tatikonda, S.: Locality sensitive outlier detection: a ranking driven approach. In: Proc. ICDE, pp. 410–421 (2011)

40. Zimek, A., Campello, R.J.G.B., Sander, J.: Ensembles for unsupervised outlier detection: Challenges and research questions. SIGKDD Explor. **15**(1), 11–22 (2013)
41. Zimek, A., Campello, R.J.G.B., Sander, J.: Data perturbation for outlier detection ensembles. In: Proc. SSDBM, vol. 13, pp. 1–12 (2014)
42. Zimek, A., Gaudet, M., Campello, R.J.G.B., Sander, J.: Subsampling for efficient and effective unsupervised outlier detection ensembles. In: Proc. KDD, pp. 428–436 (2013)
43. Zimek, A., Schubert, E., Kriegel, H.P.: A survey on unsupervised outlier detection in high-dimensional numerical data. Stat. Anal. Data Min. **5**(5), 363–387 (2012)
44. Zolotarev, V.M.: One-dimensional stable distributions. Translations of Mathematical Monographs, vol. 65. American Mathematical Society (1986)

Rare Category Exploration on Linear Time Complexity

Zhenguang Liu[1], Hao Huang[1,2]($\boxtimes$), Qinming He[1], Kevin Chiew[3],
and Yunjun Gao[1]

[1] College of Computer Science and Technology,
Zhejiang University, Hangzhou, China
{zhenguangliu,hqm,gaoyj}@zju.edu.cn
[2] State Key Laboratory of Software Engineering, Wuhan University, Wuhan, China
haohuang@whu.edu.cn
[3] Singapore Branch, Handal Indah Sdn Bhd, Johor Bahru, Malaysia
kchiew@handalindah.com.my

Abstract. Rare Category Exploration (in short as RCE) discovers the remaining data examples of a rare category from a seed. Approaches to this problem often have a high time complexity and are applicable to rare categories with compact and spherical shapes rather than arbitrary shapes. In this paper, we present FREE an effective and efficient RCE solution to explore rare categories of arbitrary shapes on a linear time complexity w.r.t. data set size. FREE firstly decomposes a data set into equal-sized cells, on which it performs wavelet transform and data density analysis to find the coarse shape of a rare category, and refines the coarse shape via an $MkNN$ based metric. Experimental results on both synthetic and real data sets verify the effectiveness and efficiency of our approach.

1 Introduction

Starting from a seed which is a known data example of a rare category, Rare Category Exploration (RCE) helps discover the remaining data examples from the same rare category of the seed. Different from classification and clustering, RCE pays more attention to the interestingness of data examples. Meanwhile, in RCE a data set is normally imbalance, i.e., the data examples from majority categories dominate the data set and the data examples of interest to users constitute a rare category. For example, fraud transactions are usually overwhelmed by millions of normal transactions.

The motivation and aims of RCE have enabled RCE to have a wide variety of applications. For example, after detecting a criminal (i.e., a data example of a rare category), RCE makes it possible to find out other members of the criminal gang by investigating this criminal's communication network. In financial security as another example, after detecting a fraud transaction, finding out the fraud transactions of the same type can help us analyze the security leaks of the system and prevent new fraud transactions.

© Springer International Publishing Switzerland 2015
M. Renz et al. (Eds.): DASFAA 2015, Part II, LNCS 9050, pp. 37–54, 2015.
DOI: 10.1007/978-3-319-18123-3_3

Given the practical importance of RCE, various solutions have been proposed with different constraints. For example, He *et al.* [5] converted RCE to a convex optimization problem and tried to represent the rare category with a hyperball. Huang *et al.* [6] formulated RCE as a local community detection problem and introduced a solution based on the compactness and isolation assumption of rare categories. Nonetheless, in many real-world data sets, rare categories may form arbitrary shapes in a high-dimensional feature space [14], i.e., the data cluster of a rare category may be long and narrow and may have holes inside or may possess concave shapes. The existing algorithms however tend to work well only on convex shaped rare categories, and either require a certain number of training data examples or assume that the rare category is isolated to all other categories. These constraints may hinder their applications on real world data sets.

Given this situation, in this paper we propose a novel algorithm FREE (**F**ast **R**are cat**E**gory **E**xploration by zooming) for RCE, which achieves linear time complexity w.r.t. data set size and is capable to handle rare categories of arbitrary shapes. More importantly, FREE does not require either any number of training data examples nor the prior knowledge of the data set. In brief, FREE is carried out by two steps, i.e., (1) finding out the coarse shape of the objective rare category from the whole feature space using coarse-grained metrics, and (2) refining the coarse shape by fine-grained metrics.

2 Related Work

The related work of RCE can be classified into three groups, i.e., (1) rare category detection, (2) imbalanced classification, and (3) the approaches to RCE which can be further classified into two subgroups, i.e., (a) optimization-based approaches and (b) community-detection-based approaches.

Rare category detection is proposed to find out at least one data example for each rare category to prove the existence of this category. The existing paradigms usually utilize the compactness characteristic to find data examples of rare categories, and can be classified into three groups, i.e., the model-based [11,12], the neighbor-based [3,4,7,8], and the hierarchical-clustering-based [16]. RCE is an intuitive follow-up action of rare category detection, i.e., after detecting a data example of a rare category, the challenge becomes how to discover the remaining data examples in the rare category [5].

Imbalanced classification aims to construct a classifier which can determine the boundary of each category in an imbalanced data set [10,15]. There are three types of methods proposed for imbalanced classification, i.e., sampling-based, ensemble-based, and adaptive-learning-based. These methods can be used for RCE by returning the data examples which are classified as from rare category. Nonetheless, since they are not specially designed for RCE, they do not perform satisfactorily due to not taking full advantage of the rare category characteristics.

Optimization-based approaches convert RCE problem to a convex optimization problem [5], and try to enclose the rare category data examples with a minimum-radius hyperball. These approaches (such as RACH [5]) can handle the scenario where the objective rare category overlaps with a majority category. Nevertheless, to build a training set, they require a certain number of labeled data examples which are usually difficult and expensive to acquire in practice, and their computational costs are high due to their high number of iterations in solving convex optimization problem.

Community-detection-based approaches transform RCE problem to a local community detection problem [6]. These approaches keep absorbing external data examples until there is no improvement of the quality of the local community. Meanwhile, they require the data examples of a rare category being isolated from other data examples. Besides, their time complexity is quadratic w.r.t. data set size because they need to construct the kNN graph of data examples.

3   Problem Statement and Assumptions

RCE can be formulated as follows [5,6].

Input: (1) An unlabeled data set $\mathbb{S} = \{\mathbf{x_1}, \mathbf{x_2}, \ldots, \mathbf{x_n}\}$, where $\mathbf{x_i} \in \mathbb{R}^d$ for $1 \leqslant i \leqslant n$ and d is the data dimension; and (2) a seed $\mathbf{s_0} \in \mathbb{O} \subset \mathbb{S}$, which is a known data example of rare category $\mathbb{O}$ that is of interest to us.

Objective: Find out the other data examples in the objective rare category $\mathbb{O}$.

For RCE, we have the following assumptions which are commonly used explicitly or implicitly by the existing work [3,4,7,8,11,16].

Assumption 1. *Data examples of rare category $\mathbb{O}$ are very similar to each other.*

In many applications this assumption is reasonable [5,6,11]. For example, network attacks utilizing the same security leak are usually similar to each other; patients of the same rare disease share similar clinical manifestations.

Assumption 2. *Data distribution of $\mathbb{O}$ is different from that of other categories.*

This assumption implies that rare category $\mathbb{O}$ does not share the same data distribution with other categories [11,12]. This is reasonable because different categories usually have different data distributions. For example, panda subspecies have quite different data distributions in fur color and tooth size comparing to giant panda [12].

Besides the assumptions, most of the existing approaches (e.g., [6–8]) also assume that rare category $\mathbb{O}$ is isolated from other categories. However, in applications like transaction fraud detection, frauds are often disguised as legal ones and often overlap with normal transactions from the majority category [3]. In this paper, we conduct study on RCE without assuming that rare categories are isolated from other categories.

Figs. 1(a) & 1(g) show two examples of data sets each of which contains three majority categories with green plotted data examples and one rare category with red plotted data examples. In Fig. 1(a), the rare category is isolated from the majority categories; whereas in Fig. 1(g), the rare category overlaps with a majority category.

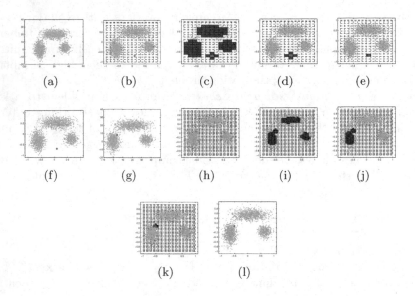

Fig. 1. Examples

4   FREE Algorithm

Throughout the paper, we use bold letters to denote vectors, normal letters to denote scalars, and calligraphic bold letters to denote sets as elaborated in Table 1. Our FREE algorithm is outlined in Algorithm 1. It takes as inputs an unlabeled data set $\mathbb{S}$, the seed data example $\mathbf{s_0}$ and parameter τ, which is the threshold of data density similarity when determining whether two nearby clusters are from the same category. The output of FREE is $\mathbb{Y}_3$ which is the set of potential data examples from rare category $\mathbb{O}$.

FREE is carried out by four phases, namely (1) phase 1, data pre-processing (lines 1–2), (2) phase 2, feature space partition (line 3), (3) phase 3, search space reducing (lines 4–5), and (4) phase 4, refinement (line 6). Phase 1 removes outliers in data set $\mathbb{S}$ (line 1) and normalizes the set (line 2), and phase 2 partitions the feature space $\mathbb{V}$ of the normalized data set $\mathbb{S}_2$ into non-overlapped and equal-sized hyper rectangles (line 3) each of which is called a *cell* corresponding to a set of data examples. In phase 3, FREE reduces the search space by (1) finding out the local cluster $\mathbb{Y}_1$ surrounding the seed using wavelet transform (line 4) and removing cells not in $\mathbb{Y}_1$, and (2) removing from $\mathbb{Y}_1$ those cells having low data density similarity as compared with the cell containing the seed (line 5). Lastly in phase 4, by adjusting its view from cells to individual data examples, FREE refines the coarse shape of the rare category and identifies each data example of the objective rare category $\mathbb{O}$ via an MkNN (mutual kNN) based metric (line 6).

The above four phases, which will be elaborated in the following subsections, allow FREE to effectively identify rare categories of arbitrary shapes from data set $\mathbb{S}$ on linear time complexity w.r.t. the data set size n.

Table 1. Symbols and meanings

Symbol	Meaning
$\mathbb{S}$	The set of unlabeled data examples
$\mathbb{S}_1$	The set of non-outlier data examples in $\mathbb{S}$
$\mathbb{S}_2$	The set of data examples after normalization of $\mathbb{S}_1$
$\mathbb{V}$	the feature space of data set $\mathbb{S}_2$
$\mathbb{V}_1$	the new feature space after performing wavelet transform on $\mathbb{V}$
$\mathbb{G}$	The set of cells generated by feature space partition on $\mathbb{V}$
$\mathbb{O}$	The objective rare category
s_0	The original seed data example
s	The seed data example after normalization
$\mathbb{C}_s$	the cell containing the seed s
$\mathbb{Y}_1$	The set of cells which are in the local cluster surrounding s
$\mathbb{Y}_2$	The set of cells that form the coarse shape of rare category $\mathbb{O}$
$\mathbb{Y}_3$	The set of data examples found for rare category $\mathbb{O}$
τ	Threshold of data density similarity
n	The number of non-outlier data examples
d	The data dimension

Algorithm 1. FREE Algorithm

Input: $\mathbb{S}$, s_0, τ
Output: $\mathbb{Y}_3$
1 $\mathbb{S}_1 =$ **remove_outliers**($\mathbb{S}$);   // phase 1
2 $[\mathbb{S}_2, s] =$ **normalize**($\mathbb{S}_1, s_0$);   // phase 1
3 $\mathbb{G} =$ **feature_space_partition**($\mathbb{S}_2, s$);   // phase 2
4 $\mathbb{Y}_1 =$ **find_local_cluster**($\mathbb{G}, s$);   // phase 3
5 $\mathbb{Y}_2 =$ **find_coarse_shape**($\mathbb{Y}_1, s, \tau$);   // phase 3
6 $\mathbb{Y}_3 =$ **refine_the_coarse_shape**($\mathbb{Y}_2, s$);   // phase 4

4.1 Phase 1: Data Pre-processing

In this phase, FREE removes outliers in the data set and normalizes the data set. This can benefit FREE in terms of saving computational cost and improving accuracy.

To remove outliers effectively and efficiently, we adopt an outlier detection algorithm proposed in [2] which has a linear time complexity.

After removing outliers, we normalize data set $\mathbb{S}_1$ by the following equation.

$$x'_j = \frac{x_j - 0.5 \times (max_j + min_j)}{0.5 \times (max_j - min_j)} \tag{1}$$

where x_j is the value of data example $\mathbf{x}$ ($\mathbf{x} \in \mathbb{S}_1$) on jth dimension, x'_j is the new value after normalization, max_j is the maximum value of $\mathbf{x} \in \mathbb{S}_1$ on jth dimension, and min_j is the minimum value of $\mathbf{x} \in \mathbb{S}_1$ on jth dimension. By Eq. (1), the distribution range of $\mathbb{S}_1$ on each dimension is normalized within the closed interval $[-1, 1]$.

4.2   Phase 2: Feature Space Partition

After data pre-processing, feature space partition is performed to partition the normalized feature space into equal-sized *cells* each of which corresponds to a set of data examples and has a list of associated statistical properties (e.g., mean and variance).

Formalization. Let $\mathbb{S}_2$ be the set of data examples after normalization, $\mathbb{A}_j$ be the domain for the jth dimension of $\mathbb{S}_2$, and $\mathbb{V} = \mathbb{A}_1 \times \mathbb{A}_2 \times \cdots \times \mathbb{A}_d$ be the d-dimensional feature space. Then each data example corresponds to a point in the feature space $\mathbb{V}$ [14].

FREE partitions $\mathbb{V}$ into equal-sized and non-overlapped hyper rectangles each of which is called a *cell* by segmenting each dimension $\mathbb{A}_j$ into m_j number of equal-sized bins [14]. Each cell $\mathbb{C}$ has the form $\langle [\ell_1, r_1), [\ell_2, r_2), \cdots, [\ell_d, r_d) \rangle$, where ℓ_j and r_j stand for the left and right interval endpoint of $\mathbb{C}$ on jth dimension respectively[1]. A data example $\mathbf{x}$ is contained in a cell $\mathbb{C}$ iff $\ell_j \leqslant x_j < r_j$ for $1 \leqslant j \leqslant d$, where x_j is the value of $\mathbf{x}$ ($\mathbf{x} \in \mathbb{S}_2$) on jth dimension. Each cell $\mathbb{C}$ has a list of statistical parameters $P(\mathbb{C})$ associated with it. In this paper, we choose *bin count* n_c, which is the number of data examples in $\mathbb{C}$, and data density of $\mathbb{C}$ as $P(\mathbb{C})$.

Definition 1. *(Populated Cell) A cell $\mathbb{C}$ is a populated cell iff its bin count n_c is above a certain threshold ϵ_1 in feature space $\mathbb{V}$.*

For a non-populated cell, the data density calculated based on the very a few data examples is meaningless, so we only consider the data density of populated cells in the remaining sections. Data density of a populated cell $\mathbb{C}$ can be measured as follows.

Definition 2. *(Data Density) The data density t of cell $\mathbb{C}$ is defined as $t = \sum_{\mathbf{x} \in \mathbb{C}} \frac{n_c}{\|\mathbf{x} - \mathbf{c}\|_2}$ where $\mathbf{c}$ defined as $\mathbf{c} = \frac{1}{n_c} \sum_{\mathbf{x} \in \mathbb{C}} \mathbf{x}$ is the cluster center of $\mathbb{C}$.*

Definition 3. *(Data Density Similarity) The data density similarity between two populated cells $\mathbb{C}_i$ and $\mathbb{C}_k$ is defined as $\min \left(\frac{t_i}{t_j}, \frac{t_j}{t_i} \right)$, where t_i and t_j are the data densities of $\mathbb{C}_i$ and $\mathbb{C}_k$ respectively.*

Bandwidth Selection. In feature space partition, the structure of cells relies on how we partition each dimension of $\mathbb{S}_2$ into non-overlapped bins, i.e., how we select the bandwidths $\mathbf{h} = (h_1, h_2, \cdots, h_d)$ for all d dimensions. Let $\mathbb{C}_s$ be the cell containing the seed, the objectives of $\mathbf{h}$ selection are (1) after feature space partition using $\mathbf{h}$, data examples of rare category $\mathbb{O}$ should be in a few cells and

---

[1] If the data distribution range on jth dimension is e, and the bandwidth of bins on jth dimension is h_j, then the jth dimension is tabulated into $e/h_j + 1$ bins if $e \mid h_j$ or $\lceil e/h_j \rceil$ bins if otherwise. Note that the bin origin is the smallest value on jth dimension and each bin is of the same bandwidth h_j.

these cells especially $\mathbb{C}_s$ should not contain too many data examples from other categories, and (2) the number of cells should not be too large; otherwise a cell may contain too few data examples.

Given the above discussion, we propose an automatic selection of $\mathbf{h}$ by two steps, i.e., (1) select the initial $h_j (1 \leqslant j \leqslant d)$ by histogram density estimation (HDE for short) on the jth dimension, and (2) modify h_j to ensure the purity of $\mathbb{C}_s$.

(1) *HDE.* HDE [13] is a non-parametric density estimation which tabulates a single dimension of a data set into non-overlapped bins. Bandwidth plays the key role in HDE. A very small bandwidth results in a jagged histogram with each data example lying in a separate bin (under-smoothed bin partition), whereas a very large bandwidth results in a histogram with a single bin (over-smoothed bin partition) [17].

For h_j selection, Scott [13] established a theoretical criterion of bandwidth selection for non-parametric density estimation, which assumes that optimal bandwidth should minimize the mean integrated squared error MISE defined as

$$\text{MISE} = \int_{-\infty}^{+\infty} \text{bias}^2\left(f(x)\right) + var\left(f(x)\right) \, dx$$

where $f(x)$ refers to the data density function, $\text{bias}^2\left(f(x)\right)$ the squared bias and $var\left(f(x)\right)$ the variance. By utilizing cross validation, MISE is estimated as [13]

$$\Gamma(h_j) = \frac{5}{6nh_j} + \frac{1}{12n^2 h_j} \sum_k (v_{k-1} - v_k)^2 \tag{2}$$

where v_k stands for the number of data examples in the kth bin.

In practice, the optimal bandwidth h_j^* that minimizes $\Gamma(h_j)$ in Eq. (2) is selected, which can avoid under-smoothed and over-smoothed bin partition. In this paper, we use h_j^* as the initial bandwidth h_j.

(2) $\mathbf{h}$ *modification.* Since the seed cell $\mathbb{C}_s$ will be used as a fingerprint cell in finding the coarse shape of rare category $\mathbb{O}$, it should not contain too many data examples from other categories. To ensure the purity of $\mathbb{C}_s$, $\mathbf{h}$ is modified as follows.

Step 1: On each dimension, select $h_j = h_j^*$ for $1 \leqslant j \leqslant d$.

Step 2: Perform feature space partition using current bandwidths $\mathbf{h} = (h_1, h_2, \cdots, h_d)$.

Step 3: Perform K-means clustering on the data examples in $\mathbb{C}_s$ to divide the data examples into K subclusters (usually K is set to 3), and choose the K initial centers as (1) the seed $\mathbf{s}$ and (2) the data examples with maximum distance to the selected centers.

Step 4: After K-means clustering, data examples in $\mathbb{C}_s$ are grouped into K subclusters. Calculate the minimum data density similarity ∂ between each two of the K subclusters. If $\partial \geqslant \tau$ (τ is similarity threshold), then it is likely that the K subclusters are from the same category, thus the current $\mathbf{h}$ is returned as the final bandwidths. Otherwise, update h_j $(1 \leqslant j \leqslant d)$ by setting $h_j = h_j \times \beta$ $(0 < \beta < 1)$ and go to Step 2.

The reason for h_j becoming smaller and smaller is that the support region of rare category $\mathbb{O}$ is usually small due to the rareness and compactness of rare category $\mathbb{O}$.

Figs. 1(b) & 1(h) show the feature space partition of the two data sets (i.e., Figs. 1(a) & 1(g)) respectively. The red plus sign in each figure shows the position of the seed **s**.

4.3   Phase 3: Search Space Reducing

Intuitively, cells containing data examples from rare category $\mathbb{O}$ should be connected to the seed cell $\mathbb{C}_s$ in location and be comparable in data density with $\mathbb{C}_s$. Given this understanding, FREE reduces search space by (1) finding out the local cluster $\mathbb{Y}_1$ surrounding seed **s** using wavelet transform and removing the cells not in $\mathbb{Y}_1$, and (2) removing from $\mathbb{Y}_1$ those cells with low data density similarity to $\mathbb{C}_s$.

Wavelet Transform. Wavelet transform is a commonly used tool for signal processing. The motivation of adopting wavelet analysis comes from the observation that the *bin count* values in $\mathbb{V}$ the feature space of normalised data set can be considered as a d-dimensional signal. One scale wavelet transform on a 1D signal **S** is illustrated in Fig. 2(a). That is, **S** is passed through two filters, the low and high pass filters **L** and **H**, and is decomposed into two frequency sub-bands, i.e., high and low frequency sub-bands D and A. The low frequency parts with high amplitude correspond to the regions where data examples are concentrated, i.e., the basic cluster structures [14], whereas the high frequency parts correspond to the regions of rapid change in the data distribution of data examples, i.e., the boundaries of clusters.

Passing a signal through filter **F** can be explained as follows. For a 1D signal **g**, the filtering process is conducted by convolving the filter **F** with **g**. Formally, let $\hat{f}_k$ be the kth coefficient of **F**, M the length of **F**, **g**′ the new signal generated by filtering and g'_i the ith component of **g**′, then

$$g'_i = \sum_{k=1}^{M} \hat{f}_k g_{i+k-\lceil \frac{M}{2} \rceil} \tag{3}$$

An adequately wavelet such as the Reverse biorthogonal 4.4 (Rbio4.4 for short) wavelet as shown in Fig. 2(b), can act as an enhancer for the cells with high

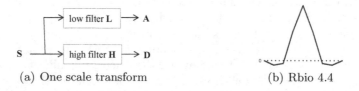

(a) One scale transform   (b) Rbio 4.4

Fig. 2. Wavelet transform

bin counts and a suppresser for the cells with low *bin counts*. Thus by applying wavelet transform with such a wavelet, the cluster structures can automatically stand out in the transformed feature space $\mathbb{V}_1$ [14].

The 1D wavelet transform in Eq. (3) can be generalized for d-dimensional feature space, where 1D wavelet transform will be applied d times, i.e., 1D wavelet transform is applied to each of the d dimensions in turn.

Definition 4. *(Significant Cell) A cell $\mathbb{C}$ is a significant cell iff its new bin count value after wavelet transform is above or equal to a threshold ϵ_2.*

(1) In the transformed feature space $\mathbb{V}_1$, each cell has its *bin count*, which is the new bin count value obtained by wavelet transform on the *bin count* values in $\mathbb{V}$. (2) In contrary to a *significant cell*, an *insignificant cell* is a cell with its bin count value in $\mathbb{V}_1$ lower than ϵ_2. Insignificant cells usually correspond to noise cells (cells with too few data examples) or cells at the boundary of a cluster. Removing them can help us clean the cluster boundary and save computational cost. (3) Threshold ϵ_2 plays the key role in determining significant cells. Instead of setting a fixed ϵ_2, we propose an automatic and data-based selection method of ϵ_2 in the next subsection.

Threshold ϵ_2 Selection

Step 1: Initialize $\mathbb{P}$ to be the set of *bin count* values in transformed feature space $\mathbb{V}_1$.

Step 2: For positive values in $\mathbb{P}$, calculate their average value a_1 and find the maximum positive value a_2.

Step 3: Using a_1 and a_2 as the initial cluster centers, perform K-means clustering with $K = 2$ to divide $\mathbb{P}$ into two clusters, one cluster of high values and the other of low values. Let ϵ_3 be the average value of the two cluster centers.

Step 4: If *bin count* value of $\mathbb{C}_s$ belongs to the high value cluster, return $\epsilon_2 = \epsilon_3$. Otherwise, remove the *bin count* values higher than ϵ_3 from $\mathbb{P}$, and go to step 2.

In step 2, cells with negative or zero values in $\mathbb{V}_1$ are ignored because these cells correspond to regions with too few data examples or local drops of data density; whereas due to the compactness characteristic of rare category $\mathbb{O}$, cells containing data examples from $\mathbb{O}$ usually correspond to local increases of data density.

In step 4, if the *bin count* value of the seed cell $\mathbb{C}_s$ belongs to the low value cluster, then there are cells with *bin count* values dramatically higher than that of $\mathbb{C}_s$. Usually these cells contain data examples in the center of a category, thus they are significant cells and will be reserved. Nonetheless, it is unnecessary to reserve them in the ϵ_2 selection procedure. So we remove these values from set $\mathbb{P}$ and return to step 2 to find the correct threshold. In Figs. 1(c) & 1(i), the blue and red cells are significant cells (the red cell is the seed cell).

In the transformed feature space $\mathbb{V}_1$, each cell forms a hyper rectangle. Letting the sides of the hyper rectangle be the sides of the cell, we have the following definitions.

Definition 5. (Adjacent Cells) *Given two cells* $\mathbb{C}_i$ *and* $\mathbb{C}_j$ *where* $i \neq j$ *with respective centers* $C_i = (c_{i1}, c_{i2}, \dots, c_{id})$ *and* $C_j = (c_{j1}, c_{j2}, \dots, c_{jd})$, *assume the side length of a cell on dimension* k *is* ℓ_k *where* $1 \leqslant k \leqslant d$. *Cells* $\mathbb{C}_i$ *and* $\mathbb{C}_j$ *are adjacent cells iff* $|c_{ik} - c_{jk}| \leqslant \ell_k$ *(i.e., either* $c_{ik} = c_{jk}$ *or* $|c_{ik} - c_{jk}| = \ell_k$) *for* $1 \leqslant k \leqslant d$.

Definition 6. (Connected Cells) *Two cells* $\mathbb{E}$ *and* $\mathbb{F}$ *are connected iff there exists a sequence of cells* $\langle \mathbb{C}_1, \mathbb{C}_2, \cdots, \mathbb{C}_j \rangle$, *where* $\mathbb{C}_1 = \mathbb{E}$ *and* $\mathbb{C}_j = \mathbb{F}$, *and* $\mathbb{C}_i$ *is adjacent to* $\mathbb{C}_{i+1}$ *for* $1 \leqslant i \leqslant j - 1$.

Definition 7. (Connected Cluster from a Cell) *The connected cluster from cell* $\mathbb{E}$ *is the set of connected cells of* $\mathbb{E}$.

Given the above definitions, the local cluster surrounding **s** is exactly the connected cluster from seed cell $\mathbb{C}_s$. In Figs. 1(d) & 1(j), the blue and red cells show the detected local cluster $\mathbb{Y}_1$ surrounding **s**. In Fig. 1(d), since rare category $\mathbb{O}$ is isolated from other categories, $\mathbb{Y}_1$ contains only the cells containing data examples in rare category $\mathbb{O}$. In Fig. 1(j), since rare category $\mathbb{O}$ overlaps with a majority category, both the clusters of the majority category and rare category $\mathbb{O}$ are reserved for further study.

After identifying local cluster $\mathbb{Y}_1$, cells not in $\mathbb{Y}_1$, including insignificant cells and significant cells unconnected to seed cell $\mathbb{C}_s$, can be removed. Thus the search space is reduced dramatically to $\mathbb{Y}_1$ to which the subsequent operations will be applied.

Data Density Analysis. Since rare category $\mathbb{O}$ may overlap with other majority categories, $\mathbb{Y}_1$ may contain both $\mathbb{O}$ and the dense part of other categories (e.g., see Fig. 1(j)). The following data density analysis will further reduce the current search space $\mathbb{Y}_1$.

Definition 8. (Peak) *A cell is a peak iff its* bin count *value in* $\mathbb{V}_1$ *is higher than that of all its adjacent cells.*

In Figs. 1(d) & 1(j), the red cells show peaks in the local cluster. Given Definition 8, FREE figures out the coarse shape of rare category $\mathbb{O}$ from local cluster $\mathbb{Y}_1$ as follows. (1) Find out peaks in $\mathbb{Y}_1$. (2) Calculate the data density similarities between seed cell $\mathbb{C}_s$ and all peaks by Definition 3[2]. (3) Divide peaks into two groups, namely group $\mathbb{M}_1$ of cells with high similarities to $\mathbb{C}_s$ and the other group $\mathbb{M}_2$ of low similarities. This is done by performing K-means clustering with $K = 2$. (4) For each cell $\mathbb{C} \in \mathbb{Y}_1$, calculate its average data

---

[2] If the data density similarity of any two peaks is not less than threshold τ, then it is very likely that all cells in $\mathbb{Y}_1$ are from the same category, and $\mathbb{Y}_1$ will be returned as the coarse shape $\mathbb{Y}_2$ of rare category $\mathbb{O}$.

density similarity to the cells in M_1 and M_2. Remove cells that is more similar to M_2 from Y_1. (5) From the remaining cells after the above four steps of selection, save the cells connected to C_s to set Y_2.

The blue cells in Figs. 1(e) & 1(k) show the result Y_2 of phase 3 for the data sets shown in Figs. 1(a) & 1(g) respectively.

4.4   Phase 4: Refinement

After the aforementioned three phases, the search space is reduced to Y_2, namely the coarse shape of rare category O. One can return the set of data examples in Y_2 as the potential set Y_3 of rare category data examples because there are very a few false negative and false positive data examples in Y_2. Nonetheless, to achieve higher accuracy, FREE continues to refine the result by zooming in its view from cells to individual data examples.

Let n' be the number of data examples in coarse shape Y_2. Due to the rareness of rare category O, $n' \ll n$. Thus fine-grained metrics can be used in this phase to refine the result obtained by Y_2. FREE adopts MkNN (mutual kNN) metric since it is capable of identifying compact clusters and can handle rare categories of arbitrary shapes.

Set Y_3 of potential data examples from rare category O is initialized as $\{s\}$. It keeps absorbing external data examples in Y_2 that are MkNN of at least one data example in Y_3 until convergence, and is returned as the potential set of rare category O.

k Selection. Parameter k plays an important role in MkNN. The objectives of k selection are (1) k should be relatively large to absorb as many data examples as possible because data examples in Y_2 have high probabilities to be from rare category O, and (2) k should not be extremely large to avoid absorbing too many *false positives*.

For k selection, FREE first performs K-means clustering on data examples in seed cell C_s to divide them into K subclusters (usually K is set to 3). Let u be the center data example in the subcluster containing seed s. We have the following claim [6].

Claim 1. *If Y_2 contains false positive data examples, then there is an abrupt increase on m-covering radius of data example u when m exceeds a threshold, where m-covering radius of u refers to the distance between u and its mth nearest neighbor.*

Proof. For the convenience of explanation, we assume that rare category O overlaps with majority category T if O is not isolated from other categories. Similar analysis applies to the case that O overlaps with a few categories.

If Y_2 contains *false positive* data examples, then there are two cases. *Case 1:* Rare category O is isolated from other categories or O overlaps with majority category T but the data density of T in the overlapped area is neglectable. For this case, the *false positive* data examples are very a few and will become local outliers in $\mathbb{Y}_2$. *Case 2:* $\mathbb{O}$ overlaps with a majority category $\mathbb{T}$ and the data density of $\mathbb{T}$ in the overlapped area is not neglectable. For this case, $\mathbb{Y}_2$ contains a few data examples from $\mathbb{T}$.

Let n_r be the number of data examples of $\mathbb{O}$ in ground truth. For case 1, the threshold is $n_r - 1$ because (1) when $m \leqslant n_r - 1$, data example $\mathbf{u}$ finds its mNN in the compact cluster of $\mathbb{O}$ with small m-covering radius due to the compactness of $\mathbb{O}$, and (2) when $m = n_r$, $\mathbf{u}$ must find its n_rth nearest neighbor from the local outliers outside $\mathbb{O}$. Thus there must be an abrupt increase in m-covering radius.

For case 2, let a be the data density of $\mathbb{T}$ in the overlapped area, b the data density of $\mathbb{O}$ and ζ the number of data examples located in the compact cluster of $\mathbb{O}$ but are actually from $\mathbb{T}$. Then the threshold is $n_r + \zeta - 1$ because (1) when $m \leqslant n_r + \zeta - 1$, data example $\mathbf{u}$ finds its mNN in the compact cluster of $\mathbb{O}$ with small m-covering radius, and (2) when $m = n_r + \zeta$, $\mathbf{u}$ must find its $(n_r + \zeta)$th nearest neighbor from outside $\mathbb{O}$. Since (i) the data density at the compact cluster of $\mathbb{O}$ is $a + b$, (ii) data density at the region of data examples from $\mathbb{T}$ is b, and (iii) a is a relatively large value due to the compactness of rare category $\mathbb{O}$, there must be an abrupt increase in m-covering radius. ∎

Following the above discussion, the optimal k value is such selected that the relative increase on m-covering radius $\frac{r(k+1)-r(k)}{r(k)}$ (where $2 \leqslant k \leqslant n'$) can be maximized [6], where $r(k)$ stands for the k-covering radius of data example $\mathbf{u}$.

If there is no abrupt increase on m-covering radius of data example $\mathbf{u}$, then $\mathbb{Y}_2$ must contain only data examples from rare category $\mathbb{O}$. Thus we can just return $\mathbb{Y}_2$ as the potential set $\mathbb{Y}_3$. The red points in Figs. 1(f) & 1(l) show the data examples in the potential set $\mathbb{Y}_3$ explored for the rare category shown in Figs. 1(a) & 1(g) respectively.

4.5   Dimension Segmenting and Pruning

When the data dimension goes too high, there may be some noisy dimensions where data examples from rare category $\mathbb{O}$ are not so compact, and the cell number may be extremely large resulting that most cells have very few data examples. In this case, our algorithm may not achieve satisfactory performance. On the contrary, if we keep only the most compact dimensions where the compactness characteristics of rare category $\mathbb{O}$ preserve, the accuracy and efficiency of FREE will be highly improved. Given this, to handle high dimensional data sets, we propose to extend the framework of FREE with dimension segmenting and pruning, i.e., iterate the four phases of RCE and remove the worst (most incompact) dimensions. The objective is to segment all dimensions and prudently prune away the most incompact dimension in each segment.

The detailed dimension segmenting and pruning process is carried out by four steps as follows. (1) Build the initial list of ordered dimensions by appending all dimensions into the list one by one, denote the list as $\mathbf{L} = (D_1, D_2, \ldots, D_d)$. (2) Divide list $\mathbf{L}$ into k segments and denote all segments as $\mathbf{G}_1 = (D_1, D_2, \ldots, D_{g_1})$, $\mathbf{G_2} = (D_{g_1+1}, D_{g_1+2}, \ldots, D_{g_1+g_2}), \ldots, \mathbf{G_k} = (D_{g_1+g_2+\ldots+g_{k-1}+1}, D_{g_1+g_2+\ldots+g_{k-1}+2}, \ldots, kD_{d-1}, D_d)$. Let ℓ_i be the number of cells in the partitioned feature subspace formed by dimensions in segment $\mathbf{G_i}$. The k segments must satisfy that ℓ_i

$(1 \leqslant i \leqslant k)$ is maximized and $\ell_i \leqslant \delta \cdot n$ where δ is a coefficient smaller than 3. (3) Run the four phases of RCE on each segment to find the potential set $\mathbb{Y}_3$ for rare category $\mathbb{O}$, sort the dimensions in each segment on an ascending (or descending) order based on the compactness measurement defined in what follows, and prune away the least compact dimension from the segment. (4) Rebuild $\mathbf{L}$ by collecting the sorted and size-reduced segments together and keeping the order of dimensions in each segment, recursively apply the same operations of steps (2) & (3) to this newly built list $\mathbf{L}$ until it cannot be further segmented.

The most compact dimension in each segment is called the *candidate dimension*. The rebuilding of dimension list in the above steps tries to avoid two *candidate dimensions* being assigned into one segment.

Compactness Measurement. The measurement for compactness of dimension j is defined as

$$w_j = \left(\sum_{\ell=1}^{d} \left(\frac{var(j)}{var(\ell)} \right)^{\frac{1}{\beta-1}} \right)^{-1} \tag{4}$$

where β is the amplification factor and is set to 1.5 in our solutions [9], and $var(i)$ is the normalized variance on the ith dimension, which is defined as

$$var(i) = \sum_{\mathbf{x}\in\mathbb{Y}_3} (x_i - o_i)^2 / \sum_{\mathbf{y}\in\mathbb{S}_2} (y_i - s_i)^2$$

where o_i and s_i are the cluster center of $\mathbb{Y}_3$ and $\mathbb{S}_2$ on the ith dimension respectively, and x_i and y_i are the values of $\mathbf{x}$ and $\mathbf{y}$ on the ith dimension respectively.

In Eq. (4), w_j measures the compactness of the data examples in $\mathbb{Y}_3$ on the jth dimension, namely, w_j will get a high value if data examples in $\mathbb{Y}_3$ are compact on the jth dimension and a low value if otherwise.

4.6 Time Complexity Analysis

Phase 1 has a linear time complexity w.r.t. data set size n. The time complexity of phase 4 is very low because the number n' of data examples in phase 4 satisfies $n' << n$. For phase 2, operations can be done by a few times of passing through all data examples with $O(n)$ time complexity; for phase 3, the most time consuming step is wavelet transform on feature space $\mathbb{V}$.

Theorem 1. *The time complexity of wavelet transform on feature space $\mathbb{V}$ is $O(dn)$.*

Proof. Wavelet transform on feature space $\mathbb{V}$ is done by applying the 1D wavelet transform to each dimension of $\mathbb{V}$ in turn. Letting a_j be the number of bins on jth dimension and z_j be the number of signals on jth dimension, we have $z_j = (\prod_{i=1}^{d} a_i)/a_j$. Let γ be the number of operations in wavelet transform and c the number of cells in $\mathbb{V}$, then the following equation holds.

$$\gamma = \lambda \times \Sigma_{j=1}^{d} z_j < \lambda \times d \times \prod_{i=1}^{d} a_i = \lambda \times d \times c$$

where λ is a small constant. In summary, $\gamma < \lambda \times d \times c$. Since in FREE, $c \leqslant \delta \times n$ always holds where δ is a coefficient, we have $\gamma < \lambda \times d \times \delta \times n$. Thus Theorem 1 is proven.   ∎

Following Theorem 1, the time complexity of phase 3 is $O(dn)$. Thus the overall time complexity of FREE is $O(dn)$.

5   Experimental Evaluation

In this section, we conduct experiments to verify the effectiveness and efficiency of FREE from three aspects, namely (1) accuracy, (2) efficiency, and (3) ability to discover rare categories of arbitrary shapes. We compare our method with RACH [5] and FRANK [6] the state-of-the-art algorithms among the existing RCE approaches of the same type. All algorithms are implemented with MATLAB 7.11 and executed on a server computer with Intel Core 4 2.4 GHz CPU and 20 GB RAM.

(1) Effectiveness Study Effectiveness study is first conducted on six UCI data sets [1] which are commonly used in the existing rare category study [3,4,7,8,12]. Specifically, the six data sets are sub-sampled to create RCE scenario. The left part of Table 2 shows the detailed information about six data sets, in which m stands for the number of categories, $\#y$ the category index of rare category ⓪ out of all categories and n_r the number of data examples in rare category ⓪.

For each data set, we first run each of three rare category detection algorithms (NNDM [3], HMS [16] and FRED [12]) to select different data examples from rare category ⓪ as three seeds, and run each of three RCE algorithms (RACH, FRANK and FREE) using the three seeds and record their average F-scores. Parameter τ in FREE is set to 0.75 for all six data sets.

The average F-score of each tested algorithm is reported in Fig. 4. The vertical axis is F-score, i.e., the harmonic mean of precision and recall, and the horizontal axis is the six data sets. Page Blocks set is in short as Page in all figures. From the figure, we can see that on all six data sets, the F-score of FREE is dramatically higher than those of other tested algorithms.

The reasons of the observations are that the rare categories in UCI data sets have different shapes and may overlap with other categories, whereas RACH tries

Table 2. Properties of UCI data sets

Data set	n	d	m	$\#y$	n_r	FREE	RACH	FRANK
Iris	104	4	3	1	5	**0.0019**	0.0028	0.0022
Glass	123	7	5	2	10	**0.0028**	0.0030	0.0037
Ecoli	288	5	8	3	4	**0.0036**	0.0039	0.0045
Page blocks	4067	8	5	4	29	**0.0231**	0.0310	0.0421
Bank	33631	2	2	2	8	**0.2334**	0.8414	2.5391
Abalone	43193	9	8	7	5	**0.5081**	723.8544	477.5160

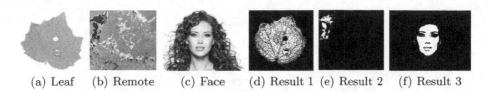

(a) Leaf   (b) Remote   (c) Face   (d) Result 1 (e) Result 2   (f) Result 3

Fig. 3. Experiments on vision data sets

Table 3. Properties of vision data sets

Data set	n (number of pixels)	d	FREE	RACH	FRANK
Leaf	$51700 = 235 \times 220$	2	**1.7039**	timeout	timeout
Remote	$60630 = 258 \times 235$	2	**1.8899**	timeout	timeout
Face	$85767 = 339 \times 253$	2	**5.6718**	timeout	timeout

to enclose the rare category data examples with a minimum-radius hyperball and FRANK requires the rare category to be isolated from other categories.

Vision Data Sets. Effectiveness study is also conducted on three vision data sets which are real-world pictures shown in Figs. 3(a)–(c). Fig. 3(a) is a picture of a leaf, Fig. 3(b) is a remote sensing picture where blue parts are lakes and green parts are land, and Fig. 3(c) is a human face picture. In the vision data sets, each picture corresponds to a data set and each pixel in the picture corresponds to a data example. The goal is to identify pixels that are similar to a given pixel. The left part of Table 3 shows the detailed information about the three vision data sets. Each data example has two dimensions (or attributes), i.e., its gray value and distance to the seed data example.

Figs. 3(d)–(f) illustrate the mining result of FREE on the three data sets respectively, where white pixels are the data examples found by FREE. In Fig. 3(d), FREE detects the leaf shape from a pixel in the green parts of the original picture; In Fig. 3(e), FREE detects the water distribution from a pixel in the blue parts of the original picture; and In Fig. 3(f), FREE detects the lady's face from a pixel within the face of the lady. We do not present the results of RACH and FRANK because they fail to obtain a result for any of the three pictures with 24 hours of the given time threshold.

From the figures, we have the following conclusions. (1) The results further verify the effectiveness of our algorithm, and show the promising applications of FREE on vision data sets. (2) Because this experiment is mainly to verify the effectiveness of FREE, we only extract two attributes for each data example. One can select different features as attributes to mine the information of interest from the pictures.

(2) Efficiency Study. The runtime of FREE, RACH, and FRANK on both UCI and vision data sets are reported in the right part of Tables 2 & 3, where the time unit is minute and the time threshold for *timeout* is 1440 minutes (24 hours). We also depict the time curve of each algorithm in Fig. 5 using these

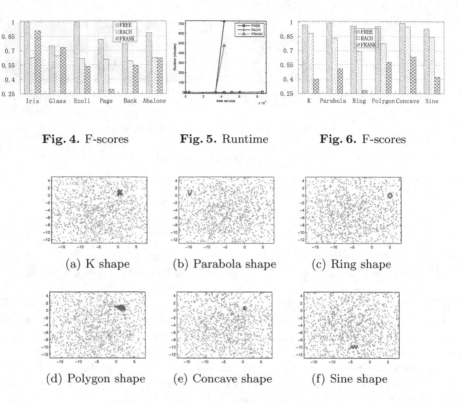

Fig. 4. F-scores   **Fig. 5.** Runtime   **Fig. 6.** F-scores

(a) K shape   (b) Parabola shape   (c) Ring shape

(d) Polygon shape   (e) Concave shape   (f) Sine shape

Fig. 7. Arbitrary shaped rare categories

data. From the figure, we can see that FREE is much efficient than other tested algorithms. For example, on Abalone set with 43193 data examples, the runtime of FREE is 0.5081 minutes, whereas FRANK takes 477.5160 minutes and RACH 723.8544 minutes.

The above observation results from that FREE has a linear time complexity w.r.t. data set size n, whereas FRANK has a quadratic time complexity and RACH has too many iterations in solving convex optimization problem.

(3) Arbitrary Shaped Rare Categories. In this experiment, we use data sets generated by our own synthetic generator and the ones used in [14] and [11]. To clearly express the arbitrary shape problem, in our settings there are only one majority category and one rare category. The data examples of the majority category are generated according to a 2D Gaussian distribution, whereas data examples of the rare category are generated randomly within a specific shape. We generate six different data sets each of which has a rare category of specific shape including K shape, parabola shape, ring shape, polygon shape, concave shape, and sine shape as shown in Figs. 7(a)–7(f). The majority category of each data set consists of 2000 data examples and the number of data examples in the rare category of different data sets varies from 60 to 190.

Fig. 6 shows the comparison results in terms of F-score on the data sets of arbitrary shapes. Note that the vertical axis is F-score and the horizontal axis is the six data sets. From the figure, we can see that the F-score of FREE is dramatically higher than those of other tested algorithms. The results further justify the effectiveness of our solution which does not *a priori* assume the shape of the rare categories.

6   Conclusion

We have proposed FREE for RCE, which achieves linear time complexity w.r.t. data set size and is able to handle rare categories of arbitrary shapes. FREE can carry out RCE tasks without certain number of training data examples or prior knowledge of the data set. This has been achieved by (1) finding out the coarse shape of the objective rare category using coarse-grained metrics, and (2) refining the coarse shape by fine-grained metrics. Extensive experiments have verified the efficiency and effectiveness of FREE.

Acknowledgments. This work was partly supported by the NSFC Grant 61472359, and the Key Science and Technology Innovation Team Fund of Zhejiang under Grant No. 2010R50041.

References

1. Asuncion, A., Newman, D.: UCI Machine Learning Repository (2007)
2. Bay, S.D., Schwabacher, M.: Mining distance-based outliers in near linear time with randomization and a simple pruning rule. In: KDD, pp. 29–38, Washington, DC, USA, August 24–27, 2003
3. He, J., Carbonell, J.: Nearest-neighbor-based active learning for rare category detection. In: Advances in Neural Information Processing Systems 20 (NIPS 2007), pp. 633–640, Vancouver, British Columbia, Canada, December 3–6, 2007
4. He, J., Carbonell, J.: Prior-free rare category detection. In: Proceedings of the SIAM International Conference on Data Mining (SDM 2009), pp. 155–163, Sparks, Nevada, USA, April 30-May 2, 2009
5. He, J., Tong, H., Carbonell, J.: Rare category characterization. In: The 10th IEEE International Conference on Data Mining (ICDM 2010), pp. 226–235, Sydney, Australia, December 14–17, 2010
6. Huang, H., Chiew, K., Gao, Y., He, Q., Li, Q.: Rare category exploration. ESWA **41**(9), 4197–4210 (2014)
7. Huang, H., He, Q., Chiew, K., Qian, F., Ma, L.: CLOVER: A faster prior-free approach to rarecategory detection. Knowledge and Information Systems **35**(3), 713–736 (2013)
8. Huang, H., He, Q., He, J., Ma, L.: RADAR: Rare category detection via computation of boundary degree. In: Huang, J.Z., Cao, L., Srivastava, J. (eds.) PAKDD 2011, Part II. LNCS, vol. 6635, pp. 258–269. Springer, Heidelberg (2011)
9. Huang, J.Z., Ng, M., Rong, H., Li, Z.: Automated variable weighting in k-means type clustering. TPAMI **27**(5), 657–668 (2005)

10. Li, S., Z. Wang, Zhou, G., Lee, S.: Semi-supervised learning for imbalanced sentiment classification. In: Proceedings of the Twenty-Second International Joint Conference on Artificial Intelligence, pp. 1826–1831 (2011)
11. Liu, Z., Chiew, K., He, Q., Huang, H., Huang, B.: Prior-free rare category detection: More effective and efficient solutions. ESWA **41**(17), 7691–7706 (2014)
12. Liu, Z., Huang, H., He, Q., Chiew, K., Ma, L.: Rare category detection on $O(dN)$ timecomplexity. In: The 18th Pacific-Asia Conference on Knowledge Discovery and Data Mining(PAKDD 2014), pp. 498–509, Tainan, Taiwan, May 13–16, 2014
13. Scott, D.W.: Multivariate Density Estimation: Theory, Practice, and Visualization. Wiley, New York (1992)
14. Sheikholeslami, G., Chatterjee, S., Zhang, A.: Wavecluster: A wavelet-based clustering approach for spatial data in very large databases. The VLDB Journal **8**(3–4), 289–304 (2000)
15. Tang, Y., Zhang, Y., Chawla, N., Krasser, S.: SVMs modeling for highly imbalanced classification. IEEE Transactions on systems, man, and cybernetics **39**(1), 281–288 (2009)
16. Vatturi, P., Wong, W.: Category detection using hierarchical mean shift. In: Proceedings of the 15th ACM SIGKDD International Conference on Knowledge Discovery and Data Mining (KDD 2009), pp. 847–856, Paris, France, June 28-July 1, 2009
17. Wand, M.P.: Data-based choice of histogram bin width. The American Statistician **51**(1), 59–64 (1997)

Probabilisstic and Uncertain Data

FP-CPNNQ: A Filter-Based Protocol for Continuous Probabilistic Nearest Neighbor Query

Yinuo Zhang[1]([✉]), Anand Panangadan[2], and Viktor K. Prasanna[2]

[1] Department of Computer Science, University of Southern California,
Los Angeles, California, USA
{yinuozha,anandvp,prasanna}@usc.edu
[2] Ming Hsieh Department of Electrical Engineering,
University of Southern California, Los Angeles, California, USA

Abstract. An increasing number of applications in environmental monitoring and location-based services make use of large-scale distributed sensing provided by wireless sensor networks. In such applications, a large number of sensor devices are deployed to collect useful information such as temperature readings and vehicle positions. However, these distributed sensors usually have limited computational and communication power and thus the amount of sensor queries should be reduced to conserve system resources. At the same time, data captured by such sensors is inherently imprecise due to sensor limitations. We propose an efficient probabilistic filter-based protocol for answering continuous nearest neighbor queries over uncertain sensor data. Experimental evaluation on real-world temperature sensing data and synthetic location data showed a significant reduction in the number of update messages.

1 Introduction

A *range query* is a test of whether a variable has its value within a specified range. When the variable is to be monitored so that a result is returned whenever its value enters the specified range, the range query is typically registered at a server and the process is called a *continuous range query*. Continuous queries, in particular continuous range queries, has attracted significant research interest with the development of wireless sensor networks and moving object databases. For example, a query to an environment monitoring sensor network could be to continuously return the identity of all temperature sensors with their readings above a certain threshold (if the temperature reading is above this limit, it could indicate a fire at the sensor location). A range query is a special case of a *non-aggregate* query as the answer only depends on the value of the object being queried (for instance, a sensor measurement or location). On the other hand, an *aggregate* query is one where accessing a single object does not provide enough information to answer the query. For example, whether vehicle v is in a specific region only depends on its own position (non-aggregate query), while

© Springer International Publishing Switzerland 2015
M. Renz et al. (Eds.): DASFAA 2015, Part II, LNCS 9050, pp. 57–73, 2015.
DOI: 10.1007/978-3-319-18123-3_4

whether v is the nearest vehicle to building b also depends on other vehicles (aggregate query). Such *continuous nearest neighbor queries* are important in location-based services. For example, a query that can continuously report the nearest ambulance to an accident site can enable rapid response to accidents.

An additional complication in practical sensor network deployments is the inherent uncertainty in the measurement process due to reasons such as sensor inaccuracy, discrete sampling intervals, and network latency. In previous work [31], we adopted the *attribute uncertainty* model to formalize the querying of uncertain data sources. The attribute uncertainty model assumes that the true value of the object being queried is within a closed region with a non-zero probability density function (PDF) for the value of interest. This region of uncertainty is an interval for the one-dimensional case, while it is a closed 2D region (e.g., circle or rectangle) for the two-dimensional case. The PDF can take on any distribution such as uniform distribution or Gaussian distribution. Figure 1 illustrates an example of attribute uncertainty for one-dimensional and two-dimensional data.

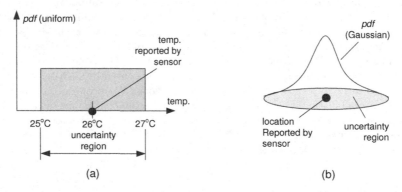

Fig. 1. Uncertainty of (a) temperature and (b) location

Queries can be issued over such uncertain data sources. Incorporating attribute uncertainty in continuous query processing results in *continuous probabilistic queries* (CPQ). In this context, "probabilistic" refers to a threshold condition defined as a probability bound that is provided with the query. For example, instead of a query requesting the identity of all temperature sensors having their readings within a specific range, a *probabilistic* query requests the identity of all sensors that are likely to have readings within that range with a probability higher than some pre-defined threshold.

As previously described, continuous queries over uncertain data can be object-independent as in the case of *continuous probabilistic range queries* (CPRQ) or be of an aggregate type, as in *continuous probabilistic nearest neighbor queries* (CPNNQ).

In [31], we comprehensively investigated object-independent queries with emphasis on range queries. We proposed a probabilistic filter protocol that reduces both communication and computational cost during query execution. Efficient concurrent query execution is also enabled with a multiple query protocol. In this paper, we extend the filter-based approach to the case of object-dependent queries over uncertain data. Such aggregate queries are more difficult to answer compared to object-independent queries since it requires accessing more information. We evaluate our approach for the special case of CPNNQ. Our main contributions in this paper are:

- Formalization of continuous probabilistic nearest-neighbor query (CPNNQ)
- A filter-based protocol to efficiently answer CPNNQs
- Evaluation of the proposed approach with comprehensive experiments on real sensor datasets

The rest of this paper is organized as follows. Section 2 presents related work. Section 3 describes the proposed data system and query model. A novel probabilistic filter protocol for CPNNQ is proposed in Section 4. In Section 5, we describe our experimental setup for evaluation and discuss the results. We conclude this paper in Section 6.

2   Related Work

In this section, we first summarize current research in efficient continuous query processing followed by progress in probabilistic query execution.

2.1   Continuous Query Processing

Efficient continuous query execution has been widely studied in the database community. Most of this work focuses on reducing the update frequency and computation load during query execution. In wireless sensor networks (WSNs), in-network processing is a common strategy to provide energy-efficient query execution in terms of communication cost [1,13]. Research has resulted in efficient protocols for a single type of query execution including aggregate query [18], top-k query [29] and spatio-temporal query [8]). In [15], a general framework was proposed to find the maximum lifetime for continuous innetwork evaluation of expression trees and can efficiently execute continuous queries in WSNs. However, in-network processing demands a large query processing capability at all sensors. In order to alleviate this requirement, [24] proposed a technique which does not require that a sensor be able to resolve a query and also supports multiple types of queries. In order to reduce transmission power, [2] also presented an optimized query routing tree which can provide a path to transmit query results to the querying node.

Most of the above works focus on the resolution of a single query. Prabhakar et al. introduced an efficient indexing framework to handle query arrival and removal for multiple query execution [23]. Xiong et al. [30] proposed an incremental algorithm for reducing query re-evaluation cost by sharing execution effort among concurrently-executing queries. Muller et al. proposed a network query approach which combines multiple requests [20]. The results for a specific user are then extracted from the results for the corresponding network query. Li et al. developed an algorithm for evaluating multiple queries, which exploit the sharing of data movement among different queries [16]. [17] provided a scalable energy-efficient multi-query processing framework which enables sharing information among different queries.

Another technique for reducing system load is *stream filter* [6,7,10,22,29], in which some query answering tasks are deployed to remote streaming sources (e.g., sensors and mobile devices). Each remote source is associated with *filter constraints* derived from a given continuous query. These constraints are used to decide whether an object needs to report its newest value to the server. Since the filter prevents all values from being sent to the server, a substantial amount of communication effort can be saved. However, data uncertainty is wide-spread in real-world applications and this issue is not addressed in existing work. In this paper, we investigate the problem of efficiently executing attribute-uncertain queries. Specifically, we develop *probabilistic filters* for continuous probabilistic nearest-neighbor queries which utilize the uncertainty information associated with sensor measurement.

2.2   Uncertainty Management in Query Execution

Chen et al. [4] studies the problem of *updating* answers for continuous probabilistic nearest neighbor queries in the server. They developed an efficient algorithm to update the answers without re-evaluating the whole query. [27,28] investigated the problem of efficiently executing continuous nearest neighbor (NN) queries for uncertain moving objects trajectories. [21] addressed probabilistic nearest neighbor queries in uncertain trajectories databases using a Markov chain model. Note that these works only handle continuous queries on the server and do not use filters to reduce communication and energy costs. Farrell et al. [11,12] proposed the notion of spatial and temporal tolerance, and examined the use of these semantics to support energy-efficient sensing of location data. The uncertainty in location value is modeled by a uniform uncertainty region; the possibility of a non-uniform PDF representing the uncertainty within the region is not considered. Moreover, the results of the queries studied in those works are not probabilistic. In this paper, we consider attribute-uncertain data and probabilistic queries.

In our preliminary work [32], we developed a filter protocol for single continuous non-aggregate query execution. In [31], we examined how to handle multiple queries efficiently. We also proposed slack filters to approximate filter regions that are hard to represent. We used range query, a typical aggregate query, as a case study in these works. In [14], we investigated efficient continuous aggregate query execution over uncertain data. However, this work mainly focus on *possible* instead of probabilistic queries (this is the special case where the probability threshold specified in the query condition is either 0 or 1). In this paper, we explore the problem of probabilistic aggregate query execution. Specifically, we will develop a filter-based protocol for continuous nearest neighbor queries over uncertain data.

3   System Architecture

In this section, we present the system model for query execution and the probabilistic query definition.

3.1   System Model

We adopt the system model defined in [31]. Figure 2 shows the system framework which has following components.

- **Uncertainty Database** stores an error model (e.g., attribute uncertainty model with region and distribution) for each type of sensor and the most recent value reported by each sensor.
- **Query Manager** receives query requests from users and evaluates them based on the data in the *uncertainty database* (e.g., [5]).
- **Filter Manager** derives *filter constraints*: the query information and data uncertainty is sent to the sensors which use this information to decide if they should report any updated value. This step reduces the energy and network bandwidth consumption.
- Each **sensor** has a *data collector* which retrieves data values (e.g., temperature or position coordinates) from external environments and a set of *filter constraints*, which are boolean expressions for determining whether the value obtained from the data collector is to be sent to the server.

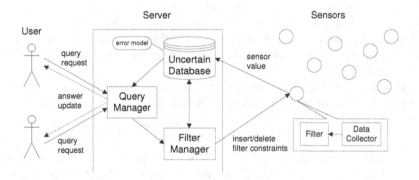

Fig. 2. System Architecture

Table 1 summarizes the symbols used throughout this paper. We describe a one-dimensional data uncertainty model (e.g., Figure 1(a)) to illustrate our techniques. However, the techniques can be directly applied to the multi-dimensional case since we can project the distance between the sensed value of a sensor and the query position to one dimension.

Table 1. Symbols used in the paper

Symbol	Description
o_i	ID of the i-th sensor, where $1 \leq i \leq n$
$v_i(t)$	Sensed value of o_i at time t
P	Probability threshold for q
$p_i(t)$	Qualification probability of o_i at time t for q
$b_i = [l_i, u_i]$	1D probabilistic filter of o_i for q_j

3.2 Continuous Probabilistic Queries

Let $o_1, \ldots, o_n$ be the IDs of n sensing devices monitored by the system. A Continuous Probabilistic Query (CPQ) [31] is defined as:

Definition 1. *Given a time interval* $[t_1, t_2]$, *a real value* $P \in (0, 1]$, *a* **CPQ** *q returns a set of IDs* $\{o_i | p_i(t) \geq P\}$ *at every time instant* t, *where* $t \in [t_1, t_2]$, *and* $p_i(t)$ *is the probability that the value of* o_i *satisfies query* q *at time* t.

We also call $[t_1, t_2]$ the *lifetime* of q. In this paper, we focus on a special case of CPQ called *continuous probabilistic nearest neighbor query* (**CPNNQ**), as defined below:

Definition 2. *Given a time interval* $[t_1, t_2]$, *a real value* $P \in (0, 1]$, *a* **CPNNQ** *q returns a set of IDs* $\{o_i | p_i^{NN}(t) \geq P\}$ *at every time instant* t, *where* $t \in [t_1, t_2]$, *and* $p_i^{NN}(t)$ *is the probability that* o_i *is the nearest neighbor of* q *at time* t.

An example of such a query is: "During the time interval $[1PM, 2PM]$, what are the IDs of sensors, whose temperature values are closest to (nearest neighbor of) 13^oC with probability $p > 0.2$, at each point of time?" Another example in 2D case is: "During the next one hour, what are the IDs of vehicles, whose probabilities of being the nearest neighbor of Staples Center are more than $P = 0.3$, at each point of time?" Notice that the answer can be changed whenever a new value is reported.

At any time t, the qualification probability of a sensor o_i for a query q can be computed by performing the following operation:

$$p_i^{NN}(t) = Pr(\forall o_j \in O, o_j \neq o_i, |o_i(t) - q| \leq |o_j(t) - q|) \tag{1}$$

where $|q - o(t)|$ is the distance between query q and sensor o's value at time t. Note that $o(t)$ represents sensor o's most recently updated value instead of sensed value at time t. However, we know that the value of sensor o is modeled using attribute uncertainty. So Equation 1 can be rewritten as:

$$p_i^{NN}(t) = \int_{n_i}^{f} Pr(|o_i(t) - q| = r)Pr(|o_j(t) - q| > r)dr \tag{2}$$

$$= \int_{n_i}^{f} pdf_i(r) \prod_{k=1 \land k \neq i}^{n} (1 - cdf_k(r))dr \tag{3}$$

where r is a variable denoting the distance to q, f is the distance between the nearest far-point among all sensor values to q, n_i is the distance between the nearest point of sensor o_i's value to q, pdf_i and cdf_i are the probability density function and the cumulative density function of o_i's value respectively. For example,

$$cdf_k(r) = \int_{n_k}^{r} pdf_k(r)dr \tag{4}$$

Basic CPNNQ Execution. A naive approach for answering a CPNNQ is to assume that each sensor's filter has no constraints. When a sensor's value is updated at time t', its new value is immediately sent to the server, and the qualification probabilities of all sensors are re-evaluated. Then, after all $p_i^{NN}(t')$ have been computed, the IDs of devices whose qualification probabilities are not smaller than P are returned to the user. The query answer is recomputed during t_1 and t_2, whenever a new value is received by the server.

However, this approach is inefficient because:

- Every sensor has to report its sensed value to the server periodically, which wastes a lot of energy and network bandwidth;
- Whenever an update is received, the server has to compute the qualification probability of each sensor with Equation 1, which can be slow.

In the next section, we will present our efficient probabilistic filter-based solution for handling continuous nearest neighbor queries.

4   The Probabilistic Filter Protocol

In this section, we present a filter-based protocol for answering continuous nearest neighbor query.

4.1   Protocol Design

Let us discuss the protocol for server side and sensor side separately.

In the server side (Algorithm 1), once a query q is registered, it comes to the initialization phase. The server requests the latest value (i.e., temperature reading) from each sensor. Based on the values and uncertainty model, the server computes a probabilistic filter for each sensor and deploy on the sensor side. How the filter is derived will be elaborated in Section 4.2. The server initializes the answer set using the uncertain database. In the maintenance phase, once the server receives an update from a sensor, it requests the latest value from each sensor. The probabilistic filters are recomputed in the server side and sent back to each sensor. Based on the latest values in the uncertain database, the server refreshes the answer set.

In the sensor side, it is quite similar to [31]. The only task is to check whether the sensed value violates its filter constraint or not. In order to do this, each sensor

```
 1  Initialization:
 2  Request data from sensors o₁, ..., oₙ;
 3  for each sensor oᵢ do
 4  │   UpdateDB(oᵢ);
 5  │   Compute filter region bᵢ;
 6  └   Send(addFilterConstraint, bᵢ, oᵢ);
 7  Initialize the answer set;
 8  Maintenance:
 9  while t₁ ≤ currentTime ≤ t₂ do
10  │   Wait for update from oᵢ;
11  │   Request data from sensors o₁, ..., oₙ;
12  │   for each sensor oᵢ do
13  │   │   UpdateDB(oᵢ);
14  │   │   Compute filter region bᵢ;
15  │   └   Send(addFilterConstraint, bᵢ, oᵢ);
16  └   Update the answer set;
17  for each sensor oᵢ do
18  └   Send(deleteFilterConstraint, oᵢ);
```

Algorithm 1. Probabilistic filter protocol for CPNN query(server side)

needs to continuously sense its value and compare with the filter constraint. The filter constraint is in the form of an interval in one dimensional case and an annulus in two dimensional case. If the sensed value is within that interval or annulus, the sensor does not need to send an update to the server. Otherwise, update is sent. Intuitively, this protocol can reduce the number of communication messages sent from sensors to server, but still provides a correct query result set.

4.2 Filter Derivation

In this section, we will present how the probabilistic filter is derived. We use one dimensional filter constraint as a case study.

For the filter constraint of o_i, it consists of two boundaries $b_i = [l_i, u_i]$. For convenience, l_i denotes the boundary near to the query point, while u_i denotes the boundary far from the query point.

Preprocessing Phase. Once the server receives the latest values and uncertainty model from all sensors, it computes the qualification probability $p_i^{NN}(t_0)$ (p_i for short in the following discussions) for each sensor o_i. Based on the qualification probabilities, the server classifies the sensors into three sets S, L (answer set) and Z.

$$S \leftarrow \{o_i | 0 < p_i < P\} \tag{5}$$
$$L \leftarrow \{o_i | p_i \geq P\} \tag{6}$$
$$Z \leftarrow \{o_i | p_i = 0\} \tag{7}$$

We now describe how to derive b_i for the sensors in each set.

Deriving u_i for S. Algorithm 2 illustrates the derivation of far boundary for sensors in set S.

---

1   $p_{diff} = min_{o_i \in S}(P - p_i)$;
2   **for** *each sensor* $o_i \in S$ **do**
3      $s_i = max(p_i - \frac{p_{diff}}{|S|-1}, 0)$;
4      $u_i = z_i(s_i)$;

Algorithm 2. Deriving u_i for S

In Algorithm 2, $z_i(x)$ is a function defined as: given that all other sensor values do not change, the qualification probability of o_i will be exactly p, when the distance from o_i's sensed value to the query point is $z_i(p)$, where $0 \leq p \leq 1$. Notice that sometimes $z_i(p)$ may not exist when p is large (i.e. if another sensor value's uncertainty region overlaps with q, o_i cannot have its qualification probability to be 1).

The intuition behind Algorithm 2 is as follows: if o_i's sensed value moves far away from query q, p_i becomes smaller. This may increase the qualification probability of other sensors in S or L since the sum of p is equal to 1. The filter boundary u_i ensures that the qualification probability of any sensor in S will not increase to a value above P. It means that no sensor in S changes its status from non-answer to answer if the filter constraint is not violated.

Deriving l_i for L. Algorithm 3 illustrates the derivation of near boundary for sensors in set L.

---

1   $p_{diff} = min_{o_i \in L}(p_i - P)$;
2   **for** *each sensor* $o_i \in L$ **do**
3      $s_i = \frac{p_{diff}}{|L|-1}$;
4      $p'_i = \{p_i | c_i = 0\}$;
5      $y_i = min(p_i + s_i, p'_i)$;
6      $l_i = z_i(y_i)$;

Algorithm 3. Deriving l_i for L

Similar to Algorithm 2, the filter boundary derived from Algorithm 3 guarantees that if o_i's value moves close to query q, all other sensors in L have no chance to switch their status from answer to non-answer. In other words, the filter constraint limits the increase on p_i so that others' qualification probability cannot decrease to a value below P.

Deriving l_i for S. The next two cases (l_i for S and u_i for L) are more complicated than the previous two. Let us imagine, if the value of a sensor $o_i \in S$ moves towards query q, it may affect the status of the sensors in L since the increase in p_i causes the decrease in the qualification probability for sensors in L. It is possible that some L sensors have their qualification probability less than P. But this never happens in the first two cases.

For convenience, we first sort all sensors. We have sensors in L as $o_1,...o_h$ in descending order of p, and sensors in S as $o_{h+1},...,o_m$ in descending order of p. c_i denotes the distance from the latest sensed value v_i to q. In order to derive l_i for S, we define two functions $Z_i(w, C[1...m])$ and $CalQP_h(C[1...m])$ as follows,

Given a sensor o_i, the distance from its sensed value to the query point is $Z_i(w, C[1...m])$ where $w = p_i \in [0,1]$ and $C[k] = |v_k - q|$.

Given a sensor o_h, its qualification probability is $CalQP_h(C[1...m])$ where $C[k] = |v_k - q|$.

Let us design the algorithm by considering the worst case. Suppose we allow p_i increase by at most δ_i. To ensure the correctness, δ_i must satisfy following conditions,

$$p'_h - \sum_{i=h+1}^{m} \delta_i \geq P \tag{8}$$

where $p'_h = CalQP_h(C[1...m])$ and $C[k] = l_i$, $k = 1...h-1$ and $C[k] = c_k$, $k = h...m$.

$$p_i + \sum_{k=h+1, k \neq i}^{m} (p'_k - p_k) + \delta_i < P \tag{9}$$

where $p'_k = CalQP_k(C[1...m])$ and $C[t] = c_t$, $t = 1...h$, i and $C[t] = u_t$, $t = h+1...m$, $t \neq i$.

Equation 8 guarantees that if all sensor values move to their inner filter boundary except o_h, then o_h's p_h is still larger than or equal to P. Equation 9 guarantees that if all S sensors values move to their outer filter boundary except o_i and o_i moves to its inner boundary, then o_i's p_i is still less than P.

As a result, the increase for p_i should satisfy

$$\delta_i \leq \frac{p'_h - P}{|S|} \tag{10}$$

$$\delta_i < P - P_i - \sum_{k=h+1, k \neq i}^{m} (p'_k - p_k) \tag{11}$$

If $P - P_i - \sum_{k=h+1, k \neq i}^{m} (p'_k - p_k) \geq \frac{p'_h - P}{|S|}$, the maximum allowed qualification probability $p_i^{max} = p_i + \delta_i$, where $\delta_i = \frac{p'_h - P}{|S|}$, and $l_i = Z_i(p_i^{max}, C[1...m])$, where $C[k] = l_k$ and $k = 1...h-1$ and $C[k] = c_k$, $k = h...m$.

If $P - P_i - \sum_{k=h+1,k\neq i}^{m} (p'_k - p_k) < \frac{p'_h - P}{|S|}$, the maximum allowed qualification

probability $p_i^{max} = p_i + \delta_i$, where $\delta_i = P - P_i - \sum_{k=h+1,k\neq i}^{m} (p'_k - p_k)$, and $l_i =$
$Z_i(p_i^{max}, C[1...m])$, where $C[k] = c_k$ and $k = 1...h, i$ and $C[k] = u_k$, $k = h + 1...m, k \neq i$.

According to Equation 10 and 11, the qualification probability of o_i can reach $p_i^{max} = p_i + \delta_i$, but no larger than p_i^{max}. Then the filter boundary can be safely set as $l_i = Z_i(p_i^{max}, C[1...m])$. The derivation is formalized in Algorithm 4.

---

1   Compute $p'_h = CalQP_h(C[1...m])$ and $C[k] = l_i$, $k = 1...h - 1$ and $C[k] = c_k$, $k = h...m$;

2   Compute $p'_k = CalQP_k(C[1...m])$ and $C[t] = c_t$, $t = 1...h, i$ and $C[t] = u_t$, $t = h + 1...m$, $t \neq i$;

3   **for** *each sensor* $o_i \in S$ **do**

4      $p_i^{max} = p_i + min(\frac{p'_h - P}{|S|}, P - P_i - \sum_{k=h+1,k\neq i}^{m} (p'_k - p_k))$;

5      $l_i = Z_i(p_i^{max}, C[1...m])$;

Algorithm 4. Deriving l_i for S

Deriving u_i for L. We omit the detailed derivation here since it is similar to the previous case (l_i for S). Algorithm 5 illustrates the derivation steps.

---

1   Compute $p'_{h+1} = CalQP_{h+1}(C[1...m])$ and $C[k] = c_k$, $k = 1...h$ and $C[k] = l_k$, $k = h + 2...m$ and $C[k] = l_k$, $k = h + 1$;

2   Compute $p'_k = CalQP_k(C[1...m])$ and $C[t] = u_t$, $t = 1...h$, $t \neq i$ and $C[t] = l_t$, $t = h + 1...m$;

3   **for** *each sensor* $o_i \in L$ **do**

4      $p_i^{min} = p_i - min(\frac{p'_{h+1} - P}{|L|}, P - P_i - \sum_{k=1,k\neq i}^{m} (p'_k - p_k))$;

5      $u_i = Z_i(p_i^{min}, C[1...m])$;

Algorithm 5. Deriving u_i for L

Deriving Filter for Z. So far, we have derived the filters for sensors in S and L. Now we investigate Z set, which is the largest set among all three in most cases. However, the filter derivation for this set is much simpler than the previous two. We first define a cut off value $cutoff = \frac{f+n}{2}$ where f is the farthest uncertainty boundary of non-zero qualification probability sensors (minimum maximum distance) and n is the nearest uncertainty boundary of zero qualification probability sensors (maximum minimum distance). The filter boundary for $o_i \in Z$ can be set as $[l_i, u_i] = [cutoff + \frac{r_i}{2}, +\infty]$ where r_i is the length of o_i's uncertain region.

The one dimensional filter constraint can be easily extended to support two dimensional data. The filter constraint region is an annulus centered at the query point with two radii as l_i and u_i.

Intuitively, with the deployment of the probabilistic filters, the updates between sensors and server can be saved compared with basic CPNNQ execution protocol.

5   Experiments

In this section, we describe the results of evaluating our protocol using a set of real temperature sensor data (Section 5.1) and a location database (Section 5.2).

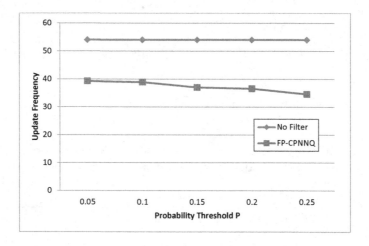

Fig. 3. Update Frequency on Temperature Data

5.1   Temperature Data

Experiment Setting. The same set of data in [31] is used for experimental evaluation. Specifically, we have 155,520 one-dimensional temperature readings captured by 54 sensors on 1st March 2004, provided by the Intel Berkeley Research lab. The temperature values are collected every 30 seconds. The lowest and the highest temperature values are $13^\circ C$ and $35^\circ C$ respectively. The domain space is $[10^\circ C, 40^\circ C]$. The uncertainty region of a sensor value is in the range of $\pm$ $1^\circ C$ [19]. By default, the uncertainty PDF is a normal distribution, with the mean as the sensed value, and the variance as 1. Also, the energy for sending an uplink message is 77.4mJ, while that for receiving a downlink message is 25.2mJ [9]. Each data point is obtained by averaging over the results of 100 random queries. Each query point is generated randomly within the domain. A query has a lifetime uniformly distributed between $[0, 24]$ hours. A query's probability threshold value, P, varies from 0.05 to 0.25. We compare our protocol with the basic protocol in which no filter is deployed. The reason we only consider no filter case is that no other work focuses on communication cost for continuous probabilistic nearest neighbor query.

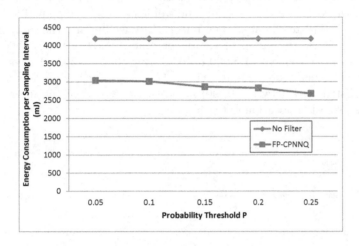

Fig. 4. Energy Consumption on Temperature Data

Experiment Results. We evaluate the probabilistic filter protocol for answering nearest neighbor query in terms of update frequency and energy consumption. As shown in Figure 3, the probabilistic filters reduce update frequency by 31% on average. When the probability threshold is 0.25, the reduction on update frequency is 36%. Similar results can be observed while evaluating energy consumption (Figure 4). The probabilistic filters reduce energy consumption on sensors by 1295 mJ per sampling interval (30 seconds) on average. The reason we have reduction in both the update frequency and energy consumption is that in some sampling periods, all sensor values stay within their corresponding filter region. As a result, no update is generated. We also observe that as the probability threshold increases, the probabilistic filters decrease the number of update messages and thus reduce energy. This is because the answer set for the queries with larger probability thresholds often has fewer qualified objects. Therefore, these answer sets are updated less frequently than that of queries with a smaller probability threshold. For example, when the probability threshold is 1.0, the answer set has at most one object.

5.2   Location Data

Experiment Setting. In this experiment, we also use the same set of location data in [31]. We simulate the movement of vehicles in a 2.0 x 2.0 km^2 European city. We use the CanuMobiSim simulator [26] to generate 5000 vehicles, which follow a smooth motion model in the streets of the city [3]. The following error model is applied to a position obtained with a (simulated) GPS device: the sensing uncertainty is obtained from a statistical error model with imprecision of 6.3m, with 95% probability [25]. The vehicles have a maximal velocity of $v_{max} = 30$m/s. The maximal sensing uncertainty is 10m and the sampling time interval is 1s. Thus, the radius of the uncertainty region of the vehicle is 40m. We simulate the movement of 5,000 objects over 90s, or 450,000 records. Each query point is generated randomly within the map. Each query has a lifetime uniformly distributed between [0, 90] seconds. Next, we present the results for expriments using this location dataset.

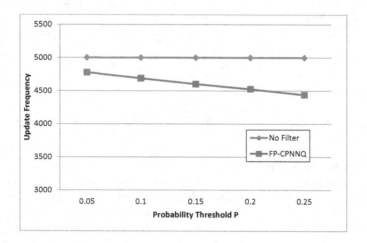

Fig. 5. Update Frequency on Location Data

Experiment Result. We also use update frequency and energy consumption as the metrics to evaluate the probabilistic filter protocol for answering nearest neighbor query over location data. As shown in Figure 5, the probabilistic filters reduce update frequency by 8% on average. For energy consumption (Figure 6), it is reduced by 30.4 J per sampling interval (1s) on average. When the probability threshold is 0.25, the reduction in both update frequency and energy consumption approaches 11%. Similar to the results on temperature data, in some sampling periods, no moving object crosses the boundary of their filter region. Update messages are not sent in these cases. However, the improvement is not as much as that for temperature data. The reason is that temperature readings, compared with vehicle locations, are relatively steady so that the filter constraints are not typically violated.

However, the performance of probabilistic filter for CPRQ [31] is much better than that for CPNNQ. This is because of the differences in their filter protocols. For CPRQ, only the sensor whose filter constraint is violated needs to send an update to the server. For CPNNQ, once a filter constraint is violated, all sensors need to send a message to the server. In the future, we will consider locality in value in order to send updates to only a subset of sensors.

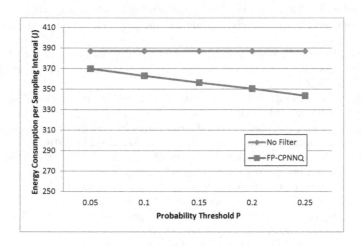

Fig. 6. Energy Consumption on Location Data

6   Conclusions

We investigated continuous nearest neighbor query execution over uncertain data. The proposed probabilistic filter protocol for processing such queries reduced communication cost and energy consumption of wireless sensors. Experimental evaluation on real-world temperature sensing data showed a reduction in the number of update messages by upto 36%. Evaluation on querying of synthetic 2D location dataset showed a reduction in both update frequency and energy consumption of 11%. The reduction in updates for continuous nearest neighbor queries is also significantly greater than that of continuous range queries.

One future direction is to extend our protocol to efficiently support multiple aggregate query execution. Another direction is to introduce tolerance in query answers in order to further reduce communication cost at sensor side.

Acknowledgments. We would like to thank Prof. Reynold Cheng (University of Hong Kong) for providing support in the early stage of this research.

References

1. Ahmad, M.B., Asif, M., Islam, M.H., Aziz, S.: A short survey on distributed in-network query processing in wireless sensor networks. In: First International Conference on Networked Digital Technologies, NDT 2009, pp. 541–543, July 2009
2. Andreou, P., Zeinalipour-Yazti, D., Pamboris, A., Chrysanthis, P.K., Samaras, G.: Optimized query routing trees for wireless sensor networks. Inf. Syst. **36**(2), 267–291 (2011)
3. Bettstetter, C.: Mobility modeling in wireless networks: categorization, smooth movement, and border effects. Mobile Computing and Communications Review **5**(3), 55–66 (2001)
4. Chen, J., Cheng, R., Mokbel, M.F., Chow, C.-Y.: Scalable processing of snapshot and continuous nearest-neighbor queries over one-dimensional uncertain data. **18**, 1219–1240 (2009)
5. Cheng, R., Kalashnikov, D.V., Prabhakar, S.: Evaluating probabilistic queries over imprecise data. In: SIGMOD (2003)
6. Cheng, R., Kao, B., Kwan, A., Sunil Prabhakar, Y.T.: Filtering data streams for entity-based continuous queries. IEEE TKDE **22**, 234–248 (2010)
7. Cheng, R., Kao, B., Prabhakar, S., Kwan, A., Tu, Y.-C.: Adaptive stream filters for entity-based queries with non-value tolerance. In: VLDB (2005)
8. Coman, A., Nascimento, M.A., Sander, J.: A framework for spatio-temporal query processing over wireless sensor networks. In: Proceedings of the 1st Workshop on Data Management for Sensor Networks, in conjunction with VLDB, DMSN 2004, Toronto, Canada, 30 August, pp. 104–110 (2004)
9. Crossbow Inc.: MPR-Mote Processor Radio Board User's Manual
10. Elmeleegy, H., Elmagarmid, A.K., Cecchet, E., Aref, W.G., Zwaenepoel, W.: Online piece-wise linear approximation of numerical streams with precision guarantees. PVLDB **2**(1), 145–156 (2009)
11. Farrell, T., Cheng, R., Rothermel, K.: Energy-efficient monitoring of mobile objects with uncertainty-aware tolerances. In: IDEAS (2007)
12. Farrell, T., Rothermel, K., Cheng, R.: Processing continuous range queries with spatio-temporal tolerance. IEEE TMC **10**, 320–334 (2010)
13. Gehrke, J., Madden, S.: Query processing in sensor networks. IEEE Pervasive Computing **3**(1), 46–55 (2004)
14. Jin, Y., Cheng, R., Kao, B., Lam, K.Y., Zhang, Y.: A filter-based protocol for continuous queries over imprecise location data. In: CIKM, pp. 365–374 (2012)
15. Kalpakis, K., Tang, S.: Maximum lifetime continuous query processing in wireless sensor networks. Ad Hoc Netw. **8**(7), 723–741 (2010)
16. Li, J., Deshpande, A., Khuller, S.: Minimizing communication cost in distributed multi-query processing. In: ICDE (2009)
17. Li, J., Shatz, S.M.: Remote query processing in wireless sensor networks using coordinated mobile objects. In: DMS, pp. 82–87. Knowledge Systems Institute (2010)
18. Madden, S., Szewczyk, R., Franklin, M.J., Culler, D.E.: Supporting aggregate queries over ad-hoc wireless sensor networks. In: WMCSA, pp. 49–58 (2002)
19. Microchip Technology Inc.: MCP9800/1/2/3 Data Sheet
20. Muller, R., Alonso, G.: Efficient sharing of sensor networks. In: MASS (2006)
21. Niedermayer, J., Züfle, A., Emrich, T., Renz, M., Mamoulis, N., Chen, L., Kriegel, H.-P.: Probabilistic nearest neighbor queries on uncertain moving object trajectories. PVLDB **7**(3), 205–216 (2013)

22. Olston, C., Jiang, J., Widom, J.: Adaptive filters for continuous queries over distributed data streams. In: SIGMOD (2003)
23. Prabhakar, S., Xia, Y., Kalashnikov, D.V., Aref, W.G., Hambrusch, S.E.: Query indexing and velocity constrained indexing: Scalable techniques for continuous queries on moving objects. IEEE Trans. Comput. **51**, 1124–1140 (2002)
24. Rahman, Md.A., Hussain, S.: Energy efficient query processing in wireless sensor network. In: AINA Workshops (2), pp. 696–700 (2007)
25. Rankin, J.: Gps and differential gps: an error model for sensor simulation. In: PLANS, pp. 260–266 (1994)
26. Stepanov, I., Marrón, P.J., Rothermel, K.: Mobility modeling of outdoor scenarios for manets. In: Annual Simulation Symposium, pp. 312–322 (2005)
27. Trajcevski, G., Tamassia, R., Cruz, I.F., Scheuermann, P., Hartglass, D., Zamierowski, C.: Ranking continuous nearest neighbors for uncertain trajectories. VLDB J. **20**(5), 767–791 (2011)
28. Trajcevski, G., Tamassia, R., Ding, H., Scheuermann, P., Cruz, I.F.: Continuous probabilistic nearest-neighbor queries for uncertain trajectories. In: EDBT, pp. 874–885 (2009)
29. Minji, W., Jianliang, X., Tang, X., Lee, W.-C.: Top-k monitoring in wireless sensor networks. IEEE Trans. Knowl. Data Eng. **19**(7), 962–976 (2007)
30. Xiong, X., Mokbel, M.F., Aref, W.G.: Sea-cnn: scalable processing of continuous k-nearest neighbor queries in spatio-temporal databases. In: ICDE (2005)
31. Zhang, Y., Cheng, R.: Probabilistic filters: A stream protocol for continuous probabilistic queries. Information Systems **38**(1), 132–154 (2013)
32. Zhang, Y., Cheng, R., Chen, J.: Evaluating continuous probabilistic queries over imprecise sensor data. In: Kitagawa, H., Ishikawa, Y., Li, Q., Watanabe, C. (eds.) DASFAA 2010. LNCS, vol. 5981, pp. 535–549. Springer, Heidelberg (2010)

Efficient Queries Evaluation on Block Independent Disjoint Probabilistic Databases

Biao Qin[✉]

School of Information, Renmin University of China, Beijing 100872, China
qinbiao@ruc.edu.cn

Abstract. Probabilistic data management has recently drawn much attention of the database research community. This paper investigates safe plans of queries on block independent disjoint (BID) probabilistic databases. This problem is fundamental to evaluate queries whose time complexity is PTIME. We first introduce two new probabilistic table models which are the correlated table and the correlated block table, and a hybrid project which executes a disjoint project and then performs an independent project in an atomic operation on BID tables. After that, we propose an algorithm to find safe plans for queries on BID probabilistic databases. Finally, we present the experimental results to show that the proposed algorithm can find safe plans for more queries than the state-of-the-art and the safe plans generated by the proposed algorithm are efficient and scale well.

1   Introduction

A diverse class of applications needs to manage large volumes of uncertain data, such as incompletely cleaned data, sensor and RFID data, information extraction, data integration, blurred data, missing data, and etc. Probabilistic data management has recently drawn much attention of the database research community [2,8,19,20]. Three probabilistic table models have been proposed. They are tuple independent model [4], block independent disjoint model [5,14] and partial block independent disjoint (p-BID) [5,14]. The p-BID model captures many representations previously discussed in the literature (e.g. p-or tables [7], ?-tables and x-relations [20]). We further introduce correlated tables and correlated block tables, which can not be classified into either tuple independent, BID or p-BID [5] table. Dalvi and Suciu's work [4] on the evaluation of conjunctive queries on tuple independent probabilistic databases showed that the class of conjunctive queries can be partitioned into "easy" queries (PTIME) and "hard" (#P-complete) queries.

The intensional method adopts lineage [20] and probabilistic reference [12,17] for query evaluation on correlated probabilistic databases. The Find-Plan algorithm [14] uses three conditions to find the safe plans of queries on probabilistic databases with the BID tuple models. Since Ré et al. [14] did not elaborate all cases, they did not prove that the Find-Plan algorithm is complete. Because project on

© Springer International Publishing Switzerland 2015
M. Renz et al. (Eds.): DASFAA 2015, Part II, LNCS 9050, pp. 74–88, 2015.
DOI: 10.1007/978-3-319-18123-3_5

BID tables has different properties from tuple independent tables, we introduce a new project, hybrid project (π^H), which executes a disjoint project [14] and then performs an independent project [14] in an atomic operation. We further introduce two new table models. Based on these findings, we propose a new algorithm to find safe plans of conjunctive queries without self-joins on BID probabilistic tables. The main contributes are as follows.

- We first introduce two new probabilistic table models which are the correlated table and the correlated block table, and a hybrid project for queries on BID tables.
- We propose an algorithm to find safe plans for queries on BID tables.
- Our experimental results show that the proposed algorithm can find safe plans for more queries than the state-of-the-art and the safe plans generated by the proposed algorithm are efficient and scale well.

The rest of this paper is organized as follows: Section 2 outlines query evaluation on probabilistic databases. Section 3 introduces two new probabilistic table models and a new probabilistic database operation. Section 4 describes an algorithm to generate safe plans for queries on BID probabilistic databases. Section 5 reports the experimental results. Section 6 discusses related work. Section 7 concludes and outlines future research.

2   Preliminaries

In this paper, we only consider conjunctive queries without self-join on probabilistic databases, unless otherwise stated. Tuple independent tables [4] and BID tables [15] are two famous uncertain data models. BID tables have the following properties: any two tuples that differ on the $\bar{K}$ attributes are independent and any two tuples that agree on the $\bar{K}$ attributes are disjoint. Moreover, $\sum_{i=1}^{n} p_i \leq 1$ in each block. Since $\bar{K} \not\to \bar{V}$ in BID tables, $\bar{K}$ is not the key of R and $R^p.Key = \{K_1, \ldots, K_n, A_1, \ldots, A_m\}$. If $\sum_{i=1}^{n} p_i = 1$ in every block of a BID R^p, we call R^p a full BID (f-BID) table. For example, T^p in Figure 1 is a tuple independent table [4] while R^p is a BID table [15]. If any two tuples in R^p have the same values of the key attributes $\{n, l\}$ then they are disjoint events; otherwise they are independent. Moreover, tA^p and tB^p are f-BID tables.

In a probabilistic database, a query answer is correct only if the resulting data and their confidences are both correct. Therefore, query evaluation on a probabilistic database includes not only data computation but also confidence computation. Safe plan [4] is based on modifying the query operators to compute probabilities rather than complex events. Select σ^e [4] acts like σ, which propagates the probabilities of tuples from the input to the output; $\bowtie^e$ [4] computes the probability of every tuple (t, t') as $p \times p'$. π^e includes disjoint project ($\pi_{\bar{A}}^D$) and independent project ($\pi_{\bar{A}}^I$) [14]. $\pi_{\bar{A}}^D$ performs as follows: If n tuples with probabilities $p_1, \ldots, p_n$ have the same value, $\bar{a}$, for their $\bar{A}$ attributes, the disjoint project will associated the answer tuple with the probability $p_1 + \cdots + p_n$. The disjoint project is correctly applied if any two tuples that share the same

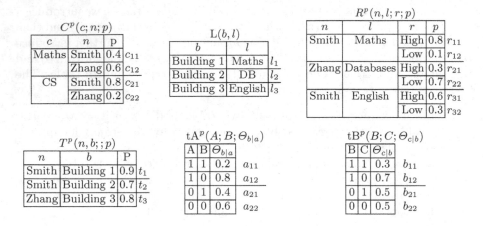

Fig. 1. A probabilistic database D^p

value of the $\bar{A}$ attributes are disjoint events. $\pi^I_{\bar{A}}$ performs as follows: If n tuples with probabilities $p_1, \ldots, p_n$ have the same value, $\bar{a}$, for their $\bar{A}$ attributes, then the independent project $\pi^I_{\bar{A}}$ will associated the answer tuple with the probability $1 - (1 - p_1)(1 - p_2) \ldots (1 - p_n)$. The independent project is correctly applied if any two tuples that share the same values of the $\bar{A}$ attributes are independent. Ré et al. [14] used independent project and disjoint project to find safe plan in BID tables. Ré and Suciu [15] further gave the definition of p-BID tables as follows.

Definition 1. *A p-BID table is a relational schema with the attributes partitioned into four classes separated by semicolons: $R(\bar{K}; \bar{K}_1; \bar{V}; p)$.*

- *R^p is $\bar{K}$-block independent if any two tuples t_i and t_j such that $t_i.\bar{K} \neq t_j.\bar{K}$ ($i \neq j$) are independent.*
- *R^p is $\bar{K}\bar{K}_1$-block disjoint if any two tuples such that $t_i.\bar{K}\bar{K}_1 = t_j.\bar{K}\bar{K}_1$ are disjoint.*
- *t_i and t_j ($t_i, t_j \in R^p$) are correlated if $t_i[\bar{K}] = t_j[\bar{K}]$ but $t_i[\bar{K}_1] \neq t_j[\bar{K}_1]$.*

We note that the p-BID table includes the BID table as a special case and the BID table further includes the tuple independent table as a special case. We can use possible world semantics to compute correct probabilities for result tuples. Let $t_1, \ldots, t_n$ be the BID tuples of $\mathcal{D}$ and W have m tuples. Then m_1 ($m_1 = n - m$) tuples do not appear in W. So the probability of W occurring is $P(W) = \Pi^m_{i=1} P(t_i) \Pi^{m_1}_{j=1} (1 - \sum P(t_j))$. If $t_i \in W$, we use its probability $P(t_i)$. Otherwise, we use the probability $1 - \sum P(t_i)$. Given a query q and a probabilistic database D, the confidence of every answer tuple t is $P(t \in q(D)) = \sum_{t \in W, W \in \mathcal{W}} P(q(W))$, where $\mathcal{W}$ is a set of all possible worlds of D.

Example 1. Assume the database shown in Figure 1 has the constraint that every teacher can work in only one building. We have the following query on the database: $q(r) = \pi^I_r(R^p)$, whose result is shown in Figure 2(a).

(b) $\pi_r^e(R^p) = \pi_r^I(R^p)$

r	p	symbolic prob.
High	0.944	$1 - (1 - r_{11})$
		$(1 - r_{21})(1 - r_{31})$
Low	0.811	$1 - (1 - r_{12})$
		$(1 - r_{22})(1 - r_{32})$

$\pi_n^H(R^p) =$

n	p	symbolic prob.
t_1^o Smith	0.99	$1 - (1 - (r_{11} + r_{12}))$
		$(1 - (r_{31} + r_{32}))$
t_2^o Zhang	1	$1 - (1 - (r_{21} + r_{22}))$

Fig. 2. (a) the answers of $\pi_r^I(R^p)$, (b) the answers of $\pi_n^e(R^p) = \pi_n^H(R^p)$

Table 1. tmp$(A; B, C; p)$

A	B	C	p	lineage
1	1	1	0.06	$a_{11}b_{11}$
1	1	0	0.14	$a_{11}b_{12}$
1	0	1	0.4	$a_{12}b_{21}$
1	0	0	0.4	$a_{12}b_{22}$
0	1	1	0.12	$a_{21}b_{11}$
0	1	0	0.28	$a_{21}b_{12}$
0	0	1	0.3	$a_{22}b_{21}$
0	0	0	0.3	$a_{22}b_{22}$

3   New Probabilistic Table Models and Operation

From Figure 2(a), we find its values for attribute r are distinct. However, their lineages are correlated by events. So we give the definition of a new kind of table as follows.

Definition 2. *Assume that a table $R_i^p(\bar{C}; p)$ has distinct values for attributes $\bar{C}$. Since their lineages are correlated by events, this kind of tables is not tuple independent, BID or p-BID. Thus they are called correlated tables.*

From the above definition, we know that the table shown in Figure 2(a) is a correlated table. The following example illustrates another new table model.

Example 2. If we perform the following query $q = tA^p \bowtie tB^p$ on BID tables shown in Figure 1, the result is shown in Table 1. We find Table 1 is a new kind of table, whose property is described in the following definition.

Definition 3. *A correlated block table (CBT) is a relational schema with the attributes partitioned into three classes separated by semicolons: $R^p(\bar{C}; \bar{V}; p)$.*

- $\sum_{i=1}^n p_i = 1$ *in each C-block.*
- R^p *is $\bar{C}$-block correlated if any two tuples t_i and t_j such that $t_i.\bar{C} \neq t_j.\bar{C}$ $(i \neq j)$ are correlated.*
- R^p *is $\bar{C}$-block disjoint if any two tuples t_i and t_j such that $t_i.\bar{C} = t_j.\bar{C}$ are block disjoint.*

Table 2. Frequently Used Notations

Symbols	Meaning
$Attr(R_i^p)$	all attributes in R_i^p
$Attr(q)$	all attributes involved in q
BL_i	a block of BID table
$R_i^p.Key$	the key of a table
$Head(q)$	the set of attributes in the output of the query q
$|Rels(q)|$	the number of relations in the query q
$\bar{X}$	a set of query variable
$\bar{x}$	a set of value for $\bar{X}$
$(\bar{X})^+$	the transitive closure under a set of FD
$\pi^I(R_i^p)$	independent project
$\pi^D(R_i^p)$	disjoint project
$\pi^H(R_i^p)$	hybrid project

From the above definition, we know that Table 1 is a correlated block table. In this paper, the symbols and their meanings are shown in Table 2. In order to perform project on BID tables and p-BID tables, We further introduce a new project operation as follows.

Definition 4. *Hybrid project is an atomic operation, which executes a disjoint project during every block and then performs an independent project across blocks in an atomic operation. For a BID table $R_i^p(\bar{K}; \bar{V}; p)$, $\pi_{\bar{A}}^H(R_i^p) = \pi_{\bar{A}}^I(\pi_{\bar{A} \cup \bar{K}}^D(R_i^p))$. For a p-BID table $R_i^p(\bar{K}; \bar{K}_1; \bar{V}; p)$, $\pi_{\bar{K}_1}^H(R_i^p) = \pi_{\bar{K}_1}^I(\pi_{\bar{K}\bar{K}_1}^D(R_i^p))$.*

If we have a query $q = \pi_n^H(R^p)$, neither $\pi_n^D(R^p)$ nor $\pi_n^I(R^p)$ can give a correct answer. While $\pi_n^H(R^p)$ gives the correct answer as shown in Figure 2(b). We give the following theorem for processing the correlated tables.

Theorem 1. *Assume we have a correlated table $R_i^p(\bar{C}; p)$. If $\bar{A} \subset \bar{C}$, $\pi_{\bar{A}}^I(R_i^p)$ is not safe.*

Proof. Let $R_j^p = \pi_{\bar{A}}^I(R_i^p)$. If $\bar{A} \subset \bar{C}$, then $\exists r \in R_j^p$ and r comes from i_1 and i_2 ($i_1, i_2 \in R_i^p$). Because i_1 and i_2 are correlated as shown in Figure 3, the confidence of r is wrong by extensional evaluation method. Hence the theorem is proved.

Assume that we project on a CBT R_i^p. If the project attributes are subset of $R_i^p.\bar{C}$, then the project is unsafe by Theorem 1. The following theorem discusses a situation whose disjoint project is safe for CBTs.

Theorem 2. *Assume we have a correlated block table $R_i^p(\bar{C}; \bar{V}; p)$. If $\bar{A} \supseteq \bar{C}$, $\pi_{\bar{A}}^D(R_i^p)$ is safe.*

Proof. For any two tuples t_1 and t_2 of R_i^p, if $t_1.\bar{C} = t_2.\bar{C}$ then they are block disjoint. Since $\bar{A} \supseteq \bar{C}$, $\pi_{\bar{A}}^D(R_i^p)$ is safe. This proves the theorem.

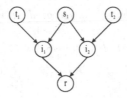

Fig. 3. The correlations among tuples

Corollary 1. *For every BID table $R_i^p(\bar{K}; \bar{V}; p)$, if $\bar{A} \not\supseteq R_i^p.\bar{K}$, then the result of $\pi_{\bar{A}}^H(R_i^p)$ is a correlated table.*

Proof. Assume that R_i^p is a BID table. From Definition 4, $\pi_{\bar{A}}^H(R_i^p)$ executes a disjoint project and then performs an independent project in an atomic operation. Assume that the result has two tuples $t_1 = (\bar{a}, p)$ and $t_2 = (\bar{b}, p)$. Then $\exists BL_i \in R_i^p$, BL_i has two tuples t_{i1} and t_{i2}, which are the source of t_1 and t_2, respectively. So t_1 and t_2 come from the same block but they can not be in a new block. Thus they are correlated. Hence, the result of $\pi_{\bar{A}}^H(R_i^p)$ is a correlated table. This proves the theorem.

4 Safe Plans of Queries on Probabilistic Databases

In this section, we introduce an algorithm to find safe plans of queries evaluation on BID tables.

4.1 The Preprocessing of the Queries

With a similar method in [13], we classify projections into simple projection which involves only one table and complex projection which involves two or more tables. We first give the following lemma to show that any execution plan for a conjunctive query without complex projections is safe, that is, it yields correct data as well as probability without requiring special treatment.

Lemma 1. *Consider a BID probabilistic database. If the query q only includes $\sigma, \bowtie$ or simple projection, then it is safe.*

Proof. By possible world semantics, $\sigma, \bowtie$, and $\pi^H(R^p)$ are all safe. So the lemma is proved.

Recall that a probabilistic database may consist both of probabilistic table names, which have an event attribute E, and deterministic relation names. Before going to the preprocessing of the queries, the attributes associated with join or $Head(q)$ in a query is considered first, where $Head(q)$ denotes the set of attributes that are in the output of the query q.

Definition 5. *[13] If an attribute is associated with join conditions or $Head(q)$ in a query q, it is called an association attribute. An association attribution set of query q is denoted by $A^2S(q)$. It includes the following attributes:*

- *If $R_i^p.A \in Head(q)$, then $R_i^p.A$ is put into $A^2S(q)$;*
- *If there is a join predicate $R_i^p.B = T_j^p.C$, then both $R_i^p.B$ and $T_j^p.C$ are put into $A^2S(q)$.*

The association attribute set of a relation R_i^p is denoted by $A^2S(R_i^p)$. It is the intersection of $A^2S(q)$ and $Attr(R_i^p)$, that is, $A^2S(R_i^p) = A^2S(q) \cap Attr(R_i^p)$. If an attribute of R_i^p cannot be inferred from $A^2S(R_i^p)$, we call it a non-association attribute. If $R_i^p.Key \not\subseteq A^2S(R_i^p)$, then R_i^p has non-association attributes. The following definition is for a query q_1 which includes only tuple independent tables.

Definition 6. *If $q = \pi^e_{A_1,\dots,A_k}(R_1^p \bowtie R_2^p \cdots \bowtie R_n^p)$ where every R_i^p is tuple independent, then q is called independent query.*

The Multiway-Split algorithm [13] makes full use of function dependency (FD) to extend the safe plan algorithm [4] for generating efficient safe plans for tuple independent probabilistic databases. We can use the Multiway-Split algorithm to find out whether an independent query is safe. We give the following definitions of 1-BID query, 2-BID query and 3-BID query.

Definition 7. *Assume that $q = R_i^p \bowtie R_j^p$, where R_i^p is a BID table and R_j^p is a tuple independent table. If $Attr(R_j^p) \supseteq Attr(R_i^p)$ and $R_i^p.\bar{K}$ joins $R_j^p.Key$, then the query q is called a 1-BID query.*

Definition 8. *Assume that $q = R_i^p \bowtie R_j$, where R_i^p is a BID table and R_j is a certain table. If $Attr(R_j) \supseteq Attr(R_i^p)$ and $R_i^p.\bar{K}$ joins $R_j.Key$, then the query q is called a 2-BID query.*

Definition 9. *Assume that $q = R_i^p \bowtie R_j^p$, where both R_i^p and R_j^p are f-BID tables. If $R_i^p.\bar{V} = R_j^p.\bar{K}$, then the query q is called a 3-BID query.*

Theorem 3. *The results of both a 1-BID query q_1 and a 2-BID query q_2 are tuple independent probabilistic tables.*

Proof. In a 1-BID query q_1, we know that $Attr(R_j^p) \supseteq Attr(R_i^p)$. Since R_j^p is a tuple independent probabilistic table, $R_j^p.Key \to Attr(R_j^p)$. Then $q_1 = R_i^p \bowtie R_j^p$ becomes a select operation as follows: for every tuple t_j in R_j^p, select at most one tuple in R_i^p. So the result of q_1 is a tuple independent probabilistic table. Similarly, we can prove the result of q_2 is also a tuple independent probabilistic table. This proves the theorem.

Theorem 4. *The result of a 3-BID query is a correlated block table.*

Proof. Assume that $q = R_i^p \bowtie R_j^p$, where both R_i^p and R_j^p are f-BID tables. Since $R_i^p.\bar{V} = R_j^p.\bar{K}$, the result of q is correlated when the result tuples have different values in $R_i^p.\bar{K}$. Since $\sum_{i=1}^n p_i = 1$ in every block, $\Pi^e_{\bar{A}}(q)$ is disjoint project if $\bar{A} \supseteq R_i^p.\bar{K}$. Thus, the result of a 3-BID query is a correlated block table. This proves the theorem.

Algorithm 1. Preprocessing($q = \pi^e_{A_1,\dots,A_k}(q_i)$)

1: push down selection predicate $R^p_i.A\theta c$;
2: get $A^2S(q)$ from q;
3: **for** (every relation R^p_i) **do**
4:     $A^2S(R^p_i) = Attr(R^p_i) \cap A^2S(q)$;
5:     **if** ((R^p_i is tuple independent) $\cap$ ($A^2S(R^p_i) \not\supseteq R^p_i.Key$)) $T^p_i = \pi^I_{A^2S(R^p_i)}(R^p_i)$;
6:     **if** (R^p_i is a BID table) $T^p_i = \pi^H_{A^2S(R^p_i)}(R^p_i)$;
7: **end for**;
8: **if** (q_j is a 1-BID sub-query or a 2-BID sub-query of q_i) **then**
9:     its result forms a tuple independent probabilistic table R^p_j;
10: **else if** (q_j is a 3-BID sub-query of q_i) **then**
11:     its result forms a correlated block table R^p_j;
12: **end if**;
13: **for** (every independent sub-query q_j) **do**
14:     **if** (Multiway-Split(q_j) is safe) its result is a probabilistic table R^p_j;
15: **end for**;
16: return $\pi^e_{A_1,\dots,A_k}(q_j)$;

The preprocess function is shown in Algorithm 1, which includes the following steps:

- We first push down selection predicate $R^p_i.A\theta c$. Based on the new query q, we derive $A^2S(q)$.
- For every tuple independent table, we project out its non-association attributes if $A^2S(R^p_i) \not\supseteq R^p_i.Key$.
- For every BID table, we project onto its association attributes.
- For both 1-BID sub-query and 2-BID sub-query, their respective results are tuple independent probabilistic tables.
- For the 3-BID sub-query, its result is a correlated block tables.
- For every independent sub-query q_j, the Multiway-Split algorithm is invoked to find its safe plan, whose result is a probabilistic table R^p_j.

We use the following example to illustrate the preprocess algorithm.

Example 3. The following query evaluates on the database shown in Figure 1.
    $q(n)$:- $C^p(c;n), T^p(n,b), L(b,l), R^p(n,l;'H')$
First, we perform the selection predicate. Then the query becomes
    $q(n)$:- $C^p(c;n), T^p(n,b), L(b,l), R^p_1(n,l,r)$
We find R^p_1 is a tuple independent table. Using Definition 5, $A^2S(q)$
$= \{n,b,l\}$ and the association attribute sets of the tables are as follows: $A^2S(C^p) =$
$Attr(C^p) \cap A^2S(q) = \{n\}$, $A^2S(T^p) = Attr(T^p) \cap A^2S(q) = \{n,b\}$, $A^2S(L) =$
$Attr(L) \cap A^2S(q) = \{b,l\}$, and $A^2S(R^p_1) = Attr(R^p_1) \cap A^2S(q) = \{n,l\}$.
    Because $A^2S(C^p) \not\supseteq C^p.Key$, C^p should project out its non-association
attribute $\{c\}$, that is, $C^p_1 = \pi^I_n(C^p)$, which is a correlated table by Corollary 1.
Similarly, R^p_1 should project out its non-association attribute $\{r\}$, that is, $R^p_2 =$
$\pi^I_{n,l}(R^p_1(n,l,r))$. Because $A^2S(T^p)$ is equal to $Attr(T^p)$, T^p need not project out

any attribute. Similarly, L need not project out any attribute either. After the pre-processing step, the query becomes:

$q(n)$:- $C_1^p(n), T^p(n,b), L(b,l), R_2^p(n,l)$

We further find q includes an independent sub-query $q_2(n)$:- $T^p(n,b), L(b,l)$, $R_2^p(n,l)$. By the Multiway-Split algorithm, q_2 has the following safe plan:

$\mathcal{P} = \pi_n^I(T^p \bowtie \pi_{n,b}^I(L \bowtie R_2^p))$

Let the result of plan $\mathcal{P}$ be table R_2^p. Then the query becomes:

$q(n)$:- $C_1^p(n), T^p(n,b), L(b,l), R_2^p(n,l)$
$\quad$:- $C_1^p(n), R_2^p(n)$

Theorem 5. *Algorithm 1 is correct and safe.*

Proof. We proves the theorem as follows.

1. Qin et al. [13] proved that simple projection is safe for tuple independent probabilistic tables. Lemma 1 has proved that simple projection is safe for BID tables. Thus Steps 5 and 6 are correct and safe.

2. If q_j is a 1-BID sub-query or a 2-BID sub-query, its result is a tuple independent probabilistic table by Theorem 3. So Step 9 is correct and safe.

3. If q_j is a 3-BID sub-query, its result is a corrected block table by Theorem 4. So Step 11 is correct and safe.

4. If q_j is an independent sub-query and safe, its result is a probabilistic table by the Multiway-Split algorithm. Otherwise, the tables involved in q_j will not be preprocessed. So Step 14 is correct and safe.

Since every step is correct and safe, Algorithm 1 is correct and safe. This proves the theorem.

From Algorithm 1, we find the time complexity for the preprocessing step is $O(n^2)$, where n denotes the number of tuples is involved in the query.

4.2   The Algorithm for Generating Safe Plans

Let q be a conjunctive query. Dalvi and Suciu [4] defined Γ as follows.

- Every FD in Γ^p is also in $\Gamma^p(q)$.
- For every join predicate $R_i.A = R_j.B$, both $R_i.A \to R_j.B$ and $R_j.B \to R_i.A$ are in $\Gamma^p(q)$.

The following is the base theorem for generating safe plans of queries evaluation on BID probabilistic databases.

Theorem 6. *Assume that* $q = \pi_{A_1,...,A_k}^e(q_i)$, $q_i = R_1^p \bowtie R_2^p ... \bowtie R_m^p$ *and* $Head(q_i) = Attr(R_1^p) \cup ... Attr(R_m^p)$. *After the preprocessing step, q is a complex projection with the following two cases.*

1. All tables are f-BID with $R_i^p.\bar{V} = R_{i+1}^p.\bar{K}$. Then $q = \pi_{A_1,...,A_k}^e(q_i)$ is safe iff $\{A_1,...,A_k\} \supseteq R_1^p.\bar{K}$.

2. At least one table R_i^p is not tuple independent table. Then, $q = \pi_{A_1,...,A_k}^e(q_i)$ is safe iff the following can be inferred from $\Gamma^p(q)$:

$$A_1,...,A_k \to Head(q_i)$$

Proof. We prove the theorem as follows.

1. Since all tables are f-BID with $R_i^p.\bar{V} = R_{i+1}^p.\bar{K}$, the result of q_i is a correlated block table. By Theorem 2, $q = \pi_{A_1,...,A_k}^e(q_i)$ is safe if $\{A_1,...,A_k\} \supseteq R_1^p.\bar{K}$.

Let the result tuples of q_i form table T_i. If $\{A_1,...,A_k\} \not\supseteq R_1^p.\bar{K}$, then any result tuple of q is derived from different blocks of T_i. Since tuples in different blocks of T_i are correlated, q is unsafe.

2. If $A_1,...,A_k \rightarrow Head(q_i)$, $\{A_1,...,A_k\}$ includes the key of every tuple independent table and all attributes of other kinds of probabilistic tables. So q is safe.

After the preprocessing step, at least one table is not tuple independent. Thus, the result of q_i is a correlated table. By Theorem 1, if $A_1,...,A_k \not\rightarrow Head(q_i)$, the plan is unsafe.

This proves the theorem.

Before presenting the algorithm for generating safe plans, we need some terminologies.

Definition 10. *After the preprocessing step, let a query be $q = \pi_{A_1,...,A_k}^e(q_j)$. Based on $\Gamma^p(q)$, the set of attributes which can be inferred from $\{A_1,...,A_k\}$ is represented by $\{A_1,...,A_k\}^+$ and we denote it by $InfAttr(q)$.*

Definition 11. *Let a query be $q = \pi_{A_1,...,A_k}^e(q_i)$. If $R_i^p \in q_i$ and $InfAttr(q) \supseteq Attr(R_i^p)$, R_i^p is covered by $InfAttr(q)$. All tables that are covered by $InfAttr(q)$ form a set called maximal coverage set (MCS(q)).*

By Definition 10, the following corollary and theorem show that we can use Theorem 6 Item 2 more flexibly.

Corollary 2. *Assume that the result of q_i is not a correlated block table after the preprocessing step. $q = \pi_{A_1,...,A_k}^e(q_i)$ is safe iff $Head(q_i) \subseteq InfAttr(q)$.*

Proof. By Theorem 6 Item 2, $q = \pi_{A_1,...,A_k}^e(q_i)$ is safe iff $A_1,...,A_k \rightarrow Head(q_i)$. Using Definition 10, $Head(q_i) \subseteq InfAttr(q)$. So $q = \pi_{A_1,...,A_k}^e(q_i)$ is safe iff $Head(q_i) \subseteq InfAttr(q)$. This proves the corollary.

Theorem 7. *After the preprocessing step, $q = \pi_{Head(q)}^e(R_1^p \bowtie R_2^p \bowtie ... \bowtie R_n^p)$ $(Head(q) \neq \{\} \wedge n > 1)$ and the result of $q_i = R_1^p \bowtie R_2^p \bowtie ... \bowtie R_n^p$ is not a correlated block table. If its $MCS(q)$ includes m $(1 \leq m \leq n)$ tables (for example $R_1^p,...,R_m^p$), then:*

  1. If m is equal to n, then the plan $\pi_{Head(q)}^e(R_1^p \bowtie R_2^p ... \bowtie R_m^p)$ is safe.

  2. If $m < n$, then the query $\pi_{Head(q)}^e(R_1^p \bowtie R_2^p ... \bowtie R_n^p)$ has no safe plan.

Proof. 1. Let $q_i = R_1^p \bowtie R_2^p ... \bowtie R_m^p$. Using Theorem 6 Item 2, $q = \pi_{A_1,...,A_k}^e(q_i)$ is safe iff $A_1,...,A_k \rightarrow Head(q_i)$. If m is equal to n, then $Head(q_i) \subseteq InfAttr(q)$. By Corollary 2, the plan is safe.

2. We prove it by contradiction. Since $MCS(q) = \{R_1^p,...,R_m^p\}$, $\overline{MCS(q)} = \{R_{m+1},...,R_n\}$. Let $q_j = R_1^p \bowtie R_2^p ... \bowtie R_m^p$ and $q_i = q_j \bowtie \pi_{\bar{A}}^e(R_{m+1}^p \bowtie ... \bowtie$

R_n^p). Assume the plan $\pi_{Head(q)}^e(q_j \bowtie \pi_{\bar{A}}^e(R_{m+1}^p \bowtie \ldots \bowtie R_n^p))$ is safe. Then using Theorem 6 Item 2, we can derive as follows.

$$Head(q) \rightarrow Attr(q_j) \cup \bar{A}$$
$$\bar{A} \rightarrow Attr(R_{m+1}^p) \cup \cdots \cup Attr(R_n^p)$$

Thus, we can further derive as follows.

$$Head(q) \rightarrow Attr(q_j) \cup Attr(R_{m+1}^p) \cup \cdots \cup Attr(R_n^p)$$
$$\rightarrow Attr(R_1^p) \cup \cdots \cup Attr(R_n^p)$$

Therefore, $|MCS(q)| = n$, which contradicts the assumption that $|MCS(q)| < n$. Hence the query has no safe plan.

The theorem is proved.

The algorithm to generate safe plans for queries on BID probabilistic databases is shown in Algorithm 2. It has the following steps.

- If q is boolean, then we use the probabilistic reference method [17].
- If q includes only tuple independent probabilistic tables, then the Multiway-Split algorithm is invoked.
- If the result of $q_i = R_1^p \bowtie \cdots R_n^p$ is a correlated block table and $\{A_1, \ldots, A_k\} \supseteq R_1^p.\bar{K}$, then the query is safe. Otherwise, the query has no safe plan.
- If $|MCS(q)|$ is equal to $|Rels(q)|$, then the query has a safe plan. Otherwise, the query has no safe plan. Here $|Rels(q)|$ denotes the number of relations involved in q.

Algorithm 2. $\text{SPlan}(q = \pi_{A_1, \ldots, A_k}^e(R_1^p \bowtie \cdots R_n^p))$

1: **if** (q is boolean) **then**
2:    return (the probabilistic inference method);
3: **end if**;
4: $q = \text{Preprocessing}(q)$;
5: **if** (q only includes tuple independent probabilistic tables) **then**
6:    return Multiway-Split(q);
7: **else if** ($q = R_1^p \bowtie \cdots R_n^p$ is a 3-BID query and $\{A_1, \ldots, A_k\} \supseteq R_1^p.\bar{K}$) **then**
8:    return $\pi_{Head(q)}^D(R_1^P \bowtie \ldots \bowtie R_k^p)$;
9: **else if** ($|MCS(q)| == |Rels(q)|$) **then**
10:    return $\pi_{Head(q)}^e(R_1^P \bowtie \ldots \bowtie R_k^p)$;
11: **else**
12:    return error("No safe plans exist");
13: **end if**;

Example 4. Continuing Example 3. We show the preprocessing result below for convenience.

$$q(n) :\text{-} C_1^p(n), R_2^p(n)$$

Using Theorem 6, $q(n)$ has a safe plan, which is $\mathcal{P} = \pi_n^I(C_1^p) \bowtie \pi_n^I(T^p \bowtie \pi_{n,b}^I(L \bowtie R_2^p))$.

However, by the Find-Plan algorithm [14], the query $q(n)$ shown in Example 3 is unsafe. We state now the soundness of our algorithm: the proof follows easily from the fact that all operations are safe.

Theorem 8. *The SPlan algorithm is complete, that is, any plan that it returns is safe and the SPlan algorithm can find a safe plan iff the query is safe.*

Proof. We prove theorem as follows. The algorithm returns safe plans in the following cases:

Case 1: If the query is boolean, the probabilistic inference method is invoked to calculate the confidence of every answer tuple.

Case 2. It returns at Line 6. The Multiway-Split algorithm is invoked to return a safe plan or return an error. The Multiway-Split algorithm is complete for tuple independent probabilistic tables.

Case 3. It returns at Line 8. Theorem 6 Item 1 ensures the plan is safe.

Case 4. It returns at Line 10. Theorem 7 ensures the plan is safe. Otherwise, the query has no safe plan.

So the SPlan algorithm is complete.

From Algorithm 2, we find the time complexity of the algorithm is $O(n^2)$, where n denotes the number of tuples is involved in the query. In this step, Algorithm 2 only generates safe plan and does not access tables in the database.

5   Experiments

We have developed a prototype for probabilistic query evaluation. Our system is implemented as a middleware [13], which can work on top of any relational database engine. We use the TPC-H benchmark [1] with databases TPC-H scale factors 0.1, 0.5 and 1. The following shorthand notations are used in the queries: Item: lineitem, Part: part, Sup: supplier, PS: partsupp, Nat: nation, Reg: region, Ord: orders and Cust: customer. In the experiments, there is no index for any table. We modified all queries by replacing all the predicates in the WHERE clause with uncertain matches. All of the following experiments are carried on the first 10 of the 22 TPC-H queries [4,13]. Dalvi and Suciu [4] had found other queries to be not very interesting for applying uncertain predicates, since most of them involves complex aggregates. The experiments were conducted on an intel core2 1.8GHz/2.0G memory and PostgreSQL as the underlying DBMS, running Windows 7.0.

Out of the 10 TPCH queries, 7 turned out to have safe plans, while $Q7$, $Q8$ and $Q10$ were unsafe. Since $Q2$, $Q3$ and $Q5$ are not satisfied with the conditions in the Find-Plan algorithm [14], they are unsafe. However, $Q2$, $Q3$ and $Q5$ are safe by the proposed algorithm. We will give their safe plans in the next subsection. Because $Q1$, $Q4$, $Q6$ and $Q9$ respective only have one probabilistic table, the proposed algorithm and the Find-Plan algorithm generate the same safe plan for them. Thus, we do not compare the performance of our algorithm with the Find-Plan algorithm.

We measured the running times for the seven queries that have safe plans, shown in Figures 4(a), 4(b) and 4(c). All times are wall-clock. The first column in the graph shows the time for running bare queries without taking into account the computation time for the uncertain predicate. The second column is the running time of the safe plan (SPlan). From those Figures, we find that the bare queries have better performance than their respective safe plans just as in [4,13]. However, we find one exception, that is, the safe plan of $Q4$ has better performance than its bare query for TPC-H scale factors 0.5 and 1. This is because the safe plan of $Q4$ adopts hash join while its bare query uses merge join. We further find that safe plans scale well on probabilistic databases. Finally, since $Q2, Q3$ and $Q5$ are complex, we further discuss their safe plans as follows.

For $Q2$, we must create a view tmp first [4]. There are two intermediate BID tables $R_i^p = \pi_{\bar{C}}^H(Part^p)$ and $R_j^p = \pi_{\bar{B}}^H(Reg^p)$. The safe plan of $Q2$ is $\mathcal{P} = R_i^p \bowtie \pi_{\bar{C}}^I(R_j^p \bowtie \pi_{\bar{A}}(PS \bowtie Sup \bowtie tmp \bowtie Nat))$, where $\bar{A}$, $\bar{B}$, $\bar{C}$ and $\bar{D}$ denote the corresponding association attribute sets. The result of $q_i = R_j^p \bowtie \pi_{\bar{A}}(PS \bowtie Sup \bowtie tmp \bowtie Nat)$ is an independent table R_k^p because $\bar{A} \supset attr(R_j^p)$.

For $Q3$, the safe plan is $\mathcal{P} = \pi_{\bar{A}}^H(Item^p) \bowtie^e \pi_{\bar{B}}^D(\pi_{\bar{C}}^D(Ord^p) \bowtie^e \pi_{\bar{D}}^D(Cust^p))$, where $\bar{A}$, $\bar{B}$, $\bar{C}$ and $\bar{D}$ denote the corresponding association attribute sets.

For $Q5$, the safe plan is $\mathcal{P} = \pi_{Head(Q5)}^I(\pi_{\bar{C}}^I(\pi_{\bar{B}}(Sup \bowtie Nat \bowtie Item) \bowtie \pi_{\bar{A}}^H(Reg^p)) \bowtie \pi_{\bar{F}}^H(\pi_{\bar{D}}^H(Ord^p) \bowtie Cust))$, where $\bar{A}$, $\bar{B}$, $\bar{C}$, $\bar{D}$, $\bar{E}$ and $\bar{F}$ denote the corresponding association attribute sets. The important part of the plan is that the result of $\pi_{\bar{F}}^H(\pi_{\bar{D}}^H(Ord^p) \bowtie Cust)$ is a BID table and the result of $\pi_{\bar{C}}^I(\pi_{\bar{B}}(Sup \bowtie Nat \bowtie Item) \bowtie \pi_{\bar{A}}^H(Reg^p)) \bowtie \pi_{\bar{F}}^H(\pi_{\bar{D}}^H(Ord^p) \bowtie Cust)$ is an independent table R_i^p because $\bar{C} \supset \bar{F}$.

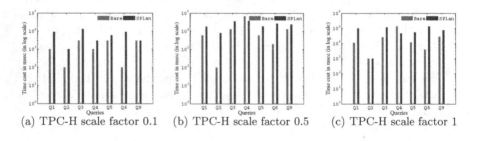

(a) TPC-H scale factor 0.1   (b) TPC-H scale factor 0.5   (c) TPC-H scale factor 1

Fig. 4. The performance of *SPlan* vs. *Bare* over BID probabilistic database

6   Related Work

Probabilistic databases have been studied for decades [2,8,19,20]. There are many uncertain data models. MystiQ [2] can handle three kinds of probabilistic tables: tuple independent tables, BID tables and p-BID tables. This paper further introduces two kinds of probabilistic tables which are the correlated table and the correlated block table.

There are two main ways to compute the result data and their confidences. One is that extensional method couples data computation and confidence computation together. Because of the rigid restrictions [3,6,10], the systems built around them can hardly handle queries flexibly. Dalvi and Suciu [4] first introduced safe plans to evaluate on tuple independent probabilistic databases. Using functional dependencies, Qin and Xia [13] further optimized Dalvi's algorithm by keeping only the necessary projections in the safe plans. For BID probabilistic tables, we propose a new project, Hybrid project (π^H), which executes a disjoint project [14] and then performs an independent project [14] in an atomic operation. A few papers [5,14] discussed safe plans of queries on BID probabilistic tables. In this paper, we introduce a new algorithm to find safe plans of conjunctive queries without self-joins on BID probabilistic databases. Our algorithm can find a safe plan for a query iff the query is safe.

The other is that there are many intensional methods to infer the confidences of answer tuples using lineage. PrDB [17] adopted probabilistic graphical model to infer the confidences. Trio [16] computed the probability of an arbitrary boolean formula of independent events, which is known to have exponential worst-case complexity. SPROUT [12] adopted its operator conf() to turn disjunctive normal form formulas into read-one formulas and then computed their probabilities on the fly. Knowledge compilation [9,11,18] can be used to compute the probabilities of query results in probabilistic databases. Sen et al. [18] proved that one only needs to test if the co-occurrence graph is a cograph to judge whether the boolean formulas of the result tuples produced by conjunctive queries without self-joins are read-once. Olteanu et al. [11] found that OBDDs can naturally represent the lineages of IQ queries and the probability of an OBDD can be computed in PTIME. Jha and Suciu [9] studied the problem of compiling the query lineage into compact representation and considered four tractable compilation targets: read-once, OBDD, FBDD, and d-DNNF.

7   Conclusions and Future Work

The field of uncertain databases has attracted considerable attention over the last few decades and is experiencing revived interest due to the increasing popularity of applications such as data cleaning and integration, information extraction, scientific and sensor databases, and etc. Dalvi and Suciu's work [4] on the evaluation of conjunctive queries on tuple-independent probabilistic databases showed that the class of conjunctive queries can be partitioned into easy queries and hard queries. This paper introduces two new probabilistic tuple models and one new probabilistic operator. Based on the above work, this paper further proposes an algorithm to find safe plans of conjunctive queries on BID tables. Thus far we have limited our discussion to safe plans of probabilistic databases. We plan to do some research work on other directions of probabilistic databases.

Acknowledgments. This work is partially funded by the National Basic Research Program of China (973 Program) under Grant No. 2012CB316205, the National Natural Science Foundation of China under Grant No. 61170012 and 61472425.

References

1. Transaction Processing Performance Council, TPC $BENCHMARK^{TM}$ H Standard Specification (revision 2.9.0)
2. Boulos, J., Dalvi, N., Mandhani, B., Mathur, S., Re, C., Suciu, D.: Mystiq: a system for finding more answers by using probabilities. In: SIGMOD, pp. 891–893 (2005)
3. Cavallo, R., Pittarelli, M.: The theory of probabilistic databases. In: VLDB, pp. 71–81 (1987)
4. Dalvi, N., Suciu, D.: Efficient query evaluation on probabilistic database. The VLDB Journal 16(4), 523–544 (2007)
5. Dalvi, N., Suciu, D.: Management of probabilistic data: foundations and challenges. In: PODS, pp. 1–12 (2007)
6. Dey, D., Sarkar, S.: A probabilistic relational model and algebra. ACM Transactions on Databases Systems 21(3), 339–369 (1996)
7. Green, T., Tannen, V.: Models for incomplete and probabilistic information. IEEE Data Engineering Bulletin 29(1), 17–24 (2006)
8. Ives, Z. G., Khandelwal, N., Kapur, A., Cakir, M.: Orchestra: rapid, collaborative sharing of dynamic data. In: CIDR, pp. 41–46 (2005)
9. Jha, A., Suciu, D.: Knowledge compilation meets database theory: Compiling queries to decision diagrams. Theory of Computing Systems 52(3), 403–440 (2013)
10. Lakshmanan, L., Leone, N., Ross, R., Subrahmanina, V.: Probview: A flexible probabilistic database system. ACM Transactions on Database Systems 22(3), 419–469 (1997)
11. Olteanu, D., Huang, J.: Secondary-storage confidence computation for conjunctive queries with inequalities. In: SIGMOD, pp. 389–402 (2009)
12. Olteanu, D., Huang, J., Koch, C.: Sprout: lazy vs. eager query plans for tuple-independent probabilistic databases. In: ICDE, pp. 640–651 (2009)
13. Qin, B., Xia, Y.: Generating efficient safe query plans for probabilistic databases. Data & Knowledge Engineering 67(3), 485–503 (2008)
14. Ré, C., Dalvi, N., Suciu, D.: Query evaluation on probabilistic databases. IEEE Data Engineering Bulletin 29(1), 25–31 (2006)
15. Ré, C., Suciu, D.: Materialized views in probabilistic databases for information exchange and query optimization. In: VLDB, pp. 51–62 (2007)
16. Sarma, A., Theobald, M., Widom, J.: Exploiting lineage for confidence computation in uncertain and probabilistic databases. In: ICDE, pp. 1023–1032 (2008)
17. Sen, P., Deshpande, A., Getoor, L.: Prdb: managing and exploiting rich correlation in probabilistic databases. The VLDB Journal 18(5), 1065–1090 (2009)
18. Sen, P., Deshpande, A., Getoor, L.: Read-once functions and query evaluation in probabilistic databases. In: VLDB, pp. 1068–1079 (2010)
19. Singh, S., Mayfield, C., Shah, R., Prabhakar, S., Hambrusch, S.: Database support for probabilistic attributes and tuples. In: ICDE, pp. 1053–1061 (2008)
20. Widom, J.: Trio: a system for integrated management of data, accuracy, and lineage. In: ICDR, pp. 262–276 (2005)

Parallel Top-k Query Processing on Uncertain Strings Using MapReduce

Hui Xu, Xiaofeng Ding$^{(\boxtimes)}$, Hai Jin, and Wenbin Jiang

Services Computing Technology and System Lab, Cluster and Grid Computing Lab,
School of Computer Science and Technology,
Huazhong University of Science and Technology, Wuhan 430074, China
{chinahui1988,xfding,hjin,wenbinjiang}@hust.edu.cn

Abstract. Top-k query is an important and essential operator for data analysis over string collections. However, when uncertainty comes into big data, it calls for new parallel algorithms for efficient query processing on large scale uncertain strings. In this paper, we proposed a MapReduce-based parallel algorithm, called MUSK, for answering top-k queries over large scale uncertain strings. We used the probabilistic n-grams to generate key-value pairs. To improve the performance, a novel lower bound for *expected edit distance* was derived to prune strings based on a new defined function *gram mapping distance*. By integrating the bound with TA, the filtering power in the Map stage was optimized effectively to decrease the transmission cost. Comprehensive experimental results on both real-world and synthetic datasets showed that MUSK outperformed the baseline approach with speeds up to 6 times in the best case, which indicated good scalability over large datasets.

1 Introduction

Due to the influence of various factors, such as the precision constraints of sensors, typographical errors, and gene mutation, uncertain strings have become ubiquitous in real life. As ranking in databases with uncertain information or attributes should concern the tradeoff between the score and the probability, existing query methods on deterministic data are no longer appropriate any more. The problem to find the top-k answers on fuzzy information has been deeply concerned for a long time. However, few literatures have dealt with top-k queries on uncertain strings, which we focus on in this paper.

Furthermore, the scale of this problem has increased dramatically in the era of big data. Regarding the limited computational capability and storage of a single machine, designing new parallel algorithms is the way to deal with the rapidly growth of both the number and length of strings. Exploiting the parallelism among a cluster of computing nodes, MapReduce [1] has become a dominant parallel computing paradigm for processing large-scale datasets, and it is already well studied and widely used in both commercial and scientific applications [2].

© Springer International Publishing Switzerland 2015
M. Renz et al. (Eds.): DASFAA 2015, Part II, LNCS 9050, pp. 89–103, 2015.
DOI: 10.1007/978-3-319-18123-3_6

However, most researches on MapReduce were focused on developing MapReduce versions of standard query algorithms [3], which is mainly dealing with queries over numerical data. To the best of our knowledge, no work has been done in the area of top-k queries over uncertain strings.

In this paper, we focus on the top-k query processing on uncertain strings based on the similarity function *expected edit distance* (EED) [4], and investigate how this problem can get benefits from the popular MapReduce framework. Since one of the most expensive operations in MapReduce is moving the data among different machines, we update the naive method by using combiners executed on the same nodes as mappers to pick up its local top-k strings. However, this method is still prohibitively expensive for large string datasets. To address this problem, we propose a semantics called *gram mapping distance* (GMD), based on which a novel lower bound for EED is derived. Finally, we propose an optimization algorithm, which seamlessly integrates the lower bound and TA algorithm to speed up. In summary, the contributions of this paper are as follows:

- We give the definition of top-k query on uncertain strings based on EED and denote the problem as EED-Top$k$ (Section 3). To solve it, we propose several MapReduce-based parallel algorithms (Section 5).
- To reduce the overall transmission cost and improve the pruning power in the Map phase, we derive a novel lower bound for EED from a new distance semantics GMD (Section 4).
- We demonstrate the superiority of our methods through extensive experiments (Section 6).

In addition, we review related work in Section 2 and conclude the paper in Section 7.

2   Related Work

Top-k query is an important data analysis tool for many applications, and it has been exhaustively studied. On the one hand, the *Fagin's Threshold Algorithm* (TA) [5] is one of the best known algorithms for answering top-k query on multi-dimensional numerical data. On the other hand, applying divide and conquer, n-gram [6] is the most widely used signature to accelerate the string search procedure. Recently, papers [7,8] proposed efficient filtering methods taking advantages of inverted n-gram indexes available for finding the top-k approximate strings on determinate strings. Without inverted n-gram index, Deng et al. [9] proposed a progressive framework by improving the traditional dynamic-programming algorithm to compute edit distance for top-k query.

When uncertainty comes into the picture, the semantics of ranking becomes ambiguous, due to the fact that both scores and probabilities of tuples must be accounted in the ranking. So existing top-k query methods on deterministic data are inappropriate for this significant challenge. Efficient top-k query processing for uncertain data in atomic data types have been discussed in [10,11] with different types of query conditions, but few literatures have dealt with uncertain

Table 1. Mainly Used Notations

Notation	Description
$|S|$	the size of string dataset S
l_s	the length of string s
μ	the average length of uncertain strings
$c[i]$	the ith character in uncertain string s
G_s	the n-gram set of string s
$\lambda_e(s_1, s_2)$	expected edit distance between s_1 and s_2
$g(s_1, s_2)$	gram mapping distance between s_1 and s_2

strings. Jestes et al. [4] focused on the similarity joins on uncertain strings using EED as measurement, and proposed two probabilistic string models to capture the fuzziness in string values in real-world applications. Then, Ge and Li [12] efficiently answered approximate substring matching problem on the character-level uncertain strings, which is different from our problem.

To deal with the rapid growth of the string number and length, parallelizing the algorithm is an obvious trend. Much recent research on MapReduce has focused on developing MapReduce versions of standard algorithms [3]. Vernica et al. [13] proposed a 3-stage MapReduce-based similarity join method. Then, a MapReduce-based string similarity joins framework, called MASSJOIN [14], was given to support both set-based similarity functions and character-based similarity functions.

Despite of the intensive efforts devoted in processing top-k queries over numerical data, or similarity joins over spatial or string datasets, to the best of our knowledge, parallel top-k query processing on uncertain strings is still a previously unresolved problem in data management.

3 Preliminaries and Problem Definition

In this section, we introduce some preliminaries in uncertain strings and top-k query. Then, we give the definition of top-k query on uncertain strings. Table 1 summarizes the mainly used notations in this paper.

3.1 Preliminaries

Uncertainty Model. In a character-level uncertain string model [4], for a string $s = c[1] \ldots c[l_s]$, each character $c[i]$ is a random variable with a discrete distribution over its possible world Ω. For $i = 1, \ldots, l_s$, $c[i] = \{(c_{i,1}, p_{i,1}), \ldots, (c_{i,\eta_i}, p_{i,\eta_i})\}$, where $c_{i,j} \in \Omega, p_{i,j} \in (0, 1]$, and $\sum_{j=1}^{\eta_i} p_{i,j} = 1$. Each character $c[i]$ instantiates into $c_{i,j}$ with probability $p_{i,j}$ independently. In this paper, we separate different choices in $c[i]$ with "$*$" and enclose each $c[i]$ with two "$|$". While, if $c[i]$ is deterministic, namely it has only one choice with probability 1, we can just write it down by itself.

Expected Edit Distance. Edit distance is one of the most widely accepted measures to determine similarities between two strings [15]. The edit distance $d(s_1, s_2)$ means the minimum number of edit operations to transform string s_1 into another string s_2. Jestes et al. [4] proposed *expected edit distance* (EED) as the similarity function between two uncertain strings. The EED between the query string q and the uncertain string s is

$$\lambda_e(s, q).= \sum_{s_i \in \Omega} p(s_i) \times d(q, s_i)$$

The probabilistic string s instantiates into one instance s_i of the possible worlds with probability $p(s_i)$, and $d(q, s_i)$ is the edit distance between s_i and q.

Probabilistic N-gram. The n-gram multi-set of the string s, denoted by G_s, is obtained by sliding a window of length n over the characters in s. Considering the uncertainty, probabilistic n-gram comes. A probabilistic n-gram of string s is $(\ell, s[\ell...\ell + n - 1], p_{i,j})$, where ℓ is the beginning position of the n-gram, and $s[\ell...\ell + n - 1]$ is the probabilistic substring of length n beginning from ℓ with probability $p_{i,j}$. Without loss of generality, before dividing s into a set of n-grams, we extend it to a new string s' by prefixing $(n - 1)$ copies of "\$" and suffixing $(n-1)$ copies of "#". If there is one uncertain character with m possible instances in s, the size of its probabilistic n-gram set will be $(l_s + m \times n - 1)$. For instance, the probabilistic 3-gram set of $s = $ "$appl|e * 0.7 * y * 0.3 * |$" is $G_s = \{(0, \$\$a, 1), (1, \$ap, 1), (2, app, 1), (3, ppl, 1), (4, ple, 0.7), (4, ply, 0.3), (5, le#, 0.7), (5, ly#, 0.3), (6, e##, 0.7), (6, y##, 0.3)\}$.

MapReduce. In the MapReduce framework, computation is carried out by using two user defined functions: Map function and Reduce function. In map phase, each map task reads the input (key, value) pairs as a list of independent records, and emits intermediate (key, value) pairs. In reduce phase, a reduce function is invoked with each distinct key and the list of all values sharing the key. The framework also allows the user to provide a combiner function that is executed on the same nodes as mappers right after the map functions have finished. The combiner allows the user to decrease the amount of data sent through the network.

3.2   Problem Definition

In this paper, we focus on the query problem on the character-level uncertain strings. Based on the scoring function EED shown above, we now formally define the top-k query on uncertain strings based on "marriage" of possible worlds and traditional top-k query semantics.

Definition 1. *(EED-Topk) Given a set of uncertain strings $S = \{s_i \mid 1 \leq i \leq |S|\}$ with possible worlds associated with probabilities on some positions, a user-appointed parameter k (the expected size of answer queue), and a query string q,*

Table 2. An Example of EED-Top$k$ with $q =$ "ACGTATGGAC" and $k = 1$

id	character-level uncertain string s	$\lambda_e(s_i, q)$
s_0	$|C * 0.5 * G * 0.5*|$ACGTATGGAC	$0.5 \times 1 + 0.5 \times 1 = 1$
s_1	ACTAGATCC$|A * 0.5 * T * 0.5*|$	$0.5 \times 6 + 0.5 \times 6 = 6$
s_2	ACGTAT$|G * 0.8 * C * 0.2*|$GAC	$0.8 \times 0 + 0.2 \times 1 = 0.2$
s_3	A$|A * 0.5 * T * 0.5*|$CCAGCAT	$0.5 \times 6 + 0.5 \times 6 = 6$
s_4	GAAG$|T * 0.7 * A * 0.3*|$TCATC	$0.7 \times 6 + 0.3 \times 6 = 6$

an EED-Topk query over S should return the k objects, whose $\lambda_e(s_i, q)$ are the global k-minimums in S. The answer of query is denoted as a k-length array $T = \{T_1, \ldots, T_k\}$. That is to say, for each $T_j \in T (1 \leq j \leq k)$, $\lambda_e(T_j, q) \leq \lambda_e(s_x, q)$, where $s_x \in S \backslash T$, $1 \leq x \leq |S|$.

Example 1. As shown in Table 2, the string set contains 5 uncertain strings. According to each string's EED with q, the answer of top-1 query is $T = \{s_2\}$.

4   A Novel Lower Bound

Since computing EEDs for every uncertain strings in the data set is prohibitively expensive, we try to give a new lower bound for EED to avoid enumerating all strings.

4.1   Gram Mapping Distance

Applying divide and conquer strategy, we propose a novel semantics called *gram mapping distance* to evaluate the similarity between the n-gram multi-sets of two strings.

Definition 2. *(Gram Mapping Distance) (GMD) Given two n-gram multi-sets G_{s_1} and G_{s_2} of the strings s_1 and s_2 with the same cardinality, the gram mapping distance between s_1 and s_2 is defined as an optimal mapping between G_{s_1} and G_{s_2}. Assume that $P : G_{s_1} \to G_{s_2}$ is a bijection. The GMD between s_1 and s_2, denoted by $\mu(s_1, s_2)$, is computed as*

$$g(s_1, s_2) = \min_P \sum_{gram_i \in G_{s_1}} \lambda_e(gram_i, P(gram_i)), \quad P : G_{s_1} \to G_{s_2}$$

Example 2. The computation of GMD is equivalent to finding an optimal mapping between two n-gram multi-sets. Similar to the work in [16], we construct a weighted matrix for each pair of n-grams from the two strings, and apply the Hungarian algorithm [17] to find the optimal mapping. As shown in Fig. 1, the strings s_1 and s_2 have different lengths with their corresponding probabilistic n-grams below them. As the definition of GMD requires that the two multi-sets must have same cardinality, it requires an additional null n-gram in the G_{s_1} before mapping.

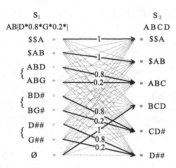

Fig. 1. Compute $g(s_1, s_2)$ between G_{s_1} and G_{s_2}. The optimal mapping is the solid line with the probability in middle, while other edges linking the n-grams are shown as dotted line. $g(s_1, s_2) = (0.8 \times 1 + 0.2 \times 1) + (0.8 \times 1 + 0.2 \times 2) + (0.8 \times 0 + 0.2 \times 1) + 1 \times 3 = 5.4$.

4.2   Lower Bound for EED

Based on GMD, a novel lower bound for EED is proposed as follows.

Lemma 1. *Given two uncertain strings s_1 and s_2, the relationship between $g(s_1, s_2)$ and $\lambda_e(s_1, s_2)$ satisfies*

$$g(s_1, s_2) \leq max\{3, (3n - 4)\} \times \lambda_e(s_1, s_2)$$

Proof. Let $E = \{e_1, e_2, \cdots, e_K\}$ be a series of edit operations that is needed to transform s_1 into s_2. Accordingly, there is a sequence of strings $s_1 = M_0 \rightarrow M_1 \rightarrow \cdots \rightarrow M_K = s_2$, where $M_{i-1} \rightarrow M_i$ indicates that M_i is the derived string from M_{i-1} by performing e_i for $1 \leq i \leq K$. Assume that there are K_1 insertion operations, K_2 deletion operations and K_3 substitution operations respectively, so we have $K_1 + K_2 + K_3 = K = \lambda_e$. Then, we analyse the detailed influence of each kind of edit operation as illustrated in Fig. 2.

According to the definition of probabilistic n-gram above, each character in string s must appear at least n times in G_s. It is clear that an edit operation will only affect at most $m \times n$ probabilistic n-grams of an uncertain string, when there is one uncertain character with m possible instances. As $\sum_{j=1}^{m} p_j = 1$, the EED between the affected $m \times n$ probabilistic n-grams and the query string's n-grams have $\lambda_e = \sum_{i=1}^{n} \sum_{j=1}^{m} p_{i,j} \times d_{i,j} \leq \sum_{i=1}^{n} max\{d_{i,j} | j \in [1, m]\}$. So we can simply consider the affected $m \times n$ probabilistic n-grams as only n n-grams with the maximum edit distance with the corresponding n-grams of the query string.

- **Insertion operation:** If we insert a character into the string M_{i-1}, as shown in Fig. 2, we see it changes from (a) to (b) with at most $m \times n$ affected probabilistic n-grams. We discuss the detailed influence in two cases: $n = 2$ and $n \geq 3$. When $n = 2$, it is clear that no more than 2 for one new derived n-gram and 1 for another one will be added to the GMD. While, when $n \geq 3$, the edit distance

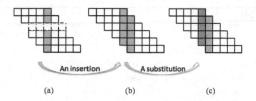

Fig. 2. From (a) to (b), we insert the pink character in front of the gray one. From (b) to (c), the blue character substitute the pink one.

will increase less than n for one new derived n-gram, 1 for both the first one and the last one n-grams, as well as 2 for the other $(n-3)$ n-grams. That's to say, if $n \geq 3$, the GMD will add no more than $1 \times 2 + 2 \times (n-3) + n = (3n-4)$. Thus, we conclude that $g(M_{i-1}, M_i) \leq max\{3, (3n-4)\}$.

- **Deletion operation:** Likewise, in Fig. 2 from (b) to (a), deleting the pink character has the same influence as inserting it.
- **Substitution operation:** As shown in Fig. 2 from (b) to (c), substitute the blue for the pink in M_{i-1} only lead to at most 1 edit distance for the affected $m \times n$ n-grams. So we have $g(M_{i-1}, M_i) \leq n$.

Above all, we conclude the following relationship:

$$g(s_1, s_2) \leq max\{3, (3n-4)\} \times K_1 + max\{3, (3n-4)\} \times K_2 + n \times K_3$$
$$\leq max\{3, (3n-4)\} \times (K_1 + K_2 + K_3)$$
$$\leq max\{3, (3n-4)\} \times \lambda_e(s_1, s_2)$$

5   Parallel Algorithms Using MapReduce

Intuitively, a naive parallel method is to enumerate all strings from the given uncertain string dataset and use MapReduce to compute the EEDs for them. Since one of the most expensive operations in MapReduce, denoted as "shuffle", is moving the data between different machines, this naive parallel method is quite time consuming as the growth of the string length, quantity and complexity.

5.1   PUSK

Fortunately, EED-topk query meets the conditions of using combiner function. So we propose a baseline parallel algorithm using the MapReduce framework as shown in Algorithm 1. The Map function computes all uncertain strings' EEDs to the query string, followed by a combiner function executed on the same nodes as mappers to pick up its local top-k strings. Then, all the local top-k strings from different nodes are sent to the Reducer, where the strings with top-k minimum EEDs from the given dataset are returned in reduce phase. We denote this basic algorithm as PUSK.

Algorithm 1. PUSK

Input: an uncertain string set $S = \{s_i \mid 1 \le i \le |S|\}$, a query string q, the size of answer set k

Output: the answer queue T: the top-k similar strings with q

1: **map**($k_1 = sid, v_1 = string$)//map phase
2:     $eed \leftarrow$ compute $\lambda_e(v_1, q)$;
3:     $output(k_2 = eed, v_2 = v_1)$;
4: **combiner**($k_2 = eed, list(v_2) = list(string)$)
5:     $count \leftarrow k$;
6:     **foreach** string in $list(v_2)$ **do**
7:         **if** $count > 0$ **then**
8:             $output(k_3 = k_2, v_3 = string)$;
9:             $count \leftarrow count - 1$;
10: **reduce**($k_3 = eed, list(v_3) = list(string)$)//reduce phase
11:     $count \leftarrow k$;
12:     **foreach** string in $list(v_3)$ **do**
13:         **if** $count > 0$ **then**
14:             $output(k_4 = k_3, v_4 = string)$;
15:             $count \leftarrow count - 1$;

Although it is very stable and almost not influenced by k, it is clear that PUSK is still brute-force and have to compute the EEDs of all the strings in the Map phase. To achieve better performance, we try to optimize the filtering power by changing the combiner function in Map phase.

5.2   MUSK

Like the original problem, our problem may be solved from scratch by applying divide-and-conquer strategy and using some TA-like algorithm. Since TA assumes that each attribute of the multidimensional data space has an index list, first of all, we design an inverted probabilistic n-gram index for uncertain strings at first. We store the index in extendible hashing and use each n-gram as the key. At the same time, we attach related string-id, frequency of n-grams, the pair of position and corresponding probability to each entries in n-gram's posting list.

By combing the lower bound as shown in Lemma 1 with the TA algorithm, we propose an optimization threshold algorithm for efficient EED-Top$k$ query processing and name it MUSK. The pseudo-code of MUSK is shown in Algorithm 2.

Fig. 3 shows an example of data flow in MUSK framework. The uncertain string dataset S is split into some smaller subsets, and each map task handles one split on the node by taking the string id sid as keys and the string content as values in parallel. To avoid too much computation and transmission cost, we optimize the filtering power in the combiner function. As both the map function and the combiner function process the input (key, value) pairs as a list of independent records, in Line 2 of Algorithm 2, we unified all the keys $k_2 = 1.0$

in the mappers to make sure that all the strings in the same split will be gathered into the same combiner as a whole after mapper. Fig. 4 shows how the combiner function works on the EED-Top1 query on uncertain string dataset as shown in Table 2.

Algorithm 2. MUSK

Input: a probabilistic string set $S = \{s_i \mid 1 \leq i \leq |S|\}$, a query string q, the size of answer queue k

Output: the answer queue T: the top-k similar strings with q

1: **map**($k_1 = sid, v_1 = string$)//map phase
2:     $output(k_2 = 1.0, v_2 = v_1)$;
3: **combiner**($k_2 = 1.0, list(v_2) = list(string)$)
4:     $G_q \leftarrow gram(q, n)$; //the n-grams of the query string q
5:     **foreach** string in $list(v_2)$ **do**
6:         $G_s \leftarrow gramIndex(string, n)$;
7:     Construct $matrix$ for TA based on EDs between G_s and G_q in ascending order;

8:     Obtain inverted list L from the $gramIndex$;
9:     $\omega = 0$;
10:    **for** (row=1; $L! = \emptyset$; row++) **do**
11:        Compute $\lambda_e(s_i, q)$ for the seen strings in current row;
12:        Update and maintain the top-k heap;
13:        ω = the summation of scores in current row;
14:        $\tau = max\{\lambda_e(s', q) | s' \in top\text{-}k\}$;
15:        **if** $\omega > \tau \times max\{3, (3n - 4)\}$ **then**
16:            All unseen strings are safely pruned;//from Lemma 1
17:            $T \leftarrow$ the pair of $(eed, string)$//early halt
18:            **foreach** pair in T **do**
19:                $output(k_3 = T.eed, v_3 = T.string)$;
20: **reduce**($k_3 = eed, list(v_3) = list(string)$)//reduce phase
21:    $count \leftarrow k$;
22:    **foreach** string in $list(v_3)$ **do**
23:        **if** $count > 0$ **then**
24:            $output(k_4 = k_3, v_4 = string)$;
25:            $count \leftarrow count - 1$;

At beginning of the filtering function, it accesses the objects row by row in the $matrix$. While some n-gram g_i is seen, find the $sids$ based on the index list of g_i and compute the $EEDs$ for these archived strings. Here, we ignore n-grams in the first and last $(n - 1)$ columns as the n-grams with \$ and # have low selectivity. In Line 13, the algorithm computes the summation of the scores in the current row as new threshold ω. It gives the minimum value of GMD for the other uncertain strings in the same split which have not been seen so far. If there is no g_i in the current row, we take the above value in the same column as its minimum value. Line 15 tests whether the current ω meets the halt condition $\omega > \tau \times max\{3, (3n-4)\}$. Here, τ is equal to the maximum of EEDs from current

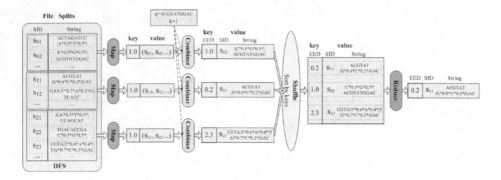

Fig. 3. An example of data flow in MUSK framework

top-k strings based on Lemma 1. Running above steps repeatedly until k objects have been found and meet the halt condition. In this case as shown in Fig. 4, when running at the second row, it meets the halt condition. The algorithm stops, and all the other unaccessed strings can be safely pruned.

	SSSSA	SSSAC	SSACG	SACGT	ACGTA	CGTAT	GTATG	TATGG	ATGGA	TGGAC	CGGAC#	GAC##	AC###	C####
	5-gram set G_q of query string q= "ACGTATGGAC"													
ED=0 ⇒	SSSSA	SSSAC	SSACG	SACGT	ACGTA	CGTAT	GTATG	TATGG	ATGGA	TGGAC	CGGAC#	GAC##	AC###	C####
ED=1 ⇒	SSSSC	SSSSC	SSACT	CACGT			GTATC	TATCG	ATCGA	TCGAC	CGAC#		AT###	A####
	SSSSG	SSSAA	SCACG										TC###	T####
ED=2 ⇒	SSSAA	SSAAC	SSAAC	SCACG	SACGT	ACGTA	CGTAT	ATGGA	TATGG	ATGGA	TGGAC	AC###	ATC##	AC###

Scan the first row and get $\omega = 0$, and $\lambda_e(s_0,q) = 1$, $\lambda_e(s_2,q) = 0.2 \Rightarrow \tau = 0.2$

If $\omega > \tau \times (3n-4)$? $\Rightarrow$ $0 > 0.2 \times (3 \times 5 - 4)$? $\Rightarrow$ NO! $\Rightarrow$ Go on!

Then scan the second row, $\omega = 1+1+1+1+0+0+1+1+1+1+1+0+1+1 = 11$, and $\tau = 0.2$

If $\omega > \tau \times (3n-4)$? $\Rightarrow$ $11 > 0.2 \times (3 \times 5 - 4)$? $\Rightarrow$ YES! $\Rightarrow$ Stop! & Return s_2

Fig. 4. An example of running with $n=5$ and $k = 1$ in the combiner

Based on the divide-and-conquer strategy, each combiner outputs the local EED-Top1 from its own split $list(k_3 = sed, v_3 = string)$. Then, for each partition, the pairs are sorted by their keys and sent across the cluster in the shuffle phase. Finally, in the reduce phase, we use only one reducer to pickup the pairs after shuffle and return the final EED-Top1 for the given uncertain string dataset.

Note that, the filtering function is an instance optimal as TA algorithm, as its buffer size is independent of the uncertain string dataset's size, and it only stores the current top-k objects, as well as the pointers to last objects accessed in the sorted list. In order to avoid the worst possible situation, like querying "xyz" from an "a/b/c/d/e" based dataset, we add one judgement at the beginning of

MUSK. If all the first elements in each column (except for the first and last $(n-1)$ columns) of the TA matrix are not less than n or user-defined maximum value of EED, the algorithm stops immediately, and return that "There's no proper result in this dataset".

6 Experimental Evaluation

In this section, we evaluate the performance of the proposed algorithms on real-world uncertain string datasets by varying different parameters. We used Hadoop 0.20.2 for the MapReduce framework implementation. All the experiments were run on a 10-node cluster. Each node had two Intel Xeon(R) E5620 2.40GHZ processors with 4 cores, 16GB RAM, 1TB disk, and installed 64-bit Linux operating systems. We set the block size of the distributed file system to 64MB.

6.1 Experimental Setup

To evaluate MUSK's scalability in terms of string length, we chose four real-world string datasets with different average string lengths standing for the short strings and the long strings, respectively. The first two datasets, *Author* and *Title*, come from author names and titles in the real data sources DBLP Bibliography[1]. We consider spelling errors and conversion between different character sets as the uncertainties. A notable difference between the two datasets is the string length. The average length of strings in *Author* is 12.8, while, the μ of *Title* is 105.08. *Query-Log*[2] contains one million distinct queries from AOL Query Log, with $\mu = 20.91$. *Protein*[3] sequences, downloaded from Uniprot, has a great deal of uncertainty [18]. The length of protein is from 52 to 687. We also generate a few synthetic datasets based on the real-world datasets above by varying some parameters of these datasets. By default, the size of generated uncertain string dataset is 20GB. After testing several experiments on the effect of n, we set $n = 3$ for short string datasets *Author* and *Query-Log*, while $n = 7$ for long string datasets *Title* and *Protein*. For each group of experimental parameters, we execute the EED-Top$k$ query for 100 times and compute the average query time.

6.2 Speedup

In order to evaluate the speedup of MUSK, we fixed the size of datasets with 20GB and varied the number of cluster nodes varying from 2 to 10 as shown in Fig. 5(a). We can see that with the increase of nodes in the cluster, the performance of our algorithm significantly improved. This is attributed to our powerful lower bound for EED, which can significantly prune dissimilar candidates and avoid computing the EEDs for all of the uncertain strings.

---

[1] http://www.informatik.uni-trier.de/$\sim$ley/db
[2] http://www.gregsadetsky.com/aol-data/
[3] http://www.uniprot.org/

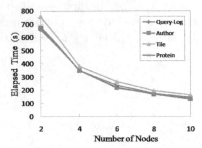 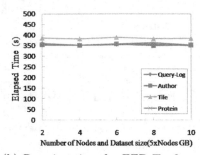

(a) Running time for EED-Top3 query using MUSK on four datasets with the number of nodes varying from 2 to 10

(b) Running time for EED-Top3 query using MUSK on four datasets with increased size from 10GB to 50GB and a cluster with 2 to 10 nodes, respectively

Fig. 5. Speedup and Scaleup

6.3   Scaleup

We evaluated the scaleup of our algorithm by increasing both the size of datasets and the number of cluster nodes with the same factor. Fig. 5(b) shows the running time for EED-Top3 queries on each of the datasets with increased size from 10GB to 50GB and a cluster with 2 to 10 nodes, respectively. A perfect scaleup could be achieved if the running time remains constant. As shown in Fig. 5(b), MUSK scales up quite well, no matter run on which string dataset.

6.4   Comparison with Baseline Method

In what follows, we compare our method MUSK with the baseline parallel algorithm PUSK for answering EED-Top$k$ queries on the four uncertain string datasets, denoted as *MUSK* and *PUSK*, respectively. The results of effects on varying parameters are shown as below.

Effects of μ. It is worth noticing that Fig. 5(b) has shown MUSK's scalability in terms of the string length by running on the four uncertain string datasets with different average string lengths from 12.80 to 302.54. With the same size of dataset, the running time is very close to each other, no matter for short strings or long strings. As shown in Fig. 6, it is clear that MUSK is much faster than PUSK and takes nearly the same run time for datasets with the same size ignoring the influence of the string length. This is reasonable, since no matter how long the uncertain string is, MUSK only need to access a few number of n-grams to get all user-wanted answers.

Effects of $|S|$. In order to analysis the effect of the database size, we generate different sizes of synthetic datasets $|S|$ from 1GB to 50GB. It is clear to notice

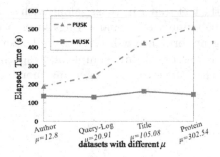

Fig. 6. Running time for different string length

from Fig. 7 that MUSK has much more stable average query time than the baseline algorithm. This is because the TA-based filtering method in the combiner stops when it meets the stopping condition, then all the other unseen strings can be safely pruned, no matter how large the dataset size is. On the contrary, the baseline method has to access all the strings in the dataset.

Effects of k. To study the query's sensitivity to user specified answer size k, a set of comparative experiments are designed by varying k from 1 to 100. As

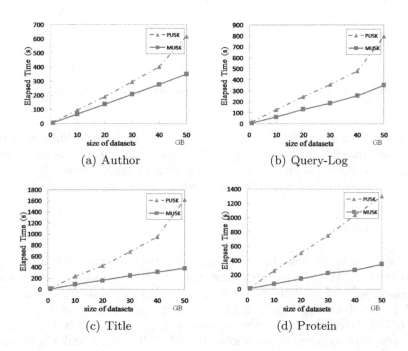

(a) Author

(b) Query-Log

(c) Title

(d) Protein

Fig. 7. Average query time for datasets with respect to varying $|S|$

shown in Fig. 8, due to the fact that τ increases along with k, it is difficult to meet the halt condition. If the number of similar strings in dataset is approximately equal to k, the filtering performance will tend to be better. The experimental results show that MUSK has lower time complexity, especially, its efficiency is improved significantly when the string is longer and more complicate.

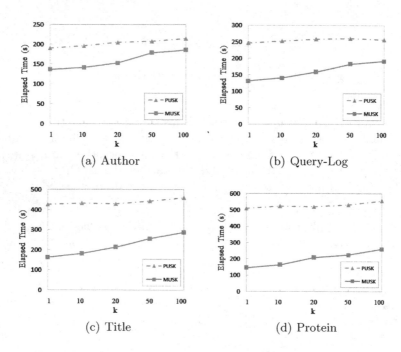

(a) Author

(b) Query-Log

(c) Title

(d) Protein

Fig. 8. Average query time for datasets with respect to varying k

From all the experiments above, the results show that our techniques significantly reduce query cost compared to the baseline parallel approach, and achieve good speedup and scalability in terms of the string length, the size of dataset, and the complexity of uncertain strings as well.

7　Conclusion

In this paper, we study how the top-k query processing on uncertain strings can get benefits from the MapReduce framework. We propose MUSK, a one-stage MapReduce-based parallel algorithm, for answering EED-Top$k$ query efficiently. We use probabilistic n-grams as signatures to produce key-value pairs in the framework. To further improve the performance, a new distance measurement GMD is proposed to drive the novel lower bound of EED. By seamlessly integrating the proposed bound with TA algorithm in the framework, we optimized the filtering power in the Map phase while not increase the transmission cost. Comprehensive experimental results show that our techniques achieve good speedup compared to the baseline parallel approach, and has good scalability performance in terms of different experimental settings.

Acknowledgments. This work was supported by the National Natural Science Foundation of China under grant 61100060 and 61472148.

References

1. Dean, J., Ghemawat, S.: Mapreduce: simplified data processing on large clusters. In: OSDI, pp. 137–150 (2004)
2. Jiang, D., Ooi, B.C., Shi, L., Wu, S.: The performance of MapReduce: An in-depth study. In: VLDB, pp. 472–483. VLDB Endowment (2010)
3. Li, F., Ooi, B.C., Tamer Özsu, M., Wu, S.: Distributed Data Management Using MapReduce. ACM Computing Survey **46**(3) (2014)
4. Jestes, J., Li, F., Yan, Z., Yi, K.: Probabilistic string similarity joins. In: SIGMOD, pp. 327–338. ACM (2010)
5. Fagin, R., Lotem, A., Naor, M.: Optimal aggregation algorithms for middleware. In: PODS, pp. 102–113 (2001)
6. Li, C., Lu, J., Lu, Y.: Efficient merging and filtering algorithms for approximate string searches. In: ICDE, pp. 257–266. IEEE (2008)
7. Kim, Y., Woo, K.-G., Park, H., Shim, K.: Efficient processing of substring match queries with inverted q-gram indexes. In: ICDE, pp. 721–732. IEEE (2010)
8. Wang, X., Ding, X., Tung, K.H., Zhang, Z.: Efficient and effective KNN sequence search with approximate n-grams. In: VLDB, pp. 1–12. VLDB Endowment (2013)
9. Deng, D., Li, G., Feng, J., Li, W.-S.: Top-k string similarity search with edit-distance constraints. In: ICDE, pp. 925–936. IEEE (2013)
10. Hua, M., Pei, J., Zhang, W., Lin X.: Efficiently answering probabilistic threshold top-k queries on uncertain data. In: ICDE, pp. 85–96. IEEE (2008)
11. Yi, K., Li, F., Kollios, G., Srivastava, D.: Efficient processing of top-k queries in uncertain databases. In: ICDE, pp. 1406–1408. IEEE (2008)
12. Ge, T., Li, Z.: Approximate substring matching over uncertain strings. In: VLDB, pp. 772–782. VLDB Endowment (2011)
13. Vernica, R., Carey, M.J., Li, C.: Efficient parallel set-similarity joins using MapReduce. In: SIGMOD, pp. 495–506. ACM (2010)
14. Deng, D., Li, G., Hao, S., Wang, J., Feng, J., Li, W.-S.: MassJoin: A MapReduce-based method for scalable string similarity joins. In: ICDE. IEEE (2014)
15. Navarro, G.: A guided tour to approximate string matching. ACM Computing Surveys **33**(1), 31–88 (2001)
16. Wang, X., Ding, X., Tung, K.H., Ying, S., Jin, H.: An efficient graph indexing method. In: ICDE, pp. 805–816. IEEE (2012)
17. Kugn, H.W.: The Hungarian method for the assignment problem. Naval Research Logistics Quarterly **2**, 83–97 (1955)
18. Bandeira, N., Clauser, K., Pevzner, P.: Shotgun Protein Sequencing: Assembly of peptide tandem mass spectra from Mixtures of Modified Proteins. Molecular and Cellular Proteomics **6**(7) (2007)

Tracing Errors in Probabilistic Databases Based on the Bayesian Network

Liang Duan[1], Kun Yue[1(✉)], Cheqing Jin[2], Wenlin Xu[1], and Weiyi Liu[1]

[1] Department of Computer Science and Engineering,
School of Information Science and Engineering, Yunnan University, Kunming, China
kyue@ynu.edu.cn
[2] Institute of Massive Computing, East China Normal University, Shanghai, China

Abstract. Data in probabilistic databases may not be absolutely correct, and worse, may be erroneous. Many existing data cleaning methods can be used to detect errors in traditional databases, but they fall short of guiding us to find errors in probabilistic databases, especially for databases with complex correlations among data. In this paper, we propose a method for tracing errors in probabilistic databases by adopting Bayesian network (BN) as the framework of representing the correlations among data. We first develop the techniques to construct an augmented Bayesian network (ABN) for an anomalous query to represent correlations among input data, intermediate data and output data in the query execution. Inspired by the notion of blame in causal models, we then define a notion of blame for ranking candidate errors. Next, we provide an efficient method for computing the degree of blame for each candidate error based on the probabilistic inference upon the ABN. Experimental results show the effectiveness and efficiency of our method.

Keywords: Data cleaning · Probabilistic database · Bayesian network · Rejection sampling · Probabilistic inference

1 Introduction

Many real word applications, such as information extraction, data integration, sensor networks and object recognition etc., are producing large volumes of uncertain data [1,2]. It is critical for such applications to effectively manage and query the uncertain data, motivating the research on probabilistic databases (PDBs) [3–5,17].

In practice, PDBs often contain errors since the data of these databases has been collected by a great deal of human effort through the consultation, verification and aggregation of existing sources. It could be worse when using the Web to extract and integrate data from diverse sources on an unprecedented scale, which the risks of creating and propagating data errors increased [4,6, 18,19]. Consequently, a user may detect an anomalous query: (1) some of the

© Springer International Publishing Switzerland 2015
M. Renz et al. (Eds.): DASFAA 2015, Part II, LNCS 9050, pp. 104–119, 2015.
DOI: 10.1007/978-3-319-18123-3_7

probabilities of the result tuples are erroneous; and (2) the tuples contributed to the incorrect output are to be found in the database. To guarantee the data quality, it is necessary to trace the errors in the input data and prevent the errors from propagating to other queries. This can be viewed as a strategy of data cleaning for improving the data quality.

For an anomalous query, it is easy to detect errors in the output by comparing the output values with the given ground truth values. This means that there are errors in the input data, but we do not know which one is not correct, and only know the output is erroneous. In this paper, we show how to trace errors in PDBs given an anomalous query.

The first step of error tracing is to find out the input data that is related to the output data. Provenance or lineage has been studied recently to describe how individual output data is derived from a certain subset of input data, so it is natural to use lineage to trace all the input data helped to produce the surprising output data [7]. Meliou et al. [8,9] proposed a method to find the causes for surprising queries and the first step is the computation of the query's lineage. However, most of current lineage-based methods make simplistic assumptions about the data (e.g., complete independence among tuples), which makes this kind of methods difficult to be used in real applications that naturally produce correlated data.

It is well known that Bayesian network (BN) is an effective framework of representing dependencies among random variables [10]. A BN is a directed acyclic graph (DAG), where nodes represent random variables and edges represent dependencies among variables. Each variable in a BN is associated with a conditional probability table (CPT) to give the probability of each state when given the states of its parents. Sen and Deshpande [11,12] provided a BN based framework that can represent probabilistic tuples and correlations. The query processing problem on this framework is casted as a probabilistic inference problem in an appropriately constructed BN. This means that the correlations among output and input data with respect to a certain query can be represented by a BN. Comparing with the lineage-based methods, this framework can describe not only how the output data is derived from input data, but also the correlations among input data. Therefore, we adopt BN as our underlying framework for probabilistic databases and construct an augmented Bayesian network (ABN) for the anomalous query to trace errors.

A query may involve large volume of input data, which is overwhelming to users. Thus, it is necessary to rank the errors in input data by their degree of contributions to output data. The notion of responsibility is first developed by Chockler et al. [13,16] to measure the degree of contributions of a cause to a result event in causal models when everything relevant about the facts of the world is known (i.e., the context is given). Meliou et al. [8,9] find causes for surprising queries by ranking candidate errors according to their responsibility after examining the lineage of the query. But the method is hampered to be carried on PDBs by two limitations: (1) the context for queries in PDBs is uncertain; and (2) the lineage cannot represent correlations among data in PDBs.

Fortunately, the notion of blame, also developed in [13], can be used when the context of causal models is unknown. Specifically, the blame of a cause is the expectation of responsibility under all of uncertain context. A context of causal models can be viewed as a possible world instance of PDBs. Therefore, inspired by the above research findings, we define the degree of blame of each node in the ABN to measure their contributions to the anomalous output. Then, we rank the candidate errors by their blame degrees for tracing the errors in input data.

Computing the blame has to find out all of possible world instances related to output data. Since the possible world instances are obtained from probabilistic inferences executed in exponential time upon the ABN, the computation of blame is exponential complexity, which is not efficient and suitable enough with respect to large scale ABNs. Thus, based on the rejection sampling [14], we propose an approximate inference algorithm to obtain the blame of nodes the ABN.

Generally speaking, our main contributions can be summarized as follows:

- We propose a method to construct an augmented Bayesian network for representing complex correlation among input data, intermediate data and output data generated in query processing given an anomalous query.
- We present a definition of blame of nodes in the augmented Bayesian networks to measure the degree of contribution of each node to the anomalous query. Then, we provide an efficient method to compute the degree of the blame and rank the candidate errors by their blame.
- We implement the proposed algorithms and make preliminary experiments to test the feasibility of our method.

The remainder of this paper is organized as follows: In Section 2, we describe the error tracing problem briefly. In Section 3, we construct an ABN for an anomalous query. In Section 4, we define the blame upon ABNs and rank candidate errors by their blame. In Section 5, we show experimental results. In Section 6, we discusses related work. In Section 7, we conclude and discuss future work.

2   Problem Statement

For a query q on a probabilistic database $D(R, P)$ where R is a set of uncertain relations and P is a BN for representing the correlation among data, the *error tracing problem* is to detect the errors which cause the anomaly that the probability $P(t)$ of result tuple t of q is not equal to the truth value $P'(t)$ [7,8].

If $P(t) = P'(t)$, then there are no errors, so we will assume that $P(t) \neq P'(t)$ for the rest of this paper. Clearly, $P(t) \neq P'(t)$ means that $P(t) > P'(t)$ or $P(t) < P'(t)$. Since $P(t) < P'(t)$ can be viewed as $P(\bar{t}) > P'(\bar{t})$, we only take the case that $P(t) > P'(t)$ into consideration. Our approach to this problem is to find all candidate errors X (i.e., tuples in D) for the anomaly, and rank them by their degree of blame.

Example 1. Figure 1(a) shows a probabilistic database D with two uncertain relations: S and T. The positive correlation (i.e., if one variable is increased, the other variable will also increased and vice versa) among uncertain data is represented by a BN, whose DAG is shown in Figure 1(b) and the corresponding CPTs are shown in Figure 1(c). For a query $q = \prod_c S \bowtie_B T$, we can obtain the result shown in Table 1. If the truth probability of $r1$ is $P'(r1) = 0$ (for ease of exposition, we assume the truth probability $P'(r1) = 0$), errors which lead to this anomaly have to be found out. It is clear that $s1.B$, $t1.B$ and $s2.B$ are the candidate errors and their degree of blame for this anomaly is shown in Table 2.

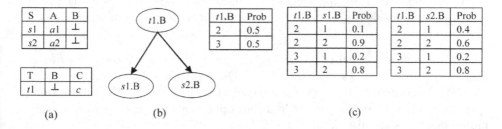

Fig. 1. (a) A small probabilistic database D with two uncertain relations. (b) DAG of a Bayesian network. (c) CPTs corresponding to the Bayesian network.

Table 1. The query result of q in D.

Query result	C	Prob
$r1$	c	0.48

Table 2. Errors with their degrees of blame.

Error	Blame
$t1.B$	1.0
$s1.B$	0.89
$s2.B$	0.55

3 Constructing Augmented Bayesian Network

In this section, we will discuss how to construct an ABN for an anomalous query on probabilistic databases.

To trace errors in D, we augment P to represent not only the correlations among the input data, but also the correlations among the input data, intermediate data and output data of an anomalous query. Constructing an ABN always has two steps: first constructing the structure of ABN and then generating the CPTs. Sen et al. [11] propose a method to construct an ABN for a query q on PDBs. Answers to query q may contain numerous result tuples, but we only concern a special tuple t which is anomalous. The ABN for tuple t, denoted by $G_t = (V_t, E_t)$, is a subgraph of the ABN $G_q = (V_q, E_q)$ for query q. Thus, we provide a method to construct an ABN specific to the anomalous result tuple t.

The structure of the ABN is constructed by the following steps:

1. For each random variable v in PDBs, add a node v to V_q.
2. Construct E_q by performing the following steps. For each operation $O^{op}(X_S, x_i)$ obtained during evaluating the operator op on the set of tuples S to generate the tuple i:
   - Add a node x_i to V_q and annotate x_i with the operator op;
   - $\forall x \in X_s$, add an edge $e(x, x_i)$ to E_q.
3. $V_t = \{v|v \in V_q, v \leadsto x_t\}$, where x_t denotes the node corresponding to the anomalous result tuple t and $v \leadsto x_t$ is true if there is a path from v_i to x_t.
4. $E_t = \{e = (v_i, v_j)|e \in E_q, v_i, v_j \in V_t\}$.

For the nodes that exist in PDBs (i.e., not generated during query processing), we copy their CPTs from the given PDBs. For the other nodes, we generate CPTs according to the operator annotated with them. For the limitation of space, we only take select, project and join operations into consideration, while more operations are discussed in [11,12].

Select: Let $\sigma_c(R)$ denote the query, where c is the predicate of the selection. For each tuple t in R, the probability in the CPT of the node corresponding to $t=1$ or $t=0$ for the case that t satisfies c or not respectively.

Project: Let $\Pi_a(R)$ denote the projection, where $a \subseteq attr(R)$ denotes the set of attributes that we want to project onto. Let r denote the result of the projecting $t \in R$, and the probability of the node corresponding to r is 1 or 0 for the case of $t.a = r.a$ or not respectively.

Join: Let $R_1 \bowtie_a R_2$ denote the join operation where R_1 and R_2 are two relations and $a \subseteq attr(R_1) \bigcap a \subseteq attr(R_2)$. Let r denote the join result of two tuples $t_1 \in R_1, t_2 \in R_2$, and the probability of the node corresponding to r is 1 or 0 for the case of that $t1.a$ equals $t2.a$ or not respectively.

Example 2. For the query $\prod_c(S \bowtie_B T)$ on the PDB presented in Figure 1, the ABN for the result tuple is shown in Figure 2. We introduce intermediates tuples produced by the join ($i1$ and $i2$) and produce a result tuple ($r1$) from the projection operation. $i1$ exists (i.e., $i1.e = 1$) only when $s1.B$ and $t1.B$ are assigned the value of 2. Similarly, $i2.e$ is 1 only when $s2.B$ and $t1.B$ are assigned the value of 2. Finally, $r1.e$ is 1 when $i1$ or $i2$ or both exist. The query result is the tuple $r1$ with the corresponding probability $P(r1.e = 1)$, which can be computed by many probabilistic inference methods, such as Enumeration algorithm and Rejection sampling [14].

4   Detecting Errors

In this section, we will discuss how to detect errors for an anomalous query based on the ABN constructed in Section 3.

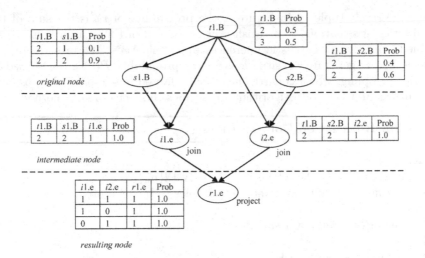

Fig. 2. ABN for query $\prod_c(S \bowtie_B T)$.

From Figure 2, we can see that there are three kinds of nodes in the ABN:

1. Original nodes, denoted as N_o, which exist in the PDBs.
2. Intermediate nodes, denoted as N_i, which are generated during query processing. These nodes are Boolean variables used to represent the existence of the intermediate tuples.
3. Resulting node, denoted as n_r, which is a Boolean variables used to represent the existence of the result tuples.

It is intuitive that errors only exist in N_o since the other nodes are generated during query processing, which is not the source of errors. Therefore, we just need detect errors in N_o.

The candidate errors for the anomalous query are all nodes in N_o, but it seems that the degrees of their contributions to the query result are not equivalent. This problem is closely related to the responsibility and blame in causal models [13], where responsibility is developed to measure the contribution of a cause to a result event when the context of the model is known. Moreover, blame is defined as the expectation of responsibility of a cause under all contexts when the context of the model is uncertain. A causal model is a directed acyclic graph and a context of the model is a set of values of each node in the model [13]. A context can be viewed as a possible world instance of PDBs. Since the number of possible worlds is exponential in the number of random variables, it is infeasible to generate all possible world instances. Fortunately, only a subset of possible world instances that satisfies the result tuple needs to be generated. We call these possible world instances as situations. We first compute the situations and then define the blame upon the ABN and rank the candidate errors by their blame.

For the result tuple $r1$ in Figure 2, the probability of $r1$ is the sum of the probabilities of situations that satisfies $r1.e = 1$. All of these situations can be obtained from the probabilistic inferences upon the ABN. We provide an efficient method to generate the situations for the probability $P(r1.e = 1)$ based on Rejection sampling, which is particularly well-adapted to sampling the posterior distributions of a BN [14]. Algorithm 1 describes the steps of this method.

Algorithm 1. Situation Computation

Input:
    G, an ABN $G = (V, E)$ where $V = N_o \bigcup N_i \bigcup \{n_r\}$;
    m, threshold of the total number of samples to be generated
Output:
    I, a vector of counts over samples
Variables:
    X, a random variable in V;
    s, the current state $\{x_1, x_2, ..., x_n\}$ where x_i is a value of $X \in V$
Steps:
$s \leftarrow$ random values of X in V
for $j \leftarrow 1$ to m **do**
    for each $X_i \in V$ **do**
        $x_i \leftarrow$ a random sample from $P(X_i | parents(X_i))$ given the values of $parents(X_i)$ in S // $parents(X_i)$ is the set of parent node values of X_i
    end for
    if s is consistent with $n_r.e = 1$ **then**
        if I contains s **then**
            $I[s] \leftarrow I[s] + 1$
        else
            insert s into $I[s]$
            $I[s] \leftarrow 1$
        end if
    end if
end for
$I[s] \leftarrow I[s]/m$
return I

For an ABN with n nodes, the complexity of Algorithm 1 is $O(mn)$.

Example 3. Three situations $I1$, $I2$ and $I3$ obtained from Algorithm 2 for the query $\prod_c (S \bowtie_B T)$ are shown in Table 3.

To measure the degree of blame of each node that induces $P(r1.e = 1) = 0.48$, we have to measure all the responsibility degrees under three situations $I1$, $I2$ and $I3$. Intuitively, we say a node has a great responsibility when changing the value of this node would affect (i.e., decrease or increase) the probability of the new situation. To describe this influence, we need some technical definitions.

Table 3. Situations for the ABN in Figure 2.

	$s1.B$	$t1.B$	$s2.B$	$i1.e$	$i2.e$	$r1.e$	Prob
$I1$	2	2	2	1	1	1	0.27
$I2$	2	2	1	1	0	1	0.18
$I3$	1	2	2	0	1	1	0.03

Definition 1. *Let X be a node of an ABN with the value domain $\{x_1, x_2, ..., x_n\}$, I be a situation and $Z = N_o \bigcup \{n_r\} - \{X\}$. The values z of Z are assigned from I, the value x_i of X is consistent with I and $\overline{x} \in \{x_1, x_2, ..., x_{i-1}, x_{i+1}, ..., x_n\}$. Changing $X = x_i$ to $X = \overline{x}$ can generate a new situation $I' = \{X = \overline{x}, Z = z\}$. The absolute value of the difference of the probabilities of I and I' is*

$$vd(X) = \mid P(I) - P(I') \mid \qquad (1)$$

If the degree of responsibility of X is greater than that of Y, then changing the value of X would make the probability $vd(X)$ be greater than changing the value of Y. Thus, we define a function $dr(X, I)$ to measure the degree of responsibility of a node X under situation I. The function $dr(X, I)$ ought to satisfy the following properties:

1. Minimal property: $dr(X, I) = 0$ when changing the value of X cannot influence the probability $P(I')$.
2. Maximal property: $dr(X, I) = 1$ when changing the value of X can make the probability $P(I') = 0$.
3. For two situations $I = \{X = x_i, Z = z\}$ and $I' = \{X = \overline{x}, Z = z\}$, if $P(I) > P(I')$ then $dr(X, I) > dr(X, I')$.

Definition 2. *The degree of responsibility of X under the situation I is defined as*

$$dr(X, I) = \lceil max\{vd(X)\} \rceil \times \frac{P(I)}{P(Z)} \qquad (2)$$

Theorem 1. *The function $dr(X, I)$ satisfies the above-mentioned properties $(1) \sim (3)$.*

Proof. Let the value domain of node X be $\{x_1, x_2, ..., x_n\}$. The value x_i is consistent with I and $\overline{x} \in \{x_1, x_2, ..., x_{i-1}, x_{i+1}, ..., x_n\}$. If changing the value of X to any other values cannot influence $P(I')$ then we can get that $P(X = x_1, Z) = P(X = x_2, Z) = ... = P(X = x_n, Z)$. Therefore, we have $dr(X, I) = \mid P(X = x_i, Z) - P(X = x_1, Z) \mid \times P(X = x_i, Z)/P(Z) = 0$.

If changing the value of X can make the probability $P(I') = 0$, then we can obtain that all values $\overline{x}$ of X such that $P(X = \overline{x}, Z = z) = 0$. Then, we have $dr(X, I) = \mid P(X = x_i, Z) - 0 \mid \times P(X = x_i, Z)/P(Z) = 1$.

For two situations $I = \{X = x_i, Z = z\}$ and $I' = \{X = \overline{x}, Z = z\}$, if $P(I) > P(I')$ then we can obtain $P(I)/P(Z) > P(I')/P(Z)$ and $\lceil max\{vd(X)\} \rceil = 1$. Therefore, we have $dr(X, I) > dr(X, I')$.

Definition 3. *The degree of blame of node X, denoted as $db(X)$, is the expectation of the degree of responsibility of X for all situations, where*

$$db(X) = \frac{\sum_{i=1}^{n}(P(I_i) \times dr(X, I_i))}{\sum_{i=1}^{n} P(I_i)} \tag{3}$$

Algorithm 2 shows the steps for computing the blame of each node in the ABN and ranking the nodes by their blame for error detection.

Algorithm 2. Error Detection

Input:
    I, set of situations;
    X, set of candidate errors //original nodes;
Output:
    X', set of errors sorted by the decreasing order of their blame degrees
Variables:
    B, a vector of blame degrees of $x \in X$
Steps:
for each x_i in X **do**
    $B[x_i] \leftarrow 0$
    for each I_j in I **do**
        $B[x_i] \leftarrow B[x_i] + \lceil max\{vd(x)\}\rceil \times P(I_j)/P(Z)$
    end for
    $B[x_i] \leftarrow B[I_j]/\sum P(I_j)$
end for
$X' \leftarrow$ sort X by the decreasing order of $B[x_i]$
return X'

For k situations, the computations of each node's degree of blame is less than $O(k)$ times. So, the complexity of computing n nodes' degrees of blame is $O(nk)$.

Example 4. Revisiting Example 3, $s1.B$, $t1.B$ and $s2.B$ are candidate errors. According to Equation(1), (2) and (3), we have

$$dr(s1.B, I1) = \lceil | 0.27 - 0.03 | \rceil \times \frac{0.27}{0.27 + 0.03} = \frac{0.27}{0.3} = 0.9$$

$$dr(s1.B, I2) = \lceil | 0.18 - 0 | \rceil \times \frac{0.18}{0.18} = 1$$

$$dr(s1.B, I3)) = \lceil | 0.03 - 0.27 | \rceil \times \frac{0.03}{0.27 + 0.03} = \frac{0.03}{0.3} = 0.1$$

$$db(s1.B) = \frac{\sum_{i=1}^{3} P(I_i) \times dr(s1.B, I_i)}{P(I1) + P(I2) + P(I3)}$$

$$= \frac{0.29 \times 0.9 + 0.18 \times 1 + 0.03 \times 0.1}{0.27 + 0.18 + 0.03} \approx 0.89$$

We can also obtain $db(s2.B) = 0.55$ and $db(t1.B) = 1$ by the same way. The degree of blame of $t1.B$ should be 1 since the joint operation $S \bowtie_B T$ requires $t1.B$ to be 2. Moreover, the value of $s1.B$ is more likely to be 2 than that of $s2.B$ when $t1.B = 1$, so $db(s1.B)$ is 0.89 and $db(s2.B)$ is 0.55.

5   Experimental Results

To verify the feasibility of the method proposed in this paper, we implemented the presented algorithms. We mainly tested the convergence and efficiency of situation computation, and then we tested the accuracy and efficiency of error detection.

5.1   Experiment Setup

—*Hardware:* For the experiments, we used MS SQL Server 2008 on a machine with 2.27GHz Intel Core i3 CPU and 2GB of RAM, running Window 7 Ultimate 32-bit operating system.
—*Data:* Our experiments used a probabilistic database with correlations represented by five classical BNs: Cancer Neapolitan, Chest Clinic, Car Diagnosis2, Alarm and HailFinder25. For each BN, we generated an original data set of 1000 tuples according to their probability distributions from Norsys [1]. To add errors into the original data sets, we randomly modified their probability distributions for one, two and three nodes to generate three kinds of test data sets. Finally, we recorded all the information of those changes as the source of errors.

5.2   Convergence and Efficiency of Situation Computation

It is pointed out that the posterior probabilities of the situations computed by an approximate algorithm for ABN's inferences are correct only if the sampling results are converged to a certain probability [14]. Thus, we tested the convergence of Algorithm 1 by recording the results upon the Chest Clinic ABN under $XRay\_result = abnormal$. Prob1 and Prob2 in Figure 3 are the probabilities obtained from Algorithm 1 and an enumeration-based algorithm. It can be seen that Prob1 and Prob2 are stable around 0.11 with the increase of the generated samples. The result shows that the probabilities returned by Algorithm 1 converge to a certain value efficiently with just about 3000 samples.

Following, we recorded the execution time of Algorithm 1 for situation computation shown in Figure 4. It can be seen that the execution time is increased linearly with the increase of samples and nearly quadratically with the increase of ABN nodes. This means that the execution time is not sensitive to the scale of the ABN. Thus, our method for situation computation is efficient.

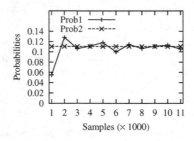

Fig. 3. Convergence of Algorithm 1

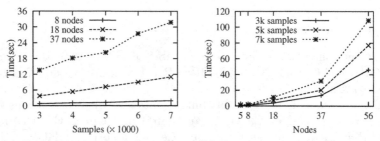

(a) Execution time with the increase of samples

(b) Execution time with increase of nodes

Fig. 4. Execution time of situation computation

5.3 Effectiveness and Efficiency of Error Detection

To test the effectiveness of our method of error detection, we ran our Algorithm 2 over the test data sets and compared the result to the source of errors using the mean average precision (MAP) as the metric of comparison [15]. The MAP is close to 1 when the result of possible errors obtained by Algorithm 2 is close to the source of errors and it means that the faulty node ranked first in the result is most likely to be erroneous.

We ran the experiments to detect one, two and three errors in three ABNs which contain 5, 8 and 18 nodes respectively. Figure 5(a) shows that the precision of the possible errors obtained by Algorithm 2 is stable above 60% with the increase of the ABN nodes. Figure 5(b) presents that the precision will be decreased slowly with the increase of the number of errors. It means that the precision of error detection is mainly determined by the number of errors in the original database. From the perspective of real applications of error detection on a database with a small number of errors, our method can work effectively.

To verify the efficiency of Algorithm 2, we recorded the execution time of error detection over five ABNs shown in Figure 6. It can be seen that the execution time is increased linearly with the increase of nodes and situations of the data sets, which guarantees the efficiency of Algorithm 2.

6   Related Work

Our work is mainly related to and unifies ideas from work on *probabilistic databases, provenance, causality* and *data cleaning*.

Probabilistic Databases. Several approaches have been proposed for managing uncertainties, such as rule-based systems, Dempster-Shafer theory, fuzzy sets, rough set, but by 1987 the probability model has been dominant [20]. Plenty of probabilistic databases have been designed to manage uncertain data where the uncertainties are quantified as probabilities, and most of those databases are based on possible world semantics. Benjelloum et al. [21] introduce a framework, called x-relation, as a representation for databases with uncertainty. X-relations can be extended to represent and query tuple-independent probabilistic data. Tuple-independent probabilistic databases are insufficient for analyzing and extracting information from practical applications that naturally produce correlated data. Wang et al. [22] introduce a probabilistic database (BAYESSTORE) that draws results from the Statistical Learning literature to express and reason about correlations among uncertain data. Jha et al. [5] propose a Markov views based framework both for representing complex correlations and for efficient query evaluation. Sen et al. [11] provide a framework based on the probabilistic graphical model that can represent not only probabilistic tuples, but also complex correlations among tuples. We adopt Sen's model as our underlying probabilistic databases in this paper.

Provenance. Different notions of provenance (also called lineage) for database queries have been studied in the past few years [24]. The most common forms of provenance describe correlations between data in the source and in the output, and can be classified into three categories: (1) explaining *where* output data came from in the input; (2) showing inputs to explain *why* an output record was produced; (3) describing in detail *how* and output record was produced [23]. Tracing the lineage of data is an important requirement for improving the quality and validity of data [7]. Galhardas et al. [25] use a data lineage to improve data

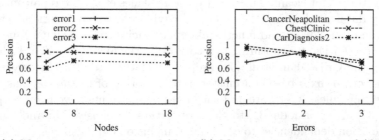

(a) Mean average precision of Algo- (b) Mean average precision of Algorithm 2 with the increase of nodes   rithm 2 with the increase of errors

Fig. 5. Precision of error detection

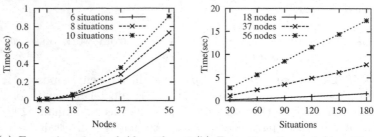

(a) Execution time of Algorithm 2 (b) Execution time of Algorithm 2
with the increase of nodes   with the increase of situations

Fig. 6. Execution time of error detection

cleaning quality by enabling users to express user interactions declaratively and tune data cleaning programs. The first step of finding causes for surprising query result proposed by Meliou et al. [8,9] is the computation of query lineages.

Causality. Causality is typically treated as the concept that either event A is a cause of event B or it is not. Halpern and Pearl [26] define a causal model, called actual causes, in terms of structural equations (i.e., a causal network, which can be presented by a directed acyclic graph), and the syntax and semantics of a language for reasoning about causality. Chockler et al. [13] extend the causality introduced by Haplern to take into account the degree of responsibility of A for B. The notion of the blame has been defined as the expected degree of responsibility of A for B when the context of the causal model is uncertain [13]. For a database query, the responsibility of each input tuple for the query output could be used for error detection [8]. Meliou et al. [9] propose a View-Conditioned Causality to trace errors in output data of a query back to the input data based on responsibility.

Data Cleaning. Data cleaning is one of the critical mechanisms for improving the data quality. Dirty data can be classified into three categories: incorrect or inconsistent data, missing data and duplicate data [28]. A variety of constraints have been studied for cleaning incorrect data. Beskales et al. [33] repair incorrect or missing data by choosing the values satisfying the given functional dependencies. Fan et al. [29] extend functional dependencies to conditional functional dependencies (CFDs) for capturing and correcting the incorrect data which does not satisfy the CFDs. Fan et al. [30] propose a method to clean incorrect data by finding certain fixes based on master data, a notion of certain regions and a class of editing rules. Aggregate and cardinality constraints have been proposed by Chen [32] and Cormode [31] to clean uncertain databases. Ma and Fan [34] extend inclusion dependencies to conditional inclusion dependencies (CINDs) to detect inconsistent data.

Statistical inference methods have been studied for cleaning missing data or correcting incorrect data when constraints are not available [35]. We [27] propose a method for missing data cleaning by adopting Bayesian network to represent and infer the probabilities of possible values of missing data. Stoyanovich et al. [36] present a framework, termed meta-rlue semi-llattices (MRSL), to infer probability distributions for missing data. Techniques for identifying duplicate data in probabilistic databases have been developed by Panse [37]. Geerts et al. [38] develop a uniform data cleaning framework for users to plug-in their preference strategies when the cleaning process involves different kinds of constraints, and a commodity data cleaning platform has been developed by Dallachiesa [39].

7   Conclusions and Future Work

In this paper, we propose an ABN-based method for tracing errors in probabilistic databases. Concentrated on the complex correlation among data, we give a method for constructing an ABN for an anomalous query to represent correlations among input data, intermediate data and output data of the query. Then, we define the blame for each node in the ABN to measure the contributions of each node to the anomalous query result. Finally, we provide an efficient method to compute the blame and rank candidate errors by their blame. However, only synthetic data test is not enough to verify the feasibility of our methods and the usability in enterprise applications.

To test our method further, we will make experiments on real life datasets. As well, we also wish to extend our method to automatically correct errors in real applications in data integration and ETL tools. These are exactly our future work.

Acknowledgments. This paper was supported by the National Basic Research (973) Program of China (No. 2012CB316203), the National Natural Science Foundation of China (Nos. 61472345, 61232002), the Natural Science Foundation of Yunnan Province (No. 2014FA023), the Program for Innovative Research Team in Yunnan University (No. XT412011), and the Yunnan Provincial Foundation for Leaders of Disciplines in Science and Technology (No. 2012HB004).

References

1. Aggarwal, C., Yu, P.: A Survey of Uncertain Data Algorithms and Applications. TKDE **21**(5), 609–623 (2007)
2. Tong, Y., Chen, L., Cheng, Y., Yu, P.: Mining frequent itemsets over uncertain databases. PVLDB **5**(11), 1650–1661 (2012)
3. Rekatsinas, T., Deshpande, A., Getoor, L.: Theodoros Rekatsinas and Amol Deshpande and Lise Getoor. In: SIGMOD, pp. 373–384. ACM (2012)
4. Buneman, P., Cheney, J., Tan, W., Vansummeren, S.: Curated databases. In: PODS, pp. 1–12. ACM (2008)
5. Jha, A., Suciu, D.: Probabilistic databases with MarkoViews. PVLDB **5**(11), 1160–1171 (2012)
6. Fan, W.: Dependencies revisited for improving data quality. In: PODS, pp. 159–170. ACM (2008)

7. Zhang, M., Zhang, X., Zhang, X., Prabhakar, S.: Tracing lineage beyond relational operators. In: VDLB, pp. 1116–1127. VLDB Endowment (2007)
8. Meliou, A., Gatterbauer, W., Moore, K., Suciu, D.: The complexity of causality and responsibility for query answers and non-answers. PVLDB 4(1), 34–45 (2010)
9. Meliou, A., Gatterbauer, W., Nath, S., Suciu, D.: Tracing data errors with view-conditioned causality. In: SIGMOD, pp. 505–516. ACM (2011)
10. Darwiche, A.: Modeling and reasoning with Bayesian networks. Cambridge University Press (2009)
11. Sen, P., Deshpande, A.: Representing and querying correlated tuples in probabilistic databases. In: ICDE, pp. 596–605. IEEE (2007)
12. Deshpande, A., Getoor, L., Sen, P.: Managing and Mining Uncertain Data. Springer (2009)
13. Chockler, H., Halpern, J.: Responsibility and blame: A structural-model approach. JAIR 22, 93–115 (2004)
14. Russell, S., Norvig, P.: Artificial Intelligence: A Modern Approach, 3rd edn. Prentice Hall (2009)
15. Jarvelin, K., Kekalainen, J.: IR evaluation methods for retrieving highly relevant documents. In: SIGIR, pp. 41–48. ACM (2000)
16. Lian, X., Chen, L.: Causality and responsibility: probabilistic queries revisited in uncertain databases. In: CIKM, pp. 349–358. ACM (2013)
17. Jin, C., Zhang, R., Kang, Q., Zhang, Z., Zhou, A.: Probabilistic Reverse Top-k Queries. In: Bhowmick, S.S., Dyreson, C.E., Jensen, C.S., Lee, M.L., Muliantara, A., Thalheim, B. (eds.) DASFAA 2014, Part I. LNCS, vol. 8421, pp. 406–419. Springer, Heidelberg (2014)
18. Liu, J., Ye, D., Wei, J., Huang, F., Zhong, H.: Consistent Query Answering Based on Repairing Inconsistent Attributes with Nulls. In: Meng, W., Feng, L., Bressan, S., Winiwarter, W., Song, W. (eds.) DASFAA 2013, Part I. LNCS, vol. 7825, pp. 407–423. Springer, Heidelberg (2013)
19. Miao, X., Gao, Y., Chen, L., Chen, G., Li, Q., Jiang, T.: On Efficient k-Skyband Query Processing over Incomplete Data. In: Meng, W., Feng, L., Bressan, S., Winiwarter, W., Song, W. (eds.) DASFAA 2013, Part I. LNCS, vol. 7825, pp. 424–439. Springer, Heidelberg (2013)
20. Dalvi, N., Suciu, D.: Management of probabilistic data: foundations and challenges. In: PODS, pp. 1–12. ACM (2007)
21. Benjelloun, O., Sarma, A., Halevy, A., Widom, J.: ULDBs: Databases with uncertainty and lineage. In: VLDB, pp. 953–964. VLDB Endowment (2006)
22. Wang, D., Michelakis, E., Garofalakis, M., Hellerstein, J.M.: BayesStore: managing large, uncertain data repositories with probabilistic graphical models. PVLDB 1(1), 340–351 (2008)
23. Cheney, J., Chiticariu, L., Tan, W.: Provenance in databases: Why, how, and where. Foundations and Trends in Databases 1(4), 379–474 (2007)
24. Green, T., Karvounarakis, G., Tannen, V.: Provenance semirings. In: PODS, pp. 31–40. ACM (2007)
25. Galhardas, H., Florescu, D., Shasha, D., Simon, E., Saita, C. A.: Improving Data Cleaning Quality Using a Data Lineage Facility. In: Workshop of DMDW, pp. (3)1–13 (2001)
26. Halpern, J., Pearl, J.: Causes and explanations: A structural-model approach. Part I: Causes. The British Journal for the Philosophy of Science 56(4), 843–887 (2005)

27. Duan, L., Yue, K., Qian, W., Liu, W.: Cleaning Missing Data Based on the Bayesian Network. In: Gao, Y., Shim, K., Ding, Z., Jin, P., Ren, Z., Xiao, Y., Liu, A., Qiao, S. (eds.) WAIM 2013 Workshops 2013. LNCS, vol. 7901, pp. 348–359. Springer, Heidelberg (2013)
28. Muller, H., Freytag, J.: Problems, methods, and challenges in comprehensive data cleansing. Professoren des Inst, Fur Informatik (2005)
29. Fan, W., Geerts, F., Jia, X., Kementsietsidis, A.: Conditional functional dependencies for capturing data inconsistencies. TODS **33**(2), 6 (2008)
30. Fan, W., Li, J., Ma, S., Tang, N., Yu, W.: Towards certain fixes with editing rules and master data. PVLDB **3**(1–2), 173–184 (2010)
31. Cormode, G., Srivastava, D., Shen, E., Yu, T.: Aggregate query answering on possibilistic data with cardinality constraints. In: ICDE, pp. 258–269. IEEE (2012)
32. Chen, H., Ku, W., Wang, H.: Cleansing uncertain databases leveraging aggregate constraints. In: Workshop of ICDE, pp. 128–135. IEEE (2010)
33. Beskales, G., Ilyas, I., Golab, L.: Sampling the repairs of functional dependency violations under hard constraints. PVLDB **3**(1–2), 197–207 (2010)
34. Ma, S., Fan, W., Bravo, L.: Extending inclusion dependencies with conditions. Theoretical Computer Science **515**, 64–95 (2014)
35. Mayfield, C., Neville, J., Prabhakar, S.: ERACER: a database approach for statistical inference and data cleaning. In: SIGMOD, pp. 75–86. ACM (2010)
36. Stoyanovich, J., Davidson, S., Milo, T., Tannen, V.: Deriving probabilistic databases with inference ensembles. In: ICDE, pp. 303–314. IEEE (2011)
37. Panse, F., Van Keulen, M., De Keijzer, A., Ritter, N.: Duplicate detection in probabilistic data. In: Workshop of ICDE, pp. 179–182. IEEE (2010)
38. Geerts, F., Mecca, G., Papotti, P., Santoro, D.: The llunatic data-cleaning framework. PVLDB **6**(9), 625–636 (2013)
39. Dallachiesa, M., Ebaid, A., Eldawy, A., Elmagarmid, A., Ilyas, I., Ouzzani, M., Tang, N.: NADEEF: a commodity data cleaning system. In: SIGMOD, pp. 541–552. ACM (2013)

Data Mining II

Mining Frequent Spatial-Textual Sequence Patterns

Krishan K. Arya[1], Vikram Goyal[1]([⊠]), Shamkant B. Navathe[2],
and Sushil Prasad[3]

[1] Indraprastha Institute of Information Technology Delhi, New Delhi, India
{krishan1241,vikram}@iiitd.ac.in
[2] Georgia Institute of Technology, Atlanta, USA
sham@cc.gatech.edu
[3] Georgia State University, Atlanta, USA
sprasad@gsu.edu

Abstract. Penetration of GPS-enabled devices has resulted in the generation of a lot of Spatial-Textual data, which can be mined or analyzed to improve various location-based services. One such kind of data is Spatial-Textual sequential data (Activity-Trajectory data), i.e. a sequence of locations visited by a user with each location having a set of activities performed by the user is a Spatial-Textual sequence. Mining such data for knowledge discovery is a cumbersome task due to the complexity of the data type and its representation. In this paper, we propose a mining framework along with algorithms for mining Spatial-Textual sequence data to find out frequent Spatial-Textual sequence patterns. We study the use of existing sequence mining algorithms in the context of Spatial-Textual sequence data and propose efficient algorithms which outperform existing algorithms in terms of computation time, as we observed by extensive experimentation. We also design an external memory algorithm to mine large-size data which cannot be accommodated in main memory. The external memory algorithm uses spatial dimension to partition the data into a set of chunks to minimize the number of false positives and has been shown to outperform the naïve external-memory algorithm that uses random partitioning.

1 Introduction

Sequential pattern mining is an important and very active research topic in the area of data mining due to its widespread use in various applications such as customer transaction analysis, web logs analytics, mining DNA sequences, mining e-bank customers financial transactions etc. In a sequential pattern mining task, a set of frequent sub-sequences is mined from a sequence database. An example of sequential pattern might be one in which the patterns shows that customers typically bought a TV, followed by the purchase of an XBOX, and then a pack of video games.

In this paper, we work on the new type of pattern mining problem called Spatial-Textual sequence pattern mining, where we mine Spatial-Textual subsequences from a set of Spatial-Textual sequences [9,10,18,22]. A Spatial-Textual

© Springer International Publishing Switzerland 2015
M. Renz et al. (Eds.): DASFAA 2015, Part II, LNCS 9050, pp. 123–138, 2015.
DOI: 10.1007/978-3-319-18123-3_8

sequence is a trajectory of locations with each location having associated with it a set of activities/ events or some other attributes. Due to the proliferation of devices enabled with GPS and Internet access, a lot of Spatial-Textual sequence data is being generated at a fast pace. Such data may be used for improving the process/ policies etc. in their respective application domains. As an example, we can see customer transactions as a Spatial-Textual sequence, where a trajectory is formed using the locations of the stores visited by the customers and a set of items purchased in a store as a set of events/ activities. Frequent Spatial-Textual sequence mined from this type of data can be useful for deciding the locations where a new shop could be opened. Another application having this type of data is related to movement of animals or birds, in which a trajectory represents a sequence of locations tracked by the monitoring system and activities at a location could represent the events related to their reproductive and migration behavior. In social networks like Foursquare, the check-in sequence of a user forms a trajectory and important words in a comment given by the user at a location constitute a set of activities for that location.

Mining Spatial-Textual frequent sub-sequential patterns is one of the major challenge due to the complexity of the data type, as we have to deal with not only two different dimensions, but also ordered data. Another challenge lies due to localization error of GPS devices in terms of defining notion of similarity of two locations, i.e. two different latitude and longitude values of locations may belong to the same actual location. To handle the first issue, we design a solution that considers one of the dimensions in the first phase of pruning the data for mining the target patterns in the second phase. Specifically we use either the spatial or the textual data first to filter the data. The remaining data is then mined for target patterns in the second phase. To handle the second issue of location similarity, we use an application defined parameter to divide the whole space into grids and consider every location within a grid cell as if its is the same location. Another way to resolve this problem may be to partition the space using an existing geographical boundary map, like zip code, city etc.

A lot of work has been done for mining general sequential patterns which is discussed in detail in Section 2. We ask two questions in the context of mining Spatial-Textual sequence data: i) how can we use existing sequence mining algorithms to mine this kind of data, and ii) can we do better than existing sequence mining algorithms? The answer to both of these questions is *YES*. To show this, we adapt an existing sequence mining algorithm called PrefixSpan[16] to mine this type of data and, we design and implement an efficient mining algorithm which outperforms the adapted PrefixSpan algorithm in terms of execution time as well as main memory requirement. We also give an external memory algorithm to mine Spatial-Textual data whereas we give a heuristic to partition the Spatial-Textual sequence database. Our experiments show that our external memory algorithm is efficient and outperforms the naïve external memory algorithms in terms of computation time.

Overall we have following contributions in this paper:

(i) We propose a flexible framework to mine a new kind of sequence pattern called Spatial-Textual subsequence pattern. The framework is flexible enough to accommodate any basic sequence mining algorithm as a baseline algorithm in the framework.

(ii) We study three different algorithms to mine patterns, namely Spatial-Textual, Textual-Spatial and Hybrid algorithm. Spatial-Textual algorithm considers spatial dimension, i.e. location sequence, in the first phase whereas Textual-Spatial uses activities in the first phase. The Hybrid version takes into account both the dimensions simultaneously and has only one phase.

(iii) We design an external memory algorithm that outperforms naïve external memory algorithm.

The remainder of the paper is structured as follows. Section 2 presents related work in the area of sequence mining. In Section 3, we formalize our problem and present the framework. Section 4 presents our proposed algorithms. In Section 5 external memory algorithm is presented. We present our experimental results in section 6. And finally we conclude our work in section 7.

2   Related Work

In this section we briefly discuss some related work done in this area. The sequential pattern mining problem was first addressed by Agrawal and Srikant [1] in which they gave the *AprioriAll* algorithm to mine frequent patterns by candidate-generation-and-test-approach. The approach is however simple but may not be efficient in mining large sequence databases having numerous patterns and/or long patterns. Subsequently, the authors in [19] proposed an algorithm which outperformed *AprioriAll* algorithm. Ayres et al.[2] used a vertical bitmap representation to count the supports and mine sequential patterns from sequence databases.

FreeSpan[11] and PrefixSpan [16] are two algorithms, which use divide and conquer approach. FreeSpan offers bi-level projection technique to improve performance. The approach given by Jian Pei et al.[16] in their subsequent work called PrefixSpan is a projection-based, sequential pattern-growth approach for the mining of sequential patterns. In this approach, a sequence database is recursively projected into a set of smaller projected databases, and sequential patterns are grown in each projected database by exploring only locally frequent fragments. PrefixSpan outperforms FreeSpan in that only relevant postfixes are projected. Giannotti et al.[7] demonstrated an approach to mine sequential patterns with temporal annotations. Another technique given by Zaki et al. [21] was SPADE algorithm. It searches the lattice formed by id-list intersections and completes mining in three passes of database scanning. MEMISP[15] is a memory-indexing approach for fast discovery of sequential patterns. It discovers sequential patterns by using a recursive find-then-index technique. Leleu et al. [14] proposed a Go-SPADE method, which extends SPADE [21] to efficiently mine sequences containing consecutive repetitions of items. It may be noted that all these works focus mainly on itemsets sequences without any spatial dimension.

Koperski and Han et al.[13] extended the concept given by Srikanta et al. to spatial databases by introducing spatial predicates like close_to and near_by. Tsoukatos and Gunopulos[20] proposed an Aprioi-based method for mining spatial regions sequences. Hwang et al.[12] presented an algorithm for mining a group of moving objects with the same movement patterns. Cao et al.[4] proposed a trajectory mining algorithm. Chung et al.[5] proposed an Apriori-based method to mine movement patterns, where the moving objects are transformed using a grid system and then frequent patterns are generated level by level. Giannotti et al.[8] focuses on a method for extracting trajectory patterns containing both spatial and temporal information. These works focus mainly on spatial and temporal dimension and do not consider Spatial-Textual data for their working.

Savasere and Navathe[17] proposed an external memory algorithm for mining association rules in large databases. They used partition-and-validation technique for mining the association rules out of the large databases. They did simple partitioning as well as random partitioning and showed that performance of external memory algorithm increases if we do random partitioning of the data. Through partition-and-validation technique of MEMISP[15] algorithm, one can mine an extremely large database in two or more database scans. However the study in [17] and [15] were limited to simple transaction database and itemset-sequences, respectively.

3 Problem Overview

Now we formalize our problem in this section.

3.1 Problem Definition

Let $I = \{i_1, i_2, \ldots, i_n\}$ be a set of items and $L = \{l_1, l_2, \ldots, l_m\}$ is a set of locations. An itemset X is a non-empty subset of I. A Spatial-Textual sequence s is an ordered list of location-itemset pairs, i.e. s= $\langle (l_i, X_i), \ldots, (l_j, X_j) \rangle$. As an example, $\langle ((20.5, 30.6), \{Bread, Butter, Beer\}), ((20.66, 30.5), \{Jam, Milk, Sauce\}), ((23,23.5), \{Bread, Milk, Coke\}) \rangle$ is a Spatial-Textual sequence. For brevity, the brackets are omitted around an itemset if the itemset has only one item.

The number of instances of location-item pairs in a sequence is called the length of the sequence. A sequence with length l is called an $l - sequence$. A sequence $\alpha = \langle a_1, a_2, \ldots, a_n \rangle$ is called a sub-sequence of another sequence $\beta = \langle b_1, b_2, \ldots b_m \rangle$ and β a super-sequence of α, denoted as $\alpha \sqsubseteq \beta$, if there exist integers $1 \leq j_1 < j_2 < \ldots < j_n \leq m$ such that $items(a_1) \subseteq items(b_{j_1})$ and $loc(a_1) = loc(b_{j_1})$, $items(a_2) \subseteq items(b_{j_2})$ and $loc(a_2) = loc(b_{j_2})$, $\ldots$,

$items(a_n) \subseteq items(b_{j_n})$ and $loc(a_n) = loc(b_{j_n})$. Here $items(p)$ returns the item-set of pair p and $loc(p)$ returns the location of pair p.

A sequence database S is a set of tuples $\langle sid, s \rangle$, where sid is a sequence-id and s a sequence.

The support of a sequence α in a sequence database S is the number of tuples in the database containing α, i.e.,

$$support_s(\alpha) = |\langle sid, s \rangle|(\langle sid, s \rangle \in S) \wedge (\alpha \sqsubseteq s)| \tag{1}$$

It can be denoted as $support(\alpha)$ if the sequence database S is clear from the context. Given a min_support (in $percentage$) as the support threshold, a sequence α is called a frequent sequential pattern in a sequence database S if $support_S(\alpha) \geq min\_support$. A sequential pattern with length l is called an $l - pattern$.

The problem here is to find frequent Spatial-Textual patterns from given Spatial-Textual sequence database S and a support threshold $min\_support$.

3.2   Framework

We propose a flexible framework to mine frequent Spatial-Textual patterns from a database of Spatial-Textual sequences. The framework is flexible in terms of allowing use of any basic sequence mining algorithm for its working, such as GSP [19], PrefixSpan [16], SPADE [21] etc. The framework is shown in figure 3.1.

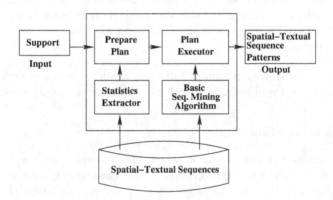

Fig. 3.1. A Framework for Mining Spatial-Textual subsequences

The framework contains a *Statistics Extractor* (SE) module which scans the Spatial-Textual sequences in the database and collects different statistics of data, such as selectivity (frequency) of the locations, selectivity of items etc. The SE module passes this information to the *Prepare Plan* (PP) module, which on the basis of statistics prepare an execution plan. We have three different algorithms namely Spatial-Textual Mining (ST-Mining), Textual-Spatial Mining (TS-Mining) and Hybrid Mining (H-Mining) algorithms supported by the framework. The PP module specifically chooses an algorithm out of the three algorithms on the basis of data statistics. The algorithm ST-Mining (TS-Mining) considers spatial dimension (textual dimension) in the first phase to prune non-candidate pairs from each sequence to produce only relevant data for the 2nd phase. The Hybrid approach takes both the dimensions simultaneously and may be applied directly or in the 2nd phase of ST-mining or TS-mining. We assume the Basic Sequence Mining Algorithm module as a repository of sequence mining algorithm such as Prefix-span, Spade etc. *Plan Executer* (PE) chooses one basic sequence mining algorithm from the repository of basic algorithms for its working. We choose Prefix-Span[16] algorithm (Baseline source code from [6]) for our study for illustrative purposes only; our framework allows use of any basic sequence-mining algorithm.

4 Algorithms

We propose three algorithms, namely Spatial-Textual mining (ST-Mining), Textual-Spatial mining (TS-Mining) and Hybrid mining (H-Mining) to mine frequent Spatial-Textual sub-sequences. The ST-Mining (TS-Mining) algorithm consists of three steps, which are:

  (i) Label locations in database S using some grid partitioning algorithm.
 (ii) Determine frequent locations (terms) for the given *min_support*.
(iii) Prune database S to S' using the obtained frequent locations (terms) and mine the filtered sequences S' to finally get the frequent Spatial-Textual sequence patterns.

The H-Mining algorithm uses simply the basic sequence mining algorithm over the whole database S and considers both the dimensions simultaneously.

4.1 Location Labeling

The Spatial-Textual sequences contain trajectories of users with associated activities on each of the location in trajectories. Two locations in trajectories may not have same values for latitudes and longitudes even if they represent the same real location due to various reasons, different localization technology, device specific characteristics etc. Therefore assigning the same location label to two locations representing the same real location needs to be fixed first. It is an important and difficult problem and is still an active research area. For our purposes we use the Grid structure with grid size being defined by passing application parameter, i.e., locations in a grid are given the same label. It may be noted that locations can also be clustered using some standard clustering algorithm like DBScan [3] to get location labels, whereas a cluster-id becomes the label of all locations in that cluster.

4.2   Algorithms

This section discusses the proposed sequence mining algorithms to mine frequent Spatial-Textual sub-sequences to be used in the framework 3.1. The first algorithm that we discuss is H-Mining algorithm given in 1. Next we discuss ST-Mining algorithm 2.

The hybrid approach first converts the data to a specific format required to run a given baseline algorithm. Basic algorithms run on one dimension, therefore the spatial-textual data is mapped to one dimension form but still preserves the actual data semantics, so that frequent Spatial-Textual sequences can be extracted back. Conversion/ mapping mainly assigns each unique pair of location and item a new symbol. The operation can be done in two steps: First, create a hash-table to map each location-item pair to a symbol (may be an integer) by scanning the sequence database. Then in the second step, the hash-table is used to map each Spatial-Textual sequence to an itemset sequence where each itemset is consisting of a set of symbols. We can then use any base sequence mining algorithm like PrefixSpan to mine frequent itemset sequences. These frequent itemset sequences then can be mapped back to the original Spatial-Textual sequences using the hash-table.

Algorithm H-Mining(*Sequences S, min_support* σ)

1   Convert Spatial-Textual sequence database S to itemset sequence form S'.

2   Call **CoreAlgo**(*Sequences S', Support* σ) over itemset sequences.

3   **return**

Procedure CoreAlgo(*Sequences S', Support* σ)

1   Mine sequences S' using chosen baseline algorithm and find each frequent sequential patterns.

2   **return**

Algorithm 1. Hybrid Mining

4.3   ST-Mining and TS-Mining Algorithms

In Spatial-Textual approach, the spatial dimension is used in the first phase to prune non-candidate location-itemset pairs. We then run basic sequence mining algorithm over this pruned Spatial-Textual sequences to find the final set of frequent Spatial-Textual sequence patterns.

The motivation for ST-Mining algorithm comes from the fact that if a spatial location is not frequent then that location with textual extension would also not be frequent. Therefore instead of working on the complete database of sequences, only trajectory (location sequence) data from the sequences database can be mined first to get frequent locations. The trajectory data being of much smaller size than the whole sequence database may provide a benefit in terms of

Algorithm ST-Mining(*Sequences S, min_support* σ)

1	Process database sequences to get frequent locations.
2	Prune Spatial-Textual sequence database using frequent locations to get pruned Spatial-Textual sequence database S'_p.
3	Call H-mining(*Sequences S'_p, min_support σ*) to get frequent Spatial-Textual sequence patterns.
4	**return**

Algorithm 2. Spatial-Textual Mining

processing time as well as total memory requirement later in the second phase due to pruning of non-candidate pairs.

Observation-1: A non-frequent location cannot appear in any of the frequent Spatial-Textual sequence.

Table 4.1 shows an example database of Spatial-Textual sequences. Each sequence has sequence-id associated with it. In this example, we have 4 Spatial-Textual sequences.To reiterate, the $l's$ are locations and the $a, b, .., ..g$ are the textual items/events/properties/attributes pertaining to those locations.

Table 4.1. Example Spatial-Textual sequence Dataset

10	$< l_1, a\ (l_2, a, b, c)(l_3, a, c)l_1, d\ (l_4, c, f) >$
20	$< (l_2, a, d)\ l_4, c(l_1, b, c)(l_7, a, e) >$
30	$< (l_1, e, f)(l_2, a, b)(l_4, d, f)l_3, c\ l_2, b >$
40	$< l_3, e\ l_6, g\ (l_2, a, f)\ l_5, c\ l_3, b\ l_4, c >$

To finally get the frequent Spatial-Textual sub-sequences, we use the frequent locations to prune out the non-candidate location-itemset pairs.

Initially, only the spatial dimension is considered from the given Spatial-Textual sequence database. For this purpose, the textual dimension i.e., the itemsets are not considered. Table 4.2 shows the spatial sequences of the given Spatial-Textual example database.

Table 4.2. Spatial sequences

10	$< l_1\ l_2\ l_3\ l_1\ l_4 >$
20	$< l_2\ l_4\ l_1\ l_7 >$
30	$< l_1\ l_2\ l_4\ l_3\ l_2 >$
40	$< l_3\ l_6\ l_2\ l_5\ l_3\ l_4 >$

After processing trajectories in Table 4.2 with *min_support* = 85%, we get the l_2 and l_4 as frequent locations.

Candidate Location-Itemset Pair (p): A pair p is a candidate location-itemset pair if its location label is present in some frequent trajectory.

In the pruning step we retain all the candidate location-itemset pairs and prune out all other non-candidate pairs. For our example case, the state of Spatial-Textual sequence database after pruning is shown in table 4.3.

Table 4.3. Pruned Spatial-Textual sequences

10	$< (l_2, a, b, c)(l_4, c, f) >$
20	$< (l_2, a, d)\, l_4, c >$
30	$< (l_2, a, b)(l_4, d, f)l_2, b >$
40	$< (l_2, a, f)(l_4, c) >$

To show that the ST-Mining algorithm would not miss any frequent Spatial-Textual sequence pattern, we prove the following lemma.

Lemma-1. Any Location-itemset pair p pruned in the pruning step is a non-candidate pair and would not contribute to any frequent Spatial-Textual sub-sequence.

Proof. Suppose a Spatial-Textual pair p contributes to a frequent Spatial-Textual pattern and is pruned by our pruning step. If p is in a frequent Spatial-Textual sequence then there should be at least k Spatial-Textual sequences, say $S' \subset S$, that contribute to the pair p. Each sequence $s \in S'$ would have at least one location-itemset pair with matching location label. In the pruning step using frequent trajectories' locations, all candidate location-itemset pairs are not pruned. Hence p should not be pruned in our pruning step. Q.E.D.

After pruning the whole Spatial-Textual sequence database, we find all the frequent Spatial-Textual sub-sequences, i.e. the sub-sequences whose occurrence frequency in the set of pruned Spatial-Textual sequences is no less than minimum support (Here $min\_support = 85\%$ is considered) by the use of any basic sequence mining algorithm. For our example, we will get $< l_2, a >$ as frequent Spatial-Textual sequence.

Till now we have discussed about the Spatial-Textual sequence data in which a user visits a sequence of locations and performs some activities at each location. But another case of the Spatial-Textual sequence data can have users visiting a sequence of locations but do perform activities very rarely. In this scenario, it may happen that the textual (itemset) sequences may become smaller in size and hence selectivity of textual dimension becomes low. Our framework in this scenario would choose TS-Mining approach where textual dimension would be used first to get the frequent items. Then using frequent textual items, the sequence database is pruned similar to the ST-Mining approach and finally the pruned sequence data is mined to get the frequent Spatial-Textual sub-sequences. The approach is symmetrical to the ST-Mining approach except that the order now is text followed by location; hence, we just present our results for the ST-Mining approach.

5 External Memory Algorithm

The size of the main memory available with a computing workstation is nowadays increasing very rapidly, however the size of data which is being generated is also increasing at a high pace. Therefore it is highly likely to have scenarios where data to be processed cannot fit in main memory. In such cases algorithms designed with the assumption of availability of complete data in main memory, cannot work. An example of this is a PrefixSpan [16] algorithm. In addition to the assumption of complete availability of data in main memory, this algorithm makes many recursive calls and requires a lot of main memory for its working.

As discussed earlier, the ST-Mining algorithm initially in its first phase runs on only spatial sequences which is smaller in size, and then in second phase it runs on pruned Spatial-Textual sequence data. It makes our algorithm to mine large datasets consisting of Spatial-Textual sequences which was not otherwise possible using the Hybrid approach. We design an external memory algorithm to mine larger size data that cannot fit in main memory and in fact will work for the scenario where even spatial sequences may also not fit in main memory. The algorithm discovers sequential patterns using a partition-and-validation approach similar to the approach proposed for frequent itemset mining in [17]. The approach, first, partitions the itemset database into a set of chunks and then each chunk is locally mined for determining local frequent itemsets. Each of the local frequent itemsets thus obtained from any chunk is then verified on the whole transaction database. The main issue with such an approach is to design a partitioning method so that number of locally frequent itemset that need to be verified is small in number. We use a similar approach for our external memory algorithm. However we propose a novel heuristic of using the spatial dimension for partitioning of data to reduce the number of patterns to be verified finally. We call this heuristic as spatial-partition. We use the ST-Mining algorithm for mining local chunks, as discussed in the previous section.

The spatial dimension is used in the following way for partitioning of data into chunks. To minimize the number of false positives, i.e., non-candidate patterns that need to be verified, we try to put most dissimilar sequences in a specific chunk. We term this rule as spatial-partition heuristic. To implement the heuristic, we use a grid structure to assign a label to each trajectory, first. A trajectory is assigned a grid label on the basis of the label of the most frequent grid that it passes through, i.e., the label of a sequence is the grid label which contains maximum number of locations of that sequence. In case of conflict for labels due to more than one grid with the same frequency, we choose one grid label arbitrarily. Though more complicated approaches can be devised for labeling, we see this simple approach to be working well for our case. However, the success of the approach depends on the grid size and distribution of locations for each grid. After labeling, these sequences are partitioned such that a partition/chunk has a minimum number of sequences with the same labels. This problem can also be modeled as a trajectory clustering problem where the distance function is defined in such a way that the two most dissimilar trajectories have the smallest distance between them so that they will be assigned the same cluster with high probability. The number of clusters obtained is equal to the number of partitions/chunks.

The steps of external memory algorithm are as follows:

(i) Generate K partitions of Spatial-Textual sequences using spatial-partition heuristic.
(ii) For each partition, generate locally frequent Spatial-Textual sequence patterns using ST-Mining algorithm.
(iii) Validate locally frequent Spatial-Textual sequence patterns obtained in the second step over complete sequential database S.

The number of partitions, K, are determined on the basis of available main memory. The chunk size should not be more than the memory required for its processing. The patterns mined from a chunk are called locally frequent spatial-textual sequence patterns. A locally frequent spatial-textual pattern is called a false pattern if it is not frequent over the complete sequence database. The main issue therefore in this approach is to cleverly partition the data so that number of false positive patterns is minimized.

6 Experiment Results

6.1 Dataset and Data Preprocessing

We conduct our experiments on two real datasets, namely Foursquare[1] and Bikely[2]. The statistics of both of these datasets have been given in Table 6.1. Foursquare dataset is prepared by crawling the check-in locations of different users along with their comments posted for a check-in place. The check-in location and words extracted from the comment constitute our location-itemset pair. Bikely dataset is also prepared by the crawling of the 100000 trajectories formed of the bicycle routes of the different users. Since the trajectories of bikely data contains only locations so we merged the comments from crawled foursquare dataset with these bikely trajectories to generate Spatial-Textual sequences. Location labels for each location are obtained by using Grid-based approach and a different grid size is chosen for each experiment. All the experiments are conducted on a workstation with Intel Duo core 3GHz CPU and 32GB memory. To generate a dataset of different sizes we sample sequences randomly from the whole dataset. The different parameters which are used for experimental study are given in Table 6.2. The min_support of 0.18% and 0.0072% are chosen for Foursquare and Bikely, respectively. We choose these support threshold values because below these values, the number of frequent patterns for the respective datasets are too many in number and both the algorithms take a lot of time. We report execution time by taking the average of the 20 iterations of each result so that the effect of the other system parameters can be minimized.

---

[1] http://www.foursquare.com
[2] http://www.bikely.com

Table 6.1. Experiment Data Statistics

	Foursquare	Bikely
Total Number of Sequences	30000	100000
Average Number of Transaction/Sequence	8	20
Average Number of items/Transaction	6	10
Total number of items(events)	50000	50000

Table 6.2. Experiment Parameters

Sr. No	Parameter	Foursquare-Values	Bikely-Values
1.	Number of Sequences	5,10,15,20,25,30K	50,75,100K
2.	Grid-Cell Size	0.1-1.0 units	0.1-1.0 units
3.	Support	0.18%	0.0072%
4.	Execution Time	milliseconds	milliseconds

6.2 Comparison of Hybrid and ST-Mining Algorithms w.r.t. Total Execution Time

Effect of Dataset Size. Figure 6.1 shows performance of Hybrid and ST-Mining algorithm for different number of sequences in Foursquare data with support value of 0.18%. The grid size is 0 in this experiment, which means that each unique location is given a different location-label.

The graph shows that the ST-Mining algorithm outperforms the H-Mining algorithm in terms of execution time. The gap in performance between Hybrid and ST-Mining algorithm increases as the data size increases. Due to lower-selectivity of location labels ST-mine algorithm has a lot of pruning on the data to be mined in the second phase. It may be noted that performance of the ST-Mining algorithm will further improve with the increase in support threshold value due to better pruning. Similar experiments are performed on Bikely data with 0 grid size and 0.0072% support value. The results are shown in figure 6.2.

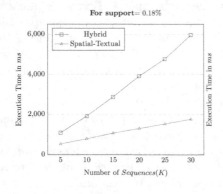

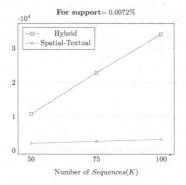

Fig. 6.1. Foursquare   **Fig. 6.2.** Bikely

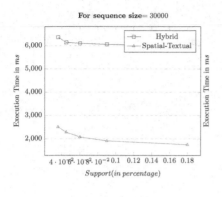

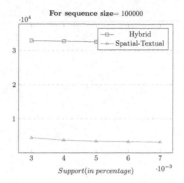

Fig. 6.3. Foursquare   **Fig. 6.4.** Bikely

Effect of Varying Support Threshold Value. Figure 6.3 shows the comparison of total execution time taken by Hybrid and ST-Mining algorithms due to change in support threshold. In this case the number of sequences and location granularity are set to constant values as given in table 6.2, i.e. location granularity as zero and number of sequences as 30000. The graph shows that the ST-Mining algorithm takes less time as compared to the Hybrid algorithm for higher support threshold values. For low support threshold ST-Mining performance is poor due to the reason of less pruning.

Figure 6.4 shows similar results for bikely dataset which contains 100000 sequences. This also shows that ST-Mining algorithm outperforms Hybrid algorithm for high support threshold values.

Effect of Varying Location Granularity. Figure 6.5 shows the comparison of total execution time taken by Hybrid and ST-Mining algorithms vis-a-vis location selectivity in terms of grid size. The location selectivity of a location is defined in terms of number of spatial-textual sequences that contain that location. Higher the number of sequences that contain a location, lower the selectivity of the location. We use support threshold value as 7.2% and number of sequences as 30000 for Foursquare. The high min_support threshold of 7.2% is chosen for this experiment to limit the number of frequent patterns. We observe too many frequent patterns at a lower support threshold value with high location granularity in our dataset.

The graph shows that ST-mine Algorithm takes less time than Hybrid algorithm in case of high selectivity of location-label. As the grid size starts increasing, the difference between the execution time of Hybrid algorithm and ST-Mining algorithm starts reducing. The reason behind this is that, because of small grid size the selectivity of location is high so frequent spatial patterns are generated less in numbers and more pruning occurs in the second phase. Figure 6.6 shows similar results for Bikely dataset with 100000 sequences and 1.2% support threshold. The min_support of 1.2% for bikely is chosen for the same reason as has been mentioned above for Foursquare data, i.e. to limit the number of frequent patterns.

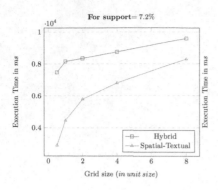

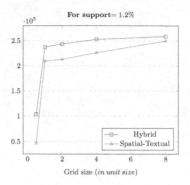

Fig. 6.5. Foursquare

Fig. 6.6. Bikely

6.3 External Memory Algorithm Performance Study

Figure 6.7, 6.8, 6.9 and 6.10 shows the results for comparison of the time taken by External-Memory algorithms, i.e., Simple Serial (SS) chunk distribution approach and dissimilar sequences (DS) chunk distribution approach. The SS approach chooses a set of spatial-textual sequences randomly to form a chunk, whereas the DS approach uses the spatial-partitioning heuristic for grouping of spatial dis-similar spatial-textual sequences for forming of a chunk. We study the performance on foursquare data for different location granularities with number of sequences as 30000 and support threshold as 7.2% and also for varying support threshold with number of sequences as 20000. The graphs 6.7, 6.8 show that DS approach takes less time than SS approach. This is due to generation of less false positive patterns that need to be verified in case of DS approach. Similar results are obtained on Bikely data for varying location granularity as shown in graph 6.9 and for varying support threshold values as shown in figure 6.10.

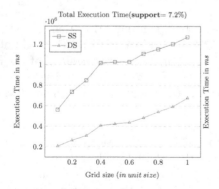

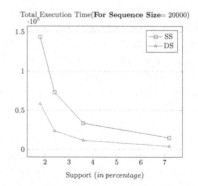

Fig. 6.7. Varying Location Granularity: Foursquare

Fig. 6.8. Varying Support Threshold:Foursquare

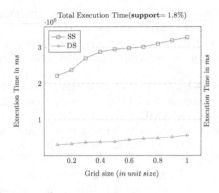

Fig. 6.9. Varying Location Granularity: Bikely

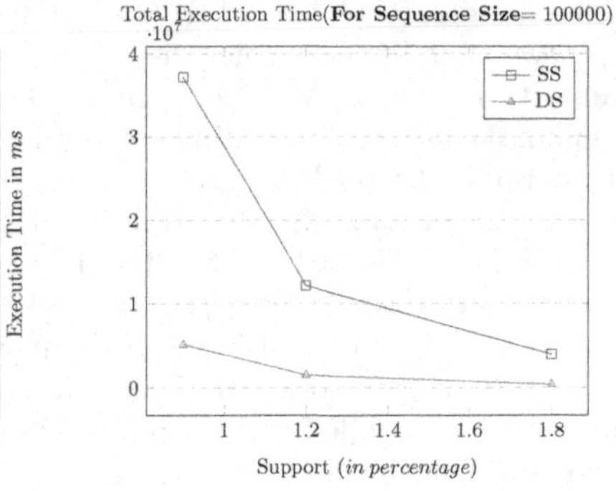

Fig. 6.10. Varying Support Threshold:Bikely

7 Conclusions and Future Work

A tremendous volumes of spatial-textual data is getting generated due to the proliferation of GPS-enabled devices and Internet access. This data can be mined for patterns that can be useful for multiple application domains. We have presented a novel yet a straightforward and simple framework to mine Spatial-Textual sequence data and have proposed three algorithms in this paper. We have also presented an external memory algorithm to mine large data that cannot fit in main memory. The heuristic proposed for external memory algorithm is simple and effective for partitioning Spatial-Textual sequence data. The experimental study conducted shows that ST-Mining and external memory algorithms actually achieve good performance. As a future work we intend to work on forming rules that would choose the best algorithm based on some properties of spatial-textual data such as location selectivity, textual terms selectivity and support threshold.

References

1. Agrawal, R., Srikant, R.: Mining sequential patterns. In: IEEE Eleventh International Conference on Data Engineering (ICDE), pp. 3–14 (1995)
2. Ayres, J., Flannick, J., Gehrke, J., Yiu, T.: Sequential pattern mining using a bitmap representation. In: Proceedings of the Eighth ACM SIGKDD International Conference on Knowledge Discovery and Data Mining, pp. 429–435. ACM (2002)
3. Birant, D., Kut, A.: St-dbscan: An algorithm for clustering spatial-temporal data. Data & Knowledge Engineering **60**(1), 208–221 (2007)
4. Cao, H., Mamoulis, N., Cheung, D.W.: Mining frequent spatio-temporal sequential patterns. In: Fifth IEEE International Conference on Data Mining, pp. 82–89 (2005)
5. Du Chung, J., Paek, O.H., Lee, J.W., Ryu, K.H.: Temporal Pattern Mining of Moving Objects for Location-Based Service. In: Hameurlain, A., Cicchetti, R., Traunmüller, R. (eds.) DEXA 2002. LNCS, vol. 2453, pp. 331–340. Springer, Heidelberg (2002)

6. Fournier-Viger, P., G.A.S.A.L.H.G.T.: SPMF: Open-Source Data Mining Platform. http://www.philippe-fournier-viger.com/spmf/ (2014)
7. Giannotti, F., Nanni, M., Pedreschi, D., Pinelli, F.: Mining sequences with temporal annotations. In: Proceedings of the 2006 ACM Symposium on Applied Computing, pp. 593–597. ACM (2006)
8. Giannotti, F., Nanni, M., Pinelli, F., Pedreschi, D.: Trajectory pattern mining. In: Proceedings of the 13th ACM SIGKDD International Conference on Knowledge Discovery and Data Mining, pp. 330–339. ACM (2007)
9. Goyal, V., Likhyani, A., Bansal, N., Liu, L.: Efficient trajectory cover search for moving object trajectories. In: Proceedings of the 2013 IEEE Second International Conference on Mobile Services, pp. 31–38 (2013)
10. Goyal, V., Navathe, S.B.: A ranking measure for top-k moving object trajectories search. In: Proceedings of the 7th Workshop on Geographic Information Retrieval, pp. 27–34 (2013)
11. Han, J., Pei, J., Mortazavi-Asl, B., Chen, Q., Dayal, U., Hsu, M.C.: Freespan: frequent pattern-projected sequential pattern mining. In: Proceedings of the Sixth ACM SIGKDD International Conference on Knowledge Discovery and Data Mining, pp. 355–359. ACM (2000)
12. Hwang, S.-Y., Liu, Y.-H., Chiu, J.-K., Lim, E.: Mining Mobile Group Patterns: A Trajectory-Based Approach. In: Ho, T.-B., Cheung, D., Liu, H. (eds.) PAKDD 2005. LNCS (LNAI), vol. 3518, pp. 713–718. Springer, Heidelberg (2005)
13. Koperski, K., Han, J.: Discovery of spatial association rules in geographic information databases. In: Advances in Spatial Databases, pp. 47–66. Springer (1995)
14. Leleu, M., Rigotti, C., Boulicaut, J.F., Euvrard, G.: Go-spade: mining sequential patterns over datasets with consecutive repetitions. In: Perner, P., Rosenfeld, A. (eds.) MLDM 2003. LNAI, vol. 2734, pp. 293–306. Springer, Heidelberg 2003
15. Lin, M.Y., Lee, S.Y.: Fast discovery of sequential patterns through memory indexing and database partitioning. J. Inf. Sci. Eng. **21**(1), 109–128 (2005)
16. Pei, J., Han, J., Mortazavi-Asl, B., Pinto, H., Chen, Q., Dayal, U., Hsu, M.C.: Prefixspan: Mining sequential patterns efficiently by prefix-projected pattern growth. In: 29th IEEE International Conference on Data Engineering (ICDE), pp. 215–224. IEEE Computer Society (2001)
17. Savasere, A., Omiecinski, E.R., Navathe, S.B.: An efficient algorithm for mining association rules in large databases. In: VLDB, pp. 432–444 (1995)
18. Saxena, A.S., Goyal, V., Bera, D.: Efficient Enforcement of Privacy for Moving Object Trajectories. In: Bagchi, A., Ray, I. (eds.) ICISS 2013. LNCS, vol. 8303, pp. 360–374. Springer, Heidelberg (2013)
19. Srikant, R., Agrawal, R.: Mining sequential patterns: Generalizations and performance improvements. In: Proceedings of the 5th International Conference on Extending Database Technology: Advances in Database Technology, EDBT 1996, pp. 3–17 (1996)
20. Tsoukatos, I., Gunopulos, D.: Efficient Mining of Spatiotemporal Patterns. In: Jensen, C.S., Schneider, M., Seeger, B., Tsotras, V.J. (eds.) SSTD 2001. LNCS, vol. 2121, pp. 425–442. Springer, Heidelberg (2001)
21. Zaki, M.J.: Spade: An efficient algorithm for mining frequent sequences. Machine Learning **42**(1–2), 31–60 (2001)
22. Zheng, K., Shang, S., Yuan, N., Yang, Y.: Towards efficient search for activity trajectories. In: 2013 IEEE 29th International Conference on Data Engineering (ICDE), pp. 230–241, April 2013

Effective and Interpretable Document Classification Using Distinctly Labeled Dirichlet Process Mixture Models of von Mises-Fisher Distributions

Ngo Van Linh[✉], Nguyen Kim Anh, Khoat Than, and Nguyen Nguyen Tat

Hanoi University of Science and Technology, 1, Dai Co Viet road, Hanoi, Vietnam
{linhnv,anhnk,khoattq}@soict.hust.edu.vn, tatnguyennguyen@gmail.com

Abstract. Document Classification is essential to information retrieval and text mining. Accuracy and interpretability are two important aspects of text classifiers. This paper proposes an interpretable classification method (DLDPvMFs) by using the Dirichlet process mixture (DPM) model to discover the hidden topics distinctly within each label for classification of directional data based on the von Mises-Fisher (vMF) distribution, which arises naturally for data distributed on the unit hypersphere. We use a mean-field variational inference algorithm when developing DLDPvMFs. By using the label information of the training data explicitly and determining automatically the number of topics for each label to find the topical space, class topics are coherent, relevant and discriminative and since they help us interpret class's label as well as distinguish classes. Our experimental results showed the advantages of our approach via significant criteria such as separability, interpretability and effectiveness in classification task of large datasets with high dimension and complex distribution. Our obtained results are highly competitive with state-of-the-art approaches.

Keywords: Variational inference · Bayesian nonparametrics · Classification · von Mises-fisher distribution

1 Introduction

Recently, as the number of online documents has been rapidly increasing, automatic text categorization is becoming a more important and fundamental task in information retrieval and text mining. The major objective of text classification system is to organize the available text documents semantically into their respective categories. This problem has attracted significant attention from lot of researchers for playing crucial role in many applications such as web page classification, classification of news articles, information retrieval etc.

Accuracy and interpretability are two important aspects of text classifiers. While the accuracy of a classifier measures the ability to correctly classify unseen

© Springer International Publishing Switzerland 2015
M. Renz et al. (Eds.): DASFAA 2015, Part II, LNCS 9050, pp. 139–153, 2015.
DOI: 10.1007/978-3-319-18123-3_9

data, interpretability is the ability of the classification to be understood by humans and supplies reason why each data instance is assigned to label. Therefore, the classifier should be represented in an intuitive and comprehensible way, such that the user can draw from this information and make an appropriate decision. However, many black-box models such as SVM or ensemble classifiers outperform more comprehensible methods like decision trees in terms of classification accuracy. This leads to a trade-off between interpretability and accuracy [1]. As a result, many researchers are interested in accuracy when the interpretability of classification is often neglected.

To interpret the way a classifier assign a document to a label, this classifier has to understand the meaning of each of labels as well as document's contents and discover words in documents which can be attributed to the document's label. Interpretable text classification in this way requires models can flexibly account for the textual patterns that underlie the observed labels. Topic modeling is a potential approach to learning hidden topics from large datasets with high dimension and complex distribution. In fact, for classification, data instances in a class usually present some special topics that are different from other classes. Therefore, to understand the meaning of each label, we can use a topic model to discover the hidden topics in respect of this label from itself label's training data. It means that each labeled document can use only topics in a special topic set associated with the document's label. Furthermore, the discovered topics for each class should be coherent and are relevant enough to interpret and can differentiate classes. And so, a new challenge in understanding the meaning of each label is to determine how many topics belong to each label.

Clearly, in some topic models, an improper estimation of the number of topics might easily mislead the classification process and result in bad classification outcomes. The Dirichlet process mixture (DPM) model is an infinite mixture model in which each component corresponds to a different topic. The DPM model can automatically determine the number of topics [2] that data instances are talking about. However, the choice of a particular generative model can affect the performance of a classification procedure. Specially, in information retrieval applications, models that use normalizing the data vectors help to remove the biases induced by the length of a document and cosine similarity is a more effective measure of similarity for analysing text documents [2–4]. The von Mises-Fisher (vMF) distribution is one of the simplest parametric distributions for directional vector data in $\mathbb{R}^d$. The mixture of vMF distributions gives a parametric model which exploits cosine similarity measure to be appropriate to high dimensional data, whereas the mixture of Gaussian distributions is similar to use euclidean distance which is not suitable to text classification. The generative model based on the vMF distributions can directly handle high dimensional vectors without dimensionality reduction, and so, can been applied for classification of texts that are high-dimensional, sparse and directional vectors [3,5].

This paper proposes an interpretable document classification method using Distinctly Labeled Dirichlet Process Mixture Models of von Mises-Fisher Distributions (DLDPvMFs) by using the DPM model to discover the hidden topics distinctly within each label. By exploiting DPM to automatically detect the number of topics within each class, our method separably learns among classes thanks to which it not only easily understands and interprets the topics of each label but also detects outliers. Moreover, our method offers the representation of instances in topical spaces to deal with visualization and dimension reduction problems. In addition, to cope with high-dimensional and sparse texts, our model generates data instances by vMF distributions. The experimental results show that our method is highly competitive with state-of-the-art approaches in classification.

The rest of the paper is organized as follows. Section 2 introduces related works. Section 3 describes our proposed approach in detail. Experimental results are discussed in Section 4 and conclusion is made in Section 5.

2   Related Works

Accuracy and interpretability are two important aspects of text classifiers. However, there are only a few works to deal with the interpretability of classification. Recently, supervised topic model incorporating document class labels such as Labeled LDA [6] applies a transformation matrix on document class labels to modify Dirichlet priors of the LDA-like models. However, Labeled LDA (L-LDA) was designed specifically for multi-label settings. Moreover, in L-LDA, the training of the LDA model is adapted to account for multi-labeled corpora by putting "topics" in 1-1 correspondence with labels and then restricting the sampling of topics for each document to the set of labels that were assigned to the document.

While labeled LDA simply defines a one-to-one correspondence between latent topics and the labels of classes and hence does not find out latent sub-topics within each class, multi-view topic model(mview-LDA) [7] and partially labeled topic models [8] discover the hidden topics within each label. Ahmed and Xing proposed a multi-view topic model for ideological perspective analysis [7]. Each ideology has a set of ideology-specific topics and an ideology-specific distribution over words. By using the hidden topics of each class which were find out in training process, their classification results improved significantly. However, the number of topics were not interested in their model and it was chosen by experiments. Similarly, Partially Labeled Dirichlet Allocation (PLDA) [8] uses the un-supervised learning machinery of topic models to discover the hidden topics within each label and interpret the hidden meaning of each class. Moreover, Partially Labeled Dirichlet Process (PLDP) [8] extends PLDA by incorporating a non-parametric Dirichlet process prior over each class's topic set, allowing the model to adaptively discover how many topics belong to each label. However, this paper aims to interpret the meaning of labels which help to understand and describe concisely the main content of each document but it does not aim to classify this document. Furthermore, in their models, they usually use multinomial distribution to generate document which is not effective to classify [9].

A more recent approach proposed by Zhu, Ahmed and Xing was maximum entropy discrimination LDA (MedLDA) [10]. They estimate model parameters which help to infer latent topical vectors of documents by integrating the max-margin principle into the process of topic learning with one single objective function. However, the quality of the topical space learned by MedLDA is heavily affected by the quality of the classifiers which are learned at each iteration of MedLDA. Moreover, due to the inappropriate use of kernels, MedLDA is not good to learn topical space.

Our work differs from [7,8] in two significant aspects. Firstly, we propose a supervised mixture model to exploit effectively label information of the training data by using the DPM model that can determine the number of topics automatically to discover the hidden topics distinctly within each label. So, class topics are coherent, relevant and discriminative and since they help us interpret class's label as well as distinguish classes. Secondly, we use generative model based on the von Mises-Fisher (vMF) distribution that can directly handle high dimensional vectors without dimensionality reduction.

3   Proposed Method

It is natural to assume that data instances in a class present some special topics or components that distinguish classes. Finding out them not only improves classification accuracy but also helps to understand and interpret the meaning of labels. In order to reinforce the separability of classes, we construct a model to discover the particular topics of each class by exploiting the data instances of each class in a separate mixture model of vMF distributions which is a suitable model for text documents [9]. Nevertheless, determining the number of topics in each class is non-trivial. Motivated by the advantages of Dirichlet process mixture model with vMF distributions [2], we propose a new model to find out automatically the number of topics that are prominent in each class.

3.1   The von Mises-Fisher (vMF) Distribution

In directional statistics, a d-dimensional unit random vector x (i.e., $x \in \mathbb{R}^d$ and $\|x\| = 1$, or equivalently $x \in \mathbb{S}^{d-1}$) is said to have d-variate von Mises-Fisher (vMF) distribution $\mathbb{R}^d$ [9] if its probability density function is given by $f(x|\mu, \kappa) = vMF(x|\mu, \kappa) = C_d(\kappa) \exp(\kappa \mu^T x)$ where $\|\mu\| = 1, \kappa \geq 0$ and $d \geq 2$. The normalizing constant $C_d(\kappa)$ is given by:

$$C_d(\kappa) = \frac{\kappa^{\frac{d}{2}-1}}{(2\pi)^{\frac{d}{2}} I_{\frac{d}{2}-1}(\kappa)} \tag{1}$$

where $I_r(\cdot)$ represents the modified Bessel function of the first kind and order. The unit vectors drawn according to vMF distribution are concentrated about μ, mean direction, which is also a unit vector. As κ, the concentration parameter, increases the distribution tends to stronger concentration around μ. In particular when $\kappa = 0, f(x|\mu, \kappa)$ reduces to uniform density on $\mathbb{S}^{d-1}$, and as

$\kappa \rightarrow \infty, f(x|\mu, \kappa)$ tends to a point density. Also, expected value $E(x) = \rho\mu$, where:

$$\rho = A_d(\kappa) = \frac{\int_{-1}^{1} t^2 e^{\kappa t}(1-t^2)^{\frac{(d-3)}{2}} dt}{\int_{-1}^{1} e^{\kappa t}(1-t^2)^{\frac{(d-3)}{2}} dt} = \frac{I_{\frac{d}{2}}(\kappa)}{I_{\frac{d}{2}-1}(\kappa)} \tag{2}$$

3.2   Dirichlet Process Mixture (DPM)

The Dirichlet process, introduced in [11], is a distribution over distributions. It is parameterized by a base distribution G_0 and a positive scaling parameter α. Suppose a random distribution G is drawn from a Dirichlet process, Dirichlet process mixture model is written as:

$$G|\{\alpha, G_0\} \sim DP(\alpha, G_0)$$

$$\eta_n|G \sim G \text{ and } X_n|\eta_n \sim P(X_n|\eta_n)$$

Sethuraman provided a more explicit characterization of the Dirichlet process in terms of a stick-breaking construction [12]. Consider two infinite collections of independent random variables, $u_t \sim Beta(1, \alpha)$ and $\eta_t \sim G_0$ for $t = \{1, 2, ...\}$. Mixing proportions $\pi_t(u)$ are given by:

$$\pi_t(u) = u_t \prod_{j=1}^{t-1} (1 - u_j) \tag{3}$$

By this construction, $\sum_{t=1}^{\infty} \pi_t = 1$ and the stick-breaking representation of Dirichlet Process mixture model is as follows:

$$G = \sum_{t=1}^{\infty} \pi_t(u)\delta_{\eta_t} \tag{4}$$

The DPM has a natural interpretation as a flexible mixture model in which the number of topics is random and grows when new data are observed. Moreover, this model estimates how many topics are needed to model the observed data. Thence, it discovers coherent and natural topics in data.

3.3   Distinctly Labeled Dirichlet Process Mixture Models of von Mises-Fisher Distributions (DLDPvMFs)

In this section, we propose a new supervised mixture model by exploiting the DPM model that can determine the number of topics automatically to discover the particular hidden topics distinctly within each label. We assume that dataset $\mathbf{X} = \{x_1, x_2, ..., x_N\}$ consisting of N d-dimensional vectors, where each vector represents a data instance. Let $\mathbf{V} = \{v_1, v_2, ..., v_N\}$ be the set of V labels for which v_i is the label of x_i. And, label v_n of x_n is generated by multinomial distribution: $v_n \sim Mult(\delta)$ where $\delta = (\delta_1, \delta_2, ..., \delta_V)$ is a proportion hyperparameter. Specially, for each class, we use a DPM model of vMF distributions to generate instances in this class. By using distinct DPM which finds out the number of topics for each class, our model will learn the particular topics that help to distinguish classes and interpret the meaning of labels. The stick-breaking construction for the DPM of vMF distributions is depicted as a graphical model in Figure 1. Let z_n be an assignment variable of the mixture topics with which the data x_n is associated. When label v_n is known, z_n is generated by multinomial distribution: $z_n \sim Mult(\pi(u_{v_n}))$. The conditional distribution of x_n given $v_n, z_n, \{\mu_1, \mu_2, ...\}$ and hyperparameter κ is $P(x_n|v_n, z_n, \kappa, \mu_1, \mu_2, ...) = vMF(x_n|\mu_{v_n,z_n}, \kappa_{v_n,z_n})$. $\mu_{v,t}$, $(t = 1, 2, ..., \infty)$ are mean direction vectors that are the particular topics of label v. Mardia and El–Atoum [13] have indentified the vMF distribution as the conjugate prior for the mean direction. So $\mu_{v,t}$ is generated by: $\mu_{v,t}|\zeta_v, \rho_v \sim vMF(\mu_{v,t}|\zeta_v, \rho_v)$ where ζ_v, ρ_v are hyperparameters of vMF distribution.

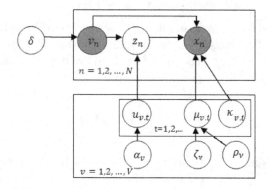

Fig. 1. Graphical Model of Distinctly Labeled Dirichlet Process Mixture Models of von Mises-Fisher Distributions (DLDPvMFs)

The complete generative model is given by:

1. The topics of each class $v \in \{1, 2, ..., V\}$ are generated as follows:
   (a) Draw $u_{v,t} \sim Beta(1, \alpha_v), t = \{1, 2, ...\}$
   (b) Draw $\mu_{v,t}|\zeta_v, \rho_v \sim vMF(\mu_{v,t}|\zeta_v, \rho_v)$
2. For each data point x_n
   (a) Draw $v_n|\delta \sim Multi(\delta)$
   (b) Draw $z_n|v_n, u_n \sim Mult(\pi(u_{v_n}))$
   (c) Draw
   $x_n|v_n, z_n, \kappa_v, \mu \sim vMF(x_n|\mu_{v_n,z_n}, \kappa_{v_n,z_n})$

3.4   Variational Inference for DPM of vMFs

Given a training set, we focus on inferring the posterior distribution that there is no direct way to compute under a DPM prior. Inference algorithms, based on

Gibbs sampling [14,15], usually are used for such problem and are guaranteed to converge to the underlying distributions. However, Gibbs sampling is not efficient enough to scale up to the large scale problems. Furthermore, it is difficult to know how many iterations are required for convergence of the Markov chain. Variational inference provides an alternative and is usually much faster Gibbs sampling. In mean-field variational methods, the true posterior is approximated by another distribution with a simpler, factored parametric form. An EM procedure is used to update the parameters of the approximate posterior and the model hyperparameters so that a lower bound on the log likelihood increases with each iteration [16].

In training process, each label v_n of instance x_n is observed. Let $\mathbf{X}^v$ and $\mathbf{Z}^v$ be instances and their assignment variables in class v. The posterior distribution is written as:

$$p(\mathbf{Z}, \mathbf{U}, \boldsymbol{\mu} | \mathbf{X}, \mathbf{V}, \alpha_{\mathbf{V}}, \zeta_{\mathbf{V}}, \kappa_{\mathbf{V}}, \rho_{\mathbf{V}})$$
$$= \prod_{v=1}^{V} p(\mathbf{Z}^v, \mathbf{U}_v, \boldsymbol{\mu}_v | \mathbf{X}^v, \alpha_v, \zeta_v, \kappa_v, \rho_v) \tag{5}$$

where $\mathbf{U}_v = \{u_{v,1}, u_{v,2}, ..., u_{v,t}, ...\}$, $\boldsymbol{\mu}_v = \{\mu_{v,1}, ..., \mu_{v,t}, ..\}$ are the hidden variables which describe particular topics of label v. It means that the each DPM of each label v are learnt independently from other classes.

We will introduce a mean-field variational method to learn DPM for class v. We assume that N_v is the number of instances that their label are v. However, in DPM models, the value of z_n is potentially unbounded. Thus the variational distribution need be truncated. The truncation level T is fixed a value and let $q(u_{v,T} = 1) = 1$. It means that the mixture proportions $\pi_{v,t}(u) = 0$ for $t > T$. The model is a full Dirichlet process and is not truncated. Only the variational distribution is truncated.

We infer latent variables and estimate hyperparameters by using mean-field variational inference method to above model. The mean-field method tries to find a distribution in a simple family that is close to the true posterior. We approximate the fully factorized family of distributions over the hidden variables:

$$q(\mathbf{U}_v, \boldsymbol{\mu}_v, \mathbf{Z}^v | \gamma_v, \tilde{\mu}_v, \tilde{\kappa}_v, \phi)$$
$$= \prod_{t=1}^{T-1} q(u_{v,t} | \gamma_{v,t}) \prod_{t=1}^{T} q(\mu_{v,t} | \tilde{\mu}_{v,t}, \tilde{\kappa}_{v,t}) \prod_{n=1}^{N_v} q(z_n | \phi_n) \tag{6}$$

and assume the factors have the parametric forms: $q(u_{v,t} | \gamma_{v,t}) = Beta(u_{v,t} | \gamma_{v,t_1}, \gamma_{v,t_2})$, $q(\mu_{v,t} | \tilde{\mu}_{v,t}, \tilde{\kappa}_{v,t}) = vMF(\mu_{v,t} | \tilde{\mu}_{v,t}, \tilde{\kappa}_{v,t})$, $q(z_n | \phi_n) = Mult(z_n | \phi_n)$ $(q(z_n = t | \phi_n) = \phi_{n,t})$. Here, $\gamma_{v,t_1}, \gamma_{v,t_2}, \tilde{\mu}_{v,t}, \tilde{\kappa}_{v,t}, \phi_n$ are the free variational parameters.

Using this factorization, a lower bound $L(\gamma_v, \tilde{\mu}_v, \tilde{\kappa}_v, \phi)$ of the log likelihood is given by:

$$
\begin{aligned}
L&(\gamma_v, \tilde{\mu}_v, \tilde{\kappa}_v, \phi) \\
&= E_q[\log P(\mathbf{X}^v|\mathbf{Z}^v, \mathbf{U}_v, \boldsymbol{\mu}_v))] + E_q[\log P(\mathbf{Z}^v|\mathbf{U}^v)] \\
&\quad + E_q[\log P(\mathbf{U}_v|\alpha_v)] + E_q[\log P(\boldsymbol{\mu}_v|\zeta_v, \rho_v)] \\
&\quad - E_q[\log q(\mathbf{U}_v|\gamma_v)] - E_q[\log q(\mathbf{Z}^v|\phi)] \\
&\quad - E_q[\log q(\boldsymbol{\mu}_v|\tilde{\mu}_v, \tilde{\kappa}_v)]
\end{aligned}
\tag{7}
$$

To optimize the lower bound of the log-likelihood, we use EM scheme to iteratively learn the model. Specifically, we repeat the following two steps, E-step and M-step, until convergence. In E-step, the lower bound is optimized with respect to each of the free parameters $\gamma_v, \tilde{\mu}_v, \tilde{\kappa}_v, \phi$ as:

$$
\gamma_{v,t_1} = 1 + \sum_{i=1}^{N_v} \phi_{i,t}
\tag{8}
$$

$$
\gamma_{v,t_2} = \alpha_v + \sum_{i=1}^{N_v} \sum_{j=t+1}^{T} \phi_{i,j}
\tag{9}
$$

$$
\phi_{n,t} \propto \exp(S_{n,t})
\tag{10}
$$

where

$$
\begin{aligned}
S_{n,t} &= \kappa_{v,t} A_d(\tilde{\kappa}_{v,t}) \tilde{\mu}_{v,t} x_n + (\Psi(\gamma_{v,t_2}) - \Psi(\gamma_{v,t_1} + \gamma_{v,t_2})) \\
&\quad + \sum_{t=1}^{T-1} (\Psi(\gamma_{v,t_2}) - \Psi(\gamma_{v,t_1} + \gamma_{v,t_2}))
\end{aligned}
\tag{11}
$$

$$
\tilde{\mu}_{v,t} = \frac{\sum_{n=1}^{N} \kappa_{v,t} \phi_{n,t} x_n + \rho_v \zeta_v}{\| \sum_{n=1}^{N} \kappa_{v,t} \phi_{n,t} x_n + \rho_v \zeta_v \|}
\tag{12}
$$

$$
\tilde{\kappa}_{v,t} = \sum_{n=1}^{N} \kappa_{v,t} \phi_{n,t} \tilde{\mu}_{v,t}^T x_n + \rho_v \tilde{\mu}_{v,t}^T \zeta_v
\tag{13}
$$

Similarly, in the M step, the lower bound is optimized with respect to each of the hyperparameters $\zeta_v, \rho_v, \kappa_v$ as:

$$
\zeta_v = \frac{\sum_{t=1}^{T} \rho_v A_d(\tilde{\kappa}_{v,t})}{\| \sum_{t=1}^{T} \rho_v A_d(\tilde{\kappa}_{v,t}) \|}
\tag{14}
$$

$$
\bar{r}_{0,v} = \frac{C_d'(\rho_v)}{C_d(\rho_v)} = -\frac{\sum_{t=1}^{T} A_d(\tilde{\kappa}_{v,t}) \tilde{\mu}_{v,t}^T \zeta_v}{T}
\tag{15}
$$

$$
\bar{r}_{v,t} = \frac{C_d'(\kappa_{v,t})}{C_d(\kappa_{v,t})} = -\frac{A_d(\tilde{\kappa}_{v,t}) \tilde{\mu}_{v,t}^T (\sum_{n=1}^{N_v} \phi_{n,t} x_n)}{\sum_{n=1}^{N} \phi_{n,t}}
\tag{16}
$$

[9] provided approximations for estimating ρ_v and κ_v:

$$\rho_v \approx \frac{d\bar{r}_{0,v} - (\bar{r}_{0,v})^3}{(1 - (\bar{r}_{0,v})^2)} \qquad (17)$$

$$\kappa_{v,t} \approx \frac{d\bar{r}_{v,t} - (\bar{r}_{v,t})^3}{2(1 - (\bar{r}_{v,t})^2)} \qquad (18)$$

The EM procedure consists of alternating E and M steps until some suitable convergence criterion is reached. After training process, in each label v, our model discovers natural topics which are represented $\tilde{\mu}_{v,t}$ and $\tilde{\kappa}_{v,t}$ for each topic. Moreover, the concentration parameter $\tilde{\kappa}_{v,t}$ expresses the concentration of instances around $\tilde{\mu}_{v,t}$. Specially, in DPM [2,16], most instances only gather in a small number of topics. It is easy to realize T_v that is the number of topics for each label v. Correspondingly, $\mathbf{T} = \sum_{v=1}^{V} T_v$ topics are identified from training set.

In test process, each unlabeled instance x_m is generated from a mixture of $\mathbf{T}$ of vMF distributions which are learnt in training process. According to [9], the probabilities of x_m to topics $\theta_m = \{\theta_1, ..., \theta_{\mathbf{T}}\}$ are determined. For classification, the label of x_m is inferred from the sum of probabilities of x_m to the topics of each label. In addition, our method find out the new representation θ of instances on topical space that is hopeful in applications as: dimension reduction, visualization data. Specially, each label is represented by the its particular topics thanks to which the discriminative property in the topical space is prominent in our method.

Algorithm 1. DLDPvMFs

Input: Set X of train data points on S^{d-1} and label set V
    Initialize randomly $\mu_0, \kappa_0, \gamma_{t_1}, \gamma_{t_2}, \tilde{\mu}_t, \tilde{\kappa}_t, \phi_n (t = 1..T; n = 1..N)$
    repeat
        {The E step}
        for $t = 1$ to T **do**
            $\gamma_{t_1} = 1 + \sum_{i=1}^{N} \phi_{i,t}$
            $\gamma_{t_2} = \alpha + \sum_{i=2}^{N} \sum_{j=t+1}^{T} \phi_{i,j}$
            $\tilde{\mu}_t = \frac{\sum_{n=1}^{N} \kappa \phi_{n,t} x_n + \kappa_0 \mu_0}{\| \sum_{n=1}^{N} \kappa \phi_{n,t} x_n + \kappa_0 \mu_0 \|}$
            $\tilde{\kappa}_t = \sum_{n=1}^{N} \kappa \phi_{n,t} \tilde{\mu}_t^T x_n + \kappa_0 \tilde{\mu}_t^T \mu_0$
            for $n = 1$ to T **do**
                Compute $\phi_{n,t}$ in (10)
            end for
        end for
        {The M step}
        Compute μ_0, κ_0 in (13),(15)
    until Convergence

---

4   Experimental Design

In this section, we describe some experiments to evaluate the advantages of our method in some aspects. We will show some clear evidences about the discriminative property in the topical space and classification effectiveness. Moreover, our method can detect outliers in learning process. Our method is compared with some prominent supervised methods.

- **mview-LDA** [7] exploits topic model to find out the topics of each label whose ideas is similar with ours. However, the number of topics within label is fixed and documents are generated by multinomial distributions.
- **MedLDA** [10] concurrently integrates the max-margin principle and topic model.

We used four high dimensional text datasets from the UCI repository data. A summary of all the datasets is shown in table 1. The balance of a dataset is defined as the ratio of the number of documents in the smallest class to the number of documents in the largest class. So a value close to (0)1 indicates a very (un)balanced dataset.

Table 1. Summary of text datasets (for each dataset, nd is the total number of documents, nw is dimension of dataset and k is the number of classes)

Dataset	nd	nw	k	Balance
hitech	2301	10080	6	0.1924
la1	3204	31472	6	0.290
la2	3075	31472	6	0.32670
ohscal	11161	11465	10	0.430

Outlier Detection

In our method, instances are automatically assigned to the topics (components) which they are almost talking about. A new instance will be considered to either assign to an old topic which is similar or generate a new topic, thanks to which our method finds out natural topics and determines the number of topics. An outlier is an instance that is different from other instances and trends to generate a new topic. Hence, there are only a few instances assigned to this topic. It is reason why our method is easy to detect outliers. It is demonstrated by the number of instances which are allocated to each topic that our method finds out. In this experiment, we use 80 percent of La2 dataset to train the model and the truncation level $T = 20$. To reduce computational complexity of algorithm, we do not estimate hyperparameters κ and fix $\kappa = 40$ in all experiments. In training process, our model detects the number of topics and the number of instances in those topics. Observing Table 2, the topics which have a few instances are not

Table 2. The number of instances allocated to each topic within each class of La2 dataset. $Topic_{ij}$ is the number of instances of topic j in class i

	Sum	topic11	topic12	topic13	topic14	topic15	topic16		
class 1	300	71	49	46	48	80	6		
	Sum	topic21	topic22	topic23	topic24	topic25	topic26		
class 2	389	44	97	89	91	52	16		
	Sum	topic31	topic32	topic33	topic34	topic35	topic36		
class 3	240	70	56	61	30	19	4		
	Sum	topic41	topic42	topic43	topic44	topic45	topic46	topic47	
class 4	724	223	99	131	90	56	106	19	
	Sum	topic51	topic52	topic53	topic54	topic55	topic56		
class 5	198	51	45	46	33	21	2		
	Sum	topic61	topic62	topic63	topic64	topic65	topic66	topic67	topic68
class 6	607	226	54	91	75	78	71	8	4

particular for classes due to which they can be considered as outliers. Outlier detection helps classification result in our algorithm is stable.

Discriminative Property in the Topical Space

The discriminative property in the topical space of a method which helps to separate classes is interested in our experiments. In order to illustrate class separation, we will visualize all instances of La2 dataset after projecting to topical space. In this experiment, the number of topics is set equally 78 in MedLDA model. In mview-LDA, the number of specific topics within each label and background topics are set equally 12 and 6. Result is shown in Fig.2. Because of finding out the specific topics of each label, the projections of DLDPvMFs and mview-LDA are discriminative than MedLDA in topical space. The projection of MedLDA confuses the dataset among classes. It is explained that the quality of the topical space learned by MedLDA is influenced heavily by the quality of the classifier which is integrated into the topic model. Moreover, due to inappropriate use of kernels, MedLDA is not good to learn topical space. Specially, DLDPvMFs detects outliers and uses vMF distributions which are suitable to document classification. Therefore, it finds out the better representation than mview-LDA.

Classification Effectiveness

We use accuracy of classification to quantify the goodness of those supervised methods. Firstly, we measure the accuracy of those methods when the number of topics increases. In mview-LDA, the number of background topics is set equally the number of classes and the number of specific topics increases to $\{6, 8, 10, 12\}$ in hitech, la2, k1b datasets and $\{4, 6, 8, 10\}$ in ohscal dataset. Accordingly, in MedLDA, the number of topics is set equally $\{42, 54, 66, 78\}$ in hitech, la2, k1b datasets and $\{50, 70, 90, 110\}$ in ohscal dataset. Whereas, by finding out the number of topics, our method is stable. In this experiment, 80 percents of dataset is used to train those models. Results (Fig.3) are shown that our method consistently achieved the best performance. In previous discussion,

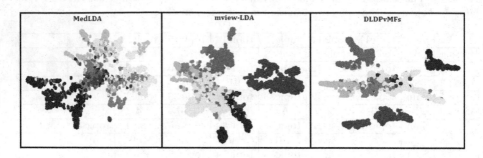

Fig. 2. Discrimination of topical space learned by MedLDA, mview-LDA, DLDPvMFs. La2 was the dataset for visualization. Data points in the same class have the same color. These embeddings were done with t-SNE [17].

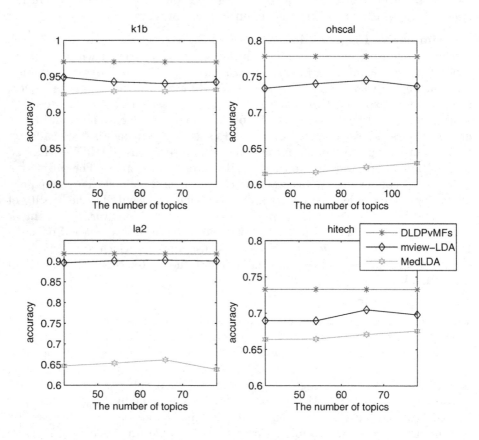

Fig. 3. Classification quality when the number of topics increases

we exposed that DLDPvMFs guarantees class separation and outliers reduction. In addition, our method exploits vMF distributions which are suitable to classify document dataset. They are reasons why mview-LDA is not as good as DLDPvMFs, albeit mview-LDA also has discriminative property among classes which are based on detecting specific topics within each class. According to above discuss, due to inappropriate use of kernels, MedLDA is not good to learn topical space which affects back classifier. Moreover, by encoding local information in each label, DLDPvMFs and mview-LDA ensure preserving inner-class local structure to help their model succeed even in cases that data points reside in a nonlinear manifold [18,19], for which MedLDA might fail. So, DLDPvMFs and mview-LDA always perform better than MedLDA.

Secondly, we examine influence on the performance of those methods when training set ratio is changed. Fig.4 is shown that the performance of our method is better when ratio increases but not many. It means that our methods is stable.

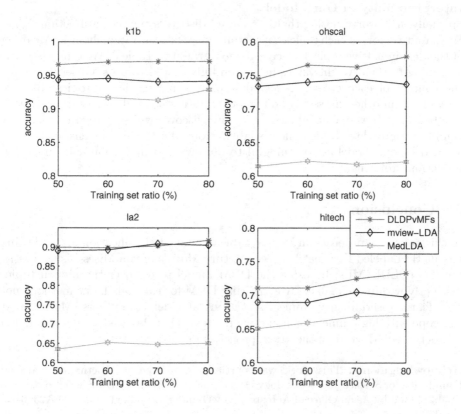

Fig. 4. Classification quality when training set ratio increases

Table 3. Meaning of the topical space which was learned by DLDPvMFs from La2. For each row, the first column shows the class label, the second column shows top terms that represent some topics in each class.

Class name	Representative topic
Entertainment	audienc, film, music, theatre, stage, product, tv, theater, televis, movie
	music, band, concert, song, perform, rock, singer, review, musician,jazz
	galleri, art, artist, review, paint, museum, viewer, charact, piece, exhibit
	...
Finance	earn, million, quarter, corpor, rose, revenu, billion, profit, incom, net
	industri, product, economi, compani, econom, research, manufactur, comput, japan, develop
	contract, business, industri, develop, product, technology, trade, merchant, equip, million
	...
Foreign	soviet, afghanistan, israel, guerrilla, rebel, military, troop, govern, soldier, army
	plane, crash, flight, airplane, pilot, bomb, wire, kill, concord, passenger
	ozone, flask, depletion, arctic, crop, roman, archaeologist, layer, research, ancient
	...

Interpretability in Our Model

Specially, it is worth noting that our model discovers the mean direction $\bar{\mu}_{v,t} \in \mathbb{R}^d$ (a unit vector) of each topic that instances are concentrated about it. Meaning of the discriminative topical space is demonstrated in Table 3. We rank all words according to their weight values in $\bar{\mu}_{v,t}$ and give top–10 words. Observationally, the content of each class is reflected well by some particular topics that are different from other classes. So, our model can understand the meaning of each of labels as well as document's contents and discover which words in documents can be attributed to the document's label. Moreover, the low-dimensional spaces learned by our model are meaningful to deal with visualization and dimension reduction problems.

5   Conclusion

In this paper, we proposes an interpretable document classification method using Distinctly Labeled Dirichlet Process Mixture Models of von Mises-Fisher Distributions (DLDPvMFs) by using the DPM model to discover the hidden topics distinctly within each label. By exploiting DPM for each label, our methods not only learns discriminative topical space but also detects outliers. Moreover, it can cope with high-dimensional and sparse text. Our obtained results are highly competitive with state-of-the-art approaches in classification.

Acknowledgments. This work was partially supported by Vietnam's National Foundation for Science and Technology Development (NAFOSTED Project No. 102.05-2014.28), and by Asian Office of Aerospace R&D under agreement number FA2386-15-1-4011.

References

1. Van de Merckt, T., Decaestecker, C.: About breaking the trade off between accuracy and comprehensibility in concept learning. In: IJCAI 1995 Workshop on Machine Learning and Comprehensibility (1995)
2. Anh, N.K., Tam, N.T., Linh, N.V.: Document clustering using dirichlet process mixture model of von mises-fisher distributions. In: 4th International Symposium on Information and Communication Technology, SoICT 2013, pp. 131–138 (2013)
3. Anh, N.K., Van Linh, N., Ky, L.H., et al.: Document classification using semi-supervived mixture model of von mises-fisher distributions on document manifold. In: Proceedings of the Fourth Symposium on Information and Communication Technology, pp. 94–100. ACM (2013)
4. Anh, N.K., Linh, N.V., Tam, N.T.: Document clustering using mixture model of von mises-fisher distributions on document manifold, pp. 146–151 (2013)
5. Gopal, S., Yang, Y.: Von mises-fisher clustering models. In: Proceedings of The 31st International Conference on Machine Learning, pp. 154–162 (2014)
6. Ramage, D., Hall, D., Nallapati, R., Manning, C.D.: Labeled lda: A supervised topic model for credit attribution in multi-labeled corpora. In: Proceedings of the 2009 Conference on Empirical Methods in Natural Language Processing, vol. 1, pp. 248–256. Association for Computational Linguistics (2009)
7. Ahmed, A., Xing, E.P.: Staying informed: supervised and semi-supervised multi-view topical analysis of ideological perspective. In: Proceedings of the 2010 Conference on Empirical Methods in Natural Language Processing, pp. 1140–1150. Association for Computational Linguistics (2010)
8. Ramage, D., Manning, C.D., Dumais, S.: Partially labeled topic models for interpretable text mining. In: Proceedings of the 17th ACM SIGKDD International Conference on Knowledge Discovery and Data Mining, pp. 457–465. ACM (2011)
9. Banerjee, A., Dhillon, I.S., Ghosh, J., Sra, S.: Clustering on the unit hypersphere using von mises-fisher distributions. J. Mach. Learn. Res. **6**, 1345–1382 (2005)
10. Zhu, J., Ahmed, A., Xing, E.P.: Medlda: maximum margin supervised topic models. The Journal of Machine Learning Research **13**(1), 2237–2278 (2012)
11. Ferguson, T.S.: A Bayesian analysis of some nonparametric problems. The Annals of Statistics **1**(2), 209–230 (1973)
12. Sethuraman, J.: A constructive definition of Dirichlet priors. Statistica Sinica **4**, 639–650 (1994)
13. Mardia, K.V., Atoum, E.S.A.M.: Bayesian inference for the von Mises-Fisher distribution. Biometrika 63, 203–206 (1976)
14. Ishwaran, H., James, L.F.: Gibbs sampling methods for stick-breaking priors. Journal of the American Statistical Association **96**(453), 161–173 (2001)
15. Neal, R.M.: Markov chain sampling methods for Dirichlet process mixture models. Journal of Computational and Graphical Statistics **9**(2), 249–265 (2000)
16. Blei, D.M., Jordan, M.I.: Variational inference for dirichlet process mixtures. Bayesian Analysis **1**(1), 121–144 (2006)
17. Van der Maaten, L., Hinton, G.: Visualizing data using t-sne. Journal of Machine Learning Research 9(11) (2008)
18. Niyogi, P.: Manifold regularization and semi-supervised learning: some theoretical analyses. Journal of Machine Learning Research **14**(1), 1229–1250 (2013)
19. Than, K., Ho, T.B., Nguyen, D.K.: An effective framework for supervised dimension reduction. Neurocomputing **139**, 397–407 (2014)

MPTM: A Topic Model for Multi-Part Documents

Zhipeng Xie[1,2(✉)], Liyang Jiang[1,2], Tengju Ye[1,2], and Zhenying He[1,2]

[1] School of Computer Science, Fudan University, Shanghai, China
[2] Shanghai Key Laboratory of Data Science, Fudan University, Shanghai, China
{xiezp,13210240017,13210240039,zhenying}@fudan.edu.cn

Abstract. Topic models have been successfully applied to uncover hidden probabilistic structures in collections of documents, where documents are treated as unstructured texts. However, it is not uncommon that some documents, which we call multi-part documents, are composed of multiple named parts. To exploit the information buried in the document-part relationships in the process of topic modeling, this paper adopts two assumptions: the first is that all parts in a given document should have similar topic distributions, and the second is that the multiple versions (corresponding to multiple named parts) of a given topic should have similar word distributions. Based on these two underlying assumptions, we propose a novel topic model for multi-part documents, called Multi-Part Topic Model (or MPTM in short), and develop its construction and inference method with the aid of the techniques of collapsed Gibbs sampling and maximum likelihood estimation. Experimental results on real datasets demonstrate that our approach has not only achieved significant improvement on the qualities of discovered topics, but also boosted the performance in information retrieval and document classification.

Keywords: Topic models · Gibbs sampling · Maximum likelihood estimation

1 Introduction

In classic topic models, such as probabilistic latent semantic analysis [8] and latent Dirichlet allocation [2], each document is represented as a mixture of topics, and each topic is represented as a probability distribution over words. To generate a document, we first draw a topic distribution independently from a prior Dirichlet distribution, and then for each word in that document, draw a topic randomly from the topic distribution and draw a word from that topic. Once the topic distribution is determined for a document, all the words in it follow the same generative procedure which is not affected by the location where a word appears in the document. In other words, each document is modeled as a whole, which is reflected in the fact that all the content of a document share the same topic distribution.

However, some documents are naturally composed of multiple named-parts, in the form of subdocuments or sections. Such documents are called multi-part documents in this paper. A typical example of multi-part documents is academic research papers, where each document is normally divided into sections such as *Abstract*, *Introduction*,

© Springer International Publishing Switzerland 2015
M. Renz et al. (Eds.): DASFAA 2015, Part II, LNCS 9050, pp. 154–168, 2015.
DOI: 10.1007/978-3-319-18123-3_10

Method, Experimental Results, and *Summary*. Logically, each section is self-existent. It is a relatively complete entity that describes the theme of the document from a specific aspect. For example, the section of *Introduction* is normally related to the motivation and related work of the paper, the section of *Method* describes the technical details of paper, while the section of *Experimental Results* may concern the performance measurements, the data used, and the comparison conclusion.

Our primary concern in this current study is taking this document-part structural information into consideration. To do this, we propose a novel topic modeling method for multi-part documents, called *Multi-Part Topic Model* (or MPTM in short). The MPTM model supposes that each topic has multiple versions (called versional topics) where each version corresponds to a specific named-part, while each part of a document is a mixture of the versional topics that corresponds to the part. Two underlying assumptions are also embodied in the model. The first one assumes that *all parts in the same document have similar topic distributions.* To enforce this assumption, we use one single Dirichlet distribution as the prior for all the parts of a document. Each document has its own Dirichlet prior. The mean parameters of the Dirichlet priors are normally different for different documents, but a common concentration parameter (also called the precision of the Dirichlet) is shared by all the Dirichlet priors, which controls how concentrated the distributions of multiple parts in the same document is around its mean. The second assumption is that *all versions of a single topic should have similar word distributions,* which is also enforced in a way similar to the first assumption. All versions of the same topic share a Dirichlet prior distribution, and the Dirichlet priors for different topics normally have different mean parameters.

By modeling document parts and versional topics separately, the proposed MPTM model allows us to judge the qualities of words and topics. A word that occurs in the top-word lists of (almost) all versions of a topic is thought of as a core word. On the other hand, if a word only appears frequently in one version of a topic, but seldom appears in other versions, it is then thought of as a word attached only to the particular version of the topic. Thus, each topic can be represented as a core-attachment structure, which facilitates the topic visualization. Similarly, a topic is thought of as stable and consistent, if it exhibits consistent probabilities across the multi-parts of documents; a topic is unstable or transient if its probabilities across the multi-parts of documents vary acutely. Accordingly, topic quality can be measured as the mean variance across the multi-parts averaged over all documents, which may help to prune unnecessary topics.

Finally, we evaluate MPTM model empirically on two real datasets. It is shown that the MPTM model not only generates topics of higher coherence than LDA, but also outperforms LDA in the tasks of information retrieval and document classification.

2   Related Work

A lot of existing work has been devoted to the incorporation of additional information into classic topic models, which can be broadly classified into three categories.

The first category of work *explores the correlation between topics*. Classic LDA model fails to model correlation between topics, because of the (nearly) independence assumptions implicit in the Dirichlet distribution on the topic proportions. To model the fact that the presence of one topic is sometimes correlated with the presence of another, [3] replaces the Dirichlet by the more flexible logistic normal distribution that incorporates a covariance structure among the components (or topics), [12] introduces the pachinko allocation model (PAM) that uses a DAG structure to represent and learn arbitrary-arity, nested and possibly sparse topic correlations, and [18] proposes a Latent Dirichlet-Tree Allcoation (LDTA) model that employs a Dirichlet-Tree prior to replace a single Dirichlet prior in LDA.

The second category pays attention to the relationships among words. The DF-LDA model [1] can incorporate the knowledge about words in the form of must-links and cannot-links using a novel Dirichlet Forest prior. Jagarlamudi et al. [9] proposes the Seeded-LDA model, allowing the user to specify some prior seed words in some topics. Chen et al. [6] proposes MC-LDA to deal with the knowledge of m-set (a set of words that should belong to the same topic) and c-set (a set of words that should not be in the same topic).

The third category focuses on the document level, to incorporate certain additional information in the topic modeling. Supervised LDA [4], DiscLDA [10], and Labeled LDA [17] try to predict the label values for input documents, based on labeled documents. TagLDA [19] extends latent Dirichlet allocation model by using a factored representation to combine the text information and tag information. Polylingual topic model [13] deals with polylingual document tuples, where each tuple is a set of documents loosely equivalent to each other, but written in different languages. It assumes that the documents in a tuple share the same tuple-specific distribution over topics, and each "topic" consists of a set of discrete word distributions, one for each language.

Our work falls into the third category, in that it makes an attempt to incorporate the information of document-part relationships into topic modeling. To the best of our knowledge, no previous work has attempted to incorporate the document-part structural information into the topic extraction problem. Our work is thus orthogonal to the previous work and complements them.

3   Multi-Part Topic Model

3.1   Generative Process

We now introduce the multi-part topic model (MPTM), an extension of latent Dirichlet allocation (LDA). Assume that there are D documents containing T topics expressed over W unique words, where each document contains P named-parts. Each document is represented as a set of P multinomial distributions over topics, where each part p of document d corresponds to one multinomial distribution over topics, denoted as $\psi_d^p = P(z|d,p)$. Each topic has multiple versions, and each versional topic is a multinomial distribution over words. For a given topic t, its versional topic corresponding to named-part p is denoted as $\varphi_t^p = P(w|t,p)$.

We first assume that *all parts within a document d should be similar in their topic distributions*, since they normally concern a common theme, and describe the theme from different aspects. In MPTM model, we enforce this assumption by requiring that all parts within a document d have their topic distributions drawn from a common prior Dirichlet distribution. The mean parameter θ_d of the Dirichlet distribution is exactly the mean of the Dirichlet distribution, which is specific to document d; while the concentration parameter s (also call precision parameter) controls how concentrated the Dirichlet distribution is around its mean θ_d, which is a hyperparameter in MPTM model.

Furthermore, we also assumed that *all versions of a topic should be similar in their word distributions*. It is enforced in MPTM model by requiring that all versions of a topic t have their word distributions drawn from a common prior Dirichlet distribution. The mean parameter ϕ_t of the common Dirichlet distribution is specific to topic t, while the concentration parameter c is also a hyperparameter that controls how concentrated the Dirichlet distribution is around its mean ϕ_t.

Table 1. Notations used

Notation	Meaning
D	the number of documents
T	the number of topics
W	the number of words in the vocabulary
P	the number of named-parts
d	a document
t	a topic
w	a word
p	a named part
ψ_d^p	the topic distribution of the part p in document d
φ_t^p	the word distribution of the version p for topic t
ϕ_t	the mean parameter of the prior Dirichlet distribution for the word distributions of versions of topic t
θ_d	the mean parameter of the prior Dirichlet distribution for the topic distributions of all parts in document d
c	the concentration hyperparameter of the prior Dirichlet distribution for the word distributions of versions of any topic
s	the concentration hyperparameter of the prior Dirichlet distribution for the topic distributions of all parts in any document

The values of s and c play an important role in our model. As we increase the value of s, all parts of a document have increasing concentration, which tends to generate similar topic distributions for those parts. As we increase the value of c, all versions of a topic have increasing concentration, which tends to get similar word distributions of those versions. When s and c go to infinity, the topic distributions of all the parts within a same topic will be constrained to be the same one, and the multi-part topic modeling method reduces to the classic topic modeling method applied on the documents. On the other hand, when s and c go to zero, there will be no constraints on the topic distributions, and the multi-part topic modeling method degenerates to the classic topic modeling methods applied on all the subdocuments where each subdocument is treated as an independent document.

The notations used in this paper are summarized in Table 1. The generative process for MPTM is given as follows:

1	For each topic $t \in \{1, \ldots, T\}$:
2	For each part $p \in \{1, \ldots, P\}$:
3	Draw $\varphi_t^p \sim Dirchlet(\phi_t, c)$
4	For each document $d \in \{1, \ldots, D\}$:
5	For each part $p \in \{1, \ldots, P\}$:
6	Draw $\psi_d^p \sim Dirichlet(\theta_d, s)$
7	For each word $w_{d,p,n}$ in part p of document d
8	Draw $z_{d,p,n} \sim Multinomial(\psi_d^p)$
9	Draw $w_{d,p,n} \sim Multinomial(\varphi_{z_{d,p,n}}^p)$

In MPTM model, the parameters include the mean vectors $\phi_t (1 \leq t \leq T)$ and the mean vectors $\theta_d (1 \leq d \leq D)$, which we treat for now as fixed quantities and are to be estimated. When the parameters are fixed, for each versional topic (t, p) (line 1), lines 2-3 draw a multinomial distribution over words φ_t^p (a versional topic) for each named part p. For each part p in each document d (lines 4-5), we first draw its multinomial distribution over topics ψ_d^p (line 6), and then generate all the words in part p of document d (lines 7-9) in the following way: for each word, a topic $z_{d,p,n}$ is randomly drawn from ψ_d^p, and then a word $w_{d,p,n}$ is chosen randomly from $\varphi_{z_{d,p,n}}^p$.

The plate notation for MPTM is given in Fig. 1. As we will see in Section 3, this model is quite powerful in improving the quality of discovered topics and boosting performance of information retrieval and document classification.

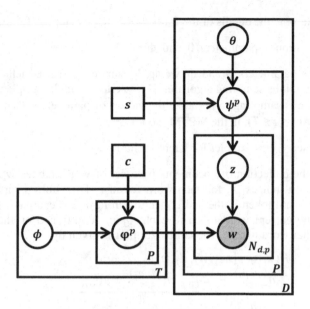

Fig. 1. Plate notation of MPTM model

3.2   Inference and Parameter Estimation

As we have described the motivation behind MPTM and its generative process, we now turn our attention to the detailed procedures for inference and parameter estimation under MPTM. In MPTM, the main parameters of interest to be estimated are the mean vectors θ_d ($1 \le d \le D$) and ϕ_t ($1 \le t \le T$) of the Dirichlet distributions. Other variables of interest include the word distributions φ_t^p of the multiple versions of a topic t, and the topic distribution ψ_d^p of the parts of a document d. Instead of directly estimating the variables φ_t^p and ψ_d^p, we estimate the posterior distribution over topics for the given observed words w, using Gibbs sampling, and then approximate φ_t^p and ψ_d^p using posterior estimates of topics for the observed words. Once φ_t^p and ψ_d^p are approximated, the parameters θ_d and ϕ_t can be estimated with a maximum likelihood procedure for Dirichlet distributions. The algorithmic skeleton for the parameter estimation in MPTM is briefly listed in Table 2.

Table 2. The framework of inference and parameter estimation for MPTM model

Step 1.	Initialize the parameters θ and ϕ
Step 2.	Sampling the hidden variables z with a collapsed Gibbs sampler
Step 3.	Update the parameters
Step 4.	Repeat the steps 2 and 3 for a fixed number of times

Next, we examine the details of the framework step by step, as follows.

Step 1. Initialization of parameters θ and ϕ

To initialize the parameters θ and ϕ, we apply standard latent Dirichlet allocation by using collapsed Gibbs sampling algorithm [7]. We use a single sample taken after 300 iterations of Gibbs sampling to initialize the values of parameters θ_d ($1 \leq d \leq D$) and parameters ϕ_t ($1 \leq t \leq T$), in the MPTM model.

Step 2. Collapsed Gibbs sampler for latent variables z

We represent the collection of documents by a set of word indices w_i, document indices d_i, and part indices p_i, for each word token i. The Gibbs sampling procedure considers each word token in the text collection in turn, and estimates the probability of assigning the current word token to each topic, conditioned on the topic assignments to all other word tokens. The Gibbs sampler is given by:

$$
\begin{aligned}
&P(z_i = t | \mathbf{z}_{-i}, w_i, d_i, p_i) \\
&\propto \frac{n_{-i,t}^{(w_i,p_i)} + c\phi_{tw_i}}{n_{-i,t}^{(\cdot,p_i)} + c} \cdot \frac{n_{-i,t}^{(d_i,p_i)} + s\theta_{d_it}}{n_{-i,\cdot}^{(d_i,p_i)} + s}
\end{aligned}
\tag{1}
$$

where the subscript "$-i$" means the exclusion of the current assignment of z_i, $n_{-i,t}^{(w_i,p_i)}$ denotes the number of times that word w_i from part p_i has been assigned to topic t, $n_{-i,t}^{(d_i,p_i)}$ denotes the number of times that a word from the part p_i of document d_i has been assigned to topic t, $n_{-i,t}^{(\cdot,p_i)} = \sum_w n_{-i,t}^{(w,p_i)}$ denotes the number of times that a word from all the part p_i has been assigned to topic t, and $n_{-i,\cdot}^{(d_i,p_i)} = \sum_j n_{-i,j}^{(d_i,p_i)}$ denotes the length of the part p_i of the document d_i.

To better understand the factors that affect topic assignments for a particular word, we can examine the two parts of Equation 1. The left part is the probability of word w_i under the part p_i version of topic t; whereas the right part is the probability that topic t has under the current topic distribution for part p_i of document d_i. Therefore, words are assigned to topics according to how likely the word in the part is for a topic, as well as how dominant a topic is in a part of a document. Clearly, the information of which part a word does occur plays an important role in determining its topic assignment.

The Gibbs sampling algorithm gives direct estimates of z for every word. Based on these estimates, the word distributions φ_t^p for part p version of topic t can be estimated from the count matrices as:

$$
\varphi_{tw}^p = \frac{n_t^{(w,p)} + c\phi_{tw}}{n_t^{(\cdot,p)} + c};
\tag{2}
$$

while topic distributions ψ_d^p for the part p of the document d can be estimated as:

$$
\psi_{dt}^p = \frac{n_t^{(d,p)} + s\theta_{dt}}{n_{\cdot}^{(d,p)} + s}.
\tag{3}
$$

Once the word distributions of all the versions for a topic and the topic distributions for all parts of a document are calculated in Equations (2) and (3), we can then update (or re-estimate) the mean parameters of the prior Dirichlet distributions in Step 3.

Step 3. How to re-estimate the parameters θ and ϕ?

Assume that a random vector, $\mathbf{p} = (p_1, \dots, p_K)$, whose elements sum to 1, follows from a Dirichlet distribution with mean vector parameter $\mathbf{m} = (m_1, \dots, m_K)$ that satisfying $\sum_k m_k = 1$ and concentration parameter s. The probability density at $\mathbf{p}$ is

$$p(\mathbf{p}) \sim \text{Dirichlet}(\mathbf{m}, s) = \frac{\Gamma(\sum_k sm_k)}{\prod_k \Gamma(sm_k)} \prod_k (p_k)^{sm_k - 1} \tag{4}$$

where the concentration parameter s, also referred to as the precision of the Dirichlet, controls how concentrated the distribution is around it mean.

In the context of MPTM model, we want to fix the concentration parameter s and only optimize the mean parameter $\mathbf{m}$ in the maximum-likelihood objective from the observed random vectors $\{\mathbf{p}_1, \dots, \mathbf{p}_N\}$. To perform this problem, we adopt the fixed-point iteration technique to compute the maximum likelihood solution [15], by iterating the following two steps until convergence:

$$\Psi(\alpha_k) = \log \bar{p}_k - \sum_j m_j^{old} \left(\log \bar{p}_k - \Psi(sm_j^{old}) \right) \tag{5}$$

and

$$m_k^{new} = \frac{\alpha_k}{\sum_j \alpha_j} \tag{6}$$

where $\log \bar{p}_k = \frac{1}{N} \sum_i \log p_{ik}$, and $\Psi(x) = \frac{d \log \Gamma(x)}{dx}$ is known as the digamma function.

The problem of finding maximum likelihood solution for mean parameter of Dirichlet distribution (with fixed concentration parameter) exists in two places of MPTM model:

For each part p of a document d, its topic distribution ψ_d^p follows from a prior Dirichlet distribution with mean parameter θ_d and concentration parameter s, given algebraically as:

$$\psi_d^p \sim \text{Dirichlet}(\theta_d, s) = \frac{\Gamma(\sum_t s\theta_{dt})}{\prod_t \Gamma(s\theta_{dt})} \prod_t (\psi_{dt}^p)^{s\theta_{dt} - 1} \tag{7}$$

For versional topic with respect to part p for a topic t, its word distribution φ_t^p follows from a prior Dirichlet distribution with mean parameter ϕ_t and concentration parameter c, given as:

$$\varphi_t^p \sim \text{Dirichlet}(\phi_t, c) = \frac{\Gamma(\sum_w c\phi_{tw})}{\prod_w \Gamma(c\phi_{tw})} \prod_w (\varphi_{tw}^p)^{c\phi_{tw} - 1} \tag{8}$$

The above fixed-point iteration technique is used in the MPTM model to estimate θ_d ($1 \leq d \leq D$) and ϕ_t ($1 \leq t \leq T$), respectively.

Step 4. Repeat the steps 2 and 3 a fixed number of times

The final step is simply to repeat the steps 2 and 3 for a fixed number of times and output the word distributions of all versional topics and the topic distributions of all parts of documents.

4   Core Words and Topic Quality

If a word appears in the top-M word lists of (almost) all versions of a topic, it is called a core word of the topic; otherwise, it is called an attached word. Thus, each version of a topic can be represented as a core-attachment structure, where the attachment represents the part-specific words.

The "core words" embodies the meaning of the topic throughout the text, they can help us understand the name of the topic clearly. While the "part-specific" attached words complement the details of the topic from different aspects, different parts may have different emphasis.

Table 3. Two exemplar core-attachment structures

Core words:	featur word relat label topic translat learn model data method
Abstract	semant paper approach propos text perform task improv languag base extract set
Introduction	approach semant task languag text sentenc work extract system tag document
Method	set term sentenc train document context exampl text tag entiti select
Experiments	tabl set train perform system evalu test baselin term select base
Summary	work approach improv system perform select achiev propos better support

Core words:	network predict social system item rate tag user recommend model method matrix
Abstract	propos effect realworld interest work novel develop review provid
Introduction	product trust work base propos person interest opinion
Method	function time set vector denot product number base group algorithm
Experiments	set data perform dataset review figur number evalu random experi paramet
Summary	represent work base propos reput trust evalu data

Let us take the IJCAI corpus as an example, where each topic has 5 versions (please refer to section 5 for the details of the IJCAI dataset). We set M=20, and define a word to be a core word for a topic if it occurs in the top-20 word lists of at least 4 versions of the topic. Table 3 illustrates the core-attachment structures of two exemplar topics in the MPTM model of IJCAI dataset.

The first example is a topic about "topic model". The core words include "topic", "model", "word", "feature", "label", etc., which well reflect the common characteristics of the topic. The words "term", "sentence", "document" appear as the attached words to the "Method" part, reflecting the technical details of the topic. The words "performance", "evaluation", "baseline", "train", and "set" are listed as the attached words to "Experiments" part. Similar analysis also applies to the second example, which is omitted here.

After analyzing the word distributions of versional topics, let us examine the topic distributions of document parts. A topic is thought of as a stable and consistent topic, if it exhibits consistent probabilities across the multi-parts of documents; otherwise, it is unstable or transient. Here, we measure the quality of a topic simply as the mean variance across the multi-parts averaged over all documents:

$$mVar(t) = \frac{1}{D} \sum_{d=1}^{D} \sum_{p=1}^{P} \left(\psi_{dt}^{p} - \frac{1}{P} \sum_{i=1}^{P} \psi_{dt}^{i} \right)^2 \tag{9}$$

In the experiment with information retrieval, it will be shown that the pruning of topics with highest mean variance can further improve the performance of MPTM.

5   Experimental Results

In this section, we evaluate the proposed MPTM model against several baseline models on two real datasets.

5.1   Data Sets

Two datasets (IJCAI and NIPS) are used in the experiments. The first dataset IJCAI is constructed by ourselves, using papers from the most recent three Proceedings of International Joint Conference on Artificial Intelligence (IJCAI) in years 2009, 2011, and 2013, because the IJCAI conferences in these three years share (almost) the same track organization, and the information of the assignments of papers to tracks can serve as external criterion for measuring the performance in information retrieval and document classification. We extracted 669 papers from 6 common tracks in total, with detailed information listed in Table 4.

The NIPS corpus contains 1740 papers published in the Proceedings of Neural Information Processing Systems (NIPS) Conferences[1] from year 1988 to year 2000.

---

[1]   The dataset is available at the NIPS Online Repository. http://nips.djvuzone. org/txt.html.

Table 4. The IJCAI Corpus

Track Name	# papers
Agent-based and Multiagent Systems	165
Constraints, Satisfiability, and Search	107
Knowledge Representation, Reasoning, and Logic	181
Natural-Language Processing	74
Planning and Scheduling	74
Web and Knowledge-based Information Systems	68
Sum	**669**

The IJCAI and NIPS papers have been preprocessed to remove the *"References"* part, to remove stop words, and to do word stemming. Each IJCAI paper is split into parts of *"Abstract"*, *"Introduction"*, *"Method"*, *"Experiments"*, and *"Summary"*; while each NIPS paper is simply split into 3 parts of equal length, called *"Head Part"*, *"Middle Part"*, and *"Tail Part"* respectively. After preprocessing, the IJCAI corpus contains 1,437,916 words with vocabulary size as 32,752, and the NIPS corpus contains 2,014,937 words with vocabulary of size 15,965.

Throughout the experiments, the MPTM models were trained using 1500 Gibbs iterations where the parameters get updated for every 300 iterations. That is, the step 2) in the algorithmic framework executes 300 iterations of collapsed Gibbs sampling, while the outer loop of steps 2) and 3) is repeated 5 times.

5.2   Topic Coherence

As indicated in [5][16][11], the perplexity measure does not reflect the semantic coherence of individual topics and can be contrary to human judges. The topic coherence measure [14] was proposed as a better alternative for assessing topic quality, which only relies upon word co-occurrence statistics within the documents, and does not depend on external resources or human labeling.   Given a topic t, if $V^{(t)} = (v_1^{(t)}, \ldots, v_1^{(M)})$ is its top-M word list, the topic coherence is defined as:

$$Coherence(t; V^{(t)}) = \sum_{m=2}^{M} \sum_{l=1}^{m-1} \log \frac{D(v_m^{(t)}, v_l^{(t)}) + 1}{D(v_l^{(t)})}$$

where $D(v)$ denotes the document frequency of word v and $D(v, v')$ denotes the number of documents containing both words v and v' . For a topic in LDA model, its top-M words are the M most probable words in the topic. In our MPTM model, because each topic has multiple versions, we define its top-M words in an intuitive manner as follows.

For a given topic t, let $\tau(\varphi_t^p, w)$ denote the position or rank of word w in the word distribution of versional topic φ_t^p . We use $cnt(t, w) = |\{p: 1 \leq p \leq P, \tau(\varphi_t^p, w) \leq M\}|$ to denote the number of versions of topic t that w occurs in its top M words, and use $sr(t, w) = \sum_p \tau(\varphi_t^p, w)$ to denote the sum of the ranks of word w in all versions of topic t. A word w is ranked before another word v with respect to a topic t, if it satisfies one of the following two conditions:

(1) $cnt(t, w) > cnt(t, v)$
(2) $cnt(t, w) = cnt(t, v)$ and $sr(t, w) < sr(t, v)$.

Accordingly, for each topic, the top-M ranked words can be calculated.

Table 5. Average Topic Coherence scores across different numbers of topics

Data Set	# Topics	LDA	MPTM	Improved Percentage
IJCAI	20	−140.7	−123.0	12.6%
	50	−210.5	−187.3	11.0%
	100	−256.6	−225.4	12.2%
NIPS	20	−154.1	−146.4	5.0%
	50	−181.6	−167.5	7.8%
	100	−210.9	−185.6	12.0%

Table 5 shows the topic coherence averaged over all topics. It can be seen that the topic coherences of MPTM models are significantly higher than those of LDA models, indicating higher quality of topics with MPTM. The improvement percentage on NIPS is less significant than IJCAI, which may be caused by the fact that the documents in NIPS are split into three parts of equal length, and it does not reflect the exact document-part relationships.

5.3   Information Retrieval

For information retrieval applications, the task is to retrieve the most relevant documents to a query document. Here we make use of the cosine similarity to measure the relevance between two documents. Mean Average Precision (MAP), for its especially good discrimination and stability, is adopted as the measure of quality to evaluate the performance of MPTM model in information retrieval.

If the set of relevant documents for a query document q_j is $\{d_1, ..., d_{m_j}\}$, and R_{jk} is the set of ranked retrieval results from the top result until you get to document d_k, then

$$\text{MAP(Q)} = \frac{1}{|Q|}\sum_{j=1}^{|Q|}\frac{1}{m_j}\sum_{k=1}^{m_j} Precision(R_{jk}).$$

To check the effects of different configurations of topic number K, parameters s, and parameter c, we have tested the MPTM model on a grid of configurations with $K \in \{5, 10, 20, 30, 50\}$, $s \in \{50, 100, 200, 400\}$, and $c \in \{50, 100, 200, 400\}$. In all the configurations, our model has consistently outperformed the LDA model. However, for different K values, the configuration at which our model obtained the best performance may vary. Without fine-tuning the parameters, we just report the MAP values with the configuration of $s = 200$ and $c = 100$, in Figure 2. Here, five-fold cross validation is conducted, where for each fold, 80% of the documents are used as the training data, and the other 20% are held-out as the query data.

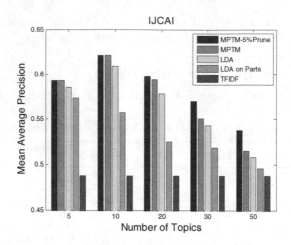

Fig. 2. Average MAP scores across different numbers of topics

In Figure 2, TFIDF method is to represent the documents using a vocabulary of 8000 words with the highest TF-IDF values; LDA on Parts method builds a LDA model by treating each part as an independent document, and then concatenate the topic distributions of all the parts of a document into a $P \times T$-dimensional representation of the document; and MPTM-5%Prune has pruned the 5% topics with highest mean variance for MPTM model.

We can observe that MPTM has achieved higher MAP values than the baseline models, and *MPTM-5%Prune* can further boost the performance of MPTM, with the aid of quality measures of topics.

5.4   Document Classification

The existence of track information associated with each document in the IJCAI corpus has also made it possible to classify a new document into the six tracks. On IJCAI corpus, five-fold cross validation is conducted as follows. At each fold, 80% of the documents are used as the training data, and the other 20% are held-out as the test data. On the training data, we train a MPTM model, with which each training or test document d can be transformed into a vector of length $T \times P$ by concatenating all the $\psi_d^p : 1 \le p \le P$. We then train a support vector machine (SVM) on the $(T \times P)$-dimensional representations of training documents provided by MPTM, and use it to classify the test documents. This SVM is compared with an SVM trained on the features provided by LDA. Both SVMs are trained with the libSVM software [5] and get optimized by a grid search with parameter ranges of $10^{-2} \le C, \gamma \le 10^4$. The mean accuracy averaged over five folds is reported in Figure 3.

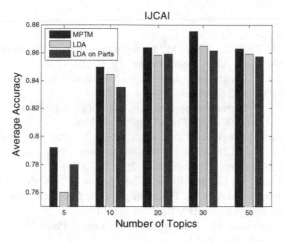

Fig. 3. Average accuracies across different numbers of topics

It can be seen from the results that the accuracy is improved in all cases, which suggests that the features provided by MPTM may be more informative in the task of document classification.

6   Conclusions

This paper proposed a novel method to exploit the multi-part composition information of documents for producing better-quality topics. To the best of our knowledge, this has not been done before. To model the multi-part documents, a novel topic model called MPTM is proposed by taking two assumptions such that *all parts within a document d should be similar in their topic distributions* and *all versions of a topic should be similar in their word distributions*. It has been manifested empirically by two datasets that MPTM has successfully produced topics of high quality, and outperformed the baseline methods in information retrieval and document classification tasks.

Finally, it is possible to remove the existence of multiple versions of topics from the MPTM model, in order to widen its applicability. For example, for a corpus where documents are labeled, it is expected to make sense to assume that the topic distributions of the documents with the same class label be drawn from a common Dirichlet prior, that is to say, to assume that the documents with same class label have similar topic distributions. Such a model can make use of the supervised information in topic modeling and may make contribution in solving the document classification task.

Acknowledgements. This work is supported by National High-tech R&D Program of China (863 Program) (No. SS2015AA011809), Science and Technology Commission of Shanghai Municipality (No. 14511106802), and National Natural Science Fundation of China (No. 61170007). We are grateful to the anonymous reviewers for their valuable comments.

References

1. Andrzejewski, D., Zhu, X., Craven, M.: Incorporating domain knowledge into topic modeling via Dirichlet Forest priors. In: ICML, pp. 25–32 (2009)
2. Blei, D.M., Ng, A.Y., Jordan, M.I.: Latent Dirichlet allocation. Journal of Machine Learning Research **3**, 993–1022 (2003)
3. Blei, D., Lafferty, J.: Correlated topic models. Advances in neural information processing systems **18**, 147–154 (2006). MIT Press, Cambridge, MA
4. Blei, D., McAuliffe, J.: Supervised topic models. (2010). arXiv preprint arXiv:1003.0783
5. Chang, C.C., Lin, C.J.: LIBSVM: a library for support vector machines. ACM Transactions on Intelligent Systems and Technology **2**(3), 27 (2011)
6. Chen, Z., Mukherjee, A., Liu, B., Hsu, M., Castellanos, M., Ghosh, R.: Exploiting domain knowledge in aspect extraction. In: Proceedings of the Conference on Empirical Methods in Natural Language Processing (EMNLP 2013) (2013)
7. Griffiths, T.L., Steyvers, M.: Finding scientific topics. Proc. Natl. Acad. Sci. U.S.A. **101**(Suppl 1), 5228–5235 (2004)
8. Hofmann, T.: Probabilistic latent semantic indexing. In: Proceedings of the 22nd annual international ACM SIGIR conference on Research and development in information retrieval, pp. 50–57 (1999)
9. Jagarlamudi, J., Daumé III, H., and Udupa, R.: Incorporating lexical priors into topic models. In: Proceedings of the 13th Conference of the European Chapter of the Association for Computational Linguistics, pages 204–213 (2012)
10. Lacoste-Julien, S., Sha, F., and Jordan, M.: DiscLDA: discriminative learning for dimensionality reduction and classification. In: Advances in Neural Information Processing Systems, pp. 89–904 (2008)
11. Lau, J.H., Baldwin, T., Newman, D.: On collocations and topic models. ACM Transactions on Speech and Language Processing (TSLP) **10**(3), 10 (2013)
12. Li, W., McCallum, A.: Pachinko allocation: DAG-structured mixture models of topic correlations. In: Proceedings of the 23rd International Conference on Machine Learning, pp. 577–584 (2006)
13. Mimno, D., Wallach, H.M., Naradowsky, J., Smith, D.A., McCallum, A.: Polylingual topic models. In: Proceedings of the 2009 Conference on Empirical Methods in Natural Language Processing, pp. 880–889 (2009)
14. Mimno, D., Wallach, H.M., Talley, E., Leenders, M., McCallum, A.: Optimizing semantic coherence in topic models. In: Proceedings of the Conference on Empirical Methods in Natural Language Processing, pp. 262–272 (2011)
15. Minka, T.: Estimating a Dirichlet distribution. Technical Report (2012). http://research.microsoft.com/en-us/um/people/minka/papers/dirichlet/minka-dirichlet.pdf
16. Newman, D., Lau, J.H., Grieser, K., Baldwin, T.: Automatic evaluation of topic coherence. In: Human Language Technologies: Proceedings of the 2010 Annual Conference of the North American Chapter of the Association for Computational Linguistics, pp. 100–108 (2010)
17. Ramage, D., Hall, D., Nallapati, R., Manning, C.D.: Labeled LDA: a supervised topic model for credit attribution in multi-labeled corpora. In: Proceedings of the 2009 Conference on Empirical Methods in Natural Language Processing, vol. 1, pp. 248–256 (2009)
18. Tam, Y.-C., Schultz, T.: Correlated latent semantic model for unsupervised LM adaptation. In: IEEE International Conference on Acoustics, Speech and Signal Processing, pp. 41–44 (2007)
19. Zhu, X., Blei, D., Lafferty, J.: TagLDA: bringing document structure knowledge into topic models. Technical Report TR-1553, University of Wisconsin (2006)

Retaining Rough Diamonds: Towards a Fairer Elimination of Low-Skilled Workers

Kinda El Maarry[✉] and Wolf-Tilo Balke

Institut für Informationssysteme, TU Braunschweig, Braunschweig, Germany
{elmaarry,balke}@ifis.cs.tu-bs.de

Abstract. Living the economic dream of globalization in the form of a location- and time-independent world-wide employment market, today crowd sourcing companies offer affordable digital solutions to business problems. At the same time, highly accessible economic opportunities are offered to workers, who often live in low or middle income countries. Thus, crowd sourcing can be understood as a flexible social solution that indiscriminately reaches out to poor, yet diligent workers: a win-win situation for employers and crowd workers. On the other hand, its virtual nature opens doors to unethical exploitation by fraudulent workers, compromising in turn the overall quality of the gained results and increasing the costs of continuous result quality assurance, e.g. by gold questions or majority votes. The central question discussed in this paper is how to distinguish between basically honest workers, who might just be lacking educational skills, and plainly unethical workers. We show how current quality control measures misjudge and subsequently discriminate against honest workers with lower skill levels. In contrast, our techniques use statistical models that computes the level of a worker's skill and a task's difficulty to clearly distinguish each worker's success zone and detect irrational response patterns, which usually imply fraud. Our evaluation shows that about 50% of misjudged workers can be successfully detected as honest, can be retained, and subsequently redirected to easier tasks.

Keywords: Crowd sourcing · Impact sourcing · Fraud detection

1   Introduction

"It's been a dream, me having my own place, paying my own rent, buying my own food. Being independent," says Martha, a Samasource Kenyan worker[1]; one of the many faces behind the international taskforce ready to work through crowd sourcing platforms. The social model of *Impact Sourcing*, first implemented by the social enterprise Digital Divide Data (DDD)[2] back in 2001 has been adopted by many

---

[1] http://www.samasource.org/impact/
[2] http://www.digitaldividedata.com/about/

© Springer International Publishing Switzerland 2015
M. Renz et al. (Eds.): DASFAA 2015, Part II, LNCS 9050, pp. 169–185, 2015.
DOI: 10.1007/978-3-319-18123-3_11

companies and crowd sourcing platforms like Samasource[3], RuralShores[4], etc. The new Impact Sourcing industry aims at hiring people at the bottom of the income pyramid to perform small, yet useful cognitive and intelligent tasks via digital interfaces, which ultimately promises to boost the general economic development [1]. However, the mostly anonymous, highly distributed and virtual nature of the short-term work contracts also carry the danger of being exploited by fraudulent workers: By providing incorrect (usually simply random) answers, they compromise the overall result quality and thus not only directly hurt the respective task provider, but in the long run also all honest workers in dire need of employment.

Anecdotic evidence can be drawn from our own research work on crowd sourcing as reported in [2]: By completely excluding workers from just two offending countries, where the number of clearly fraudulent workers seemed to be much higher than average, the overall result correctness in our experiments instantly increased by about 20%. In particular, correctness for a simple genre classification task for movies increased from 59% to 79% using majority vote for quality assurance. Of course, this crude heuristics was bound to exclude many honest workers too.

Although typical quality control measures for crowd sourcing like gold questions or majority votes promise to mitigate this problem, they also face serious problems: firstly, they are only applicable for factual tasks, which hampers creative task design and thus overall benefit. Secondly, they incur additional costs for the task provider and thus reduce the readiness to crowd source tasks, and finally (and probably worst for impact sourcing) they exclude honest and willing workers that may not have been provided tasks on their individual skill levels. Indeed, the bottom of the income pyramid encompasses a heterogeneous set of workers, who according to their skill level provide responses that are: either *sufficiently good* by non-expert, diligent workers with higher skill levels, or *mixed* by honest workers with lower skill levels, as well as by unethical workers who exploit the system for financial gains. Yet, according to Samasource 92% of its taskforce are unemployed or underemployed, i.e. for the crowd, every incoming living wage counts and contributes to a better standard of living.

On the other end of the spectrum, crowd sourcing emerges as an unparalleled solution [3] to companies with intelligent digital tasks, to name but a few: text translation, image tagging, text sentiment analysis. Generally speaking, through the collective intelligence of the diligent workers, high quality responses could be attained. Naively, such high quality can be assured by manually cutting out the unethical workers through submitted response checks. However, this instantly invalidates the core gains attained through crowd sourcing, and becomes both costly and time consuming. Consequently, unethical workers are further encouraged to submit low quality results, and it becomes a question of automatically detecting such workers for an improved overall quality [4].

As argued above, some common practices are 1) injecting a set of questions whose answers are already known, so-called gold questions, within each Human Intelligent Task (HIT), 2) Employing Majority vote to filter out workers who often fail to agree

---

[3] http://www.samasource.org/
[4] http://ruralshores.com/about.html

with the majority, or 3) adopting a reputation based system, where workers' history is recorded, and a reputation score is assigned to each worker by combining e.g., the requestor's satisfaction level, the ratio of completed to aborted HITS, etc. Unfortunately, such practices can heavily misjudge honest, yet less skilled workers.

Example 1 *(Collecting motion picture ratings):*
Given a dataset of movies, the crowd is asked to classify each movie as either PG, or PG-13. Assume a skewed distribution where 90% of the movies are PG, and only 10% are PG-13. Assume worker A simply tags all movies as PG, ultimately he/she will only have a 10% error rate. On the other hand, consider worker B who's actually checking each movie. Worker B can easily exhibit similar or even higher error rates, because he/she perceives some movies according to his/her standards as PG-13. Although worker A is obviously a spammer, in a reputation based system he/she would be given a higher reputation score than worker B, since more gold questions were answered correctly.

Our results in initial experiments indeed indicate that even with datasets where answer possibilities are evenly distributed, unethical workers can still through random guessing surpass honest workers with lower skill levels. Based on these insights, in this paper we design a method that abstracts each worker's gold questions' responses to a skill-graded response vector, and then zooms in on their individual success zone for quality control. The success zones are individually defined by each worker's skill level and bounded by questions' whose difficulty levels are well within the worker's skill level. We can compute both parameters through psychometric *item response theory* (IRT) models: in particular, the Rasch model [5]. The underlying assumption is that honest workers should exhibit a clear tendency to correctly answer tasks within their difficulty levels. Moreover, failing to answer easy tasks, yet correctly answering more difficult ones would indicate fraud. We develop three techniques that are designed to detect irrational response patterns in success zones. The contributions of this paper can be summarized as follows:

- We show how gold questions and reputation-based systems can be *bypassed by unethical workers*, using real-world dataset in a laboratory-based study.
- We present a framework for distinguishing between unethical and honest workers with lower skill levels in *a fair, yet reliable* manner.
- We extensively test our framework in practical crowd sourcing experiments and demonstrate how honest workers with lower skills levels can be indeed detected and redirected to skill-fitted tasks.

The rest of the paper is structured as follows: In section 2, we give an overview of related work. In section 3, we motivate our problem with a case study, then start describing our framework in section 4, by presenting the underlying statistical Rasch model and illustrating how it can be used to identify workers' success zone. In section 5, we introduce three techniques that aim at recognizing irrational patterns in success zones. This is backed up by a laboratory-based experiment that offers ground-truth and a real-world crowd sourcing experiment in the evaluation section. Finally, the last section gives a summary and an overview of future work.

2   Related Work

Crowdsourcing provides both a *social* chance that indiscriminately reaches out to poor, yet diligent workers, as well as an affordable digital solution for companies with intelligent digital business problems like e.g., web resource tagging [6], completing missing data [7], sentiment analysis [8], text translation [9], information extraction [10], etc. But as with every chance, the challenge of acquiring high quality results, which is compromised by unethical workers, must be overcome. In this section, we give an overview of the current crowd sourcing quality control measures, as well as a brief overview of the Rasch Model and its related work in crowd sourcing.

A rich body of research has examined many different techniques to mitigate the quality problem in crowdsourcing. Aggregation methods aim at improving the overall quality through redundancy and repeated labeling. Through assigning several workers to the same task, an aggregation of their responses help identify the correct response. Such aggregation methods include: basic Majority decision, which has been shown to have severe limitations [11]. This was further developed by Dawid and Skene [12] to take the response's quality based on the workers into consideration, through applying an expectation maximization algorithm. Other approaches that considered such error rates relied on: Bayesian version of the expectation maximization algorithm approach [13], or a probabilistic approach that takes into account both the worker's skill and the difficulty of the task at hand [14]. A further step was taken in [15] with an algorithm separating the unrecoverable error rates from recoverable bias. Manipulating monetary incentives have also been investigated, yet proves tricky to implement, where low paid jobs yield sloppy work, and high paid jobs attract unethical workers [16].

Other techniques focus on trying to eliminate unethical workers on longer time scales like constantly measuring performance with injected gold questions or employing the workforce via reputation-based systems. Even when the procurement of gold questions is feasible and not too expensive, the question of how many gold questions to include immediately materializes [17]. On the other hand reliably computing the workers' reputation poses a real challenge, and many reputation approaches have been investigated whether it's based on a reputation model [18-19], on feedback and overall satisfaction [19], or on deterministic approaches [20], etc.

Our work is tightly related to the IRT paradigm [21] in psychometrics, which enables us to focus on the workers' capabilities. More specifically, we employ the Rasch models [5], which computes the expected response correctness probability of a worker to a given task, the task's difficulty, and the ability of the worker. This helps us address a principal concern of Impact sourcing: distinguishing honest workers with low skill levels from unethical workers. So far, most research have focused on one or two of those aspects, with the exception to the work of Whitehill in [14], where they presented GLAD – a generative model of labels, abilities and difficulties. The presented model is close to our work as it's also based on IRT. They obtain the maximum likelihood estimates of these three parameters through utilizing an Expectation-Maximization approach (EM) in an iterative manner for results aggregation. GLAD's robustness wavers when faced with unethical workers, especially when they constitute more than 30% of the task force [22]. Our perspective is however focused on detecting irrational response patterns to be able to distinguish diligent workers with lower skill levels.

Other works, considered only one aspect, namely the worker's ability. Dawid and Skene [12] utilized confusion matrices, which offered an improved form of the redundancy technique. However, as pointed out and addressed by Ipeirotis [23], this underestimates the quality of workers who consistently give incorrect results. In contrast, in [24], the workers' ability together with the inference of correct answers are investigated. The downside however, is that it overlooks the varying difficulties of the task at hand, which should influence the workers' abilities. In our model, the correctness probability of a worker for a given task isn't static, but varies according to the task's difficulty and the worker's ability. Furthermore, it's measured across the different tasks' difficulty level.

3   Motivational Crowd Sourcing Laboratory-Based Study

To acquire a basic ground truth dataset, we conducted a small-scale laboratory-based experiment, where a total of 18 workers volunteered for the study. Given a set of 20 multiple choice questions, the volunteers were asked to answer the questions twice. In the first round, they would randomly select answers in any fashion. In the second round, they were asked to truthfully consider the questions before answering.

In this paper we formulate our HITs over an American standardized test for college admission with medium difficulty level: the Graduate Record Examination (GRE) dataset, which was crawled from graduateshotline.com. A GRE test is typically made up of five sections. We extracted the questions corresponding to the verbal section, namely the verbal practice questions. The task then is to select the right definition of a given word. For each question, four definitions are given, and the worker should select the correct definition corresponding to the word in question.

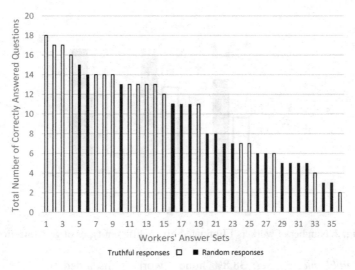

Fig. 1. Truthful versus Random Responses

Figure 1 sorts all workers' answer sets according to the respective total number of correct answers achieved over the 20 questions. Although no worker got all 20 answers right, it comes as no surprise that truthful answers tend to be more correct than random answers: in the random response round, the workers had on average 40% correct answers, while in the truthful response round, the workers had on average 58.6% correct answers. Furthermore, one can clearly see that even though the dataset is in no way biased, random responses at times produced better overall results than workers who actually tried to truthfully answer the questions.

3.1   Unethical Workers Versus Reputation-Based Systems

Consider the top ten workers getting the most correct answers in Figure 1. In a reputation based system, the worker at rank 5 (scoring 15 correct answers) would be given a higher reputation score than workers on ranks 6 to 9, who scored only 14 correct answers. Yet, here three workers at least tried to answer correctly. In biased datasets, (e.g. the movie dataset from example 1), experienced unethical workers' could easily improve their reputation score even more by choosing proper answering schemes.

3.2   Unethical Workers Versus Gold Questions

Upon setting up 40% gold questions (i.e. 8 gold questions out of the 20 questions) and a 70% accuracy level threshold (i.e. workers scoring less than 70% correct answers in the gold questions are eliminated), we considered the following three gold question set scenarios as depicted in Figure 2:

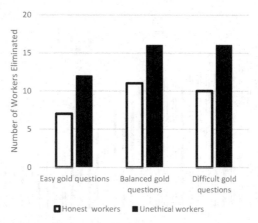

Fig. 2. Number of workers Eliminated given different types of gold questions

1. *Easy gold question set:* 38.8% honest workers discarded (i.e. 7 workers) and 66.67% unethical workers eliminated (i.e. 12 workers).

2. *Balanced gold question set, addressing all difficulty levels*: 61% honest workers discarded (i.e. 11 workers are eliminated) and 88% random workers eliminated (i.e. 16 workers are eliminated).
3. *Difficult gold question set*: 55.5% honest workers discarded[5] (i.e. 10 workers) and 88.8% unethical workers eliminated (i.e. 16 workers).

Although gold questions are more biased to penalize unethical workers, still this bias is relatively low, and a significant number of honest workers are penalized, too. Moreover, unethical workers can with a higher chance bypass an easy set of gold questions (33.33% unethical workers kept) than they are to bypass either a balanced or a difficult set of gold questions (around 12% unethical workers kept).

[5]Typically, it's to be expected that the number of honest workers eliminated by the difficult set of gold questions would be higher than that with the balanced gold, as it would impose a higher skill threshold on the employed workers. However, due to the relatively small number of volunteers, the 6% difference which is in fact only 1 worker more that was eliminated by the balanced gold question can be misleading and should be considered as an artifact of the experiment and not generalized. For the rest of our experiments, we use a balanced set of gold questions.

4   Identifying Success Zone Bounds

In this section we provide a short overview of the underlying statistical *Rasch Model* (RM), and illustrate how we employ it to 1) abstract workers' responses to a skill-graded vector, and 2) identify each worker's success zone for quality control.

4.1   The Rasch Model

Crowdsourcing inherently involves many human factors, accordingly our attention is drawn to the science assessing individual's capabilities, aptitudes and intelligence, namely, psychometrics and its IRT classes, in particular the RM. Simply put, the RM computes the probability P_{ij} that a worker's $w_i \in W$ response to a given task $t_j \in T$ is correct as a function of his ability θ_{w_i} and the difficulty of the task β_{t_j}. Assuming a binary setting, where a worker's response $x_{ij} \in [0,1]$ is known (with 0 meaning an incorrect and 1 a correct response), RM's dichotomous case can be employed. Basically, both the RM's parameters: a worker's ability θ and a task's difficulty β are depicted as latent variables, whose difference yields the correctness probability P.

Definition 1: (Rasch Model for Dichotomous Items) given a set of workers $W= \{w_1, w_2, ..., w_n\}$, where $|W| = n$, and a HIT $T= \{t_1, t_2, ..., t_m\}$, where $|T| = m$. Assume $w_i \in W$ and $t_j \in T$, then correctness Probability P_{ij} can be given as follows

$$P_{ij} = P_{ij}(x_{ij} = 1) = \frac{exp(\theta_{w_i} - \beta_{t_j})}{1 + exp(\theta_{w_i} - \beta_{t_j})}$$

This can also be reformulated, where the distance between θ_{w_i} and β_{t_j} is given by the logarithm of the odds ratio, also known as the log odd unit *logit*.

$$log\left(\frac{P_{ij}}{1 - P_{ij}}\right) = \theta_{w_i} - \beta_{t_j}$$

Such a logit scale enforces a consistent valued unit interval that is meaningful between the locations of the workers and the questions when they're drawn on a map [25]. So whereas a worker w_i has a 50% chance of answering a question within his/her exact ability, this success probability increases to 75% for questions that is 1 logit easier and similarly drops to 25% for questions that is 1 logit more difficult. The difficulty of a question with a logit value of 0 is average. A negative logit value implies an easy β_{t_j} and a low θ_{w_i} and vice versa. Accordingly, the correctness probability of a worker's response will be high when his/her ability exceeds the corresponding task's difficulty. We exploit this to detect irrational response patterns in section 5.3.

Objectively Estimating θ and β Parameters.
Naively, β can be determined by observing the proportion of incorrect responses, while θ_{w_i} could be defined by where a worker w_i stands in the percentile among all the other workers. In fact, the practice of using the raw score (i.e. total number of correct responses) to estimate the worker's ability is quite ubiquitous [25]. However, this fails to capture the reliability of responses and raises many questions as can be seen in the following example:

Example 2:
a) Assume that the same HIT is assigned to two different groups of workers. The majority of the first group's workers are unethical, while the majority of the second is honest. A non-objective measure of β would label questions assigned to the first group as difficult, because the majority gave incorrect responses. Simultaneously, the same questions would have a lower β for the second honest group of workers.

b) Assume two HITs with different β, where the first has a low difficulty level and the second has a high difficulty level. An unobjective measure of θ would equally treat the workers lying in the same percentile of each HIT, irrespective of the difficulty of the questions they're handling.

Accordingly, objectivity is needed. Simply put 1) θ measurements should be independent of T, and 2) β measurements should be independent W. In RM, the "objective" measurement of (θ, β) is emphasized [26].

Numerous ways for estimating the Rasch model's parameters (θ, β) exist. We employ the conditional maximum likelihood estimation (CML) for estimating β, as it leads to minimal estimation bias with well-defined standard error. For estimating the θ, join maximum likelihood estimation (JML) is used, which proved to be robust against missing data, as well as an efficient estimation method [27].

These parameters can only be estimated on workers' responses to gold questions, where all responses can be judged to be correct or not. In the evaluation section, we illustrate that 40% gold question suffice for the RM to correctly approximate each of the (θ, β) parameters (i.e. for a HIT of 20 questions, 8 gold questions are needed).

4.2   Skill-Graded Vectors and Success Zones

In a perfect setting, a worker who's able to correctly answer questions of a certain difficulty level, should be able to answer all less difficult questions. Similarly, failing to submit a correct response to a question of a certain difficulty would imply the worker's incompetency in answering more difficult questions. Different workers will have different skill levels and a fair judgment of a worker should be based on how well they do on tasks within their skill level. To that end, we reorder each worker's gold question response vectors, such that the questions are ordered ascendingly in terms of their difficulty using the RM's estimated β parameter. At which point, in an ideal setting, we can observe the following workers' response vectors as illustrated in.Table 3 Next, we zoom in on the set of questions within the worker's ability (i.e., success zone). Using the RM's estimated θ parameter, we can define the bound of the success zone.

Table 1. (Ideal) Skill-graded Response Matrix

Workers \ Tasks	1	2	3	4	5	6	7	8	9
A	1	1	1	0	0	0	0	0	0
B	1	1	1	1	1	0	0	0	0
C	1	1	1	1	1	1	1	0	0

Definition 2: (Skill-graded Vector - SV) for w_i assigned to a HIT with golden questions $T= \{t_1, t_2, ..., t_m\}$, $|T| = m$, with corresponding $\beta = \{\beta_{t_1}, \beta_{t_2}, ..., \beta_{t_m}\}$. Worker w_i submits the following corresponding response vector $RV_{w_i} = \{x_1, x_2, ..., x_m\}$.
$$SV_{w_i} = \{\{x_a, x_b, ..., x_z\} \mid \beta_{t_z} > \cdots > \beta_{t_b} > \beta_{t_a}\}, \text{ where } |SV_{w_i}| = |T| = m$$

Definition 3: (Success Zone - SZ) given a Skill-graded Vector $SV_{w_i} = \{x_1, x_2, ..., x_m\}$ for worker w_i to a set of gold questions $|T| = m$.
$$SZ_{w_i} = \{\{x_a, x_b, ..., x_h\} \mid h \leq m \wedge \beta_{t_{h+1}} > \theta_{w_i}\}$$

5   Rational and Irrational Patterns in Success Zones

In reality, the SV matrix shown in Table 1, also known as a perfect Guttmann scale [28], is more plausible in theory, and is rather the exception than the rule. Observing the workers in reality, different response patterns can be seen, which are trickier to understand and handle.Table 2 illustrates a more realistic SV matrix, the shaded cells represent the success zones of each worker. Workers will sometimes miss out on easy questions within their ability $(\theta_{w_i} > \beta_{t_j})$ e.g. worker A responds incorrectly to task 3.

And sometimes they may as well successfully respond to difficult questions beyond their ability ($\theta_{w_i} < \beta_{t_j}$) e.g. worker C answers task 8 correctly. Some of these different response patterns can be explained [26] as seen in Table 3.

Table 2. (Realistic) Skill-graded Response Matrix

Workers \ Tasks	1	2	3	4	5	6	7	8	9
A	1	1	0	1	1	1	1	0	0
B	1	1	1	0	1	0	0	1	0
C	1	0	1	1	1	0	0	1	0

Accordingly, there will be unexpected false or correct responses in a worker's SV. These discrepancies are however of greater or lesser importance and impact the overall rationality of the response pattern depending on the number of their occurrences and their location in SV. We only focus on discrepancies within success zones, while SV entries that are outside of the success zone are: 1) incorrect and workers shouldn't be penalized for or, 2) correct and will be often attained through guessing.

Next we present three techniques that focus on success zones and aim at recognizing irrational patterns within these zones: 1) Skill-adapted Gold questions, 2) Entropy-based elimination, and 3) Rasch-based elimination.

Table 3. Observed response patterns

Response Pattern	Response Vector
Lucky-guessing	A few unexpected correct responses to difficult tasks
Carelessness	A few unexpected incorrect responses to easy tasks
Rushing	Unexpected error near the end for tasks with difficulty levels within a worker's abilities
Random guessing	Unexpected correct and incorrect responses irrespective of the question's difficulty
Constant response set	Same submitted response over and over again

5.1   Skill-Adapted Gold Questions

This technique is similar to using gold questions with a certain accuracy level (i.e. workers not achieving the required accuracy level are eliminated), except that gold questions don't provide a fair basis for judging workers with varying skill levels. The new technique also applies the gold question's accuracy level, but strictly on the success zone i.e. it discards all workers failing to achieve the required accuracy level on tasks within their skill level. Accordingly an irrational pattern in a worker's success zone can be defined as follows:

Definition 4: *(Skill-adapted gold questions)* given a success zone $SZ_{w_i} = \{x_1, x_2, ..., x_h\}$ of a worker w_i and a required accuracy level A, all workers are eliminated for which holds:

$$\sum_{j=0}^{h} x_j \bigg/ |SZ_{w_i}| < A$$

5.2   Entropy-Based Elimination

We utilize the *Shannon Entropy* [29] to measure the randomness of responses within the success zone. Following Shannon's entropy definition, a response pattern with high entropy indicates randomness, while rational response pattern will have low entropy. According to our experimental results on our dataset, entropy values higher than 0.89 indicated randomness. Accordingly an irrational pattern in a worker's success zone can be defined as follows:

Definition 5: *(Entropy-based Elimination)* given a success zone $SZ_{w_i} = \{x_1, x_1, ..., x_h\}$ of a worker w_i and an entropy threshold=0.89, all workers are eliminated for which holds:

$$-\sum_{j=0}^{h} p(x_j) \log_2 p(x_j) > 0.89$$

5.3   Rasch-Based Elimination

One shortcoming of both the skill-adapted gold questions and the entropy-based elimination technique, is that none of them take into consideration the point at which an incorrect response is given. For instance, the following two response vectors (101011) and (111100) get the same entropy of 0.9182. Following RM's definition of having higher correctness probability the easier the task is relative to a worker's skill ability, then failing on the second task in the first response pattern should be penalized more than failing on the fifth task in the second response pattern. Using the correctness probability P_{ij} that is estimated by the Rasch model, we add up the correctness probabilities for task entries that are correctly answered and penalize falsely answered tasks by subtracting its corresponding correctness probability. We then define an irrational response pattern as follows:

Definition 6: *(Rasch-based Elimination)* given a success zone $SZ_{w_i} = \{x_1, x_1, ..., x_h\}$ and the corresponding computed correctness probabilities P = $\{P_{i1}, P_{i2}, ..., P_{ih}\}$ of a worker w_i and a required accuracy level A, all workers are eliminated for which holds:

$$\sum_{j=0}^{h} (-P_{ij} + (x_j * 2P_{ij})) < A$$

6   Experimental Results

In this section we evaluate the proposed framework and extensively evaluate the efficiency of the three proposed techniques: 1) Skill-adapted Gold Questions, 2) Entropy-based elimination and 3) Rasch-based elimination.

In section 3, we evaluated the impact of the type of gold questions used (easy, balanced and difficult). Next, we investigate how many gold questions should be injected, which would allow the rasch model to correctly approximate each worker's skill level, and their correctness probability. Similar to the motivational case study in section 3, we use the verbal section from the GRE dataset to create our HITS for evaluation purposes.

The open source eRm package for the application of IRT models in R is utilized, where correct responses are coded as 1 and incorrect ones are coded as 0 in a worker's success zone. The eRm package uses conditional maximum likelihood CML estimation as it maintains the concept of specific objectivity [5], [30].

6.1   Gold Question Set Size

We experiment with different gold question set sizes (κ) and examine how it impacts the reliability of the RM's approximated parameters (θ_{w_i} worker's skill level and P_{ij} worker's correctness probability). This ultimately allows the inference of a heuristic for gold question set size κ, which would permit a good a reliable approximation of the parameters that are used in identifying a worker's success zone. Starting off with $\kappa = 20\%$ (i.e. 4 gold questions), we incrementally upsurge the size till 80%. We designed the different sets to be balanced in terms of questions' difficulty. Figure 3

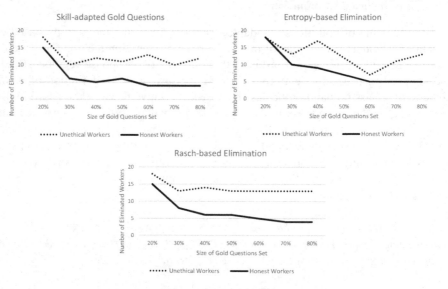

Fig. 3. Impact of varying κ

illustrates the impact of the different κ sizes on the three techniques in terms of: 1) number of eliminated unethical workers and 2) number of misjudged honest workers.

As κ increases, the three techniques can more efficiently judge honest workers and the number of misjudgments decrease. Moreover, for the three techniques the number of misjudged honest workers curves tend to converge on average when $\kappa = 40\%$.

Both the skill-adapted gold questions and the Rasch-based elimination approaches are optimistic compared to the Entropy-based elimination, which maintains on average a higher rate of eliminated unethical workers. Interesting to note is that convergence is much smoother and pronounced with the number of misjudged ethical workers curve than that of the unethical workers. This can be attributed to the implicit random response nature of the unethical workers i.e. for every set of gold questions that are of the same size, the unethical workers' performance would vastly fluctuate. This is especially prominent with the pessimistic entropy-based elimination approach.

For the rest of our experiments, we design our HITS to constitute 40% gold questions.

6.2   Ground Truth-Based Evaluation

We use the dataset from the laboratory-based study in section 3, which is made up of 20 questions, 40% out of which are gold questions (i.e. 8 questions). This gives us ground truth, where responses corresponding to the first random round constitutes unethical workers, while the second truthful response round constitutes honest workers. Accordingly, we can precisely evaluate how each of the proposed techniques fare in terms of detecting and distinguishing between irrational and rational response patterns in success zones and accordingly eliminate the workers fairly.

Figure 4 depicts the number of eliminated workers by each of the: a) skill-adapted gold questions, b)entropy-based elimination, c)Rasch-based elimination techniques, and compares them to the standard gold questions technique. For the gold question technique, we use a balanced set (i.e. set of gold questions target all difficulty levels) with an accuracy level set to 70% (i.e. workers must provide 70% correct answers, otherwise they're eliminated).

As can be seen, the three proposed techniques perform significantly better than solely relying on gold questions. Both the Skill-adapted gold questions and the Rasch-based elimination retain 50% more of the honest workers. Whereas gold questions misjudge 61% of the honest workers, both the skill-adapted gold questions and the Rasch-based elimination misjudge only 22% and 33% honest workers, respectively. Moreover, in support of our earlier findings, the elimination ratio of unethical to honest workers for the skill-adapted gold questions and the Rasch-based elimination is around 40%, designating them as more optimistic techniques. On the other hand, the entropy-based elimination is more rigid and pessimistic with a 52% ratio, yet still much better than simple gold questions. So while both the Skill-adapted gold questions and Rasch-based elimination can more efficiently recognize honest workers, entropy-based elimination tends to more efficiently recognize unethical workers. In summary, all three techniques exhibit better ratios than gold questions.

A closer look on the eliminated honest workers by skill-adapted gold questions techniques ascertains the justification of these elimination, where eliminated workers either: a) failed in answering all gold questions or b) failed in having a skill level higher than the easiest question (e.g. worker ability θ_{w_i} = -1.24 and the easiest question's difficulty level β_{t_1} = -1.21).

Furthermore, examining the different computed entropies for both honest and unethical workers, we observed that entropies higher than 0.89 indicated randomness in the responses. Accordingly, by setting the entropy threshold to 0.89, 94% unethical workers are eliminated, and 50% ethical honest are misjudged.

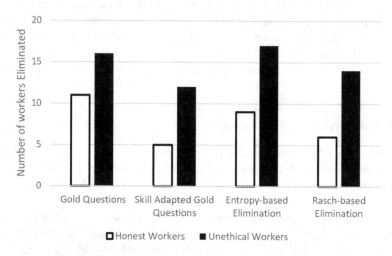

□ Honest Workers   ■ Unethical Workers

Fig. 4. Number of Ethical/Unethical Workers Eliminated

6.3   Practical Crowd Sourcing Experiments on Real World Data

In this experiment, we evaluate the efficiency of our techniques in real world crowd sourcing experiments. We used CrowdFlower as a generic crowd sourcing platform. Following the results of sections 3 and 6.1, we designed a balanced set of Gold questions. Each HIT consisted of 28 questions. We overrode the gold question policy of CrowdFlower, allowing workers to participate in the job, even if they did not meet the required accuracy level. No restraints were set to geography, and skills were set to minimum as defined by the platform. We collected 1148 judgments from 41 workers incurring a total cost of 35$. Table 4 illustrates how many workers were eliminated by each technique.

The results closely reflect our results on synthetic data. Using the platform's static accuracy levels, 32% of the workers would have been eliminated (13 workers). Both the Skill-adapted gold questions and Rasch-based Elimination tend to retain 50% of the workers originally eliminated by gold questions. On the other hand, the Entropy-based elimination retains its pessimism, discarding 26% of the workers.

Table 4. Percentage of Eliminated Workers

Gold Questions (70% accuracy level)	Skill-Adapted Gold Questions	Entropy-based Elimination	Rasch-based Elimination
32% (13 workers)	14% (6 workers)	26% (11 workers)	16.6% (7 workers)

Note that unlike the laboratory-based experiment, no ground-truth is available. Accordingly, to measure the efficiency of the techniques, we evaluate the set of workers who would have been discarded by static gold questions, yet retained by our techniques, namely: the Skill-Adapted Gold Questions and the Rasch-based Elimination. We provided each of those workers with HITS of 10 gold questions that corresponded to their skill level, at which point they provided on average 74% correct answers (i.e. providing only 2 incorrect answers in a HIT), which even beats the initial accuracy level threshold set by static gold questions.

7   Summary and Outlook

In this paper, we addressed the central question of a fair choice of workers that concerns impact sourcing: distinguishing between honest workers, who might just be lacking educational skills, and unethical workers. Indeed, our laboratory study shows how current quality control measures (namely gold questions and reputation-based systems) tend to misjudge and exclude honest and willing workers who may just not have been provided tasks on their individual skill levels. Yet, impact sourcing has to be fair by providing tasks for honest and willing workers that are well within their skill level, while avoiding unethical workers to keep up result quality constraints.

Accordingly, we developed a framework to promote fair judgments of workers. At its heart we deployed the Rasch model, which takes into account both, the level of a worker's skill and the task's difficulty. This aids in distinguishing each worker's success zone. We then presented three advanced, yet practical techniques for detecting irrational patterns within success zones: 1) Skill-adapted gold questions, 2) Entropy-based Elimination and 3) Rasch-based Elimination, with the first and third techniques being optimistic and the second more pessimistic in nature. Both laboratory-based and real world crowd souring experiments attested to the efficiency of the techniques: about 50% of misjudged honest workers can be successfully identified, and subsequently be redirected to skill-fitted tasks.

So far, the Rasch model can be applied on Gold Questions to compute tasks' difficulties, in future work we will expand our framework to majority vote scenarios, too. Due to the redundant nature of majority voting, here interesting financial gains can be realized. Of course, this also needs the computation of task difficulties of non-gold questions, which would allow for a dynamic adaptive task allocation fitting the corresponding workers' skills. Thus, skill ontologies and competence profiles are bound to be of major importance.

References

1. Gino, F., Staats, B.R.: The microwork solution. Harvard Business Review **90**(12) (2012)
2. Selke, J., Lofi, C., Balke, W.-T.: Pushing the boundaries of crowd- enabled databases with query-driven schema expansion. In: 38th Int. Conf. on Very Large Data Bases (VLDB), pp. 538–549 (2012)
3. Howe, J.: The Rise of Crowdsourcing. The Journal of North **14**(14), 1–5 (2006)
4. Zhu, D., Carterette, B.: An analysis of assessor behavior in crowdsourced preference judgments. In: SIGIR 2010 Workshop on Crowdsourcing for Search Evaluation, no. Cse, pp. 17–20 (2010)
5. Rasch, G.: Probabilistic Models for Some Intelligence and Attainment Tests. Nielsen & Lydiche (1960)
6. Finin, T., Murnane, W., Karandikar, A., Keller, N., Martineau, J., Dredze, M.: Annotating named entities in twitter data with crowdsourcing. In: CSLDAMT 2010 Proceedings of the NAACL HLT 2010 Workshop on Creating Speech and Language Data with Amazon's Mechanical Turk, pp. 80–88 (2010)
7. Lofi, C., El Maarry, K., Balke, W.-T.: Skyline queries in crowd-enabled databases. In: Proceedings of the 16th International Conference on Extending Database Technology, EDBT/ICDT Joint Conference (2013)
8. Kouloumpis, E., Wilson, T., Moore, J.: Twitter sentiment analysis: the good the bad and the OMG! In: International AAAI Conference on Weblogs and Social Media, pp. 538–541 (2011)
9. Callison-Burch, C.: Fast, cheap, and creative: evaluating translation quality using amazon's mechanical turk. In: EMNLP 2009: Proceedings of the 2009 Conference on Empirical Methods in Natural Language Processing, vol. 1(1), pp. 286–295 (2009)
10. Lofi, C., Selke, J., Balke, W.-T.: Information Extraction Meets Crowdsourcing: A Promising Couple. Proceedings of the VLDB Endowment **5** (6), 538-549 (2012)
11. Kuncheva, L.I., Whitaker, C.J., Shipp, C.A., Duin, R.P.W.: Limits on the majority vote accuracy in classifier fusion. Journal: Pattern Analysis and Applications, PAA **6**(1), 22–31 (2003)
12. Dawid, A.P., Skene, A.M.: Maximum likelihood estimation of observer error-rates using the EM algorithm. Journal of Applied Statistics, 20–28 (1979)
13. Raykar, V.C., Yu, S., Zhao, L.H., Valadez, G.H., Florin, C., Bogoni, L., Moy, L.: Learning From Crowds. The Journal of Machine Learning Research **11**, 1297–1322 (2010)
14. Whitehill, J., Ruvolo, P., Wu, T., Bergsma, J., Movellan, J.: Whose Vote Should Count More: Optimal Integration of Labels from Labelers of Unknown Expertise. Proceedings of NIPS **22**(1), 1–9 (2009)
15. Ipeirotis, P.G., Provost, F., Wang, J.: Quality management on amazon mechanical turk. In: Proceedings of the ACM SIGKDD Workshop on Human Computation, p. 3 (2010)
16. Kazai, G.: In Search of Quality in Crowdsourcing for Search Engine Evaluation. In: Clough, P., Foley, C., Gurrin, C., Jones, G.J., Kraaij, W., Lee, H., Mudoch, V. (eds.) ECIR 2011. LNCS, vol. 6611, pp. 165–176. Springer, Heidelberg (2011)
17. Qiang Liu, A.I.: Mark steyvers, "scoring workers in crowdsourcing: how many control questions are enough? In: Proceedings of NIPS (2013)
18. El Maarry, K., Balke, W.-T., Cho, H., Hwang, S., Baba, Y.: Skill ontology-based model for quality assurance in crowdsourcing. In: UnCrowd 2014: DASFAA Workshop on Uncertain and Crowdsourced Data, Bali, Indonesia (2014)

19. Ignjatovic, A., Foo, N., Lee, C.T.L.C.T.: An analytic approach to reputation ranking of participants in online transactions. In: IEEE/WIC/ACM Int. Conf. Web Intell. Intell. Agent Technol., vol. 1 (2008)
20. Noorian, Z., Ulieru, M.: The State of the Art in Trust and Reputation Systems: A Framework for Comparison. Journal of Theoretical and Applied Electronic Commerce Research **5**(2) (2010)
21. Traub, R.E.: Applications of item response theory to practical testing problems, vol. 5, pp. 539–543. Erlbaum Associates (1980)
22. Quoc Viet Hung, N., Tam, N.T., Tran, L.N., Aberer, K.: An Evaluation of Aggregation Techniques in Crowdsourcing. In: Lin, X., Manolopoulos, Y., Srivastava, D., Huang, G. (eds.) WISE 2013, Part II. LNCS, vol. 8181, pp. 1–15. Springer, Heidelberg (2013)
23. Wang, J., Ipeirotis, P.G., Provost, F.: Managing crowdsourced workers. In: Winter Conference on Business Intelligence (2011)
24. Batchelder, W.H., Romney, A.K.: Test theory without an answer key. Journal Psychometrika **53**(1), 71–92 (1988)
25. Bond, T.G., Fox, C.M.: Applying the Rasch Model: Fundamental Measurement in the Human Sciences. Journal of Educational Measurement **40**(2), 185–187 (2003)
26. Karabatsos, G.: A critique of Rasch residual fit statistics. Journal of Applied Measures. **1**(2), 152–176 (2000)
27. Linacre, J.M.: Understanding Rasch measurement: estimation methods for Rasch measures. Journal of Outcome Measurment **3**(4), 382–405 (1999)
28. Guttman, L.: A basis for scaling qualitative data. Journal of American Sociological Review, 139–150 (1944)
29. Shannon, C.E.: A mathematical theory of communication. SIGMOBILE Mobile Computing and Communications Review **5**(1) (2001)
30. Rasch, G.: On specific objectivity: An attempt at formalizing the request for generality and validity of scientific statements. Journal of Danish Yearbook of Philosophy **14** (1977)

Spatio-temporal Data II

Skyline Trips of Multiple POIs Categories

Saad Aljubayrin[1][(✉)], Zhen He[2], and Rui Zhang[1]

[1] Department of Computing and Information Systems, The University of Melbourne,
Melbourne, Australia
`aljubayrin@su.edu.sa, rui.zhang@unimelb.edu.au`
[2] Department of Computer Science and Computer Engineering, Latrobe University,
Melbourne, Australia
`z.he@latrobe.edu.au`

Abstract. In this paper, we introduce a new interesting path finding problem, which is the Skyline Trips of Multiple POIs Categories
($STMPC$) query. In particular, given a road network with a set of Points
of Interest (POIs) from different categories, a list of items the user is planning to purchase and a pricing function for items at each related POI;
find the skyline trips in term of both trip length and trip aggregated
cost. This query has important applications in everyday life. Specifically,
it assists people to choose the most suitable trips among the skyline trips
based on two dimensions; trip total length and trip aggregated cost. We
prove the problem is NP-hard and we distinguish it from existing related
problems. We also proposed a framework and two effective algorithms to
efficiently solve the $STMPC$ query in real time and produce near optimal
results when tested on real datasets.

Keywords: Path finding · Skyline paths · Skyline trips · Trip planing

1 Introduction

Over the past few decades spatial databases have been studied extensively, resulting in significant outcomes in areas such as spatial indexing, path finding and
data modelling [5,8,11,21,26,29]. In this paper we focus on the path finding field
and introduce a new interesting path finding problem, which is the Skyline Trips
of Multiple POIs Categories $STMPC$ query. In particular, given a road network
graph G, source s and destination d, a set of n POIs categories $C = \{c_1, c_2, ...c_n\}$
with a set P of POIs in each category $c_i = \{p_1, p_2, ...p_n\}$, a list O of items the
user is planning to purchase $O = \{o_1, o_2, ...o_n\}$ and a pricing function $f(o_i)$ for
items at each related POI p_i; find the skyline trips $Sky(T) = \{t_1, t_2, ...t_n\}$ that
each starts at s, pass through at least one POI p_i from each related category
and ends at d, thus $t_i = \{s, p_1 \in c_1, p_2 \in c_2, ...p_n \in c_n, d\}$. $Sky(T)$ is the set
of trips that are not dominated by any possible other trip in term of both trip
length and trip cost aggregated from POIs creating the trip.

Since the Trip planing query (TPQ) studied in [19] is a special case of our
problem, we would illustrate it first and distinguish it from the typical shortest

M. Renz et al. (Eds.): DASFAA 2015, Part II, LNCS 9050, pp. 189–206, 2015.
DOI: 10.1007/978-3-319-18123-3_12

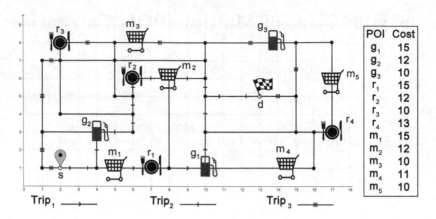

Fig. 1. Motivating example

path query. TPQ is a generalization of the Traveling Salesman problem (TSP) [9] and more inclusive than normal shortest path query. Specifically, a TPQ usually starts from a source point and pass through a number of POIs in particular order and possibly restricted by some constraint (e.g. time, distance). In contrast, a typical shortest path query aims at finding the shortest path between source and destination by finding the smallest aggregated weight of road segments connecting the query points. For example, TPQ can be finding a path from a user's office that pass by an ATM, supermarket, restaurant and ending the trip at the user's home. While a shortest path query can be finding the path between two targets (e.g. ATM and supermarket) in a trip. The cost at each POIs is not considered in the TPQ, thus, there is only one optimal trip, which is the shortest trip that passes through at least one POI from each category.

The main difference between the TPQ query and the query studied in this paper is that, the *STMPC* query involves another optimality dimension, which is the trip total cost aggregated from POIs creating the trip. Specifically, while the TPQ query only finds the shortest trip, our query uses the trips length and cost to find the set of skyline trips. This results in the possibility of having multiple skyline optimal trips. We illustrate the *STMPC* query in the following example.

Figure 1 shows an example of a road network with source s and destination d and a number of POIs from three different categories; gas stations, restaurants and supermarkets. Based on the prices of items the user is planning to purchase and their quantity, we define the cost at each POI. When the user at s wants to visit one POI from each category on her way to d, there can be a large number of possible trips which all include one POI from each category. For instance, $Trip_1 = \{s, m_1, r_1, g_1, d\}$ has the shortest distance (15) with expensive cost (45), while $Trip_3 = \{s, r_3, m_3, g_3, d\}$ has the lowest cost (30) but with quite long distance (27). In addition, there can be a trip that is both short (17) and has a low cost (36) i.e. $trip_2 = \{s, g_2, r_2, m_2, d\}$ .Therefore, the user can choose the most preferable trip amongst the skyline trips, which are based on two dimensions;

trip distance and trip total cost. In this example the trips $Trip_1$, $Trip_2$ and $Trip_3$ are the skyline trips because they are not dominated by any other possible trip.

Motivated by scenarios such as the previous example, we formulated the *STMPC* query and proposed two effective algorithms to efficiently answer the query in real time and produce near optimal results. Our first algorithm is generally based on defining a new weight for each POI, where this weight is a combination of both POI cost and POI distance from query points. Next, we perform a number of iterations, where in each iteration we change the weight of the two dimensions (cost and distance) in order to get the skyline candidates from each category. Our second algorithm does not only consider the distance between each POI and the query end points, it also considers clustered POIs using a suitable data structure (e.g. Quadtree, Rtree). This results in more accurate skyline trips especially if some POIs are clustered in geographical spots.

In this paper we also proposed a framework to estimate the network distance between POIs and query points. This is because using the Euclidean distance in road network does not usually provide accurate measurement while using the exact network distance between all POIs and query points would be an expensive task at the query time. Our framework is based on precomputing the network distance between POIs and a group of geographical spots in the network and then using this distance to estimate the actual network distance.

We make the following contributions:

1. We introduce the *STMPC* query, which is a novel path finding problem and has important applications in everyday life.
2. We proposed two interesting heuristic algorithms to solve the *STMPC* query and produce near optimal results.
3. We proposed an offline framework to estimate the network distance between POIs and predefined geographical regions, which can contain the query points. Our framework shows superior results compared to the Euclidean distance estimation.
4. We perform extensive experiments to evaluate the effectiveness and efficiency of the proposed algorithms, and the results are summarized as follows:
   (a) In term of effectiveness, our algorithms produce up to 0.99 optimal results based on our optimality indicter. The accuracy of the distance estimated by our framework is between 0.96 and 0.99 compared to actual network distance.
   (b) In term of efficiency, our algorithms answer *STMPC* queries in real time and up to four orders of magnitude faster than the baseline solution.

The reminder of the paper is organized as follows. Related work is discussed in Section 2. Section 3 presents the preliminaries and problem definition. Sections 4 details the proposed efficient heuristics and the distance estimation framework. The experimental results are shown in Section 5. Finally we conclude the paper in Section 6.

2   Related Work

Related work can be categorised into two categories; road network skyline query related problems and trip planing query related problems. We are unaware of any attempt to investigate the problem of finding the skyline trips of multiple POIs categories within either categories.

First, most existing studies on skyline queries (e.g.[3,17,25,27]) have focused on efficiently finding the skyline points for datasets in the traditional database systems. Only a few studies considered the skyline concept in spatial database systems, specifically, in road networks. Deng in [7], proposed to solve the "Multi-source Skyline Query", which aims at finding skyline POIs in road network based on the attribute of the POI (e.g. price) and the aggregated distance between the POI and multiple locations. For example, find the skyline hotels that are both cheap and close to the beach, the university and the botanic garden. The setting of this problem is different from ours in that it only assumes one POI category (e.g. hotels), while we consider multiple POIs categories and also consider the total trip distance aggregated from travelling between POIs from different categories. Therefore, the solution of Deng is not applicable to our problem.

Some other studies in road network skyline query [18,20,28] focus on finding the skyline paths considering multiple path attributes such as distance, driving time, gas consumption, number of traffic lights, etc. They assume different paths would have different values at each attribute and thus their goal is to find the set of skyline paths to allow the user to choose the most preferable path. Again, this problem is different than ours in that it does not include any POIs nor the distance between them, hence, their solution is not applicable.

Other road networks skyline studies such as [12–14], consider continuous skyline queries for POIs in road network. They continuously search for the skyline POIs for a moving object considering both the attributes of the POIs (e.g. price, rating) and their distance to the moving object. In these studies, the outcome is not a complete trip consisting of POIs from multiple categories, instead, it is a set of skyline POIs from the same category and hence, their solutions are not applicable to our problem.

Second, most trip planing studies [4,16,19,24] have one optimal trip that answer their queries, while we consider a set of skyline trips. The TPQ [19] discussed in the introduction can be considered as a special case of our problem. This is because TPQ does not consider the cost at POIs while constructing the trip. Therefore, applying the TPQ solution to our problem would only return one trip from the skyline trips, which is the one with the lowest distance. The optimal sequenced rout query studied in [16,24] aims to solve the TPQ with order constraint, is a special case of the TPQ and thus a special case of our problem. Finally, the "The multi-rule partial sequenced route query" studied in [4] is similar to both the optimal sequenced route query and the TPQ in that, it may involve some order constraint.

Although some of the above mentioned studies might seem similar to the $STMPC$ query, their solutions are not applicable. This is because our problem is mainly inherited from three major queries types, which are multi-dimensional skyline queries, nearest neighbour queries and shortest path queries. On the other hand, most of the studies discussed in this section are only inherited from two queries types.

There are some existing studies on the nearest neighbor or range queries [1, 10, 30, 31], which retrieve the set of objects with the smallest distance to the query point. However, the *STMPC* and TPQ queries uses the aggregated distance between query points and POIs to find the trip with the lowest total distance.

3   *STMPC* Query

We first formalize the *STMPC* query and present the baseline algorithm, and then we prove it is NP-hard. Table 1 summarizes our notation.

Table 1. Frequently Used Symbols

Symbol	Explanation
s	The source of a STMPC query.
d	The destination of a STMPC query.
P	A set of POIs.
C	A set of POIs categories.
O	A set of items the user wants to purchase.
t	A trip from s to d through one POI from each category.
$Dis(p_i, p_j)$	The network distance between p_i and p_j.
p^c	The cost at a POI p.
p^d	The aggregated distance from a POI p to both s and d.
t^c	A trip total cost.
t^d	A trip total distance.
p^p	The priority value of a POI p.

3.1   Problem Definition

Giving a road network graph G, source s and destination d, a set C of POIs categories $C = \{c_1, c_2, ...c_n\}$ with a set P of POIs in each category $c_i = \{p_1, p_2, ...p_n\}$, a list O of items the user is planning to purchase $O = \{o_1, o_2, ...o_n\}$. Each item in the user list o_i can be associated with different costs at different related POIs. Let $t_i = \{s, p_1 \in c_1, p_2 \in c_2, ...p_n \in c_n, d\}$ be a trip that starts from s and passes at least through one POI from each category and finishes at d. We use $t_i{}^d$ to donate the total distance of a trip, which is the sum of the network distances between s, the trip POIs and d in the travelled order; $t_i{}^d = Dis(s, p_1) + Dis(p_1, p_2) + ... + Dis(p_n, d)$. We use $t_i{}^c$ to donate the total cost of a trip, which is the sum of the cost at each POI in the trip $t_i{}^c = p_1{}^c + p_2{}^c + ... + p_i{}^c$. For example the trip distance of $trip_1$ in Figure 1 is $t_1{}^d = Dis(s, m_1) + Dis(m_1, r_1) + Dis(r_1, g_1) + Dis(g_1, d) = 15$, while the cost of the same trip is $t_1{}^c = m_1{}^c + r_1{}^c + g_1{}^c = 45$.

We can consider any trip t_i that starts at s, pass at least through one POI from each category and ends at d as a valid trip because it answers the user query. However, different trips have different distance and cost values, hence a trip with a short distance such as $trip_1$ may have a high cost and vice versa. Therefore, we leave it up to the user to decide the relative importance of trip distance travelled and cost by returning the skyline trips. The problem of the *STMPC* query is defined as follow:

Definition 1. *Skyline Trips of Multiple POIs Categories (STMPC) Query: given a road network graph G, source s and destination d, a set C of POIs categories with a set P of POIs in each category, a list O of items the user is planning to purchase and a pricing function f(o_i) for items at each related POI, the STMPC query finds the skyline trips Sky(T) that each starts at s, passes through at least one POI from each related category and ends at d such that any trip t ∈ Sky(T) are not dominated by any other trip t' in term of both trip distance t^d and trip cost t^c, i.e., ∀t', $t^d \leq t'^d \vee t^c \leq t'^c$.*

Based on the above definition, a straightforward solution is as follow. First, compute all the possible POI permutations, where only one POI from each category is chosen based on the list of items the user wants to purchase. Next, add s and d to the beginning and end of each found permutation in order to construct valid trips. Finally, compute cost and network distance for each constructed trip and perform a skyline query to find the skyline trips. The problem with this solution is that, it is extremely slow and only applicable for very small dataset sizes. For example, it takes more than 24 hours to find the skyline trips when applied only to 40 POIs.

3.2   STMPC NP-Hard

The *STMPC* query can be considered as a generalization of some known path finding problems such as TPQ [19] and the travelling salesman problem (TSP) [9]. We will show in the following theorems that these two problems are special cases of the *STMPC* problem.

Theorem 1. *The metric travelling salesman problem with defined start and end points is a special case of the* STMPC *query.*

Proof. According to definition 1, when we simplify the *STMPC* problem to assume that, there is only one POI in each category, all trips will have the same cost (e.g. $t_1{}^c = t_2{}^c = \ldots = t_n{}^c$) while visiting POIs in different order may results in trips with different distances (e.g. $t_1{}^d \neq t_2{}^d \neq \ldots \neq t_n{}^d$), thus, there will only be one skyline trip, which is the trip with the minimum distance. According to the TSP definition, this version of the *STMPC* problem is an instance of the TSP. Therefore, TSP is a special case of the *STMPC* problem.

Theorem 2. *The trip planning query (TPQ) is a special case of the* STMPC *query.*

Proof. According to definition 1, when we simplify the *STMPC* problem to assume that all POIs from the same category have the same cost (e.g. $p_1{}^c \in c_1 = p_2{}^c \in c_2 = ... = p_n{}^c \in c_n$), all trips will have the same cost (e.g. $t_1{}^c = t_2{}^c = ... = t_n{}^c$). While visiting different POIs in different orders may results in trips with different distances (e.g. $t_1{}^d \neq t_2{}^d \neq ... \neq t_n{}^d$), thus, there will only be one skyline trip, which is the trip with the minimum distance. According to the TPQ definition, this version of the *STMPC* problem is a an instance of the TPQ. Therefore, TPQ is a special case of the *STMPC* problem.

Corollary 1. *The Skyline Trips of Multiple POI Category query* STMPC *is NP-hard.*

Proof. According to [9] and [19] the problems TSP and TPQ are proven to be NP-hard. Therefore, since the problems TSP and TPQ are special cases of the *STMPC* problem (theorems 1 and 2), the *STMPC* query is NP-hard.

The aim of the *STMPC* query is to find a set of optimal trips, which are the skyline trips in regard to two quality dimensions; trip cost and distance. Based on theorem 2, the TPQ is a simpler version of the *STMPC* query, where only the shortest optimal trip is queried. This means, finding each skyline trip is at least as hard as the TPQ.

4   Proposed Heuristics

In this section we present our proposed heuristics algorithms and a network distance estimation framework. For ease of understanding, we first describe a simple Euclidean distance based solution, which we call Weighted POIs Algorithm (WPOIs). Next we detail the distance estimation framework, which is used by both algorithms to estimate the network distance instead of using the Euclidean distance. Finally, we cover the second algorithm; Clustered Weighted POIs Algorithm (CWPOIs), which is an improved cluster based version of the first algorithm.

4.1   Weighted POIs Algorithm (WPOIs)

Here we present an efficient algorithm to find the skyline trips for multiple POIs categories based on two dimensions (trip cost and trip distance). The WPOIs algorithm is divided into two stages; POIs nomination stage and trip construction stage. It works by repeatedly iterating through these two stages, where the outcome of each iteration is a skyline trip candidate. In the first stage of each iteration, every POI category nominates one POI as the most superior POI in the category. In the trip construction stage of each iteration, we use s, d and the nominated POI from each category to construct a trip using a greedy approach.

POI Nomination Stage: Before we start illustrating the process of this stage, we need to define new properties for both POIs and iterations. First, for each POI, we define two properties, which are POI aggregated distance p^d and POI cost p^c. As mentioned in Section 3 ,the property p^c represents the POI expected cost based on the items the user wants to purchase, while the property p^d represents the POI aggregated Euclidean distance from both s and d. Second, for each iteration, we define two dependant weighting values w^c and w^d, which are the cost weight and the distance weight, respectively, where always $w^c + w^d = 1$. These two weights represent the importance of cost and distance when nominating a POI from each category. We also define a third property for each POI, which is the POI priority value p^p. The priority value p^p is simply the weighted sum of the POI cost and distance, thus, $p^p = w^c p^c + w^d p^d$. However, before finding the value of p^p, we first need to normalise the POIs cost range to match their distance range. Data normalisation is discussed in [22]. The value of p^p represents the total weight of a POI in each iteration.

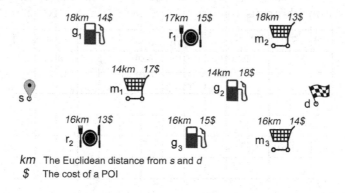

Fig. 2. WPOIs algorithm example

The main idea of this stage is to vary the weights of the cost and distance (w^c, w^d) during every iteration in order to nominate the POI with the lowest p^p from each category. For example in Figure 2, when $w^c = 1$, thus $w^d = 0$ $(1 - w^c)$, the priority value p^p of the three gas stations (g_1, g_2, g_3) are $(18 * 0 + 14 * 1 = 14, 14 * 0 + 18 * 1 = 18, 16 * 0 + 15 * 1 = 15)$ respectively. Therefore, the gas station category nominates the POI with the lowest p^p, which is g_1. Similarly m_2 and r_2 are nominated. On the other hand when $w^d = 0.5$ and thus $w^c = 0.5$, the nominated POIs from each category are is g_3, m_3, r_2 and so on. The order of the POIs is not important at this stage of the algorithm because the trip will be formed in the trip construction stage.

Based on definition 1, if a trip t_a consists of the cheapest POI from each category (e.g. $\min(p_i{}^c)$), then, t_a is a skyline trip because it is the cheapest. We could also expect that, if a trip t_b consists of POIs with the least aggregated distance from s and d (e.g. $\min(p_i{}^d)$), then, t_b is a skyline trip because it is the shortest. In the previous process, when $w^c = 1$, only POIs with the lowest cost will be nominated resulting in the cheapest trip (skyline). Similarly, when $w^d = 1$ the POIs of the shortest trip (skyline) will be nominated. The two skyline trips t_a and t_b are the most extreme skyline points on each dimension, thus, based on definition 1, if $t_a \neq t_b$, then, all other skyline points are between t_a and t_b.

As mentioned at the beginning of this stage, the variables w^c and w^d are dependant because $w^c + w^d = 1$. Therefore, a possible approach of trying to get all the skyline trips is to vary the cost weight w^c between 1 and 0 for a number of iterations. In every iteration, we get the best POI from each category, which together could form a skyline trip. A straightforward technique of achieving this is to have a fixed number of iterations (e.g. 100) where we gradually change the cost weight w^c between 1 and 0 in every iteration. For example the values of w^c in the 100 iterations are $(1, 0.99, 0.98...0)$ as a results the values of w^d are $(0, 0.01, 0.02...1)$. However, this approach has two main disadvantages. First, it is possible that a POI category nominates the same POI for different iterations. For example in Figure 2, the gas station g_1 is nominated when the cost weight is $1 \geq w^c \geq 7.5$. This is because the p^p of g_1 is the lowest compared to other gas stations in the specified range. Second, it is also possible to miss some skyline trips if they exist between two fractions with a small difference between them (e.g. $w^c = 0.751$ and $w^c = 0.752$). As a result, the straightforward technique is inefficient. We use an efficient iterating technique that is based on the following theorem.

Theorem 3. *If $a > b$ and p is nominated when $w^c = a$ and also nominated when $w^c = b$, then p is nominated when $a \geq w^c \geq b$.*

Proof. Since $p^p = w^c p^c + (1 - w^c) p^d$ and p is nominated based on the p^p of a and b, $p_a{}^p$ and $p_a{}^p$ respectively, where $p_a{}^p \geq p_a{}^p$. let $w^c = x$, thus, $p_x{}^p = xp^c + (1 - x)p^d$. if $a \geq x \geq b$ then, $p_a{}^p \geq p_x{}^p \geq p_a{}^p$ and hence p is nominated when $w^c = x$.

Based on the above theorem, we do not need to fix the number of iterations (e.g. 100). Instead, for each POI category, we compare the nominated POI for the two extreme weight of w^c, where $w^c = 0$ and $w^c = 1$. If the same POI is nominated, we do not need check any other value of w^c. Otherwise, we compare the nominated POI of the middle weight (e.g. $w^c = 0.5$) to both extreme values and so on. For example in Figure 2, when $w^c = 1$ the Restaurant category nominates r_2 because $r_2{}^p < r_1{}^p$ ($13 < 15$) and also nominates r_2 when $w^c = 0$ because $r_2{}^p < r_1{}^p$ ($16 < 17$). Therefore, we do not to iterate between 1 and 0 in order to look for a POI with less p^p in the Restaurant category.

Trip Construction Stage: As mentioned at the beginning of this subsection, the outcome of the POI nomination stage is a set of skyline candidate trips. Each consists of one nominated POI from each category. In this stage, we focus on the order of the nominated POI in order to achieve a trip that starts from s, pass through the nominated POIs and ends at d with the minimum distance. As explained in Section 3, this stage of the algorithm is NP-hard problem by itself because it is a special case of the TSP. Therefore, we use the same greedy technique used to solve the TPQ and TSP problems. It works by visiting the nearest neighbour of the last POI added to the trip starting from s and ending at d, where the Euclidean distance is used in the nearest neighbour search. In each iteration, we use the greedy technique to form the general shape of a candidate trip (POIs order) regardless of the real network distance between them.

Once all candidate trips are formed, we can use any shortest path algorithm to create the final trips using the network distance. Next, we perform a skyline query over the candidate trips using their costs and network distances as the two dimensions in order to prune any dominated trip. Finally, we get the set of skyline trips $Sky(T)$.

4.2   Distance Estimation Framework

In the previous illustration of the WPOIs algorithm, we used the Euclidean distance to estimate the aggregated distance p^d from each POI to both s and d. We also used the Euclidean distance in the greedy solution to find the next nearest neighbour of POIs forming a trip. The disadvantage of using the Euclidean distance in road networks is that, it does not reflect the actual network distance [2,15,23]. In contrast, the exact network distance is computationally expensive when computed online and hard to store when computed off-line. Therefore, we propose a network distance estimation framework to be used instead of the Euclidean distance in the WPOIs and CWPOIs algorithms. The main idea of this framework is to precompute and store the average network distance between every POI and specific geographical regions in the road network. It estimates the network distance between a POI and any network vertex by retrieving the stored distance between the POI and the region containing the queried vertex. Dividing the network into multiple geographical regions can be done using any suitable multi-dimensional space data structure (e.g. Quadtree, Rtree). In the settings of this study we use Quadtree for its simplicity.

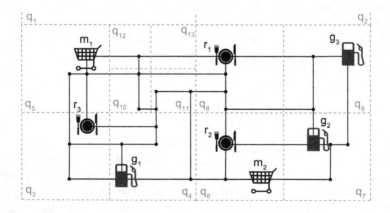

Fig. 3. WPOIs algorithm example

The pre-computation process starts by indexing the network vertices into the Quadtree with suitable density level at each leave node. Next, for each POI we perform a single source shortest path search (Dijkstra's [8]) to find the distance from that particular POI to all network vertices. Based on these distances, we measure and store the average distance to each of the geographical regions (Quadtree leave nodes). For example in Figure 3, the average network distances between every POI $\{g_1, g_2, .., r_1, ..m_3\}$ and each of the Quadtree leave nodes $\{q_1, q_2, ...q_{12}\}$ are stored. At the online stage, the distance between a network vertex and a POI is estimated by retrieving the stored average distance between the POI and the geographical square containing the vertex. For example in Figure 3, we can estimate the network distance between g_1 and r_1 by retrieving the stored distance between g_1 and the Quadtree leave node q_8.

The required time to pre-compute the distances and the memory consumption of storing the distances are highly sensitive to the number of POIs and the number of Quadtree leaf nodes, which is based on the node density. However, since the framework is computed offline, the time consumption is not critical. Moreover, the memory consumption can be well managed by using the right density level in each Quadtree leaf node as will be shown in Section 5. The higher time and memory cost of pre-computing the distances is well justified by the more accurate network distance estimations.

The distance estimation framework is more suitable for static road networks. However, in order to make it applicable on time-dependent road networks [6], we can use the road network historical data to store different traveling times between POIs and the geographical regions for different times of the day.

4.3   Clustered Weighted POIs Algorithm (CWPOIs)

The CWPOIs algorithm is an improved version of the first algorithm illustrated in Section 4.1. The problem with the WPOIs algorithm is that, it only considers the aggregated distance between a POI and the query points at the nomination stage, regardless of the distance to POIs from other categories. This could result in missing some of the skyline trips candidates when their POIs are clustered far from query points. For example, the WPOIs algorithm may miss a skyline trip $t_1 = \{g_x, r_x, m_x\}$ (POIs located in the same location e.g. mall) when there is a another trip $t_2 = \{g_y, r_y, m_y\}$ (POIs located in different locations) with the same cost and worse total distance $t_2{}^d > t_1{}^d$. This is because the distance p^d of each of the three POIs $(g_x{}^d, r_x{}^d, m_x{}^d)$ is large when considered individually, thus, the t_1 POIs are dominated by the POIs of t_2 (i.e. $g_x{}^d > g_y{}^d, r_x{}^d > r_y{}^d, m_x{}^d > m_y{}^d$). Therefore, in order to overcome this obstacle we propose the CWPOIs algorithm, which considers both the aggregated distance from query points p^d and the distance between clustered POIs.

Similar to the first algorithm, the CWPOIs algorithm is based on the framework illustrated in Section 4.2 to estimate the network distance. However, it also uses the framework to cluster POIs in geographical regions (Quadtree nodes). The main idea of this algorithm is to define two properties for each Quadtree node, which are the node lowest cost n^c and the node average distance n^d. This is only applicable for nodes containing at least one POI from each related category. The first property n^c can be defined by the aggregated cost of the POI with the lowest cost p^c from each category located within the geographical area of a node n. The second property n^d is the aggregated average distance between a node n and the query points s and d, which is obtained from the pre-computation of the framework.

The process of the CWPOIS algorithm starts by defining the values for the n^c and n^d properties for each applicable Quadtree node, which can be leaf or non-leaf node. Next, we use these two values to perform a skyline query over the Quadtree nodes, where the outcome of this step is a set of non-dominated Quadtree nodes. Then, we apply the first algorithm (WPOIS) at each of the skyline Quadtree nodes to find the skyline trips candidates in each geographical region. Finally, we measure the network distance for each of the candidate trips and perform a skyline query to return the final skyline trips.

Based on the above process, the CWPOIs algorithm finds the skyline geographical regions and apply WPOIs algorithm to each of them individually. This results in separating the nomination competition (first stage of WPOIs algorithm) performed for POIs in a clustered region from other regions, thus, returning more accurate skyline trips at the final stage.

5   Experimental Study

In this section we evaluate the effectiveness and efficiency of the proposed framework and algorithms. We conducted the experiments on a desktop PC with 8GB RAM and a 3.4GHz Intel$^{(R)}$ Core$^{(TM)}$ i7 CPU. The disk page size is 4K bytes. We use the London road network dataset extracted from Open Street Map[1], which contains 515,120 vertices and 838,702 edges. We also extracted the locations of 11,030 POIs in London classified into 12 different categories.

We vary parameters such as the number of POIs categories, POIs cardinality within each category, and Quadtree density level to gain insight into the performance of the framework and the algorithms in different settings. The detailed settings are given in the individual experiments.

5.1   Framework Evaluation

As discussed in Section 4.2, the purpose of the framework is to provide a better estimation of the network distance than using the Euclidean distance. We validate the framework in term of both effectiveness and efficiency.

Framework Accuracy: First, we compare the accuracy of the distance estimated by our framework denoted as *Frame-Dis*, to the Euclidean distance estimation denoted as *Eu-Dis*. An intuitive way to find the accuracy ratio AR for a

---

[1] http://metro.teczno.com/#london

distance estimation method *Est-dis*, which can be either *Frame-Dis* or *Eu-Dis*, is to compare it to the actual network distance denoted as *Act-Dis* as follow:

$$AR = \frac{Act\text{-}Dis - (|Act\text{-}Dis - Est\text{-}Dis|)}{Act\text{-}Dis}$$

For example, when $Act-Dis = 37km$ and $Frame-Dis = 37.5km$, the accuracy indicator $AR = 0.98$. The accuracy of our framework is highly sensitive to the density level at each geographical region. The less dense the Quadtree cell, the more accurate distance estimation. We vary the maximum density level inside each Quadtree node from 0.01% to 0.001% of the total network vertices. We compare the average accuracy ratio AR of both *Frame-Dis* and *Eu-Dis* when running 1000 queries from random POIs to random network vertices. It can be seen from Figure 4a that, the accuracy ratio of *Frame-Dis* increases up to 0.99 as the density level decreases. This is because the size of geographical regions decrease and hence, the difference between the average distance from a POI to a region and the actual distance from the POI to any vertex within that region decreases. In all density levels, our framework estimates network distance more accurately than the Euclidean distance.

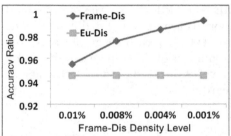
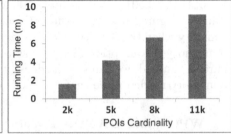

(a) Distance Accuracy   (b) Construction Running Time

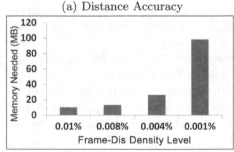
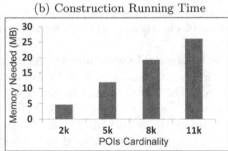

(c) Memory Consumption (density level)   (d) Memory Consumption (cardinality)

Fig. 4. Framework Evaluation

Framework Efficiency: The framework efficiency is evaluated using two metrics; the time taken to construct the framework and its memory consumption.

Framework construction running time: the running time of the framework is highly effected by the total number of POIs from all categories. This is because we need to find the distance to all network vertices in order to measure the average distance from each POI to every geographical region. Figure 4b illustrates the increase of framework construction time as the total number of POIs increases. When all the POIs are used, the framework is precomputed in less than 10 minutes. The graph construction running time is not affect by Quadtree density level and thus the number of the geographical locations. This is because there will be a path finding query for each POI regardless of the number of Quadtree nodes.

Framework memory consumption: the main purpose of the framework is to precompute and store the average distance between every pair of a POI and a geographical region. Therefore, the number of stored distances is nm, where n is the number of POIs and m is the number of geographical regions. Figure 4c illustrates the memory space needed to store the precomputed distance for different density levels in each geographical region when all of the 11,030 POIs are used. When the density level decreases from 0.01% to 0.001%, the number of geographical regions increase and thus more space is needed to store the distances. At the 0.001% density, only 100 MB is needed to store the precomputed distances. Figure 4d shows different memory consumptions for different POIs cardinality when the density level is fixed to 0.004% in each geographical region. The memory consumption increases as the number of POIs increases. Based on Figures 4c and 4d, we can see the density level of the used data structure affects the memory needed more than the POIs cardinality.

5.2 WPOIs and CWPOIs Effectiveness Evaluation

In this subsection, we validate the effectiveness of our proposed algorithms in finding skyline trips. However, we need to fist discuss how to measure the optimality of our results. Finally, we measure the effectiveness of our algorithms.

Effectiveness Measurement: As discussed in Section 3, the baseline optimal solution is extremely slow as it takes more than 24 hours to only process 40 POIs dataset. There is no straightforward way to measure the optimality of a set of points compared to the optimal skyline set. Therefore, we propose a formula to compare the results of our algorithms to the optimal results and provide an optimality metric Opt, defined as follows:

$$Opt = 1 - \sum_{x \in Sky} = \frac{minDis(x, tp)}{minDis(x, dp)}$$

Where, Sky is the set of optimal skyline points, $minDis(x, y)$ is the minimum graph distance required for a point y to dominate point x, tp is the point with the minimum $minDis$ from the examined points, and dp is the best case point

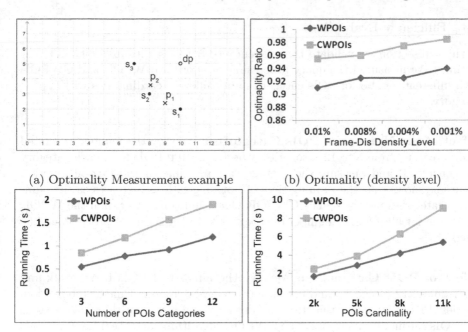

(a) Optimality Measurement example   (b) Optimality (density level)

(c) Running Time (categories number)   (d) Running Time (cardinality)

Fig. 5. Algorithms Evaluation

that is dominated by all optimal skyline points. For example in Figure 5a, when $Sky = \{s_1, s_2, s_3\}$ are the optimal skyline points, the best case point should have the coordinates $(10, 11)$ as shown in the figure to be the best point dominated by all skyline points. In addition, the $minDis(s_1, p_1)$ is 0.5 because the point p_1 needs to move down on the y axis by 0.5 in order to dominate s_1 and so on.

Effectiveness Experiments: We use the optimality metric Opt to evaluate the performance of our algorithms (WPOIs, CWPOIs) compared to the optimal results for 20 different queries. We also vary the density level of the framework to reflect different performance levels. It can be seen from Figure 5b that, the optimality for both algorithms increase as the density level decreases due to higher network distance estimation. In addition, although both algorithms have high optimality metric value(over 0.9), the CWPOIs outperforms the WPOIs for all density levels. This is because CWPOIs algorithm considers clustered POIs when finding skyline trips while WPOIs algorithm only considers the distance between a POI and the query points.

5.3   Efficiency Evaluation

In this subsection we validate the efficiency of both algorithms under different settings. We measured the query processing time of both WPOIs and CWPOIs with different number of POIs categories and different cardinalities within each category.

Effect of the Number of POIs Categories: We vary the number of related POI categories from 3 to 12 categories, where only 100 POIs from each category are considered and measure the average running time of 100 random queries. Figure 5c shows that the running time of both algorithms increases as the number of categories increases. The CWPOIs algorithm run slower than the WPOIs algorithm for all different number of categories, which is due to processing clustered POIs.

Effect of POIs Cardinality: We vary the number of POIs between 2k and 11k from all 12 categories and measure the average running time of 100 random queries. Figure 5d shows that the running time of both algorithms increases as the POIs cardinality increases. The CWPOIs algorithm can take up to 9 seconds to answer a query when all POIs from all 12 categories are used. However, in reality it is not common for a user to plan to visit 12 different POI categories in one trip. In addition, the running time can be tolerated considering the near optimal results obtained by both algorithms for an NP-hard problem. Both algorithms are more than four orders of magnitude faster than the baseline solution.

6   Conclusion and Future Work

We proposed a new path finding problem, *STMPC*, which finds the skyline trips of multiple POI categories between two points based on cost and distance. We define the problem in road network settings and proved it to be NP-hard. We proposed an independent framework to estimate the network distance. This framework is based on precomputing and storing the distances between POIs and some geographical regions in the network. We also proposed two interesting heuristic algorithms, which are WPOIs and CWPOIs algorithms. The CWPOIs algorithm considers clustered POIs when nominating skyline candidate trips. As shown in the experiments section, both algorithms return skyline trips that are close to optimal trips within reasonable running time when processing a large real dataset. Our algorithms are four orders of magnitude faster than the naive optimal solution.

For future work we will consider solving the same problem when there are multiple quality dimensions (e.g. distance, cost, rating, number of stops ... etc). We might also investigate improving the memory consumption of the proposed framework by only storing network distance between different geographical regions.

Acknowledgments. The author Saad Aljubayrin is sponsored by Shaqra University, KSA. This work is supported by Australian Research Council (ARC) Discovery Project DP130104587 and Australian Research Council (ARC) Future Fellowships Project FT120100832.

References

1. Ali, M.E., Tanin, E., Zhang, R., Kulik, L.: A motion-aware approach for efficient evaluation of continuous queries on 3d object databases. The VLDB Journal The International Journal on Very Large Data Bases **19**(5), 603–632 (2010)
2. Aljubayrin, S., Qi, J., Jensen, C.S., Zhang, R., He, Z., Wen, Z.: The safest path via safe zones. In: ICDE (2015)
3. Borzsony, S., Kossmann, D., Stocker, K.: The skyline operator. In: Proceedings of the17th International Conference on Data Engineering, pp. 421–430. IEEE (2001)
4. Chen, H., Ku, W.S., Sun, M.T., Zimmermann, R.: The multi-rule partial sequenced route query. In: Proceedings of the 16th ACM SIGSPATIAL International Conference on Advances in Geographic Information Systems. p. 10. ACM (2008)
5. Comer, D.: Ubiquitous b-tree. ACM Computing Surveys (CSUR) **11**(2), 121–137 (1979)
6. Demiryurek, U., Banaei-Kashani, F., Shahabi, C.: Efficient K-Nearest Neighbor Search in Time-Dependent Spatial Networks. In: Bringas, P.G., Hameurlain, A., Quirchmayr, G. (eds.) DEXA 2010, Part I. LNCS, vol. 6261, pp. 432–449. Springer, Heidelberg (2010)
7. Deng, K., Zhou, X., Shen, H.T.: Multi-source skyline query processing in road networks. In: IEEE 23rd International Conference on Data Engineering, ICDE 2007, pp. 796–805. IEEE (2007)
8. Dijkstra, E.W.: A note on two problems in connexion with graphs. Numerische Mathematik **1**(1), 269–271 (1959)
9. Dorigo, M., Gambardella, L.M.: Ant colonies for the travelling salesman problem. BioSystems **43**(2), 73–81 (1997)
10. Eunus Ali, M., Zhang, R., Tanin, E., Kulik, L.: A motion-aware approach to continuous retrieval of 3d objects. In: IEEE 24th International Conference on Data Engineering, ICDE 2008, pp. 843–852. IEEE (2008)
11. Guttman, A.: R-trees: a dynamic index structure for spatial searching, vol. 14. ACM (1984)
12. Huang, X., Jensen, C.S.: In-Route Skyline Querying for Location-Based Services. In: Kwon, Y.-J., Bouju, A., Claramunt, C. (eds.) W2GIS 2004. LNCS, vol. 3428, pp. 120–135. Springer, Heidelberg (2005)
13. Huang, Y.K., Chang, C.H., Lee, C.: Continuous distance-based skyline queries in road networks. Information Systems **37**(7), 611–633 (2012)
14. Jang, S.M., Yoo, J.S.: Processing continuous skyline queries in road networks. In: International Symposium on Computer Science and its Applications, CSA 2008, pp. 353–356. IEEE (2008)
15. Jensen, C.S., Kolářvr, J., Pedersen, T.B., Timko, I.: Nearest neighbor queries in road networks. In: Proceedings of the 11th ACM International Symposium on Advances in Geographic Information Systems, pp. 1–8. ACM (2003)
16. Kanza, Y., Levin, R., Safra, E., Sagiv, Y.: Interactive route search in the presence of order constraints. Proceedings of the VLDB Endowment **3**(1–2), 117–128 (2010)

17. Kossmann, D., Ramsak, F., Rost, S.: Shooting stars in the sky: An online algorithm for skyline queries. In: Proceedings of the 28th International Conference on Very Large Data Bases, pp. 275–286. VLDB Endowment (2002)
18. Kriegel, H.P., Renz, M., Schubert, M.: Route skyline queries: A multi-preference path planning approach. In: 2010 IEEE 26th International Conference on Data Engineering (ICDE), pp. 261–272. IEEE (2010)
19. Li, F., Cheng, D., Hadjieleftheriou, M., Kollios, G., Teng, S.-H.: On Trip Planning Queries in Spatial Databases. In: Medeiros, C.B., Egenhofer, M., Bertino, E. (eds.) SSTD 2005. LNCS, vol. 3633, pp. 273–290. Springer, Heidelberg (2005)
20. Mouratidis, K., Lin, Y., Yiu, M.L.: Preference queries in large multi-cost transportation networks. In: 2010 IEEE 26th International Conference on Data Engineering (ICDE), pp. 533–544. IEEE (2010)
21. Nutanong, S., Zhang, R., Tanin, E., Kulik, L.: The v*-diagram: a query-dependent approach to moving knn queries. Proceedings of the VLDB Endowment **1**(1), 1095–1106 (2008)
22. Pyle, D.: Data preparation for data mining, vol. 1. Morgan Kaufmann (1999)
23. Shahabi, C., Kolahdouzan, M.R., Sharifzadeh, M.: A road network embedding technique for k-nearest neighbor search in moving object databases. GeoInformatica **7**(3), 255–273 (2003)
24. Sharifzadeh, M., Kolahdouzan, M., Shahabi, C.: The optimal sequenced route query. The VLDB Journal **17**(4), 765–787 (2008)
25. Sharifzadeh, M., Shahabi, C.: The spatial skyline queries. In: Proceedings of the 32nd International Conference on Very Large Data Bases, pp. 751–762. VLDB Endowment (2006)
26. Shipman, D.W.: The functional data model and the data languages daplex. ACM Transactions on Database Systems (TODS) **6**(1), 140–173 (1981)
27. Tan, K.L., Eng, P.K., Ooi, B.C., et al.: Efficient progressive skyline computation. VLDB **1**, 301–310 (2001)
28. Tian, Y., Lee, K.C., Lee, W.C.: Finding skyline paths in road networks. In: Proceedings of the 17th ACM SIGSPATIAL International Conference on Advances in Geographic Information Systems, pp. 444–447. ACM (2009)
29. Vranken, W.F., Boucher, W., Stevens, T.J., Fogh, R.H., Pajon, A., Llinas, M., Ulrich, E.L., Markley, J.L., Ionides, J., Laue, E.D.: The ccpn data model for nmr spectroscopy: development of a software pipeline. Proteins: Structure, Function, and Bioinformatics **59**(4), 687–696 (2005)
30. Zhang, R., Lin, D., Ramamohanarao, K., Bertino, E.: Continuous intersection joins over moving objects. In: IEEE 24th International Conference on Data Engineering, ICDE 2008, pp. 863–872. IEEE (2008)
31. Zhang, R., Qi, J., Lin, D., Wang, W., Wong, R.C.W.: A highly optimized algorithm for continuous intersection join queries over moving objects. The VLDB Journal The International Journal on Very Large Data. Bases **21**(4), 561–586 (2012)

Keyword-Aware Dominant Route Search for Various User Preferences

Yujiao Li, Weidong Yang$^{(\boxtimes)}$, Wu Dan, and Zhipeng Xie

School of Computer Science, Shanghai Key Laboratory of Data Science,
Fudan University, Shanghai, China
liyujiaocs@hotmail.com, {yweidong,wudan0425}@gmail.com,
xiezp@fudan.edu.cn

Abstract. Route search has been studied extensively. However existing solutions for route search are often insufficient in offering users the flexibility to specify their requirements. Recently, a new kind of keyword-aware optimal route (KOR) query is proposed for finding an optimal route such that it covers a set of user-specified keywords, it satisfies a budget constraint (e.g., time), and the total popularity of the route is maximized. For example, a user may search for the route passing through *cinema* and *bookstore* within a travel time of 3 hours, which has the highest total rating score. KOR only returns one result regardless of user preferences:however some may care more about *bookstore*, while others think *cinema* more important. Apparently, it is not user-friendly for users to specify their own preference explicitly. To meet the various user preferences, this paper presents a new route search query called Keyword-aware Dominant Routes (KDR) query, which returns all possible preferable routes, covering a set of keywords under a specified budget constraint. We devise a novel algorithm DR for efficiently finding the exact result of a KDR query and a heuristic algorithm FDR to approximately answer a KDR query. Experimental results show that our algorithms are efficient.

1   Introduction

Route search has been an important problem that has application in online map services and location based services. Recently, keyword route search has been studied for route planning [1,2,3], which allows users to specified their interests for route search. In the Trip Plan Query (TPQ)[1], the user can specify a set of keyword category, and the TPQ retrieves the shortest route between the specified source and destination that covers all the keywords. To further improve the flexibility for users to specify their requirements for route search, the keyword-aware optimal route (KOR) query is proposed, which aims to find an optimal route such that it covers a set of user-specified keywords, satisfies a budget constraint (e.g., time), and the total popularity (or ratings) of the route is maximized. An example KOR query is to search for the route passing through

© Springer International Publishing Switzerland 2015
M. Renz et al. (Eds.): DASFAA 2015, Part II, LNCS 9050, pp. 207–222, 2015.
DOI: 10.1007/978-3-319-18123-3_13

cinema and *bookstore* within a travel time of 3 hours, which has the highest total rating score. Here we assume that each Point of Interest (PoI) is associated with a user rating score, which is available in many websites (e.g., TripAdvisor and FourSquare). The KOR query only returns one optimal result by treating all the keyword equally. However, a user may have different preferences toward the query keywords—a user may think *cinema* more important, and the other may think *bookstore* more important for her route search. The KOR query fails to capture the different user preferences.

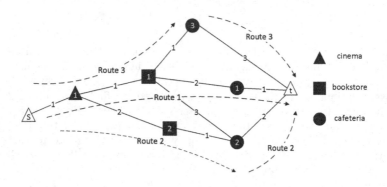

Fig. 1. Finding a preferable route from s to t

We further illustrate the problem using an example. Suppose that Lucy wants to find a route in a city such that "the route starts from her hotel and ends at the airport in the city, passing by a *cinema*, a *cafeteria* and a *bookstore*, the total time spending on the road is within 6 hours, and she expects that the locations covered by the route have the highest rating."

As shown in Fig. 1, s (her hotel) and t (the airport) are start location and target location, respectively. The value on each node represents the rating of the location while the value on each edge represents the time cost of traveling from one location to another. And the distinct shapes of each node indicate the different kinds of keywords. From the query by Lucy, we know that the start location is hotel, target location is airport, and keywords are *cinema, cafeteria* and *bookstore*.

If Lucy cares more about *bookstore*, Route 2 will be the best choice for her; If Lucy regards *cafeteria* more important, Route 3 is preferable. However, the KOR query only returns Route 2 as the result route that has the maximum value for the sum of all the ratings of the places covered by the route.

To address the problem of the KOR query, one straightforward solution is to request users to enter their preferences to each of her query keywords. This, however, is not user friendly because it puts the burden on the users for specify their preferences, which may not always be easy.

In contrast, we propose to find both routes for Lucy, namely Route 2 (the *bookstore* enjoys larger popularity) and Route 3 (*cafeteria* is the most important). Then Lucy can choose one according her preference. To meet the need, we propose a new type of query, called keyword-aware dominant route (KDR) query. The KDR query can be defined in two steps for easy understanding: (1) We consider the set of feasible routes in which each route covers a set of specified query keywords, and satisfies a specified budget constraint (e.g., travel time from the source to target); and (2) KDR returns a subset of dominant routes from all the feasible routes such that each returned dominant route is not dominated by any other feasible route in terms of the ratings of places. Intuitively, we can guarantee that the most preferable route for a user with any preferences on the ratings of places covering keywords will be in our result set.

Similar to the original KOR query, answering the KDR query is also NP-hard. In this paper, we propose an exact algorithm called DR, which only considers the nodes covering the query keywords and skips the irrelevant nodes in the road network to avoid unnecessary cost of time and space. Based on DR, we also propose a more efficient heuristic algorithm FDR to answer a KDR query approximately. FDR assumes that people seldom turn backward during their journey.

The contributions of our study are as follows:

- We introduce the notion of keyword-aware route dominance and formally define a new type of route search query, called KDR query.
- We propose an efficient exact algorithm called DR, and a heuristic algorithm called FDR.
- We conduct experiments to study the efficiency of our approaches.

The rest of the paper is organized as follows. The problem is formally defined in Section 2. In Section 3 the algorithm DR and the algorithm FDR are presented. The experimental results are presented in Section 4. Section 5 surveys the related work. Section 6 concludes this paper.

2   Keyword-Aware Dominant Route (KDR) Query

In this section, we define the proposed keyword-aware dominant route query.

To ease the presentation, we use a directed graph $G = (V, E)$ to represent the road network [5], where V is the set of vertices (nodes) representing locations and E is the set of edges (arcs). Our discussion can be easily extended to undirected graphs. Each edge in E, denoted as (v_i, v_j), is associated with a cost value $b(v_i, v_j)$ which can be travel duration, travel distance or travel cost. Each node in V is associated with a set of keywords represented by $v.\psi$, and each node is also associated with a rating score $o(v)$, which indicates the popularity of the location.

Route $R = (v_0, v_1, ..., v_n)$ is a path that goes through v_0 to v_n sequentially, following the relevant edges in G . The **budget score**(BS) of a route R is defined as sum of the cost of its edges, i.e. $BS(R) = \sum_{i=1}^{n} b(v_{i-1}, v_i)$. The keyword set that route R covers can be denoted by $\psi(R) = \bigcup_{v \in R} v.\psi$.

A Keyword-aware Dominant Route (KDR) query is represented as a quadruple $Q = < s, t, \psi, \Delta >$, where s is the source location, t is the target location, ψ is a set of keywords, and Δ is a budget limit. Before we explain the semantic meaning of the KDR query, we first introduce the definition of feasible route.

We denote $\mathbf{R}_{s,t}$ as the set of all paths between the start location s and target location t. A route in $\mathbf{R}_{s,t}$ is called a **feasible route** if its budget score is under the specified budget limit Δ and covers all the query keywords in ψ. The set of all feasible routes are denoted by $\mathbf{FR}_Q$. We next explain the notion of keyword-aware route dominance based on objective vector of each route.

Definition 1. *(Objective Vector). Given a route $R = (v_0, v_1, ..., v_n)$ and a keyword set ψ, the objective vector of R denote the value of route with respect to ψ. Each element in the objective vector of route R is the maximum objective value (i.e., the rating score) of the nodes covered by each keyword, i.e., $OV(R) = (o_1, o_2, ...o_k)$, $o_i = \max \{o(v) | v \in R, v.\psi = \psi[i]\}$ where k is the size of the query keyword set ψ, $\psi[i]$ is the i-th keyword of ψ according to alphabetic order.*

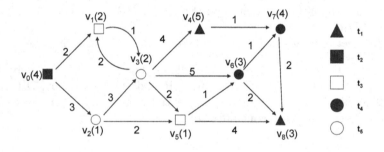

Fig. 2. Example of G

Fig. 2 shows an example of the graph G. We consider five keywords $t_1, \cdots, t_5$, each represented by a distinct shape, respectively. For simplicity, each node contains a single keyword with an objective score (rating score) inside a bracket. And on each edge, there is a number which represents the budget value. For example, given the route $R = (v_0, v_2, v_5, v_8)$ and $\psi = \{t_1, t_2\}$, we have $OV(R) = (3, 4)$ and $BS(R) = 3 + 2 + 4 = 9$.

Definition 2. *(Dominance Relation between Feasible Routes). Given a KDR query $Q = < s, t, \psi, \Delta >$, feasible routes R_p and R_q whose objective vectors are $OV(R_p) = (p_1, p_2, ..., p_k)$ and $OV(R_q) = (q_1, q_2, \cdots, q_k)$ respectively. We say*

that R_p is dominated by R_q, denoted by $R_p \prec R_q$ iff 1) $p_i \leq q_i$ holds for all $i = 1, \ldots, k$, and there exists $j \in \{1, \ldots, k\}$ such as $p_j < q_j$; or 2) and the nodes that contribute on objective vector in R_p and R_q are the same, i.e.
$V_p = \{u|o(u) = max\{o(v)|v \in R_p, v.\psi = u.\psi\}, u \in R_p\}$ is the same as
$V_q = \{u|o(u) = max\{o(v)|v \in R_q, v.\psi = u.\psi\}, u \in R_q\}$ and $BS(R_p) > BS(R_q)$.

The KDR query is to find the dominant routes set from feasible routes.

Definition 3. *(Keyword-aware Dominant Route). Given a KDR query $Q =<$ $s, t, \psi, \Delta >$, a dominant route R_p is a feasible route which satisfies there exists no route $R_q \in$ **FR**$_Q$ such as $R_p \prec R_q$ .We denote by* **DR**$_Q$ *the set of dominant routes for query Q. We denote the set of non-dominant paths for Q by* $\overline{\textbf{DR}_Q} =$ **FR**$_Q$ − **DR**$_Q$.

In the example graph in Fig. 2, given a query $Q =< v_0, v_8, \{t_3, t_4\}, 8 >$, the dominant routes are $R_1 = (v_0, v_1, v_6, v_8)$ and $R_2 = (v_0, v_5, v_7, v_8)$ with objective vector $OV(R_1) = (2, 3)$ and $OV(R_2) = (1, 4)$, and budget score $BS(R_1) = 9, BS(R_2) = 8$, respectively.

Lemma 1. *The problem of answering the KDR query is NP-hard.*

Proof. This problem can be reduced from the generalized traveling salesman problem, which is NP-hard. The general traveling salesman problem is to find a path with starting and ending at two specified nodes such that it goes through each group once in the graph, where nodes are divided into groups. If we set all the objective score in KDR to the same value, it is equivalent to the generalized traveling salesman problem.

3 Algorithm

We propose an exact algorithm in Section 3.1 and a heuristic algorithm in Section 3.2.

3.1 DR Search Algorithm

Traditional solutions to route search problem in road network are usually based on breadth-first or depth-first traversal on graph. They start search from the start location and extend to target literally with adjacent nodes. After the traversal, it selects the non-dominated ones of all the feasible routes as the answer to the query[17,18]. However, this kind of search is computationally prohibitive since the huge solution space in the spatial graph.

To this end, we propose an efficient route search algorithm, called DR. In DR algorithm, we add the shortest path from the start to the end as an initial path and iteratively refine the route by inserting nodes containing uncovered keywords. Using designed pruning strategies, the algorithm searches all the routes with different keyword node combination. After all the feasible routes are generated, we compute the exact path of the remaining feasible routes.

Before introducing the prune strategies, we will give some definitions.

Definition 4. *(Keyword Node). Given a KDR query $Q =< s, t, \psi, \Delta >$, we use $\mathbf{K}_\psi$ to denote the **keyword node set** of Q, that is,each node in $\mathbf{K}_\psi$ is called a **keyword node**.*

Each route R can be divided into segments by keyword node. For example, $R = (s, v_1, v_2, v_3, v_4, v_5, t)$ with v_3 as a keyword node can be divided into $R_1 = (s, v_1, v_2, v_3)$ and $R_2 = (v_3, v_4, v_5, t)$.

Definition 5. *(Keyword-node Route). Given a KDR query $Q =< s, t, \psi, \Delta >$, with keyword node set $\mathbf{K}_\psi$, If each of the segments of a route divided by keyword node is a shortest path from its start to its end, we call the whole route a **keyword-node route**. All the keyword-node routes of query Q comprise the **Keyword-node Route set**, denoted by $\mathbf{KR}_{s,t,\psi}$.*

Clearly, if a route is confirmed to be a dominant route, it must be a keyword-node route. A Keyword-node Route is fixed once we specify the keyword node passed by in the route. Thus route $R = (s, v_1, v_2, v_3, v_4, v_5, t)$ can be denoted by a Keyword-node Sequence (KN-sequence) such as $KS = [s, v_3, t]$.

Definition 6. *(Potential Route). Given a KDR query $Q =< s, t, \psi, \Delta >$, A route is called a potential route if and only if $R \in \mathbf{KR}_{s,t,\psi}$, $\psi(R) \subset \psi$, and $BS(R) \leq \Delta$.*

We use $d(v_i, v_j)$ to represent the shortest distance between node v_i and v_j which is preprocessed before querying [6]. For a keyword-node route R whose KN-sequence is$KS = [s, v_1, v_2, ...v_k, t]$. Its budget score is computed by Eq. 1 and its objective vector is computed by Eq. 2.

$$BS(KS) = BS(R) = \sum_{i=0}^{k} d(v_i, v_{i+1}) \tag{1}$$

$$OV(KS) = (o_1, o_2, \ldots o_k), o_i = \max\{o(v)|v \in KS, v.\psi = \psi[i]\} \tag{2}$$

The budget score of KN-sequence KS is the sum of the shortest distance between any two contiguous keyword nodes, which is the same as the budget score of R. The objective vector of KN-sequence KS is computed similar to route R, and we select objective score which is the maximum for each keyword among all the nodes objective score in KS.

For example, given a query $Q =< v_0, v_8, \{t_3, t_4\}, 8 >$, R is an initial route whose KN-sequence is $[v_0, v_8]$. The route represented by KN-sequence $[v_0, v_1, v_8]$ is a potential route whose objective vector is $(2, 1)$.

Definition 7. *(Route Refinement Operation $\otimes$). Route refinement $R \otimes v$ is to insert a keyword node v to a Keyword-node route R with the least budget score by adjusting the order of keyword nodes in KN-sequence. For a route R belongs to KR, whose KN-sequence is $KS = [s, v_1, v_2, ...v_k, t]$ and a keyword node v, $v \in K, \forall 0 \leq i \leq k, KS' = [s, v_{n_1}, v_{n_2}, ...v_{n_i}, v, v_{n_{i+1}}...v_{n_k}, t]$ and $R \otimes v = argmin_{R'} BS(R')$ (i=0 means node s)*

After some route refinements, a potential route R may become a feasible route.

Definition 8. *(Potential Route Ancestor). For route R and R' of keyword node route, we say route R is the parent of R' if $R' = R \otimes v$ and we say R is the ancestor of R' if R' can be generated by several route refinements in R, $R' = R \otimes v_1 \otimes v_2 \otimes \ldots \otimes v_k$.*

We illustrate the route refinement operation with an example of generating route R of KN-sequence $[v_0, v_1, v_6, v_8]$. From the KN-sequence $[v_0, v_8]$, we first generate $[v_0, v_1, v_8]$ by inserting keyword node v_1. Then from $[v_0, v_1, v_8]$, we generate $[v_0, v_1, v_6, v_8]$ by inserting v_6. We say that $[v_0, v_8]$ is the parent of $[v_0, v_1, v_8]$ and the ancestor of $[v_0, v_1, v_6, v_8]$.

Given a KDR query Q, each result route needs to match all the keywords. The KN-sequence of the initial route consists of start point and end point, and then we insert the keyword nodes uncovered one by one to refine the route until all keywords are covered. During the refinement, we omit the intermediate nodes in the shortest path of any two sequenced nodes. After obtaining the KN-sequence of dominate routes, we calculate the nodes located between any two keyword nodes u and v using existing shortest path algorithm.

To reduce the calculation in the procedure of a routes refinement, we must seek for some reasonable and effective prune strategies.

One hard constraint of our KDR problem is the budget limit—if a route violates the limit, then it should be deleted and should not be expanded anymore because any of its descendants will violate the budget limit too. It can be proven as follows.

Lemma 2. *If $R' = R \otimes v$ and $R', R \in KR$, we can get $BS(R) \leq BS(R')$.*

Proof. Denote the KN-sequence of R and R' as $KS = [s, v_1, \ldots, v_k, t]$ and $KS'' = [s, v_{n_1}, \ldots v_{n_i}, v, v_{n_{i+1}}, \ldots, v_{n_k}, t]$, respectively. Since R is the shortest route passing by $v_1, \ldots, v_k$, route R' with $KS'' = [s, v_{n_1}, \ldots v_{n_i}, v_{n_{i+1}}, \ldots, v_{n_k}, t]$ satisfies $BS(R'') = BS(KS'') = \sum_{i=0}^{k} d(v_{n_i}, v_{n_{i+1}}) \geq BS(R)$. The budget score of R' can be computed as follows

$$BS(R') = \sum_{j=0}^{k} d(v_{n_j}, v_{n_{j+1}}) - d(v_{n_i}, v_{n_{i+1}}) + d(v_{n_i}, v) + d(v, v_{n_{i+1}}) \geq BS(KS'')$$

So, we can get $BS(R) \leq BS(R')$

A potential route will be dominated if its objective vector is dominated by the objective vector of a feasible route on matched keywords of the potential route and the objective vector of the feasible route has the maximum scores in the whole graph on unmatched keywords. We have the following lemma.

Lemma 3. *For potential route R, if there exists a feasible route R' such that R' dominates R on the set of keywords $\dot{\psi} = \psi(R)$ and $OV(R')$ has the maximum scores in the whole graph on $\ddot{\psi} = \psi - \dot{\psi}$, then R is dominated by R'.*

Algorithm 1. DR Algorithm($Q = < s, t, \psi, \Delta >, G = (V, E)$)

---

1 Initialize a queue Q and a array F for KN-sequence of feasible routes;
2 $Q.enqueue(R = [s, t])$;
3 **while** Q *is not empty* **do**
4     $R \leftarrow Q.dequeue$;
5   **if** $|R.\psi| < |\psi|$ **then**
6   **forall the** *node u which contains uncovered keyword in R* **do**
7           $\hat{R} = R \otimes u$;
8   **if** $BS(\hat{R}) < \Delta$ *and* $FdominatePR(F, \hat{R})$ *is false* **then**
9              $Q.enqueue(\hat{R})$;

10   **else**
11   **if** $FdominateFR(F, R)$ *is false* **then**
12   add R to F;
13   remove from F the routes that are dominated by R
14   compute the exact route of each route in F;

Proof. Obvious from Definition 2.

Based on these lemmas, algorithm DR is described in Algorithm 1. In algorithm DR, we insert the root KN-sequence $[s, t]$ into the queue Q, which organizes all the KN-sequences of potential routes during search process. While Q is not empty, we iteratively get a potential route and expand it with an unmatched keyword node (lines 6–9). If the expanded route satisfies the budget constraint and is not dominated by a feasible route ($FdominatePR(F, R)$ is false), we insert the expanded route into Q (lines 8–9). Otherwise, if the route is a feasible route, we use function $FdominatePR(F, R)$ to check whether route R is dominated by a route in F. We filter out all the routes in F that are dominated by R (lines 13–14).

After the expansion and refining of routes, we output the final dominant routes by computing the exact path of each remaining route R in F. The search process takes $O(n^{2k} + n^k k!)$, where n is the maximum number of nodes that a keyword matches, and k is the number of query keywords, which is usually small.

Theorem 1. *The DR algorithm finds the exact answer for any KDR query.*

Proof. Since a dominate route must be a keyword-node route, the whole search space equals to the permutation of keyword nodes without any pruning strategy. In spite of the pruning strategies we employ (Lemma 2 and Lemma 3), the search space of DR algorithm is the combination of the keyword nodes such that the potential routes having the least budget scores among the routes with the same keyword nodes should be considered. As to Definition 2, the routes whose budget scores are not the least can be dominated and such routes must be deleted after their refinement.Hence, we complete the proof.

3.2   FDR Algorithm

Based on the the DR algorithm, we develop the FDR algorithm, which is a heuristic algorithm.FDR is inspired by the observation that people normally do not move backward during their journeys.

FDR just expands current routes which match part of query keywords by following a forward direction toward the target node, i.e., each expanding will make the last node of the route closer to the destination. Specifically, when we choose a node containing an unmatched keyword to expand the current route, the selected node should make the route after expanding becomes closer to the target node. With the search strategy, FDR would explore few keywords nodes than does the DR algorithm, but may miss result routes. We call the algorithm FDR (Forward Expanding DR).

Given a KDR query $Q = <s, t, \psi, \Delta>$, for a keyword-node route R whose KN-sequence is $[s, v_1, v_2, \ldots, v_n]$, we say it is a **partial route** if $v_n \neq t$. We say the expansion from v_i to v_{i+1} obeys the forward expansion when $d(v_{i+1}, t) - d(v_i, t) \leq \theta$, $\theta \geq 0$. Here $\theta = 0$ defines a strict forward expansion, i.e., after the expansion, the distance from v_{i+1} to t must be smaller than or equal to the distance from v_i to t; A positive θ defines a weak forward expansion, which allows a constrained backward under length limit θ.

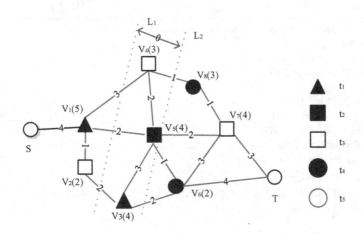

Fig. 3. Example of Forward Route Expansion

As shown in Figure 3, given a target node as node t, for a partial route R that ends at v_5, the two parallel lines L_1 and L_2 seperate the whole graph to three parts. If $\theta = 0$, nodes in the left of L_2 violate the direction $v_5 \rightarrow T$ constraint, while nodes in the right of L_2 are qualified to be expanded. L_1 indicates the situation while $\theta = 2$.

For a keyword-node route R whose KN-sequence is $[s, v_1, v_2, \ldots, v_n]$, we use notation $R \oplus v$ to represent the forward route expansion and the KN-sequence of the expanded route is $[s, v_1, v_2, \ldots, v_n, v]$.

Based on the forward expansion mechanism, we choose a new unmatched keyword node and insert it to the end of the current route iteratively until the route becomes a feasible route. For example, in Figure 3, given KDR $Q =< s, t, \{t_1, t_2, t_3\}, 15 >$, the partial route R with KN-sequence $[s, v_1, v_5]$ whose last expanded node is v_5. Only v_7 can be used to expand R if $\theta = 0$.

Based on the forward expansion mechanism, we have another pruning strategy:

Lemma 4. *(Partial Route Domination). For two partial routes R_1 and R_2 we say R_1 dominates R_2, if the last expanded node of R_1 and that of R_2 are the same, $R_1.\psi = R_2.\psi$, $OV(R_2) \prec OV(R_1)$, and $BS(R_1) \leq BS(R_2)$.*

For example, consider two partial routes R_1 and R_2 in Figure 3. Their KN-sequences are $[s, v_1, v_4, v_5]$ and $[s, v_1, v_2, v_5]$, respectively, and their last expanded node is V_5, and they match the same set of keywords, $OV(R_2) \prec OV(R_1)$, and $BS(R_1) = 9$ is smaller than $BS(R_2) = 10$. Thus, no matter which nodes are inserted to expand R_1 and R_2 , the final feasible route of R_2 must be dominated by the final route of R_1.

Algorithm 2. FDR Algorithm($Q =< s, t, \psi, \Delta >$,$G = (V, E)$)

---

1 Initialize a queue Q,a array F for KN-sequence of feasible routes;
2 Array labels={};
3 $Q.enqueue(R = [s])$;
4 **while** Q *is not empty* **do**
5 $\quad$ $R \leftarrow Q.dequeue$;
6 $\quad$ **if** $|R.\psi| < |\psi|$ **then**
7 $\quad\quad$ $v \leftarrow$ lastnode(R);
8 $\quad\quad$ filter(labels(v), R);
9 $\quad\quad$ **foreach** *node u that contains uncovered keyword in R and isForward(v, u)=true* **do**
10 $\quad\quad\quad$ $\hat{R} = R \oplus u$;
11 $\quad\quad\quad$ **if** $BS(\hat{R}) < \Delta$ *and* $FdominatePR(F, \hat{R})$ *is false* **then**
12 $\quad\quad\quad\quad$ $Q.enqueue(\hat{R})$;

13 $\quad$ **else**
14 $\quad\quad$ **if** $FdominateFR(F, R)$ *is false* **then**
15 $\quad\quad\quad$ add R to F;
16 $\quad\quad$ remove from F the routes that are dominated by R;
17 $\quad$ compute the exact route of each route in F;

The pseudo code of FDR is outlined in Algorithm 2 based on the DR framework.

In Algorithm 2, according to the forward expansion mechanism, FDR expands a potential route by an unmatched keyword node at the end of the current route. Similar to the DR algorithm, we iteratively get a potential route from Q, and expand it with a keyword node by following the forward expansion mechanism. If the expanded route satisfies the budget limit and is not dominated by a feasible route ($FdominatePR(F, R)$ is false), we add it to Q (line 12). Otherwise, if the route is a feasible route, we use function $FdominateFR(F, R)$ to check whether it is dominated by another one; if not, the route is appended to F. In addition, we check if any route in F can be dominated by the new feasible route R and remove such routes from F (line 16). The worst time complexity of FDR is the same as that of DR. But in practice, FDR checks much fewer nodes.

4   Experiment

To evaluate the performance of the proposed algorithms for KDR query, we use real-world road-network data extracted from the OpenStreetMap[1] database. We focus on the city of Shanghai, and we obtained 6,050 POIs with information on their latitudes and longitudes. These POIs are labeled with nine categories including eating, drink- ing, leisure place, etc. Three datasets (denoted by Node2000, Node4000, Node6000) are generated from the data of Shanghai, which contain 2,000 POI nodes, 4,000 POI nodes and 6,000 POI nodes, respectively. The distance serves as the budget score of the edge. To assign rating values for different POIs, we randomly generate it from a uniform distribution over $\{1,2,3,4,5\}$. Those rating values are used as the objective scores, which will be maximized in the KDR query.

We study the performance of our algorithms by varying budget limit, the number of keyword nodes, the number of query keywords, and the size of dataset. Four query sets are generated by varying budget limit Δ from 10,000 to 50,000, varying number of keyword nodes $|\mathbf{K}_\psi|$ from 100 to 500, varying number of keywords from 2 to 5 and varying dataset. The default values of some parameters are as follows: $\Delta = 20,000$, $|\mathbf{K}_\psi|=200$ and $|\psi| =3$ .The default dataset is Node4000. Each set comprises 50 queries. The starting and ending locations are selected randomly. The query keywords are generated randomly while making sure the requirement of $|\mathbf{K}_\psi|$ is satisfied.

All algorithms were implemented in VC++ and run on an Intel(R) Core(TM) i7-3770 CPU @3.40GHZ.

4.1   Efficiency

This set of experiments is to study the efficiency of the proposed algorithms with variation of the budget limit, number of keyword nodes, number of query keywords and the size of dataset.

---

[1] www.openstreetmap.org

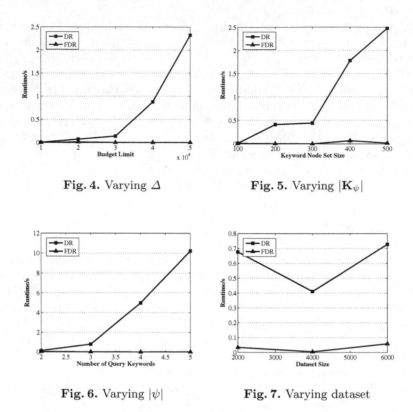

Fig. 4. Varying Δ **Fig. 5.** Varying $|\mathbf{K}_\psi|$

Fig. 6. Varying $|\psi|$ **Fig. 7.** Varying dataset

Varying the Budget Limit Δ. Fig. 4 shows the runtime on Node4000 dataset with the variation of Δ. At each Δ, the average runtime is reported over 10,000, 20,000, 30,000, 40,000, and 50,000 with a number of keywords being 3.The runtime of DR and FDR grows when budget limit increases as a larger limit corresponds to more keywords nodes. Compared to DR, the runtime of FDR increases much more slightly because the pruning strategy in direction reduces the complexity of FDR to be linear to while DR is exponential.

Varying the Size of Keyword Node Set. Fig. 5 shows the runtime of the queries of different $|\mathbf{K}_\psi|$.With the increase of keyword nodes, DR slows down and FDR runs slightly slower. This is because the complexity of DR is exponential to the size of cover size of each keyword while FDR prunes more routes.

Varying the Number of Query Keywords. Fig. 6 shows the runtime of queries with different numbers of keywords ranging from 2 to 5. The runtime of DR and FDR increases because more permutations are needed when the number of nodes increases and the keyword covering size remains the same.

Varying the Dataset. Fig. 7 shows the runtime of queries on different datasets, Node2000, Node4000, Node6000. We do not see a clear tendency of the effect of the different datasets. A possible reason is that the time is mainly affected by factors such as query keywords, Δ, start and end points, etc.

Summary. These results show that our exact method DR is efficient when the number of keywords and the number of keyword nodes are small. But when the size of keywords or the number of keyword nodes are relatively large, approximate method FDR should be employed, which can usually find routes for users to choose. Both algorithms are less affected by dataset size because their computation complexity is irrelevant to size of dataset.

4.2 Number of Returned Routes

We report the number of returned routes of the two algorithms DR and FDR.

Varying the Budget Limit Δ. Figure 8 shows the size of results of the two algorithms on Node4000 dataset with the variation of Δ. The number of keywords is set to 3. The size of result route set of DR and FDR grows when Δ increases, which permits more permutations of keyword nodes.

Varying the Number of Query Keywords. Figure 9 shows the size of route set of queries with different numbers of keywords ranging from 2 to 5. The number of returned routes of FDR decreases significantly as the querying keyword size increases. FDR may fail to return routes when the number of keywords is large.

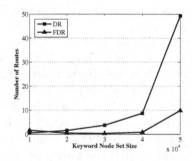   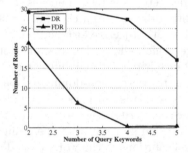

Fig. 8. Varying Δ **Fig. 9.** Varying $|\psi|$

5   Related Work

Li et al. proposed Trip Plan Query (TPQ) in spatial databases [1]. In a TPQ, the user specifies a set of categories. The TPQ retrieves the shortest route that starts at the source point, passes through at least one point from each category, and ends at the specified destination. It is a NP-hard problem. The optimal sequenced route (OSR) query studied by Sharifzadeh et al.[2] adds a constraint on the sequence imposed on the types of passing by locations without specifying the destination. Several queries based on OSR have been proposed such as multi-type nearest neighbor (MTNN) query and multi-rule partial sequenced route (MRPSR) query [7,3]. Considering the difference between keywords on nodes and query keywords, Yao et al.[15] propose the multi-approximate-keyword routing (MARK) query.

Cao et al. studied a problem called the keyword-aware optimal route (KOR) problem and use IR-Tree as index structure [4,8]. KOR is the most similar problem to KDR. Our KDR query improves upon KOR as we discussed in Introduction and the algorithms for KOR cannot be used to answer the KDR query.

Skyline queries[12] aim to find skyline objects which are superior to other object at least in one attributes, i.e., skyline queries tend to find objects which cannot be dominated by other objects. Various types of spatial queries have been extensively studied in the database literature. For example, Branch-and-Bound Skyline algorithm[13] is a progressive optimal algorithm for the general skyline query. Huang and Jensen[14] study the problem of finding locations of interest which are not dominated with respect to two attributes: the network distance to a single query location and the detour distance from the predefined route through the road network. The authors in [16] proposed two algorithms for the spatial skyline query, the R-tree-based B_2S_2 and the Voronoi-based VS_2. Multi-preference path planning [19] computes skylines on routes in a road network which concentrating on preference like distance, driving time, the number of traffic lights, etc. Our KDR query is partly inspired by the skyline query, but is significantly different from the existing skyline queries.

In the multi-objective formulation, several parameters can be associated with each edge, which allows the possibility of incorporating various criteria, such as cost, distance, time and reliability[9],etc. Multi-objective shortest path problem (MSPP) is to search non-dominated paths in the network (also called Pareto optimal[10] which is not possible to find a better value for one criterion without letting the other criteria get worse with a specified source location. MSPP is hard to solve. Hansen [9] prove that there might exist an exponential number of non-dominated solutions in the worst case. The solution that computes the complete non-dominated set includes approaches of labelling algorithm [9]. The procedures are based on ranking paths [10] and Mote's algorithm [11] The solution of MSPP cannot solve our problem since it focuses on the sum or min-max function where our KDR is kind of max-max function. And adding keyword to multi-criteria path computation is a novel generalization of MSPP with budget constraint.

6 Conclusion

In this paper, we propose the notion of keyword route dominance and define keyword-aware dominant route (KDR) query which is to find a dominant route set to meet various user preference. We devise two algorithms, i.e., DR and FDR. Results of empirical studies show that the two proposed algorithms are capable of answering KDR queries efficiently, and FDR is a good approximation of DR.

Acknowledgments. This work was partially supported by the Polar Research Program of China under Grant CHINARE2014-04-07-06, Public science and technology research funds projects of ocean under Grant 201405031.

References

1. Li, F., Cheng, D., Hadjieleftheriou, M., Kollios, G., Teng, S.-H.: On trip planning queries in spatial databases. In: Medeiros, C.B., Egenhofer, M., Bertino, E. (eds.) SSTD 2005. LNCS, vol. 3633, pp. 273–290. Springer, Heidelberg (2005)
2. Sharifzadeh, M., Kolahdouzan, M., Shahabi, C.: The optimal sequenced route query. VLDBJ **17**(4), 765–787 (2008)
3. Chen, H., Ku, W.-S., Sun, M.-T., Zimmermann, R.: The multi-rule partial sequenced route query. In: GIS (2008)
4. Cao, X., Chen, L., Cong, G., Xiao, X.: Keyword-aware optimal route search. PVLDB **5**(11), 1136–1147 (2012)
5. Cormen, T., Leiserson, C., Rivest, R., Stein, C.: Introduction to Algorithms. The MIT Press (1997)
6. Floyd, R.W.: Algorithm 97: Shortest path. Communications of the ACM **5**(6), 345 (1962)
7. Ma, X., Shekhar, S., Xiong, H., Zhang, P.: Exploiting a Page-Level Upper Bound for Multi-Type Nearest Neighbor Queries. In: Proceedings of the 14th ACM International Sym-posium on Geographic Information Systems (ACM-GIS) (2006)
8. Cao, X., Cong, G., Jensen, C.S., Ooi, B.C.: Route skyline queries: A multi-preference path planning approach. In: SIGMOD (2011)
9. Hansen, P.: Bicriterion path problems, in Multiple Criteria Decision Making Theory and Application, vol. 177. Springer Verlag, Berlin (1978)
10. Martins, E., Santos, J.: The labeling algorithm for the multiobjective shortest path problem, Departamento de Matematica, Universidade de Coimbra, Portugal, Tech. Rep. TR-99/005 (1999)
11. Mote, J., Murthy, I., Olson, D.L.: A parametric approach to solving bicriterion shortest path problems. Eur. J. Oper. Res. **53** (1991)
12. Börzsönyi, S., Kossmann, D., Stocker, K.: The Skyline Operator. In: ICDE, pp. 421–430 (2001)
13. Papadias, D., Tao, Y., Fu, G., Seeger, B.: Progressive Skyline Computation in Database Systems. ACM Trans. Database Syst. **30**(1), 41–82 (2005)
14. Huang, X., Jensen, C.S.: In-route skyline querying for location-based services. In: Kwon, Y.-J., Bouju, A., Claramunt, C. (eds.) W2GIS 2004. LNCS, vol. 3428, pp. 120–135. Springer, Heidelberg (2005)
15. Yao, B., Tang, M., Li, F.: Multi-approximate-keyword routing in gis data. In: GIS (2011)

16. Sharifzadeh, M., Shahabi, C.: The Spatial Skyline Queries. In: VLDB (2006)
17. Goldberg, A.V., Harrelson, C.: Computing the shortest path: A* search meets graph theory. In: SODA (2005)
18. Jensen, C.S., Kolrvr, J., Pedersen, T.B., Timko, I.: Nearest neighbor queries in road networks. In: Proceedings of the 11th ACM International Symposium on Advances in Geographic Information Systems (ACM-GIS), pp. 1–8 (2003)
19. Kriegel, H., Renz, M., Schubert, M.: Route skyline queries: a multi-preference path planning approach. In: ICDE (2010)

Spatial Keyword Range Search on Trajectories

Yuxing Han[1]([✉]), Liping Wang[1], Ying Zhang[2], Wenjie Zhang[3],
and Xuemin Lin[1,3]

[1] Shanghai Key Lab for Trustworthy Computing, East China Normal University,
Shanghai, China
yxhan@student.ecnu.edu.cn, lipingwang@sei.ecnu.edu.cn
[2] University of Technology, Sydney, Australia
[3] The University of New South Wales, Sydney, Australia
Ying.Zhang@uts.edu.au, {zhangw,lxue}@cse.unsw.edu.au

Abstract. With advances in geo-positioning technologies and ensuing location based service, there are a rapid growing amount of trajectories associated with textual information collected in many emerging applications. For instance, nowadays many people are used to sharing interesting experience through Foursquare or Twitter along their travel routes. In this paper, we investigate the problem of spatial keyword range search on trajectories, which is essential to make sense of large amount of trajectory data. To the best of our knowledge, this is the first work to systematically investigate range search over trajectories where three important aspects, *i.e.*, spatio, temporal and textual, are all taken into consideration. Given a query region, a timespan and a set of keywords, we aim to retrieve trajectories that go through this region during query timespan, and contain all the query keywords. To facilitate the range search, a novel index structure called IOC-Tree is proposed based on the inverted indexing and octree techniques to effectively explore the spatio, temporal and textual pruning techniques. Furthermore, this structure can also support the query with order-sensitive keywords. Comprehensive experiments on several real-life datasets are conducted to demonstrate the efficiency.

1 Introduction

The proliferation of GPS-enabled devices such as smartphones and the prosperity of location-based service have witnessed an unprecedented collection of trajectory data. Latest work on spatio-temporal trajectories includes travel time estimation [21], trajectory compression [18], route recommendation [19], frequent path finding [12], etc. In addition to spatio-temporal trajectory, semantic trajectories [2] which combine textual information with each spatial location also attract great research attention in recent years. A large amount of semantic trajectories are generated from location-based social networking services (LBSNs), such as Foursquare and Twitter. Representative work includes pattern mining [22] and activity trajectories search [23].

M. Renz et al. (Eds.): DASFAA 2015, Part II, LNCS 9050, pp. 223–240, 2015.
DOI: 10.1007/978-3-319-18123-3_14

Motivation. While significant efforts have been devoted to exploiting trajectory dataset, to the best of our knowledge, none of the existing work considers three critical aspects (*spatio, temporal, textual*) of the trajectory data at the same time during the range search processing. Previous studies either consider the spatio-temporal properties (e.g., [4,16]), or only explore spatial and textual aspects (e.g., [6,23]) of the trajectories. However, we stress that, in many real-life scenarios, three aspects of the trajectories are considered by users at the same time to retrieve desirable results. In particular, users may only be interested in the trajectory points (e.g., activities) within a particular region (e.g., a nearby area or a suburb) during a time period (say, last week or recent 10 days). Meanwhile, as keywords tagged on the trajectory points carry rich information such as activities involved and users' personal experiences, it is natural to choose the trajectories based on query keywords.

Motivated by the above facts, in this paper we study the problem of **S**patial **K**eyword **R**ange search on **T**rajectories (*SKRT*) which retrieves meaningful trajectories based on the spatio-temporal and keyword constraints. Specifically, the spatio-temporal constraint is captured by a spatial region and a timespan while a set of query keywords is used to express user's interests. A trajectory will be retrieved if the trajectory points satisfying spatio-temporal constraint contain all query keywords. Below is a motivating example.

Example 1. A travel recommendation system has collected a number of user trajectories as shown in Fig 1. Suppose a tourist wants to plan a trip in a nearby region (dotted circle with Fig 1) to enjoy wonderful local *flower* and *pizza*. Undoubtedly, it is beneficial to the tourist if the system can provide some relevant trajectory data in the last one month. In the example, trajectory R_3 is not the appropriate candidate since there is no keyword *pizza* in the query range although it does contain pizza. Similarly, R_4 does not satisfy keyword constraint because *pizza* is not covered. Therefore, only R_1 and R_2 are retrieved for this tourist.

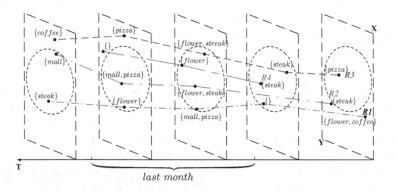

Fig. 1. Motivation Example

Challenges. In many applications, the number of trajectories might be massive, calling for efficient indexing techniques to facilitate trajectory range search proposed in this paper. The key challenge is how to effectively combine the spatio, temporal and keywords features of the objects such that a large number of non-promising trajectories can be pruned. As observed by Christoforaki et al. [5], the number of keywords issued from users in real life is generally small, typically 2-5, and hence higher priority is given to keyword feature. In this paper, we adopted *inverted index* technique and hence the query processing follows a keyword-first-pruning strategy. Only a few relevant trajectories containing query keywords will be loaded during query processing. Another advantage of inverted index is that the query with order-sensitive keywords can be naturally supported by choosing inverted lists in order. Regarding each keyword (i.e., inverted list), we also need effective spatial-temporal index structure to organize related trajectory points. As we observe that the spatial and temporal distributions of the real-life trajectories might be highly skewed for each keyword, we use *octree* [14] to organized trajectory points in each inverted list. Note that octree is a three-dimensional analog of quadtree [8], which is self-adaptive to the spatial and temporal distributions of the trajectory points. As the nodes of octrees can be encoded based on *morton code* [15], thus two nodes with higher spatio-temporal proximity are likely to be assigned to the same page in the secondary storage. As a result, the number of I/Os during query processing could be reduced due to principle of locality. Furthermore, to expedite the process of spatio-temporal search, *signature* technique is utilized to prune non-promising trajectories without loading trajectory points resident on the disk.

Contribution. Our main contribution can be summarized as follows:

- This is the first work to investigate the problem of spatial keyword range search where both spatio-temporal and keyword constraints are considered.
- We proposed a novel structure, namely IOC-Tree, to effectively organize trajectories with keywords.
- We also proposed an efficient algorithm to process spatial keyword range search on trajectories.
- Comprehensive experiments on real-life datasets demonstrate the efficiency of our techniques.

Organization. The remainder of the paper is organized as follows: Section 2 gives a formal problem definition and the related work is also reported. Section 3 presents the IOC-Tree structure. Efficient spatial keyword range search algorithm is proposed in Section 4, followed by experimental evaluation in Section 5. We conclude the paper in Section 6.

2   Preliminary

In this section, we first provide a formal definition of problem we study in this paper, then give a brief review of related work. Table 1 summarizes the notations used throughout the paper.

Table 1. List of Notations

Notations	Explanation
Tr	a trajectory with keywords
$SubTr(i,j)$	a sub-trajectory of Tr
$\psi(Q.R)$	the diameter of query range
$\mathbb{V}$	keyword vocabulary
w	a keyword in $\mathbb{V}$
h	maximal depth of IOC-Tree
ψ	split threshold of a node in IOC-Tree
m	number of query keywords

2.1   Problem Description

In this paper, a *trajectory* Tr is represented as a time-ordered sequence of location points with keywords: $\{(t_1, p_1, \phi_1),\ (t_2, p_2, \phi_2),...,(t_n, p_n, \phi_n)\}$, where t_i is the timestamp, p_i is the location comprised of *latitude* and *longitude*, ϕ_i is the set of keywords.

Definition 1. *A **sub-trajectory** $\{(t_i, p_i, \phi_i),\ (t_{i+1}, p_{i+1}, \phi_{i+1}),...,(t_j, p_j, \phi_j)\}$ where $1 \leq i \leq j \leq n$, is a part of a trajectory. We denote above sub-trajectory as $SubTr(i,j)$.*

A sub-trajectory is a consecutive part of a trajectory and it can have only one point. The concept of sub-trajectory is given because one trajectory may enter and leave a particular area multiple times.

Definition 2. *Spatial Keyword Range search on Trajectories (SKRT) Q consists of a spatial **region** R, a **timespan** $T = [t_s, t_e]$ and a set Φ of **keywords** ($= \{k_1, k_2, ..., k_m\}$). We call a trajectory Tr **satisfies** query Q if we could find sub-trajectories of Tr, $SubTr(i_1, j_1)$, $SubTr(i_2, j_2)$, ..., $SubTr(i_t, j_t)$, which locate within region during query timespan $[t_s, t_e]$, and collectively contain query keywords, i.e., $\Phi \subseteq (\cup_{x=i_1}^{j_1} \phi_x) \cup (\cup_{x=i_2}^{j_2} \phi_x)... \cup (\cup_{x=i_t}^{j_t} \phi_x)$.*

Essentially, $SKRT$ consists of three constraints: spatial constraint R, temporal constraint T, and keyword constraint Φ. A trajectory will be retrieved if the trajectory points within the spatio-temporal range (i.e., satisfying both spatial and temporal constraints) cover all the query keywords. Following is an example of $SKRT$ based on trajectories in Fig 2.

Example 2. In Fig 2, there are four trajectories R_1, R_2, R_3, R_4 with 17 points where p_{ij} represents the point in R_i with its timestamp t_j. Notice that in practical applications, points of trajectories are usually not collected at the same timestamp. Assume an $SKRT$ Q is given as follows: $Q.R$ is the space within dotted circle depicted in Fig 2, $Q.T$ is $[t_s, t_e]$ where $t_1 < t_s < t_2$, $t_3 < t_e < t_4$, and $Q.\Phi = \{a, c\}$.

Fig 2 shows that trajectory R_1 and R_2 are the results returned by $SKRT$, for R_3 is eliminated due to spatio-temporal constraint and R_4 are eliminated due to keyword constraint.

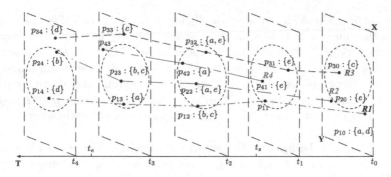

Fig. 2. Example of $SKRT$

Problem Statement. Given a database D of trajectories and an $SKRT$ query Q, we aim to retrieve all of the trajectories which **satisfy** the query Q from D.

2.2 Related Work

To the best of our knowledge, there is no existing work studying the problem of $SKRT$ proposed in this paper. Below, we introduce two categories of closely related existing work.

Spatial Keyword Search. Due to huge amounts of spatio-textual objects collected from location-based services, Spatial Keyword (SK) search has been widely studied. One of the most important queries is the top-k spatial keyword search, which aims to find k objects which have the most spatial proximity and contain all the query keywords. Many efficient index structures have been proposed such as inverted R-tree [24] and IR^2-tree [7]. In addition to top-k spatial keyword search, many interesting query variants are proposed such as direction-aware spatial keyword search (DESKS) [9] and collective spatial keyword search (CoSKQ) [11]. Nevertheless, the temporal information is not considered in the above work. Recently, some recent work on SK search also consider the temporal constraint. In [13], Magdy et al. proposed a system called *Mercury* for real-time support of top-k spatio-temporal queries on microblogs, which allow users to browse recent microblogs near their locations. In [17], Skovsgaard et al. proposed an index structure that extends existing techniques for counting frequent items in summaries and a scalable query processing algorithm to identify top-k terms seen in the microblog posts in a user-specified spatio-temporal range. However, these techniques are especially designed for spatio-textual objects which are inherently different from the trajectory data.

TkSK [6] and ATSQ [23] are the two most relevant work to our problem. The TkSK proposed by Cong et al. [6] is comprised of a user location and a keyword set, and returns k trajectories whose text descriptions covering the keyword set with the shortest match distance. The authors developed a hybrid index called B^{ck}-*tree* to deal with text relevance and location proximity between the query and trajectories. The ATSQ studied by Zheng et al. [23] finds k distinct trajectories have the smallest minimum match distance with respect to query locations with their query activity. A hybrid grid index called GAT was proposed to prune the search space by location proximity and activity containment simultaneously. As shown in our empirical study, although we can extend the above techniques to support temporal pruning by further organize the trajectory points with B+ tree according to their timestamps, the performance is not satisfactory.

Historical Spatio-Temporal Trajectory Indexing. There has been considerable related work on storing and querying historical spatio-temporal trajectories. Prior work proposed TB-tree [16] to solve the problem of range query over spatio-temporal trajectories. The main idea of TB-tree indexing method is to bundle segments from the same trajectory into leaf nodes of the R-tree. The MV3R-tree [20] is a hybrid structure that uses a multi-version R-tree (MVR-tree) for time-stamp queries and a small 3D R-tree for time-interval queries. This structure has been proved to outperform other historical trajectory index structures.

SETI [4] is a grid-based index which partitions the spatial space into grids and then index temporal information within each spatial partition based on R*-tree. PIST [3] is also grid-based which focuses on indexing trajectory points. It utilizes a quad-tree like data structure to partition points into a variable-sized grid according to the density of data. Since in our problems, trajectories have extra textual data than traditional ones, structures mentioned above can't be straightforwardly extended to solve our problem.

3   Inverted Octree

In this section, we introduce a new indexing structure, namely Inverted Octree (IOC-Tree), to effectively organize the trajectories with textual information. Section 3.1 provides the overview of our IOC-Tree structure. Section 3.2 introduces the detailed data structure, followed by the index maintenance algorithms in Section 3.3.

3.1   Overview

As discussed in Section 1, we follow the keyword-first-pruning strategy because we observe that the query keyword usually has the lowest selectivity (i.e., highest pruning capability) compared with spatial and temporal factors. Therefore, we adopted *inverted index* technique such that only the trajectory points associated with at least one query keyword will be involved in the range search. For each keyword in the vocabulary, a corresponding octree is built to organize the relevant trajectory points. The spatio-temporal space (i.e., 3-dimensional space) is recursively divided into cells (nodes) in an adaptive way, and trajectory points are kept on the leaf nodes of the octree. Moreover, we apply the *3D morton code* technique to encode the leaf nodes of the octree, and the non-empty leaf nodes are organized by an auxiliary disk-based one dimensional structure (e.g., B+ tree) where the morton code is the key of the nodes. In this way, trajectory points with high spatio-temporal proximity are likely to be resident in the same page on the disk. We also employ the *signature* technique to keep trajectory identification information at high level node so that some non-promising trajectories can be pruned earlier without invoke extra I/O costs. For the purpose of verification, we also keep the exact trajectory information for each non-empty leaf node of octrees.

3.2 IOC-Tree Structure

In our IOC-Tree, we have an octree OC_w for each keyword $w \in \mathbb{V}$ [1]. In octree, each non-leaf nodes have eight children, and a simple illustration of octrees is depicted in Fig 3(a). Construction of an octree starts from treating the whole spatio-temporal three-dimensional space as a root node, and then recursively divides space into eight subspaces as children nodes if the node has sufficient objects, *i.e.*, spatio-temporal points. As shown in Fig 3(a), the space is first partitioned into 8 nodes, and only one of them is further divided. In this way, the octree can effectively handle the skewness of the trajectory points.

Next, we will explain how to generate morton code for each leaf node of octree. In [15], a 3-dimensional space will be recursively divided into 8^{h-1} cells where h is the depth of the partition, and the morton code of each cell is assigned based on its visiting order. Nevertheless, in this paper, the octree is constructed in an adaptive way and hence the leaf nodes may resident on different levels. Thus, the morton code of each leaf node is assigned as follows. Let h denote the maximal height of the octree, we assume the space is evenly partitioned into 8^{h-1} *virtual cells* and the morton code of each virtual cell is immediate. Then the morton code of a leaf node v corresponds to the minimal code of virtual cells covered by v. Fig 3(b) illustrates an example where the maximal depth of the octree is 3. Each circle represents a leaf node of the octree built from Fig 3(a) and each of them is assigned a morton code according to our approach. The morton code of Node #24 is 24 as it is the smallest codes in the virtual cells it contains. We remark that we do not need to materialize the virtual cells in our implementation, the morton code can be assigned based on its split sequence. Details are omitted due to space limitation.

In order to efficiently prune non-promising nodes when searching octrees, for each node v, we also maintain a *signature* to summarize the identifications of a set of trajectories that go through the corresponding spatio-temporal region of v. In particular, a signature of a node v is a bit vector, and each trajectory ID will be mapped onto one bit by a hash function. Then its i-th bit will be set 1 if there *exists* a trajectory point within in v (i.e., point satisfying spatio-temporal constraint regarding v) and its ID is mapped to the i-th position. Otherwise, the i-th bit is set to 0. As shown in Section 4.1, the non-promising trajectories may

---

[1] Note that as the frequencies of the keywords follow the *power-law* distribution in real-life, and there is a large portion of low frequency keywords. In our implementation, we simply keep trajectory points of low frequency keyword in one disk page, instead of building the related octree.

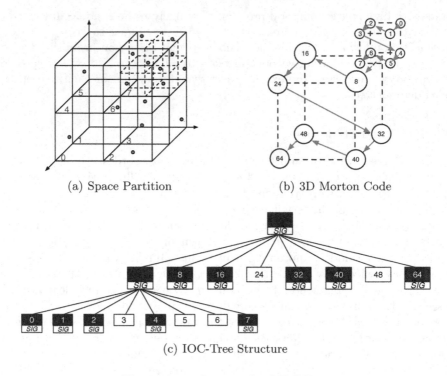

(a) Space Partition   (b) 3D Morton Code

(c) IOC-Tree Structure

Fig. 3. A simple example of IOC-Tree

be pruned at high level of octree with the help of node signatures, and hence significantly reduce the I/O costs. Note that although the number of trajectory points is very huge, the number of trajectories is usually one or two orders of magnitude smaller, which makes the storage of *signature* feasible.

Fig 3(c) demonstrates the structure of an octree built from Fig 3(a), where nodes that contain trajectory points (i.e., non-empty nodes) is set *black* and the rest nodes (empty nodes) are set *white*. Each leaf node is labeled by its morton code, and a signature is maintained to summarize the trajectory IDs within the node. In addition to the octree structure, we also keep the trajectory points in each black leaf node. Each trajectory point is recorded as a tuple $(pID, tID, lat, lng, time)$, where pID is the point ID, tID is corresponding trajectory ID, and $(lat, lng, time)$ is the spatio-temporal value of this point. The non-empty leaf nodes from octrees will be kept on the disk by one dimensional index structure (e.g., B+ tree), which are ordered by their morton codes. However, *signature* only keeps a rough description for trajectories and will only be helpful for the pruning non-promising nodes. Therefore, for each leaf nodes of octrees, the exact information to explain the trajectories that they contain (i.e., trajectory IDs) need to be kept on the disk respectively. These information will also be organized by a B+ tree with the morton code of the corresponding leaf node as the key.

3.3 IOC-Tree Maintenance

The insertion of a trajectory includes two steps. Firstly for every point p from a trajectory Tr, it will be assigned into the corresponding octrees based on the keywords it contains. A leaf node of the octrees will be split if it contains more than ψ points and does not reach the maximal depth h. Meanwhile, for every octree node along the inserting path of p, the bit of its *signatures* mapped by Tr will be set to 1. As to deletion of a particular trajectory, we should remove all its points from their corresponding octrees along with some possible merges of the nodes. A bit in signature of a node from octrees will be reset correspondingly. Moreover, for leaf nodes in both cases of insertion and deletion, the related exact trajectory information on the disk also need to be updated.

Table 2. Distribution of Trajectory Points

Node#	0	1	2	3	4	5	6	7
Points	p_{11}	p_{10}	p_{30}	p_{20}	p_{23}	p_{12}	p_{22}	p_{42}
		p_{31}	p_{41}		p_{24}	p_{13}	p_{32}	p_{43}
					p_{33}	p_{14}	p_{34}	

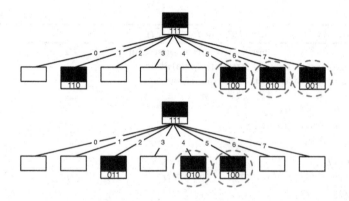

Fig. 4. IOC-trees' Construction

Example 3. For the sake of brevity, we build 2-level octrees regarding the case in Fig 2, where 17 trajectory points are distributed in different subspaces as shown in Table 2. Assume that each *signature* has three bits, and trajectory R_1 is mapped to the first position, R_2 and R_3 are mapped to the second position, R_4 is mapped to the third position. Fig 4 demonstrates the inverted octrees OC_a and OC_c where a node is set black if it contains points and white otherwise. The *signatures* are only kept for black nodes. As a matter of fact, white nodes are not reserved in our implementation. As an illustration of *signature*, for *node#1* in OC_a, its *signature* is 110 because only trajectories R_1 (p_{10}) and R_3 (p_{31}) go through the corresponding region. Notice the exact trajectory information for each non-empty leaf nodes are stored on the disk and they are not shown in Fig 4.

4   Algorithm for Query Processing

In this section, we present efficient algorithms for processing $SKRT$ and one of its variant query with the assumption that trajectory data are organized by IOC-Tree. Section 4.1 will give a specific description of algorithm for $SKRT$. As an extension, a variant query of $SKRT$ called $SKRTO$ and its brief processing algorithm will be introduced in Section 4.2.

4.1   Algorithm for Processing $SKRT$

In our algorithm, octree nodes are divided into three types with regard to spatio-temporal query range: one locate outside the range, one locate totally in the range which are called *fully-covered* nodes, and the other intersect with range which are called *partially-covered* nodes. we employ a set L to keep octree nodes that are being processed. Moreover, sets of candidate nodes CN_i^f and CN_i^p are employed to record *fully-covered* and *partially-covered* candidates from related octree OC_i respectively. $\mathcal{TR}(CN)$ denotes a set of trajectories that the nodes in CN contains. Besides, CT^f contains candidate trajectories that appear in the *fully-covered* nodes for all the query keywords, while CT^p contains the rest candidate trajectories based on sets of candidate nodes.

Algorithm 1. Algorithm Outline For $SKRT$

Input: Q: an $SKRT$ query with three constraints $(Q.R, Q.T, Q.\delta)$, OC:
        inverted octree, D: trajectory database
Output: $\mathcal{A}$: set of trajectories from D satisfy Q
1 $\mathcal{A} \leftarrow \emptyset; L \leftarrow \emptyset; CT^f \leftarrow \emptyset; CT^p \leftarrow \emptyset;$
2 **foreach** $k_i \in Q.\delta$ **do**
3   put $root(OC_i)$ into L;
4     $CN_i^f \leftarrow \emptyset; CN_i^p \leftarrow \emptyset;$
5 $\textbf{\textit{Prune}}(Q, L);$
6 $CT^f \leftarrow \cap_{i=1}^m \mathcal{TR}(CN_i^f); CT^p \leftarrow \cup_{i=1}^m \mathcal{TR}(CN_i^p) + \cup_{i=1}^m \mathcal{TR}(CN_i^f);$
7 $CT^p \leftarrow CT^p - CT^f;$
8 $\textbf{\textit{Verification}}(Q, \mathcal{A});$
9 return $\mathcal{A};$

The basic outline of algorithm for processing $SKRT$ is illustrated in Algorithm 1. The main idea is to prune as many trajectories as possible based on the spatio, temporal and textual information by using IOC-Tree structure. In Line 3, the root nodes of related octrees are put into the set L. Sets of different types of candidate nodes for each query keywords are initialized in Line 4. The next step is to explore the nodes in L to prune nodes that does not satisfy spatio-temporal constraint (Line 5). After that, we determine the different sets of candidate trajectories based on the exact trajectory information (Line 6-7). Finally, we validate each candidate trajectory and put right ones into the result set $\mathcal{A}$ (Line 8).

We proceed to give more details of procedures invoked in Algorithm 1.

Procedure *Prune* prunes non-promising nodes based on the inverted octree and deals with nodes level by level among the related octrees. In each level, nodes that don't satisfy spatio-temporal constraint will be firstly pruned by *STRange-Filter*. During the process of *STRangeFilter*, we only explore *black* nodes (i.e., non-empty nodes) and keep a one-bit flag for each node that satisfy spatio-temporal constraint to indicate a *fully-covered* or *partially-covered* node. Line 3-7 perform a *signature* test to prune nodes that definitely share no trajectories with nodes from other octrees. Among the nodes that survive from the *signature* test, leaf nodes will be inserted in the corresponding sets of candidate nodes (Line 8-13), and the non-*white* children nodes of non-leaf nodes will be put into L (Line 14-17). The last step is to retrieve candidate trajectories from the disk for each query keyword based on different types of candidate nodes (Line 19-20).

Procedure *Verification* aims at further validating candidate trajectories and inserting trajectories that pass all the tests into result set $\mathcal{A}$. Firstly, we believe all the trajectories in CT^f are appropriate ones which can be easily proved (Line 1). For each trajectory Tr in CT^p, a set Ω of query keywords that Tr doesn't have within *fully-covered* nodes are identified (Line 2-4). Then according to each keyword k_j in Ω, only if Tr appears in $\mathcal{TR}(CN_j^p)$ will it be loaded from the corresponding cell on the disk to verify spatio-temoral constraint (Line 5-7).

Example 4. Consider the $SKRT$ problem given in Fig 2. Since the query keyword set $Q.\phi$ contains two keywords a and c, only OC_a and OC_c which we have built in Fig 4 need to be explored. The nodes marked with dotted circle signify the ones within query spatio-temporal range in both trees. $node\#7$ in OC_a is pruned because it does not pass the *signature* test. According to containment relationship with query range, after procedure *Prune*, we have $CN_a^f = CN_c^f = \{node\#5\}$, $CN_a^p = \{node\#6\}$ and $CN_c^p = \{node\#4\}$. Table 2 shows that only trajectory R_1 goes through $node\#5$, and trajectory R_2 and R_3 go through $node\#4$ and $node\#6$. Therefore, $\mathcal{TR}(CN_a^f) = \mathcal{TR}(CN_c^f) = \{R_1\}$ and $\mathcal{TR}(CN_a^p) = \mathcal{TR}(CN_c^p) = \{R_2, R_3\}$. After intersecting $\mathcal{TR}(CN_a^f)$ and $\mathcal{TR}(CN_c^f)$, we get the first qualified trajectory $R1$. Nevertheless, the trajectories in $CT^p = \{R_2, R_3\}$ obtained from $\mathcal{TR}(CN_a^p)$ and $\mathcal{TR}(CN_c^p)$ still have to be verified. To do that, we need to load $node\#4$ and $node\#6$ from the disk, where we find that R_2 has points p_{22} and p_{23} that contain a and c in the query range respectively, while R_3 doesn't. This implies that R_2 is another qualified trajectory while R_3 is not. Finally we get the answer set $\mathcal{A} = \{R_1, R_2\}$.

4.2   Extension for Query with Order-Sensitive Keywords

As mentioned before, one advantage of inverted octree is that it can also support the query with order-sensitive keywords ($SKRTO$). The definition of $SKRTO$ has similar constraints with $SKRT$, except that query keywords from $Q.\Phi$ should be satisfied by a trajectory in chronological order.

In the sequel, we denote the earliest timestamp of a node v as $v.t_{start}$ and the latest timestamp as $v.t_{end}$. Due to space limits, we just highlight two important

---

Procedure Prune(Q, L)

---

1 **while** $L \neq \emptyset$ **do**
2   ***STRangeFilter***$(Q.R, Q.T, L)$;
3   **foreach** $k_i \in Q.\delta$ **do**
4   SIG_i = bitwise-OR of signatures of nodes $v \in L$ from OC_i;
5   **foreach** *node* $v \in L$ *from* OC_i **do**
6   **foreach** SIG_j *where* $j \neq i$ **do**
7   ***SignatureCheck***(v, SIG_j);
8   **foreach** *node* $v \in L$ *that survive from the signature test* **do**
9   Suppose v comes from OC_j;
10   **if** v *is a fully covered leaf node* **then**
11   add v into CN_j^f;
12   **else if** v *is a partially covered leaf node* **then**
13   add v into CN_j^p;
14   **else if** v *is non-leaf node* **then**
15   **foreach** *child node* v' *of* v **do**
16   **if** v' *is not a white node* **then**
17   put v' into L;
18   delete v from L;
19 **foreach** $k_i \in Q.\delta$ **do**
20   determine $\mathcal{TR}(CN_i^f)$ and $\mathcal{TR}(CN_i^p)$;

Procedure Verification$(Q, \mathcal{A})$

---

1 $\mathcal{A} \leftarrow CT^f$;
2 **foreach** *trajectory* $Tr \in CT^p$ **do**
3   find out a keyword set $\Psi = \{k_i | Tr \in \mathcal{TR}(CN_i^f)\}$;
4      $\Omega \leftarrow Q.\Phi - \Psi$;
5   **foreach** $\mathcal{TR}(CN_j^p)$ *where* $k_j \in \Omega$ **do**
6   **if** $Tr \in \mathcal{TR}(CN)_j^p$ *and* ***LoadAndJudge***$(Tr, \mathcal{TR}(CN_j^p))$ **then**
7             $\mathcal{A} \leftarrow \mathcal{A} \cup Tr$;

---

techniques to modify Algorithm 1 to answer $SKRTO$. One is a pruning technique during the process of procedure *Prune*. In each level, we will visit nodes of related octrees in order of query keywords. If there is a node v from OC_i whose latest timestamp is not larger than all the earliest timestamp of qualified nodes in OC_j $(j < i)$, then this node can be pruned safely. The other aspect is how to deal with trajectories in CT^f in procedure *Verification* because they may not satisfy order-sensitive keyword constraint. Instead, we should find a node sequence $v_1, v_2, ..., v_m$ such that $v_i \in CN_i^f$ $(i \leftarrow 1$ to $m)$ and $v_{j-1}.t_{end} \leq v_j.t_{start}$ $(j \leftarrow 2$ to $m)$. After obtaining such a cell sequence, we can guarantee trajectories

Table 3. Dataset Statistics

	LA	NY	TW
#trajectory	31,553	49,022	214,834
#location	215,614	206,416	1,287,315
#tag	3,175,597	3,068,401	28,645,905
#distinct-tag	100,843	89,665	1,852,141

Table 4. Experimental Settings

	$\|Q.\phi\|$	$\|Q.T\|$ (month)	$\delta(Q.R)$(km)
LA,NYC	2, 3, 4, 5	1, 2, 3, 4, 5	10, 20, 30, 40, 50
TW	2, 3, 4 ,5	0.5, 1, 1.5, 2, 2.5	5, 10, 15, 20, 25

from intersection of trajectory set contained by v_i ($i \leftarrow 1$ to m) satisfy $SKRTO$ query. In this way, a considerable number of node access can be avoided.

5 Experiments

We conduct comprehensive empirical experiments in this section to evaluate CPU and I/O performance between our proposed algorithm and two baselines for both $SKRT$ and $SKRTO$ query.

5.1 Experimental Setup & Datasets

All the algorithms including baselines are implemented in C++ and the experiments are performed on a machine with Intel i5 CPU (3.10GHz) and 8GB main memory, running Windows 7. The raw datasets are all stored in binary files with page size 4096 bytes. Notice that the number of I/Os is considered as the number of accessed pages in different algorithms.

Three real trajectory datasets are used, two of which are from check-in records of Foursquare within areas of Los Angeles (LA) and New York (NY) [1], the third one is from geo-tagged tweets (TW) [10] collected from May 2012 to August 2013. For all three datasets, records belonging to the same user are ordered chronologically to form a trajectory of this user. The frequent and meaningful words are collected from the plain text in each record. Table 3 summaries important statistics of three datasets.

For different datasets, different set-ups are designed as shown in Table 4. The default settings of the parameter are underlined. For each experimental set, we generate 50 queries and report the average running time and accessed pages. We randomly pick up several trajectories from datasets and generate queries by selectively choosing keywords and setting reasonable query range and timespan. For the inverted octree in all the experiments, the maximal depth h is set to 5 and the split threshold ψ is set to 80.

5.2   Baselines

Two baselines extended from techniques from [6] and [23] respectively are proposed for comparison and validation of our proposal algorithm. Both baselines need extra order examination when dealing with $SKRTO$ query.

B^{ck}-tree The original B^{ck}-tree [6] is designed to solve spatial keyword problem on trajectories. It divides the spatial region into quad cells, and builds a B+ tree based on the cell division. Three elements including *wordID, cellID, posting list*, are contained by leaf entries of the B+ tree, among which the first two are the keys and *posting list* is a sequence of trajectories that go through cell *cellID* and contain word *wordID*. To incorporate temporal information, some modification are made on the posting lists, *i.e.*, trajectories are sorted by the timestamp of its point which locates in the corresponding cell and contains the corresponding word. In general, B^{ck}-tree firstly prunes trajectories by spatio-textual constraint, and then by temporal constraint.

GAT The second baseline is a natural extension of Grid Index for Activity Trajectories (GAT)[23]. We only divide a grid when necessary in a way similar to how we divide spatio-temporal space. For each grid, we construct a B+ tree to index trajectory points based on their timestamps and then build an inverted index of points for every keyword in this grid to record textual information. Generally, **GAT** adopts a strategy by pruning trajectory on spatio, temporal and textual constraint in sequence.

5.3   Performance Evaluation

Varying Number of Query Keywords $|Q.\phi|$. In the first set of experiments, we vary $|Q.\phi|$, the number of query keywords, to compare CPU and IO cost on $SKRT$ query among three algorithms. Fig 5 shows that as $|Q.\phi|$ gets larger, the running time of IOC-Tree doesn't grow as fast as other two baselines on all three datasets. Especially in the case of dataset TW, IOC-Tree spends much less time than GAT and B^{ck}-tree when $|Q.\phi|$ becomes larger. The reason can be revealed when we carefully check Fig 6, which shows IO cost result. While the number of accessed pages by GAT and B^{ck} become larger as $|Q.\phi|$ increases, that of IOC-Tree decreases conversely. In the case of TW, IOC-Tree outperforms two baselines for nearly one order of magnitude when $|Q.\phi|$ is 5. This is due to that given a fixed spatio-temporal search space, more nodes of the octree can be pruned when more query keywords are involved during process of octrees' exploration and *signature* test. Therefore, the whole running time can be greatly reduced although more keywords may incur more examination on trajectories.

Varying Query Timespan $|Q.T|$. Then we proceed to investigate the effect of $|Q.T|$, the query timespan, on the performance of algorithms by plotting the time cost of $SKRT$ and $SKRTO$ on different datasets in Fig 7 and Fig 8. Apparently, longer query timespan means larger spatio-temporal search space, which results in longer running time. Generally, $SKRTO$ query requires more runtime cost because of the high computation of verification on order-sensitive keywords.

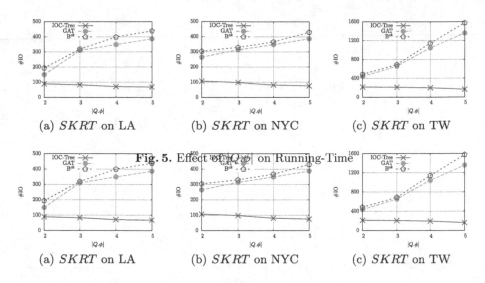

Fig. 5. Effect of $|Q.\phi|$ on Running-Time

(a) *SKRT* on LA   (b) *SKRT* on NYC   (c) *SKRT* on TW

Fig. 6. Effect of $|Q.\phi|$ on #*IO*

However, IOC-Tree maintains a relatively slow growth of time cost and outperforms two baselines under all cases which is benefited by fast nodes' pruning and morton code's utilization. In the case of dataset TW, the gap between IOC-Tree and two baselines becomes even larger. The reason is that trajectories of TW contain more tags averagely, and IOC-Tree adopts a keyword-first-pruning strategy; as the query timespan gets larger, GAT and B^{ck}-tree have much more candidates to verify while IOC-Tree has pruned a lot in the first place.

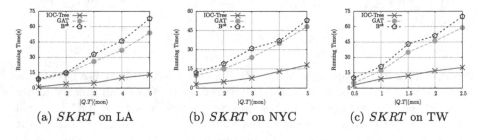

(a) *SKRT* on LA   (b) *SKRT* on NYC   (c) *SKRT* on TW

Fig. 7. Effect of $|Q.T|$

Varying Diameter of Query Range $\delta(Q.R)$. Finally we study the effect of $\delta(Q.R)$, the diameter of query spatial range, on the running time of *SKRT* and *SKRTO* among three algorithms. The experimental results are demonstrated in Fig 9 and Fig 10. Similar to varying $|Q.T|$, larger diameter of query spatial range naturally involves larger search space. Therefore, more cells and trajectories will be identified as candidates. As expected, the performance of all algorithms

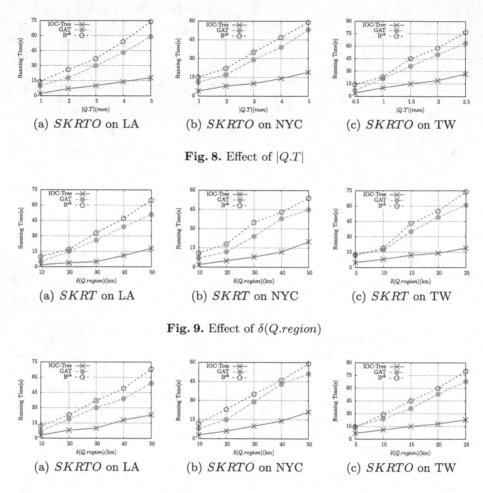

Fig. 8. Effect of $|Q.T|$

Fig. 9. Effect of $\delta(Q.region)$

Fig. 10. Effect of $\delta(Q.region)$

degrades regarding the increase of $\delta(Q.R)$. Between two baselines, GAT performs better than B^{ck}-tree since given a fixed spatial space, GAT can quickly locate the quad cells and explore the corresponding B+ tree. However, IOC-Tree still has superior performance among all the algorithms. This is mainly because it takes fully advantage of spatio-temporal proximity and thus a considerable IO cost can be saved.

6   Conclusion

We study spatial keyword range search on trajectories, which takes spatio, tempral and textual properties of trajectories into consideration. To efficiently solve spatial keyword range search on trajectories ($SKRT$) and its variation with order-sensitive keywords ($SKRTO$), we design a novel index structure named

IOC-Tree with signature to organize trajectory data and propose an efficient algorithm for query processing. Extensive experiments on real datasets confirm the efficiency of our techniques.

Acknowledgments. The work is supported by NSFC61232006, NSFC61321064, ARC DE140100679, ARC DP130103245, ARC DP150103071 and DP150102728.

References

1. Bao, J., Zheng, Y., Mokbel, M.F.: Location-based and preference-aware recommendation using sparse geo-social networking data. In: Proceedings of the 20th International Conference on Advances in Geographic Information Systems, pp. 199–208. ACM (2012)
2. BOGORNY, V., ALVARES, L.O., Kuijpers, B., Fernandes de Macedo, J. A., MOELANS, B., and Tietbohl Palma, A.: Towards semantic trajectory knowledge discovery
3. Botea, V., Mallett, D., Nascimento, M.A., Sander, J.: Pist: an efficient and practical indexing technique for historical spatio-temporal point data. GeoInformatica **12**(2), 143–168 (2008)
4. Chakka, V.P., Everspaugh, A.C., Patel, J.M.: Indexing large trajectory data sets with seti. Ann Arbor 1001, 48109–2122 (2003)
5. Christoforaki, M., He, J., Dimopoulos, C., Markowetz, A., Suel, T.: Text vs. space: efficient geo-search query processing. In: Proceedings of the 20th ACM international conference on Information and knowledge Management, pp. 423–432. ACM (2011)
6. Cong, G., Lu, H., Ooi, B. C., Zhang, D., Zhang, M.: Efficient spatial keyword search in trajectory databases (2012). arXiv preprint arXiv:1205.2880
7. De Felipe, I., Hristidis, V., Rishe, N.: Keyword search on spatial databases. In: 2008 IEEE 24th International Conference on Data Engineering. ICDE 2008, pp. 656–665. IEEE (2008)
8. Finkel, R.A., Bentley, J.L.: Quad trees a data structure for retrieval on composite keys. Acta informatica **4**(1), 1–9 (1974)
9. Li, G., Feng, J., and Xu, J. Desks: Direction-aware spatial keyword search. In: 2012 IEEE 28th International Conference on Data Engineering (ICDE), pp. 474–485. IEEE (2012)
10. Li, G., Wang, Y., Wang, T., Feng, J.: Location-aware publish/subscribe. In: Proceedings of the 19th ACM SIGKDD International Conference on Knowledge Discovery and Data Mining, pp. 802–810. ACM (2013)
11. Long, C., Wong, R.C.-W., Wang, K., and Fu, A.W.-C.: Collective spatial keyword queries: a distance owner-driven approach. In: Proceedings of the 2013 International Conference on Management of Data, pp. 689–700. ACM (2013)
12. Luo, W., Tan, H., Chen, L., and Ni, L. M. Finding time period-based most frequent path in big trajectory data. In: Proceedings of the 2013 International Conference on Management of Data, pp. 713–724. ACM (2013)
13. Magdy, A., Mokbel, M. F., Elnikety, S., Nath, S., He, Y.: Mercury: a memory-constrained spatio-temporal real-time search on microblogs. In: 2014 IEEE 30th International Conference on Data Engineering (ICDE), pp. 172–183. IEEE (2014)

14. Meagher, D.J.: Octree encoding: a new technique for the representation, manipulation and display of arbitrary 3-d objects by computer. Electrical and Systems Engineering Department Rensseiaer Polytechnic Institute Image Processing Laboratory (1980)
15. Morton, G.M.: A computer oriented geodetic data base and a new technique in file sequencing. International Business Machines Company (1966)
16. Pfoser, D., Jensen, C.S., Theodoridis, Y., et al.: Novel approaches to the indexing of moving object trajectories. In: Proceedings of VLDB, pp. 395–406. Citeseer (2000)
17. Skovsgaard, A., Sidlauskas, D., Jensen, C.S.: Scalable top-k spatio-temporal term querying. In: 2014 IEEE 30th International Conference on Data Engineering (ICDE), pp. 148–159. IEEE (2014)
18. Song, R., Sun, W., Zheng, B., Zheng, Y.: Press: a novel framework of trajectory compression in road networks (2014). arXiv preprint arXiv:1402.1546
19. Su, H., Zheng, K., Huang, J., Jeung, H., Chen, L., Zhou, X.: Crowdplanner: a crowd-based route recommendation system. In: 2014 IEEE 30th International Conference on Data Engineering (ICDE), pp. 1144–1155. IEEE (2014)
20. Tao, Y., Papadias, D.: The mv3r-tree: A spatio-temporal access method for timestamp and interval queries
21. Wang, Y., Zheng, Y., Xue, Y.: Travel time estimation of a path using sparse trajectories. In: Proceeding of the 20th SIGKDD Conference on Knowledge Discovery and Data Mining (2014)
22. Zhang, C., Han, J., Shou, L., Lu, J., La Porta, T.: Splitter: Mining fine-grained sequential patterns in semantic trajectories. Proceedings of the VLDB Endowment **7**, 9 (2014)
23. Zheng, K., Shang, S., Yuan, N.J., Yang, Y.: Towards efficient search for activity trajectories. In: 2013 IEEE 29th International Conference on Data Engineering (ICDE), pp. 230–241. IEEE (2013)
24. Zhou, Y., Xie, X., Wang, C., Gong, Y., Ma, W.-Y.: Hybrid index structures for location-based web search. In: Proceedings of the 14th ACM International Conference on Information and Knowledge Management, pp. 155–162. ACM (2005)

TOF: A Throughput Oriented Framework for Spatial Queries Processing in Multi-core Environment

Zhong-Bin Xue[1,2], Xuan Zhou[1,2(✉)], and Shan Wang[1,2]

[1] MOE Key Laboratory of DEKE, Renmin University of China,
Beijing 100872, China
[2] School of Information, Renmin University of China, Beijing 100872, China
{zbxue,swang}@ruc.edu.cn, xuan.zhou@outlook.com

Abstract. In this paper, we develop a Throughput Oriented Framework (TOF) for efficient processing of spatiotemporal queries in multi-core environment. Traditional approaches to spatial query processing were focused on reduction of query latency. In real world, most LBS applications emphasize throughput rather than query latency. TOF is designed to achieve maximum throughput. Instead of resorting to complex indexes, TOF chooses to execute a batch queries at each run, so it can maximize data locality and parallelism on multi-core platforms. Using TOF, we designed algorithms for processing range queries and kNN queries respectively. Experimental study shows that these algorithms outperform the existing approaches significantly in terms of throughput.

Keywords: Multi-core · Large update rates · Real-time response · High throughput · Batch query · Spatial-temporal database

1 Introduction

With the advance of mobile devices, communication technologies and GPS systems, the technology of Location Based Service (LBS) has attracted widespread attentions. Many Web companies, such as Google, Baidu, Twitter and Facebook, are nowadays providing location related information service for a growing number of users online. Location based spatial queries, which allow users to search nearby Points of Interest (POIs), are one of the core techniques for LBS applications.

In the past few years a series of algorithms have been proposed to handle location based spatial queries [1][2][3][10][15][16]. These algorithms are usually designed to provide fast response time and high update performance. Spatial indexes, such as B-tree and R-tree, are widely used to achieve the goal. However, the landscape has changed recently, due to the dramatic growth of users and available digital contents. As suggested by some recent investigations [4], for many LBS applications, throughput is more crucial, especially when confronted

© Springer International Publishing Switzerland 2015
M. Renz et al. (Eds.): DASFAA 2015, Part II, LNCS 9050, pp. 241–256, 2015.
DOI: 10.1007/978-3-319-18123-3_15

with a large number of Web users, where both queries and data updates arrive at a high rate.

Fortunately, the development of hardware technologies provide new means to meet the performance needs of LBS. Nowadays, the capacity of main memory is up to 6 TB per server [5]. The dramatically increasing of main memory size has allowed us to store the entire workloads in memory. Meanwhile, the number of cores in multi- and many-core processors is steadily increasing. With the speed of in-memory computing, query latency is no longer a major concern for spatial queries. The increase of cores allows us to concurrently process an increasing number of queries, which is helpful in achieving high throughput. However, most existing approaches of spatiotemporal query processing do not utilize the new features effectively. This urges us to reconsider the framework of spatiotemporal data processing and create new algorithms that can make the best of modern architecture to achieve maximum throughput.

In this paper, we present a Throughput Oriented Framework (TOF) for efficient execution of massive concurrent queries over a large quantity of moving objects. TOF is a snapshot-based framework. In each snapshot, TOF handles spatial queries and object updates through batch processing, so it avoids maintaining sophisticate index structures. TOF's design makes full use of data locality to improve the cache hit ratio. It also aims to maximize intra and inter parallel between queries and give full play to multi-cores processors. In this framework, we designed two algorithms for handling range queries and kNN queries respectively. We conducted a thorough experimental study on real world data sets. The results show that our algorithms can achieve high throughput with response time guarantees.

The remainder of the paper is organized as follows. Section 2 describes the related work about main-memory based spatiotemporal indexes; Section 3 shows the problem formulation, our proposed throughput oriented framework, grid index method, and two main-memory algorithms to handle moving object range queries and kNN queries respectively; Section 4 reports our experimental evaluation that compare our approach against existing ones, utilizing spatial indexes; Finally, we conclude the paper in last section.

2   Related Work

There has been a significant body of work for dealing with spatiotemporal queries. In this section, We present the related work which employs memory-resident indexes.

Range Query. Range query which computes the objects that lie within a given spatial range. It is one of the fundamental problems in the field of moving object management. A serious of main-memory indexes and algorithms have been proposed to deal with it, such as MOVIE, TwinGrid, and PGrid, etc.

MOVIE [8][10] is a snapshot-based algorithm, which uses frequently building short-lived throwaway indexes. In MOVIE, during the ith snapshot, termed as time frame T_i, an index I_i is created. I_i is a read-only index, which is used to answer the queries in T_i. The incoming updates are collected in a dedicated update buffer, B_i. When a new time frame T_{i+1} starts, a new index I_{i+1} is created based on B_i and I_i. After I_{i+1} being built, the buffer B_i and index I_i are destroyed. In MOVIE, a linearized kd-trie [11] is used for indexing. In order to simulate a pointer-based index structure, each node of the kd-trie is assigned by a unique identifier, which is based on a space-filling curve, such as Hilbert curve [10] or z-curve [12][13][14]. Similar to TOF, MOVIE adopts batch updates. However, its query processing scheme is not tailored to modern hardware, such that no tactic is applied to improve cache efficiency or multi-core parallelism. Its index creation is also a costly operation.

TwinGrid [15] is another index structure that based on data snapshot and used for querying moving objects. TwinGrid applies a grid based index, in which two grid structures are maintained. One grid structure is called read-store, which is used for answering incoming query. The other one is write-store, which is used to deal with incoming update. Unlike MOVIE, TwinGrid uses a copy-on-write mechanism to periodically merge the write-store and the read-store to keep the read-store fresh. Some evaluation [15] indicates that TwinGrid runs faster than MOVIE. Based on TwinGrid, PGrid [16] is proposed. PGrid is an improved version of TwinGrid, which contains only one grid structure. Parallel queries and updates operate on the same grid structure. PGrid uses a variant lock protocol to resolve the read and write contention on the grid structure. For both query and update, PGrid can respond quickly.

The above main-memory approaches put a lot of weights on query latency. Throughput is their secondary concern. In contrast, TOF regards query latency less important, as most real-world applications are satisfied with sub-second response time, which is relatively easy to achieve through modern hardware. Throughput appears a more crucial issue, as LBS needs to deal with a large number of queries simultaneously. If a higher latency can be tolerated, the throughput of the existing approaches can be improved significantly.

kNN Query. Main-memory kNN query processing has been mainly investigated in the field of continuous queries. A lot of work leverage main-memory grid indexes. Examples include YPK-CNN, SEA-CNN, and CPM, etc.

YPK-CNN[9] and CPM [18] apply a query-and-refinement approach to handle kNN queries. Firstly, they use range query to identify k objects as a candidate result set. Then in the refinement phase, the algorithms use the farthest distance between the query and the objects in the candidate set as the radius, and evaluate all the objects within this radius to obtain the final results. The region within this radius is normally called **refinement region**. To keep the results up-to-date, YPK-CNN defines a new refinement region based on the maximum distance between the query point and the current locations of previous kNN objects. If moving objects that move out of the refinement region are more than those that move into the region, then the query has to be recomputed from scratch. Otherwise, the k objects in the search region that lie closest to q form the new result. SEA-CNN [17] introduced an idea of shared execution of multiple kNN queries over moving POIs.

Similar to the existing work on range queries, most algorithms used for kNN queries also aim to minimize response time rather than throughput. Our proposed TOF aims to achieve improvement on throughput. Nevertheless, it guarantees to meet the general requirement on response time.

3   Throughput Oriented Framework (TOF)

This section presents TOF, a throughput oriented framework for handling spatiotemporal queries on modern hardware. We first outline the problem formulation. Then we present the framework. After that, we introduce two main memory based algorithms for handling range queries and kNN queries respectively, which exploit the multi-core architecture to maximize the throughput.

3.1   Problem Formulation

Consider a setting in which a data set of N moving objects, be it mobile phone users or vehicles, in two-dimensional space within the domain $|X| * |Y|$, where $|X|$(resp. $|Y|$) represents the number of different positions in the horizontal (resp. vertical) dimension. A grid structure is used to index the two-dimensional space. Each rectangle in grid is called a cell. We formally define the moving objects and queries as follows:

Definition 1. *A moving object is modeled as a point and defined as $o_i = \{ OID, \langle x, y \rangle \}$, where OID is a unique key that identifies the moving object, and x and y are the coordinates.*

We assume all the relevant information about the moving objects is maintained at a central server. Whenever an object moves to a new location, an update message is sent to the central server.

Definition 2. *A range query is modeled as a rectangle and defined as $q_r = \{QID, [x_{low}; y_{low}] * [x_{high}; y_{high}]\}$, where QID is a unique key that identifies the query, and $[x_{low}; y_{low}]$(resp.$[x_{high}; y_{high}]$) represents the lower-left (resp. upper-right) corner of the rectangle.*

Obviously, the results of the query are supposed to contain the complete set of OIDs, whose objects fall in this rectangle. One or more grid cells could be fully or partially covered by a query. Objects in the partially covered cells should be checked to determine if they are within the query range, whereas objects in the fully covered cells are the results and do not need to be checked.

Definition 3. *A kNN query is defined as $q_{knn} = \{ QID, \langle x, y \rangle, k \}$, where QID is a unique key that identifies the query, x and y are the coordinates of the query, and k is the required number of the nearest neighbors.*

Given a kNN query q on a set of moving objects O, the task is to ensure that the query's result set Q_r, which is a subset of O, always satisfies the following conditions: $| Q_r | = $ k and for $\forall o_i \subseteq Q_r, \forall o_j \subseteq O - Q_r, d_E(q, o_i) \leq d_E(q, o_j)$, where $d_E(q, o_i)$ is the Euclidean distance between the query q and the object o_i.

TOF uses a snapshot based approach. The position of a moving object o_i at snapshot t is defined as $loc_t(o_i) = (x_t, y_t)$, where x_t and y_t are the coordinates of o_i at snapshot t. In each snapshot, all the moving objects, such as taxis, which are equipped with GPS, send their locations to the central server. After accumulating all the updates, at the end of a snapshot, TOF performs the updates altogether. In each snapshot, all queries submitted to the central server are also cached in the memory. In the following snapshot, all the queries are executed in a batch on the updated locations. Specifically, a range query submitted by a user at time t_i is computed based on the locations of moving objects at t_i. But the results are returned only at time t_{i+1}. We have $t_{i+1} - t_i < \triangle t$, where $\triangle t$ is a predefined time interval, which determines the response time of the system.

Based on the foregoing, we focus on maximizing the throughput of processing the range and kNN queries (i.e. processing as many queries as possible within a fixed time interval) using the modern main-memory and multi-core platforms.

3.2   The Framework

As shown in Figure 1, the processing model of TOF is divided into three stages: preprocessing stage, executing stage and dispatching stage. The stages work as a pipeline approach and use a queue buffer to communicate with each other. The outputs of former stage become the inputs of latter stage.

Preprocessing Stage. The task of the first stage is to handle the incoming queries and moving object updates. As we mentioned earlier, TOF is a snapshot-based framework. At *Snapshot i*, TOF caches the incoming updates and queries into the object buffer OB_i and query buffer QB_i. At the end of the time interval, TOF applies all the updates on the data in parallel, and forwards the updated data and queries to the executing stage. Meanwhile, it starts to prepare *snapshot i + 1*, and collects OB_{i+1} and the query buffer QB_{i+1}. In order to achieve a proper trade-off between response time and throughput, TOF uses a variable query window mechanism. TOF decides the query window size, the time

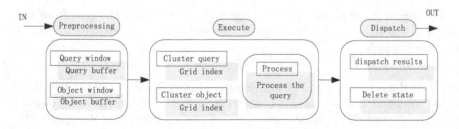

Fig. 1. Processing model of TOF

interval of a snapshot, according to the QoS. As the query response time meets the requirements in the Service-Level Agreement (SLA), TOF would maximize the number of queries to be handled in each time interval.

Executing Stage. In this stage, TOF first creates grid indexes on the object buffer and the query buffer coming from the pre-processing stages. The grid indexes are created based on the spatial coordinates of the objects and the queries. Multi-core parallelization is performed to maximize the throughput. Based on the index, TOF performs query processing. Queries and objects that fall into the same cell of the grid are always processed together. This helps TOF achieve high data locality and thus cache efficiency. Queries close to one another are likely to share computation, such that performance can be further improved. Moreover, queries and objects in different cells can be processed in parallel, which gives full play to multi-core processors.

Dispatching Stage. In this stage, TOF dispatches the query results to the clients and all intermediate data are deleted, including the grid structures.

To summarize, TOF attempts to make the best of new hardware and enable thousand fold parallelism in query processing.

3.3   Grid Indexing

In this part, we show how TOF maps moving objects to a grid index.

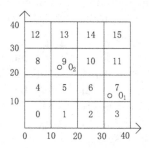

Fig. 2. Grid index of moving object

TOF uses a conventional encoding method, which orders the cells from bottom left to top right, as shown in Fig. 2. To map a moving object to its cell, we apply the function

$$cell = x/L + y/L * N,$$

where L is the cell length and N is the number of cells in x-ray. Our grid index uses an array to store the cells. For instance, in Fig. 2, the two moving objects, O_1 (1, 31, 11) and O_2 (2, 13, 22) are mapped to Cell 7 and Cell 9 respectively.

During indexing, multiple threads can be used to assign data to cells simultaneously. However, this would introduce contention, which would severely impair the performance. TOF adopts a lock free mechanism, in which the data is scanned twice. In the first round, the data is divided into M parts and scanned by M threads in parallel. The first scan generates a histogram for each cell and obtains the number of moving object in each cell. Then, a contiguous memory space is allocated to store the objects. As the number of moving object in each cell is already known, we can pre-compute the exclusive location where each thread writes its output. During the second scan, all threads read and write their own data independently. Contention is completely avoided. The same approach has been applied in [6].

When constructing the gird index, TOF uses an array to store the data in each cell. The pointer structure is avoided to optimize the performance. When the number of cells is large, writing into the arrays concurrently can cause cache thrash. In this case, TOF applies the strategy of Radix Clustering [7] to improve cache efficiency. Basically, the grid index can be constructed in two steps. First, a coarse grid with much fewer cells is first created. Then, each cell in the coarse grid is further divided into more cells to form the required grid index.

3.4   Algorithms for Handling Range Queries and kNN Queries

In this part, we describe the algorithms for handling range queries and kNN queries in TOF. Our algorithms are performed in the execution stage of TOF. Both algorithms work in two phases: grid index building phase and processing phase. During the first phase, they build grid structures to index the moving objects and queries according to their coordinates. In the processing phase, they perform the query according to the grid index.

Range Query. For range query, in the grid index building phase, a range query, which is a rectangle as mentioned in Section 3.1, could cross multiple cells in a grid index. As shown in Fig. 3, TOF replicates each range query and assigns a replica to each cell overlapping with the query. Accordingly, the moving objects are assigned to the grid index too. The index building procedure follows which is introduced in Section 3.3. After the indexing, the objects only need to be compared against the queries in the same cell to generate the complete results.

In the processing phase, the queries and objects in the same cells are compared against each other to decide if an object falls in each query range. This can be regarded as a join operation. To conduct the join, we can choose bucket-chaining-join or nested-loop-join. Bucket-chaining-join was first proposed in [20]. Compared to nested-loop-join, its main advantage is to reduce the comparison cost. Our experiments show that the comparison between an object and a query is the most computation intensive task in the join operation. Given a moving object $o_i(OID, (x, y))$ and a range query q_j (QID, $((x_{min}, y_{min}), (x_{max}, y_{max})))$, the follow expression is used to perform the comparison :

$$if((x >= x_{min}) \&\& (x <= x_{max}) \&\& (y >= y_{min}) \&\& (y <= y_{max})) (1)$$

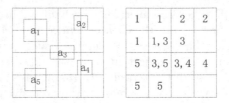

Fig. 3. Partitioning (left) and assignment (right) for query

Depending on the results of the expression, the program decides whether to add the object to the result set of the query. This will lead to a branch prediction failure. For main memory programs, branch prediction failure could result in significant performance decline.

To achieve improved performance, TOF tries to utilize the SIMD feature and the prefetching mechanism of modern CPUs. Figure 4 illustrates the SIMD

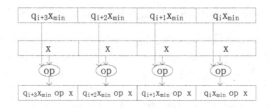

Fig. 4. SIMD mechanism for moving objects range query

mechanism for range query processing. TOF computes four queries and one object each time. As shown in Fig. 4, each time TOF compares between the x from one moving object and the x_{min} from four queries through a single SIMD execution. After all the dimensions are computed, the algorithm integrates the outcome into final results. To facilitate this mechanism, column-store is used to store the data and the queries, such that each dimension is stored in a single array.

As data comparison in bucket-chaining-join is not contiguous, SIMD and the pre-fetching mechanism cannot be used efficiently. The nested-loops-join can benefit from the SIMD more significantly. Therefore, in this paper, we propose to use the SIMD based nested-loops-join.

kNN Query. In kNN query processing, each query is a point instead of a range, as mentioned in Section 3.1. In the index building phase, both queries and objects are mapped to a grid index. The procedure follows which is introduced in Section 3.3. After the indexing, for the queries in each cell, we compute their common refinement region, which determines the scope of the objects the queries need to compare with.

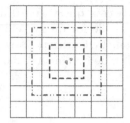

Fig. 5. Expansion method for kNN query

In the processing phase, we use the cell as the basic processing unit. Given a cell c_i, we call the cells surrounding c_i the 1st adjacent circle of c_i. For example, in Fig. 5 the cells that are connected by a dotted line belong to first adjacent circle of the cell that q located. The cells surrounding the lth adjacent circle form the $(l + 1)$th adjacent circle. TOF uses a refinement-and-searching approach to perform kNN search, similar to that used by YPK-CNN. Firstly, we use the histogram of the object cells built in the grid index building phase to calculate the refinement region for corresponding query cell. Given a query cell, suppose its Mth adjacent circle contains at least k moving objects. Then, it can be proven that the kNN objects must be within the $M + \lceil(\sqrt{2} - 1) * M + \sqrt{2}\rceil$th adjacent circle of the query cell. Thus, this circle defines the refinement region of the query cell. In the searching phase, TOF only needs to compare the queries in the cell against the objects in its refinement region to obtain the final results. To further speedup the process, we apply a pruning strategy, that is, if the farthest distance between the query and the k objects is less than the distance between the query and a cell, we can skip the cell. This refinement-and-searching approach minimizes the comparison conducted in kNN query processing.

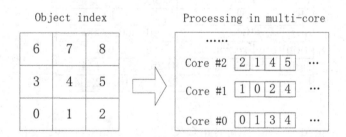

Fig. 6. Data locality of kNN query on multi-core processing

To parallelize the processing phase on multi-cores, TOF assigns different query cells to different threads. This is illustrated in Figure 6. We assume that for each query cell we should expand one level to find the final results. Each thread processes a query cell. As shown in Fig. 6, thread 0 is responsible for the query cell 0. It needs to access query cell 0 and the object cells 0,1,3,4 accordingly. Thread 1 is responsible for query cell 1, so it needs to access the object cells 0,1,2,3,4,5, and so on. TOF tries to assign adjacent query cells to the threads, so that data locality can be maximized and cache miss can be reduced.

4   Performance Study

We conducted several sets of experiments to study the properties of the proposed algorithms for range query and kNN query processing over spatiotemporal data. Before reporting the findings, we describe the experimental setting first.

4.1   Experimental Setup

We used the real Germany road network to generate the data sets. The Germany road network contains 380 million vertices and 400 million road segments. The data sets are produced using MOTO [8], which is an open-source moving object trace generator and based on Brinkhoff's moving-object generator [19]. MOTO follows a network-based object placement approach, where objects are placed and navigated (to a random destination) in a given road network.

To obtain realistic skew and to stress test the indexing techniques, the generator was slightly modified, so that half of the objects are placed in five major German cities according to the number of inhabitants in those cities. The queries are also distributed in those cities accordingly. This ensures that the most update-intensive regions are also the most queried ones.

Table 1. Workload configuration

Parameter	Values
objects, $*10^6$	5, 10, 20, 40
queries, $*10^3$	5, 50, 500, 5000
monitored region, km^2	Germany, 641 * 864
range query size, km^2	0.25, 0.5, 1, 2, 4, 8
k	2, 4, 8, 16, 32, 64

We implemented our algorithms in C/C++ and compiled with g++ under the maximum optimization level. All experiments ran on a 32-core hyper-threaded machine (4 Intel E5-2670 @2.6GHz) with 256GB RAM running SUSE 11 (64-bit). Caches are shared by all threads in an entire chip. In the following part, all the experiments were conducted on 5 million objects.

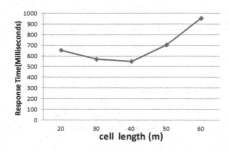

Fig. 7. Optimal grid cell size

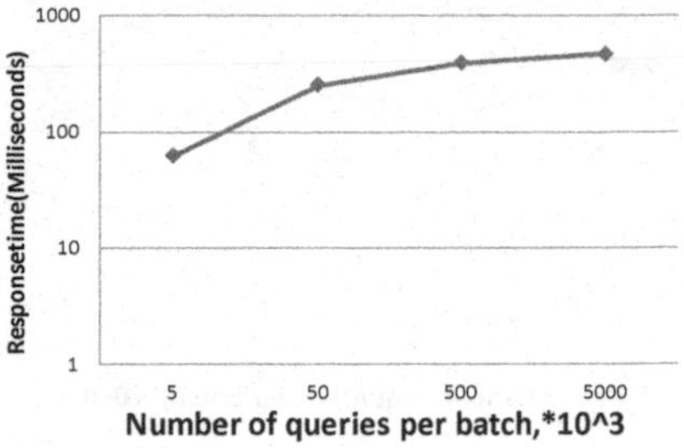

Fig. 8. Response time vs Increasing number of queries

4.2   Range Query Performance

In this part, we show the performance of our range query processor and compare it against PGrid[16]. To the best of our knowledge, PGird is the fastest among all the existing in-memory approaches to process moving object range queries. We mainly focus on the throughput.

Grid cell size is an important parameter that affects the performance of the algorithm. To obtain the optimal value, we tested 5 million objects and 5 million queries with different cell sizes. We varied the grid cell size from 20 to 60 meters. When the cell length is longer, there will be more objects in each cell. When the cell length is shorter, for range queries, there will be more query duplicates. Both could increase the comparison cost. The results are shown in Figure 7. We can see that 40m seems to be the optimal cell size. Thus, in the following experiments, we choose 40m as the cell size.

TOF needs to guarantee that its response time meets the requirements of applications. The way to tune response time is to change the window size – the number of queries processed in each time interval. Figure 8 shows how the response time varies with the window size. As we can see, if the number of queries processed in each round is controlled within a certain range, TOF is able to achieve sub-second response time easily, which satisfies most real-world applications.

Figure 9 shows the throughput comparison between TOF and PGrid when changing the query window size. With the increase of query numbers, the throughput increase dramatically for TOF. TOF uses a batch processing to maximize the throughput. So in a certain degree, the more queries, the more throughput for TOF. Whereas, PGrid uses one-by-one query processing mechanism. When the queries coming too fast, PGrid cannot process the incoming queries on time. Users have to queue up for the response.

Figure 10 shows the throughput comparison between TOF and PGrid when changing the number of objects. When the number of moving objects increase, there will be more objects in each cell, which increases the comparison cost. For TOF, due to the SIMD mechanism used in join processing, it performance is steady with the increasing of moving objects. For PGrid, it used a grid index, for

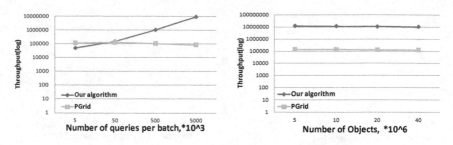

Fig. 9. Throughput vs Increasing number of queries

Fig. 10. Throughput vs Increasing number of objects

each query it only need to scan the data in corresponding cell, so its throughput is steady too.

Figure 11 compares the throughput of our algorithm against that of PGrid by varying the query range size. When the query range size increases, more cells will be processed for each query. Therefore, the throughput decreases gradually with the increasing query range size for both TOF and PGrid. Clearly, TOF outperforms PGrid by almost two orders of magnitude. This is attributed to the throughput oriented design of TOF, which apply batch processing to maximize the computation sharing among queries and uses the grid index to achieve high data locality. To prevent contention between threads, PGrid uses a lock mechanism, which is costly and prevents it from scaling well on multi-cores.

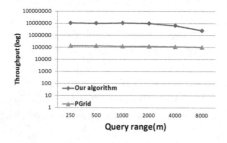
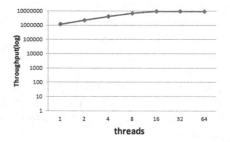

Fig. 11. Throughput vs Increasing query range size

Fig. 12. Throughput vs Increasing number of threads

Figure 12 shows how TOF scale with increasing number of hardware threads in a multi-core environment. In the TOF framework, a lock free mechanism is used to index the moving objects, making query processing scalable. As we can see, more threads are used, a higher throughput can be achieved by the system. The throughput reaches its maximum of 10M/s, when the number of threads is 32, which is exactly the number of hardware threads of the platform.

4.3   kNN Query Performance

In this part, we show the performance of kNN query processing of TOF and compare it against a kd-tree based main-memory algorithm, named RKNN [21].

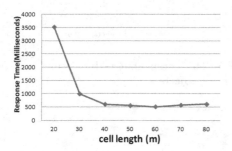
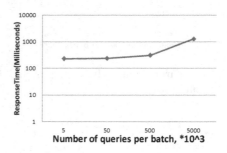

Fig. 13. Optimal grid cell size

Fig. 14. Response time vs Increasing number of queries

Similar to range query processing, grid cell size is an important parameter for kNN query processing. Our first set of experiments aimed to study how the grid cell size affects the performance. The experiments were conducted on 5 million objects and 5 million queries. We varied the grid cell size from 20 to 80 meters. From figure 13, we can see that when the cell length is too short, such as 20 meters, the response time is quite high. This is because most of the time was used to calculate the M level regions for each query cell. On the contrary, when the cell length is too long, there will be too many objects in each cell, which would increase the comparison cost. Figure 13 shows that 60m is the optimal cell length. In the following experiments, we used 60m as the cell length.

We also conducted experiments to see how the response time varies with the window size in kNN processing. Figure 14 shows the results. As we can see, TOF is able to achieve reasonable response time on kNN, as long as the window size is configured appropriately.

Figure 15 shows the throughput comparison between the two algorithms when varying the query window size. The RKNN algorithm builds a kd-tree index over objects, then using the index to search for the results. In query processing, RKNN executes one query at a time. TOF uses grid to index the moving objects and uses a batch query mode to get the results. TOF enables computation sharing among queries. Its design enables better data locality and scalability. Thus, it is able to achieve much better throughput than RKNN (an order of magnitude better).

Figure 16 shows the throughput comparison between the two algorithms when increasing the number of objects. TOF uses a batch query mode, which makes full use of data locality to enhance the throughput. With the increase of moving objects quantity, the throughput decreases in both algorithms. But our algorithm's performance is an order of magnitude better than RKNN.

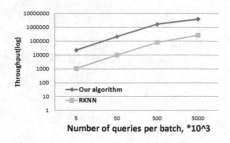

Fig. 15. Throughput vs Increasing number of queries

Fig. 16. Throughput vs Increasing number of objects

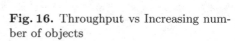

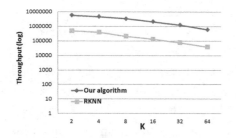
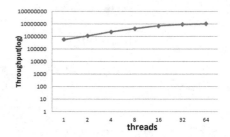

Fig. 17. Throughput vs Varying query region

Fig. 18. Throughput vs Increasing number of threads

Figure 17 plots the performance of both algorithms when k changes. With the increase of k, the throughput decreases for both algorithms. This is because increasing k would increase the region to search for top k objects. Nevertheless, TOF achieves a much higher throughput performance than the RKNN algorithm.

Figure 18 shows how TOF scales in a multi-core environment when performing kNN query processing. Similar to the results on range queries, TOF scales almost linearly with the number of hardware threads, owing to it throughput oriented design.

5   Conclusions

In this paper, we present a throughput oriented framework (TOF) that utilizes various feature of new hardware for efficient processing location-dependent spatial queries. We found that it is necessary to reconsider the design of traditional spatiotemporal query engines when throughput instead of query latency becomes the primary requirement. This was proven by our experimental evaluation, as TOF can achieve an order of magnitude of performance improvement over existing approaches.

References

1. Biveinis, L., Saltenis, S., Jensen, C.S.: Main-memory operation buffering for efficient R-tree update. In: 33rd International Conference on Very Large Data Bases, pp. 591–602. Vienna, Austria (2007)
2. Yiu, M.L., Tao, Y., Mamoulis, N.: The B^{dual}-Tree: indexing moving objects by space filling curves in the dual space. In: 34rd International Conference on Very Large Data Bases, pp. 379–400. New Zealand (2008)
3. Zhang, J., Zhu, M., Papadias, D.: Location-based spatial queries. In: 2003 ACM SIGMOD International Conference on Management of Data, pp. 443–454. California (2003)
4. Chen, Y.J., Chuang, K.T., Chen, M.S.: Spatial-temporal query homogeneity for KNN object search on road networks. In: 22nd ACM International Conference on Information Knowledge Management, pp. 1019–1028. ACM (2013)
5. http://ark.intel.com/zh-cn/products/75258/Intel-Xeon-Processor-E7-8890-v2-37_5M-Cache-2_80-GHz
6. Cagri, B., Gustavo, A., Jens, T., Özsu, M.T.: Main-Memory Hash Joins on Modern Processor Architectures. In: IEEE Transactions on Knowledge and Data Engineering, IEEE Press (2014)
7. Manegold, S., Boncz, P., Kersten, M.: Optimizing main-memory join on modern hardware. In: IEEE Transactions on Knowledge and Data Engineering, pp. 709–730. IEEE Press (2002)
8. Dittrich, J., Blunschi, L., Vaz Salles, M.A.: Indexing Moving Objects Using Short-Lived Throwaway Indexes. In: Mamoulis, N., Seidl, T., Pedersen, T.B., Torp, K., Assent, I. (eds.) SSTD 2009. LNCS, vol. 5644, pp. 189–207. Springer, Heidelberg (2009)
9. Yu, X., Pu, K.Q., Koudas, N.: Monitoring k-nearest neighbor queries over moving objects. In: 21st International Conference on Data Engineering, pp. 631–642. IEEE Press, Tokyo (2005)
10. Dittrich, J., Blunschi, L., Salles, M.A.V.: MOVIES: indexing moving objects by shooting index images. Geoinformatica 15(4), 727–767 (2011)
11. Harizopoulos, S., Liang, V., Abadi, D.J., Madden, S.: Performance tradeoffs in read-optimized databases. In: 32nd International Conference on Very Large Databases, pp. 487–498. VLDB Endowment, Seoul (2006)
12. Hilbert, D.: Ueber die stetige Abbildung einer Line auf ein Flichenstck. Mathematische Annalen 38. 3, pp. 459–460. IEEE (1891)
13. Orenstein, J.A., Merrett, T.H.: A class of data structures for associative searching. In: 3rd ACM SIGACT-SIGMOD Symposium on Principles of Database Systems, pp. 181–190. ACM (1984)
14. Tropf, H., Herzog, H.: Multidimensional Range Search in Dynamically Balanced Trees. ANGEWANDTE INFO (2), pp. 71–77. IEEE (1981)
15. Šidlauskas, D., Ross, K.A., Jensen, C.S., Šaltenis, S.: Thread-level parallel indexing of update intensive moving-object workloads. In: Pfoser, D., Tao, Y., Mouratidis, K., Nascimento, M.A., Mokbel, M., Shekhar, S., Huang, Y. (eds.) SSTD 2011. LNCS, vol. 6849, pp. 186–204. Springer, Heidelberg (2011)
16. idlauskas, D., altenis, S., Jensen, C.S.: Parallel main-memory indexing for moving-object query and update workloads. In: 2012 ACM SIGMOD International Conference on Management of Data, pp. 37–48. ACM, Scottsdale (2012)
17. Xiong, X., Mokbel, M.F., Aref, W.G.: SEA-CNN: scalable processing of continuous k-nearest neighbor queries in spatio-temporal databases. In: 21st International Conference on Data Engineering, pp. 675–686. IEEE Press, Tokyo (2005)

18. Mouratidis, K., Hadjieleftheriou, M., Papadias, D.: Conceptual partitioning: an efficient method for continuous nearest neighbor monitoring. In: 2005 ACM SIGMOD International Conference on Management of Data, pp. 634–645. ACM, Baltimore (2005)
19. Brinkhoff, T.: A framework for generating network-based moving objects. GeoInformatica **6**(2), 153–180 (2002)
20. Manegold, S., Boncz, P., Kersten, M.: Optimizing main-memory join on modern hardware. IEEE Transactions on Knowledge and Data Engineering, 709–730. Springer (2002)
21. http://www.cs.umd.edu/mount/ANN/

Query Processing

Identifying and Caching Hot Triples
for Efficient RDF Query Processing

Wei Emma Zhang[1]($\boxtimes$), Quan Z. Sheng[1], Kerry Taylor[2], and Yongrui Qin[1]

[1] School of Computer Science, The University of Adelaide,
Adelaide, SA 5005, Australia
{wei.zhang01,michael.sheng,yongrui.qin}@adelaide.edu.au
[2] CSIRO, Canberra, ACT 2601, Australia
kerry.taylor@csiro.au

Abstract. Resource Description Framework (RDF) has been used as a general model for conceptual description and information modelling. As the growing number and volume of RDF datasets emerged recently, many techniques have been developed for accelerating the query answering process on triple stores, which handle large-scale RDF data. Caching is one of the popular solutions. Non-RDBMS based triple stores, which leverage the intrinsic nature of RDF graphs, are emerging and attracting more research attention in recent years. However, as their fundamental structure is different from RDBMS triple stores, they can not leverage the RDBMS caching mechanism. In this paper, we develop a *time-aware frequency* based caching algorithm to address this issue. Our approach retrieves the accessed triples by analyzing and expanding previous queries and collects most frequently accessed triples by evaluating their access frequencies using *Exponential Smoothing*, a forecasting method. We evaluate our approach using real world queries from a publicly available SPARQL endpoint. Our theoretical analysis and empirical results show that the proposed approach outperforms the state-of-the-art approaches with higher hit rates.

Keywords: Caching · Query expansion · Exponential smoothing · RDF

1 Introduction

The Resource Description Framework (RDF), originally proposed for the Semantic Web to represent machine understandable data, has been increasingly used as a general model for conceptual description and information modeling. For example, knowledge management applications such as DBpedia[1] and Freebase[2] offer large collections of facts about entities and their relations with RDF-based representations. The RDF model is more expressive than the relational data model, yet increases the complexity of data querying and processing. The SPARQL Protocol and RDF Query Language (SPARQL) is the W3C standard query language

---

[1] http://dbpedia.org/
[2] https://www.freebase.com/

© Springer International Publishing Switzerland 2015
M. Renz et al. (Eds.): DASFAA 2015, Part II, LNCS 9050, pp. 259–274, 2015.
DOI: 10.1007/978-3-319-18123-3_16

to retrieve information modelled in RDF from either local or publicly accessible triple stores. It is an SQL-like deductive query language that provides a graph pattern based query syntax. Since the number of publicly available RDF datasets grows rapidly and the volume of the data is increasing, it becomes essential for efficient querying and processing of large scale RDF datasets. In recent years, many techniques have been designed for this purpose including index building [12,16], query optimization [3,11,15], and caching [7,8,17]. Among these, caching focuses on speeding up queries based on the principle that frequently used data is kept in memory (i.e., cached), while subsequent queries first access the data in memory and then if needed, access the data stored in slower and cheaper hardware components (e.g., disks). If the requested data is cached (also called *cache hit*), the result will be returned immediately without accessing slower hardware. This can accelerate the process of querying. Caching techniques for relational databases have been developed for over 40 years and many algorithms have been proposed, e.g., LRU [2], LRU-k [13] and ARC [9]. Triple stores often adopt relational databases as their underlying management systems and utilize the caching mechanism provided by these database systems.

Meanwhile, in recent years, non-relational based triple stores are emerging [18,19] and attracting research interest. However, the caching algorithms designed for relational databases are not directly applicable to these non-relational based triple stores. A number of research efforts have been devoted to addressing this issue, such as caching query results [8] or caching the result of subgraphs of a query pattern graph [17]. However, the solution in [8] relys on the assumption that subsequent queries are exactly the same to the previous cached ones, and work in [7] assumes that subsequent queries are very similar. Work in [17] ignores the order of joins over the subgraphs. Thus if subsequent queries that contain the same subgraphs is decomposed in different orders, limited contribution of cached results will be given.

In this paper, we concentrate on solving the above issues by developing a *time-aware frequency* based caching algorithm which leverages the idea of a novel approach recently proposed for main memory databases in Online Transaction Processing (OLTP) systems [6]. More specifically, we cache the complete triples that have been frequently accessed by previous queries. We draw on the notion of query expansion in getting access logs used in this algorithm. We first extract the accessed triples by rewriting the original queries according to query pattern analysis and executing the rewritten queries. Then we use a smoothing method to estimate the access frequencies of the accessed triples. The triples with the highest estimated access frequencies are considered "hot" and will be cached. Other triples are considered "cold". When a new query arrives, it will be performed on the union of triples both in cache and on disk. As the access cost to cold triples is much higher than the access cost on hot triples, the overall querying speed depends on how much proportion of the frequently accessed triples are in cache, i.e., the hit rate achieved by the cache. It should be noted that filtering cold data to reduce the querying time is not the focus of this paper. We will concentrate on solving the very first key issues in this research area, i.e., how to identify and cache hot RDF data. Our caching approach can be applied in a number of application scenarios. For example, it can be adopted as a server-side caching approach for a publicly accessible endpoint or as a component of local non-RDBMS based triple stores. To the best of our knowledge, this is the first work that takes advantage of frequency-based caching algorithms in a non-RDBMS based triple store. In a nutshell, the main contributions of this work are summarized as follows:

- We develop a time-aware frequency based caching algorithm for non-RDBMS based triple stores. Our approach adopts a smoothing method, which is widely used for economic forecasting and recently introduced into main memory databases, to evaluate frequently accessed triples.
- We utilize the techniques in the fields of query analysis and query expansion to retrieve triples accessed by previously performed queries. When analyzing and expanding queries, we consider not only conjunctive relationships among triple patterns, but also patterns such as UNION and OPTIONAL.
- We perform our experiments on a widely accessed SPARQL endpoint. Both theoretical analysis and empirical results show that our approach outperforms state-of-the-art approaches with higher hit rates.

The remainder of this paper is structured as follows. We introduce some background to this paper and briefly discuss the main tasks for hot data management in Section 2. In Section 3, we describe the details of our methodology. Experiments are presented in Section 4. Section 5 overviews the related work. Finally, we conclude this paper in Section 6 and discuss some future research directions.

2   Preliminaries

In this section, we will discuss the background context for this work. We will then introduce the main tasks for hot RDF data management.

2.1   RDF and SPARQL Queries

The Resource Description Framework (RDF) presents a statement as a triple in the form of $(subject, predicate, object) \in UB \times U \times UBL$ where U, B, and L denote the sets of URIs, Blank nodes and Literals, respectively. Each triple represents a relationship between two resources and can be considered as an edge ($predicate$) connecting two vertices ($subject$ and $object$). All the interconnected triples in an RDF dataset form a directed graph.

The SPARQL Protocol and RDF Query Language (SPARQL) is a pattern matching-based RDF query language which contains a number of graph patterns. A SPARQL graph pattern expression is defined recursively as follows [8,14]:

(i) A valid triple pattern T is a graph pattern,
(ii) If P_1 and P_2 are graph patterns, then expressions $(P_1 \text{ AND } P_2)$, $(P_1 \text{ UNION } P_2)$ and $(P_1 \text{ OPTIONAL } P_2)$ are graph patterns,

(iii) If P is a graph pattern and R is a SPARQL condition, then the expression (P FILTER R) is a graph pattern.

A *Basic Graph Pattern* (BGP) is a graph pattern when it is represented by the conjunction of multiple triple patterns.

2.2   Exponential Smoothing

The *Exponential Smoothing* (ES) is a technique to produce a smoothed data presentation, or to make forecasts for time series data, i.e., a sequence of observations [5]. It can be applied to any discrete set of repeated measurement and is currently widely used in smoothing or forecasting economic data in the financial markets. Equation 1 shows the simplest form of exponential smoothing. This equation is also regarded as *Simple Exponential Smoothing* (SES).

$$E_t = \alpha * x_t + (1 - \alpha) * E_{t-1} \tag{1}$$

where E_t stands for smoothed observation of time t, x_t stands for the actual observation value at time t, and α is a smoothing constant with $\alpha \in (0, 1)$. From this equation, it is easy to observe that SES assigns exponentially decreasing weights as the observation becomes older. In our work, we will adapt this equation to estimate record access frequencies (details in Section 3.3). The reason behind our choice of SES is its simplicity and effectiveness [6].

2.3   Main Tasks for Hot RDF Data Management

Hot RDF data management generally involves two main tasks:

- *Identifying hot data*: This task focuses on efficiently identifying hot and cold data in RDF datasets, and updating the cache. Identifying hot data essentially requires pre-computation of previously accessed data, which can be done off-line. During the querying time, subsequently accessed records will be added and processed. They will also contribute to further estimation of the hot data. Due to the limitation of the cache size, old cached triples might need to be replaced by newly accessed triples. In this paper, we focus on proposing techniques to identify hot data and the cache replacement.
- *Filtering cold data*: This task aims at exploiting accurate filters to reduce unnecessary accesses to the cold data. As the query result is the union of results from the hot data (in cache) and the cold data (stored on disk), it is necessary to filter the redundant cold data when performing queries. As mentioned before, we will not focus our discussions on this task in the paper, which will be part of our future work.

3   The Methodology

Figure 1 illustrates our approach, which consists of three major steps, namely *data acquisition and pre-processing*, *access log extraction*, and *caching*. Section

3.1 discusses our experience in pre-processing a set of real-world query logs to extract SPARQL queries. Section 3.2 focuses on generating the access log, which is formed by triples that are accessed during query processing. Section 3.3 introduces our techniques in caching hot data based on the accessed triples.

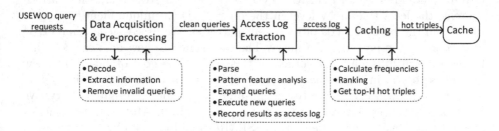

Fig. 1. Steps for Identifying and Caching Hot Triples

3.1 Data Acquisition and Pre-processing

In this paper, we use a real world dataset to showcase the RDF data acquisition and pre-processing. Specifically, we analyze parts of the DBpedia 3.9 query logs provided by USEWOD 2014 data challenge[3]. These logs are formatted in Apache Common Log Format. The log files were collected from late 2013 to early 2014 and each log contains requests received by DBPedia's SPARQL endpoint for one day. Each request contains requesters' anonymized IP address and a timestamp in addition to the actual query. The annoymized IP is encoded to an ID which represents a unique query requester from a same IP address. Request date is the time that the query is performed and the timestamps are in an hourly resolution. The query content is encoded with HTML URL encoding.

Table 1. Query Form Distribution of DBpedia Queries

Query Forms	SELECT	ASK	DESCRIBE	CONSTRUCT
Percentage in Total Queries(%)	98.07	1.51	0.29	0.13

In order to extract queries, we process the original query by decoding, extracting interesting values (IP, date, query string), identifying SPARQL queries from query strings and removing duplicated and invalid queries. Here, invalid queries include all incomplete queries, queries in languages rather than English, queries with syntax errors according to SPARQL1.1 specification and queries generate no result. Our work only focuses on SELECT, as shown from our own analysis of the USEWOD data set (see Table 1) more than 98% of queries are SELECT queries.

---

[3] http://usewod.org/

3.2   Access Log Extraction

After obtaining the valid queries, we expand them to extract the access log. There are three main steps. We first decompose a query into patterns and record patterns' information in a pattern table containing the pattern feature (UNION, GROUP, OPTIONAL, FILTER, BIND, VALUES, MINUS etc.), the pattern content, the level (hierarchy level in the query), ID (a unique identifier in the table) and the parentID (ID of its parent pattern). We use Apache Jena TDB[4] to extract all the patterns of a query. Then we generate new queries according to the pattern information recorded in this table. The goal of generating new queries is to obtain triples that a query accesses. In other words, the results of new queries will contribute to getting final results of the original query. We focus on discussing this step in the rest of this section. After the new queries are generated, we execute them on the SPARQL endpoint and get all accessed triples that are related to the final returned results and record them in the access log. This step is straightforward so we will not discuss in details.

Let $Q_s = (S, P)$ be the original query where S is the SELECT expression and $P = P_1 \oplus ... \oplus P_n$ is the query pattern where $\oplus \in \{$UNION, GROUP, OPTIONAL, FILTER, BIND, VALUES, MINUS$\}$. P_i can be finally decomposed to triple patterns with modifier and constraint expressions: $P_i = \{T_1, T_2, ..., T_k, E_1, E_2, ..., E_n\}$.

Let Q_e be the expanded query and

$$Q_e = \begin{cases} (S_i, P), & \text{if } T_i \text{ is not from UNION pattern} \\ (S_i, P_e), & \text{if } T_i \text{ is from UNION pattern } P_j \end{cases}$$

where S_i is modified from T_i by building T_i as a selection expression, P_e is obtained by replacing P_j with P_j' in P, where P_j' is obtained by removing T_i from P_j. To build S_i, we choose the triple patterns in the pattern table one by one. It simply adds "SELECT" before the triple pattern T_i and add "as ?s", "as ?p" and "as ?o" after each component of T_i. Thus the results are in a uniform style and easily merged together. For the pattern of new queries, we analyze the type of triple pattern T_i's parent, the graph pattern P_j and its position and relation with other patterns. Then we can decide whether it is the same as the query pattern in the original query (P, if P_j is not UNION), or needs to be modified by removing the triple pattern T_i (P_e, if P_j is UNION). More specifically, we develop four solutions for generating new queries according to the characteristics of different query patterns.

- *UNION.* As each part of UNION contributes to the result separately, each graph pattern in it will generate a new query, whose query pattern is built by removing this graph pattern from the pattern of the original query.
- *GROUP, OPTIONAL.* Each triple pattern in GROUP and OPTIONAL produces a new query. As GROUP and OPTIONAL contribute to the final query result by working with other patterns, the pattern of the new queries are the same as the pattern of the original query.

---

[4] http://jena.apache.org/index.html

```
Query 1:
    SELECT * WHERE {
        {?city rdfs:label 'Evry'@en.}
        UNION
        {?alias dbpedia2:redirect ?city; rdfs:label 'Evry'@en.}
        OPTIONAL {?city dbp:abstract ?abstract}
        FILTER (langMatches(lang(?abstract),'en'))
    }
    LIMIT 10
```

Fig. 2. Example Query

- *FILTER, BIND, VALUES, MINUS.* FILTER is an additional constraint and usually cannot be further decomposed into BGPs or triple patterns. Thus we keep it unmodified in the new queries. BIND and VALUES are assignments that cannot be decomposed into BGPs or triple patterns either, we also keep them unmodified in the new queries. Hence, FILTER, BIND and VALUES will not generate new queries. MINUS can be decomposed into BGPs or triple patterns, but as we only consider the triples related to the final result, we ignore the processing of triple patterns in MINUS.
- *LIMIT, OFFSET.* Regarding to modifiers, we only take LIMIT and OFFSET into consideration by simply appending them to the new queries.

We illustrate our approach with Query 1 in Figure 2, which retrieves certain information in English about city "Evry". The graph patterns for Query 1 are:

P_{GROUP1} := {?city rdfs:label 'Evry'@en.}
P_{GROUP2} := {?alias dbpedia2:redirect ?city.
                 ?alias rdfs:label 'Evry'@en.}
P_{UNION} := P_{GROUP1} UNION P_{GROUP2}
$P_{OPTIONAL}$:= {?city dbp:abstract ?abstract}
P_{FILTER} := FILTER langMatches(lang(?abstract),'en')

Curly braces delimit a graph pattern that can be further decomposed into triple patterns. According to our four solutions discussed above, the triple patterns that can generate new queries are:

T_1 := ?city rdfs:label 'Evry'@en
T_2 := ?alias dbpedia2:redirect ?city
T_3 := ?alias rdfs:label 'Evry'@en
T_4 := ?city dbp:abstract ?abstract

T_1 and $T_2.T_3$ (. represents AND) are two parts of UNION, thus they can produce one new query with pattern removing T_1 from the original query and other two new queries with pattern removing $T_2.T_3$. For the latter case, since T_2 and T_3 belong to a GROUP, they can generate two new queries respectively that only differ with the SELECT expression. T_4 is OPTIONAL, and it can build a new query just by using itself in a SELECT expression and other part is the same as the original query. For the modifier LIMIT, we simply append it at the end of the new queries. Thus four new queries will be generated. Figure 3 gives the new query generated from T_1. To avoid naming conflict in the SELECT expression in the new queries, we rename all the variables by labeling them with sequential integers (i.e. ?var1, ?var2). In Figure 3, the variables in the new query follow this renaming rule. Other new queries as well as the pattern table of Query 1 can be found in our website[5].

```
SELECT DISTINCT (?var0 AS ?s) (rdfs:label AS ?p) ("Evry"@en AS ?o)
WHERE {
    { ?var0  rdfs:label "Evry"@en }
    OPTIONAL { ?var0 dbp:abstract ?var2 }
    FILTER langMatches(lang(?var2), "en")
}
LIMIT 10
```

Fig. 3. New Query for T1

3.3 Caching

In this section, we discuss how we use the access log to realize caching as well as cache replacement. We modify the SES equation according to our requirement and process the access log in a forward way to compute access frequencies for each record in the log (Section 3.3.1). According to the estimation, we rank the frequencies and consider those with the highest values as hot, which will be cached. During the query processing, a new estimation will be made both for triples that are in the estimation record and for new triples (Section 3.3.2). To update the cache, we implement two cache replacement approaches (Section 3.3.3).

3.3.1 Modified Simple Exponential Smoothing (MSES)

As introduced in Section 2.2, *Exponential Smoothing* (ES) is a smoothing method that considers the estimation according to time, which meets the requirement of caching the most frequently and *recently* used triples. In our approach, we exploit *Simple Exponential Smoothing* (SES) to estimate access frequencies of triples. Here x_t represents whether the triple is observed at time t, thus it is either 1 if an access for a triple is observed; or 0 otherwise. Therefore we can modify the Equation 1 as:

$$E_t = \alpha + E_{t_{prev}} * (1 - \alpha)^{t_{prev} - t} \qquad (2)$$

where t_{prev} represents the time when the triple is last observed and $E_{t_{prev}}$ denotes the previous frequency estimation for the triple at t_{prev}. α is a smoothing constant with value between 0 and 1. The accuracy of ES can be measured by its

---
[5] http://cs.adelaide.edu.au/~wei/sublinks/projects/SemanticCaching/
PatternsTableAndNewQueries.pdf

standard error. For a record with true access frequency p, it can be shown that the standard error for SES is $\sqrt{\alpha p(1-p)/(2-\alpha)}$ as indicated in a recent study [6]. The authors of [6] do not give the derivation of this standard error. In this paper, we give our derivation according to Equation 2 and provide a theoretical proof that SES achieves better hit rates than the most used caching algorithm LRU-2. The variance of the estimation E_t is [10]:

$$Var(E_t) = \sigma^2 \alpha^2 (\sum_{i=0}^{t-1}(1-2\alpha)^{2i}) \tag{3}$$

As

$$\sum_{i=0}^{t-1}(1-2\alpha)^{2i} = \frac{1-(1-\alpha)^{2t}}{1-(1-\alpha)^2} \tag{4}$$

we can get

$$Var(E_t) = \sigma^2 \frac{\alpha}{2-\alpha}[1-(1-\alpha)^{2t}] \tag{5}$$

When $t \to \infty$, we get

$$Var(E_t) = \sigma^2 \frac{\alpha}{2-\alpha} \tag{6}$$

Because the observation value can only be 0 or 1, it follows a *Bernoulli Distribution* (or *Two-point Distribution*), and the frequency p is actually the possibility of the observation being 1. So, the standard error for this Bernoulli distribution is: $\sigma = \sqrt{p(1-p)}$ [4]. Thus we can get the standard error of the estimation SE as:

$$SE(E_t) = \sqrt{Var(E_t)} = \sqrt{p(1-p)}\sqrt{\frac{\alpha}{2-\alpha}} = \sqrt{\frac{\alpha p(1-p)}{(2-\alpha)}} \tag{7}$$

We also give the measurement of standard error for the most commonly used caching algorithm LRU-2 here as we compare our approach with LRU-2 in Section 4. In a recent work in [6], the authors present a probability model for evaluating the distribution of LRU-2 estimation and find that it follows a geometric distribution. Thus its standard error is:

$$SE(LRU-2) = \sqrt{\frac{1-p}{p^2}} \tag{8}$$

By comparing Equation 7 and 8, it is easy to observe that $SE(LRU-2)$ is always bigger than $SE(E_t)$. Our evaluation in Section 4 also empirically proves this theoretical derivation.

3.3.2   Identify and Cache Hot Triples

We use a forward algorithm to identify hot triples. This algorithm works as follows. It scans the access log from a beginning time to an ending time. A parameter H represents the number of records to be classified as "hot".

The output is the H hot triples that will be cached. When encountering an access to a triple at time t, the algorithm updates this triple's estimation using Equation 2. When the scan is completed, the algorithm ranks each triple by its estimated frequency and returns the H triples with the highest estimates as the hot set.

The forward algorithm has three main advantages. Firstly, it is simple as we only need to choose a starting time and then calculate the new estimation using Equation 2 when a triple is observed in the log again. Secondly, the algorithm enables us to update the estimation and the cache immediately after a new query is executed based on the previously recorded estimation. Thirdly, this algorithm implements an incremental approach that helps identify the *warm-up* stage and the *warmed* stage of the cache.

However, this algorithm also has several drawbacks. Specifically, it requires storing the whole estimation record which is a large overhead. Furthermore, the algorithm also consumes a significant amount of time when calculating and comparing the estimation values. To solve these issues, we consider improving the algorithm in two ways. One possible solution is that we do not keep the whole records. Instead we just keep a record after skipping certain ones. This is a naive sampling approach. We vary the sampling rate but it turns out that the performance of this sampling approach is not desirable (see Section 4). The other possible approach is that we maintain partial records by only keeping those within a specified range of time. Assume $last\_access\_time$ is the time a record is last observed. We find the earliest $last\_access\_time$: $t_{earliest}$ in the hot triples and only keep the estimation records whose $last\_access\_time$ is later than $t_{earliest}$. Thus we only keep estimation records from $t_{earliest}$ to the access time of currently processed triple. The update of $t_{earliest}$ will be discussed in next section.

3.3.3   Cache Replacement

We provide two ways for cache replacement based on the two possible improved forward algorithms discussed in Section 3.3.2, namely the *full-records based replacement* and the *improved replacement*.

In the full-records based replacement, each time when a new query is executed, we examine the accessed triples using MSES. If they are in the cache, we update the estimation for each triple. Otherwise, we record the new estimations. We keep the estimation records for all accessed triples. When the top H estimations are changed, the cache will be updated to the new top H hot triples.

In the improved replacement, we only keep estimation records from $t_{earliest}$ to the access time of the current processing triple. Algorithm 1 describes the details on updating the cache by using part of the estimation records. The input of the algorithm is the whole estimation records, triples in cache and a new estimated triple. Line 1-4 initialize variables that represent the latest access time, the earliest access time, the minimal estimation and the maximum estimation in cached triples (i.e., the hot triples). Then the algorithm gets partial estimation records that are within the time range between $t_{earliest}$ and the access time of the last processed triple in the log (Line 5). If the new estimated triple is in

---

Algorithm 1. Algorithm for Improved Caching Replacement

    Data: $Records$, $cachedTriples$, $newAccTriple$
    Result: Updated $cachedTriples$
    begin

1        $t_{latest} \longleftarrow max(last\_acc\_time, cachedTripples)$
2        $t_{earliest} \longleftarrow min(last\_acc\_time, cachedTripples)$
3        $est_{max} \longleftarrow max(est, cachedTripples)$
4        $est_{min} \longleftarrow min(est, cachedTripples)$
5        $Records \longleftarrow getPartialRecords(t_{latest}, t_{earliest})$
6   **if** $newAccTriple$ **in** $cachedTriples$ **then**
7            $calculateNewEstimation()$
8   Update est, last_acc_time in $Records$
9            $t_{latest} \longleftarrow newAccTriple.last\_acc\_time$
10          $t_{earliest} \longleftarrow getEarliest(cachedTriples)$
11   remove from $Records$ the records with last_acc_time **less than** $t_{earliest}$

        else
12   **if** $newAccTriple$ **in** $Records$ **then**
13             $calculateNewEstimation()$
14   Update est, last_acc_time in $Records$
15   **if** est **between** est_{min} and est_{max} **then**
16   remove from $cachedTriples$ the records with est_{min} with minimum last_acc_time
17                $addToCached(newAccTriple)$
18                $t_{latest} \longleftarrow newAccTriple.last\_acc\_time$
19                $t_{earliest} \longleftarrow getEarliest(cachedTriples)$
20   remove from $Records$ the records with last_acc_time **less than** $t_{earliest}$

          else
21             $addToRecords(newAccTriple)$

---

the cache, it shows a cache hit, and the algorithm updates the new estimation calculated by Equation 2 and the *last_access_time* of this triple in estimation records. It calculates the new $t_{earliest}$ and t_{latest} if the new estimated triple holds the previous $t_{earliest}$ and t_{latest}. If $t_{earliest}$ is changed, estimation records with *last access time* earlier than $t_{earliest}$ will be removed (Line 6-11). If the new estimated triple is not in the cache, which is a cache miss, the algorithm checks whether the new estimated triple is in the estimation records. If so, it updates its estimation and *last_access_time* in records. In addition, the cache needs to be updated if the estimation of the new estimated triple is in the range of (est_{min}, est_{max}). This means it becomes a new hot triple that should be placed into the cache. When the cache is updated, new $t_{earliest}$ and t_{latest} will be calculated, and the estimation records outside the time range will be removed (Line 15-20). If the new estimated triple is not in the estimation records, it needs to be added to the records (Line 21).

4   Experiments

We designed three experiments to verify the effectiveness of our approach: i) comparison of the hit rates between our proposed approach and the state-of-the-art algorithms; ii) space overhead comparison between the full record based forward MSES and the improved forward MSES; and iii) time overhead comparison between the full record based forward MSES and the improved forward

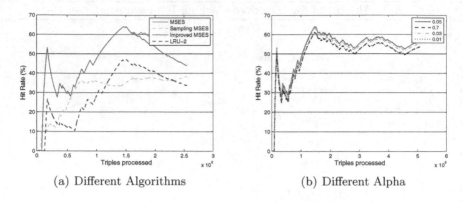

(a) Different Algorithms   (b) Different Alpha

Fig. 4. Hit Rate Comparisons

MSES. We conducted the three experiments on a PC with 64-bit Windows 7 Operation System, 8GB RAM and 2.40GHZ intel -i7-3630QM CPU using Java SE7 and Apache Jena-2.11.2.

Hit Rate Comparison. To evaluate the performance of our approach, we implemented various algorithms including the forward MSES, an Improved MSES, the Sampling MSES, and LRU-2. LRU-2 is a commonly used page replacement algorithm which we implemented based on record rather than page. Figure 4(a) shows the hit rates achieved by these algorithms. It should be noted that in the experiment, the caching size was set to 20% of the total access log and α was set to 0.05 for MSES and its variants. As the *Exponential Smoothing* has only one parameter α, the choice for α would affect the hit rate performance. However, as per our experiments on different values for α, the hit rates differ only slightly and a value of 0.05 shows better performance (see Figure 4(b)). We chose 20% as the caching size because it is neither too large (e.g., > 50%) to narrow the performance differences among algorithms, nor too small (e.g., < 10%) leading to inaccurate performance evaluation due to insufficient processed data. From the figure, we can see that the MSES and Improved MSES have the same hit rate until they have processed about 1.8 million RDF triples, after which MSES has a higher hit rate than Improved MSES. This is because MSES maintains the estimations for all processed records while the Improved MSES only keeps part of the estimations. The changing point denotes that from which, the Improved MSES maintains partial volume of estimation records. From the figure, we can also see that the Sampling MSES does not perform well. This figure only shows the hit rate of sampling MSES with the sampling rate of 50%, which is expected to have a high hit rate. The LRU-2 algorithm has the lowest hit rate of all the algorithms. The hit rates of all algorithms start from 0 and reach their first peak at certain points, and then go down and up. The direction to the first peak shows the warm-up stage and the rest of the lines are the warmed stage. This illustrates that we exploit an incremental approach, which includes a warm-up stage to calculate the hit rate.

Storage Overhead. This experiment compares the two implementations of our cache replacement algorithms. As discussed before, MSES performs better than the Improved MSES. However, it consumes more storage space to maintain the estimation records for all processed triples. It also takes a longer time to check the cache. Figure 5 shows the performance comparison between MSES and the Improved MSES. Figure 5(a) shows the maximum space consumption for each algorithm. The columns are classified into 4 groups which represent the percentage of hot triples to all accessed triples. In each group, the left column represents the maximum space used by MSES, including the hot triples and the estimation records (we have processed 5,095,807 triples in this experiment). The middle column represents the space usage of the Improved MSES that also includes the hot triples and the estimation records. The right column represents the size of the hot triples. From this figure, we can see that the Improved MSES consumes less space.

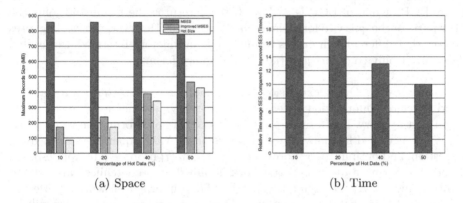

(a) Space   (b) Time

Fig. 5. Performance Comparisons Between MSES and Improved MSES

Time Overhead. Figure 5(b) shows the relative average hit checking time MSES uses compared to Improved MSES. We also used 10%, 20%, 40%, 50% as our testing hot data ratios. MSES takes much longer time than MSES does to check if there is a cache hit. Specifically, when the hot triple ratio is 10%, MSES takes almost 20 times longer than the Improved MSES. Even when the hot triple ratio grows to 50%, the gap is still close to 10 times. The more percentage of the data is cached, the narrower the gap will be. This is because the number of estimations checked will be closer for the two algorithms when the percentage of data cached grows high.

5   Related Work

In this section, we review existing efforts in semantic caching and query expansion that are related to our research in this paper.

Semantic Caching. Semantic caching involves techniques that keep previously fetched data from past queries. In this way, if subsequent queries use the same data, results can be returned immediately. The recent approaches include caching complete query results [8] and caching the result of subgraphs of a query pattern graph [17]. The work in [8] is the first step towards semantic caching, in which both the triple query result caching and the application object caching are implemented. It essentially builds a proxy layer between an application and a SPARQL endpoint. The proxy layer caches the query-result pairs. When a new query arrives, this approach first checks if the query is already in the cache. If so, it will answer the query using the cached result without accessing the triple store. In this approach, only subsequent queries that are the same as cached ones can result in a cache hit. However, real world queries vary and even if they access the same triples, they might not return the same results, e.g., different queries may be interested in different parts of a triple. [17] provides an approach that caches intermediate result of basic graph patterns in SPARQL queries. For a new query, it first decomposes the query into BGPs and checks if the result of any BGP or join of BGP is cached. If so, it returns the result and joins the other parts of the query to form the final query result. However, the work ignores the fact that different join orders would result in different intermediate results. Moreover, it only considers join (i.e., AND) query form.

Query Expansion. Query expansion (also called *query relaxation*) aims at discovering related information based on a user query and increasing recall when executing the query. It is a research topic in the field of information retrieval [1] and is recently used by many research works on SPARQL queries, e.g., caching, suggestion, and optimization. Lorey and Naumann [7] propose a template-based method to expand a query and the work is aimed at pre-fetching data related to future queries. By modifying the query, related information will be perfected and cached for the subsequent queries. Thus, these subsequent queries will be answered without accessing the triple store. It is based on the assumption that subsequent queries will access relevant resources to the cached query results. However, this assumption is based on the application scenario, that only considers queries requested by the same agent. Note that the highest cache hit rate of their approach is 39%, which is much lower than the hit rate of our approach (64%, see Figure 4(a)).

In a very recent work by Levandoski et al. [6], the authors present an approach for identifying hot and cold data in main memory of OLTP systems. They introduce a forecasting method that is commonly used in the finance field: *Simple Exponential Smoothing* (SES). It is shown that the approach achieves a higher hit rate than the best existing caching techniques. Our work is inspired by this work and we leverage the idea of using SES as the caching algorithm because our research problem shares similar requirements with their work in two aspects. Firstly, our caching is record based, rather than traditional page-based caching algorithms. Thus the classical algorithms are not applicable. Secondly, the potential applications of our approach are in non-RDBMS based triple stores, which require relatively less computing and analyzing ability but efficient query answering, and this is similar to OLTP systems.

6 Conclusion and Future Work

In this paper, we analyze the caching problem for non-RDBMS based RDF triple stores and develop a caching method that is based on the estimation of triples' access frequencies. We extract real world queries from publicly accessible SPARQL endpoints and expand these queries according to an analysis of query patterns. We exploit expanded queries to obtain accessed triples, which are recorded as the access log in our caching process. Then we extend a smoothing method, *Simple Exponential Smoothing*, to estimate the access frequencies of the accessed triples. The triples with the highest estimation are cached. We also develop several cache replacement algorithms to update the cache so that only triples with the highest estimates are kept in the cache. Our experimental studies with a real-world dataset reveal that the sampling does not work well as expected and our proposed Improved MSES achieves a higher hit rates than LRU-2, the most commonly used caching algorithm in relational databases. This shows that our time-aware frequency based caching approach performs well and has great potential to improve the query processing performance in non-RDBMS triple stores.

There are several directions for the future work. In this paper, we only consider results-related triples as accessed triples. Although our approach is able to return results (in case of cache hit) immediately, the querying process is yet finalized because many other triples (i.e., cold triples that are related to the query) are also accessed without contributing to the final result. In order to improve the performance of the whole query answering process, we will work on how to filter these related cold triples. Moreover, the forward algorithm for processing the access log is comparatively slower than the backward algorithm proposed in a recent work in [6]. We will investigate the possibility of designing a backward algorithm for our scenario. Finally, we plan to integrate the pattern caching into our approach to further enhance the hit rate performance.

References

1. Carpineto, C., Romano, G.: A Survey of Automatic Query Expansion in Information Retrieval. ACM Computing Survey **44**(1), 1 (2012)
2. Denning, P.J.: The Working Set Model for Program Behaviour. Communications of the ACM **11**(5), 323–333 (1968)
3. Huang, J., Abadi, D.J., Ren, K.: Scalable SPARQL Querying of Large RDF Graphs. The VLDB Endowment (PVLDB) **4**(11), 1123–1134 (2011)
4. Johnson, N.L., Kemp, A.W., Kotz, S.: Univariate Discrete Distributions (2nd Edition). Wiley (1993)
5. Jr., E.S.G.: Exponential Smoothing: The State of The Art-Part II. International Journal of Forecasting **22**(4), 637–666 (2006)

6. Levandoski, J.J., Larson, P., Stoica, R.: Identifying hot and cold data in main-memory databases. In: Proc. of 29th International Conference on Data Engineering (ICDE 2013), pp. 26–37. Brisbane, Australia, April 2013
7. Lorey, J., Naumann, F.: Detecting SPARQL query templates for data prefetching. In: Cimiano, P., Corcho, O., Presutti, V., Hollink, L., Rudolph, S. (eds.) ESWC 2013. LNCS, vol. 7882, pp. 124–139. Springer, Heidelberg (2013)
8. Martin, M., Unbehauen, J., Auer, S.: Improving the Performance of Semantic Web Applications with SPARQL Query Caching. In: Aroyo, L., Antoniou, G., Hyvönen, E., ten Teije, A., Stuckenschmidt, H., Cabral, L., Tudorache, T. (eds.) ESWC 2010, Part II. LNCS, vol. 6089, pp. 304–318. Springer, Heidelberg (2010)
9. Megiddo, N., Modha, D.S.: ARC: a self-tuning, low overhead replacement cache. In: Proc. of the Conference on File and Storage Technologies (FAST 2003). San Francisco, California, USA, March 2003
10. Movellan, J.R.: A Quickie on Exponential Smoothing. http://mplab.ucsd.edu/tutorials/ExpSmoothing.pdfa/
11. Neumann, T., Weikum, G.: Scalable join processing on very large RDF graphs. In: Proc. of the International Conference on Management of Data (SIGMOD 2009)
12. Neumann, T., Weikum, G.: The RDF-3X Engine for Scalable Management of RDF Data. The VLDB Journal 19(1), 91–113 (2010)
13. O'Neil, E.J., O'Neil, P.E., Weikum, G.: The LRU-K page replacement algorithm for database disk buffering. In: Proc. of the International Conference on Management of Data (SIGMOD 1993), pp. 297–306. Washington, D.C., USA, May 1993
14. Pérez, J., Arenas, M., Gutierrez, C.: Semantics and Complexity of SPARQL. ACM Transactions on Database Systems 34(3) (2009)
15. Stocker, M., Seaborne, A., Bernstein, A., Kiefer, C., Reynolds, D.: SPARQL basic graph pattern optimization using selectivity estimation. In: Proc. of the 17th International World Wide Web Conference (WWW 2008), pp. 595–604. Beijing, China, April 2008
16. Yan, Y., Wang, C., Zhou, A., Qian, W., Ma, L., Pan, Y.: Efficiently querying RDF data in triple stores. In: Proc. of the 17th International World Wide Web Conference (WWW 2008), pp. 1053–1054. Beijing, China, April 2008
17. Yang, M., Wu, G.: Caching intermediate result of SPARQL queries. In: Proc. of the 20th International World Wide Web Conference (WWW 2011), pp. 159–160. Hyderabad, India, March 2011
18. Zeng, K., Yang, J., Wang, H., Shao, B., Wang, Z.: A Distributed Graph Engine for Web Scale RDF Data. The VLDB Endowment (PVLDB) 6(4), 265–276 (2013)
19. Zou, L., Mo, J., Chen, L., Özsu, M.T., Zhao, D.: gStore: Answering SPARQL Queries via Subgraph Matching. The VLDB Endowment (PVLDB) 4(8), 482–493 (2011)

History-Pattern Implementation for Large-Scale Dynamic Multidimensional Datasets and Its Evaluations

Masafumi Makino, Tatsuo Tsuji[(⊠)], and Ken Higuchi

Information Science Depatment, Faculty of Engineering,
University of Fukui, Bunkyo 3-9-1, Fukui City 910-8507, Japan
{makino,tsuji,higuchi}@pear.fuis.u-fukui.ac.jp

Abstract. In this paper, we present a novel encoding/decoding method for dynamic multidimensional datasets and its implementation scheme. Our method encodes an n-dimensional tuple into a pair of scalar values even if n is sufficiently large. The method also encodes and decodes tuples using only *shift* and *and/or* register instructions. One of the most serious problems in multidimensional array based tuple encoding is that the size of an encoded result may often exceed the machine word size for large-scale tuple sets. This problem is efficiently resolved in our scheme. We confirmed the advantages of our scheme by analytical and experimental evaluations. The experimental evaluations were conducted to compare our constructed prototype system with other systems; (1) a system based on a similar encoding scheme called history-offset encoding, and (2) PostgreSQL RDBMS. In most cases, both the storage and retrieval costs of our system significantly outperformed those of the other systems.

1 Introduction

In general, an n-dimensional data tuple can be mapped to the n-dimensional coordinate of a multidimensional array element. The coordinate can be further uniquely mapped to its position in the array by calculating the addressing function of the array. However, in a dynamic situation where new attribute values can emerge, a larger array is necessary to cover the new values, and the positions of the existing array elements must be recalculated according to the new addressing function.

An extendible array (e.g., [11]) can extend its size along any dimension without relocating any existing array elements. History-offset encoding [14] is a scheme for encoding multidimensional datasets based on extendible arrays. If a new attribute value emerges in an inserted tuple, a subarray to hold the tuple is newly allocated and attached to the extended dimension. A tuple can be handled with only two scalar values, *history value* of the attached subarray and *position* of the element in the subarray regardless of the dimension n. Dynamic tuple insertions/deletions can be performed without relocating existing encoded tuples.

Many of the tuple encoding schemes, including history-offset encoding, use the addressing function of a multidimensional array to compute the position. However, there are two problems inherent in such encodings. First, the size of an encoded result may exceed the machine word size (typically 64 bits) for large-scale datasets.

© Springer International Publishing Switzerland 2015
M. Renz et al. (Eds.): DASFAA 2015, Part II, LNCS 9050, pp. 275–291, 2015.
DOI: 10.1007/978-3-319-18123-3_17

Second, the time cost of encoding/decoding in tuple retrieval may be high; more specifically, such operations require multiplication and division to compute the addressing function, and these arithmetic operations are expensive. To resolve these two problems without performance degradation, we present a *history-pattern* encoding scheme for dynamic multidimensional datasets and its implementation scheme called History-Pattern implementation for Multidimensional Datasets (HPMD). Our encoding scheme ensures significantly smaller storage and retrieval costs.

Our scheme encodes a tuple into a pair of scalar values *<history value, pattern>*. The core data structures for tuple encoding/decoding are considerably small. An encoded tuple can be a variable length record; the *history value* represents the extended subarray in which the tuple is included and also represents the bit size of the *pattern*. This approach enables an output file of the encoded results to conform to a sufficiently small sequential file organization. Additionally, our scheme does not employ the addressing function, hence avoiding *multiply* and *divide* instructions. Instead, it encodes and decodes tuples using only *shift* and *and/or* register instructions. This makes tuple retrieval significantly fast and further provides an efficient scheme for handling large-scale tuples whose encoded sizes exceed machine word size.

In this paper, after history-offset encoding is outlined, our history-pattern encoding is presented. Next HPMD and an implementation scheme for large-scale tuple sets is described. Then the retrieval strategy using HPMD is explained. Lastly the implemented HPMD is evaluated and compared with other systems.

2   History-Pattern Encoding

2.1   Preliminary (Extendible Array and History-Offset Encoding)

As preliminary, we first introduce an extendible array model employed in *history-offset* encoding[14].   Fig. 1 is an example of a two-dimensional extendible array.

An n-dimensional extendible array A has a *history counter* h, *history table* H_i, and a *coefficient table* C_i for each extendible dimension i ($i = 1,..., n$). H_i memorize the extension history of A. If the current size of A is $[s_1, s_2,..., s_n]$, for an extension that extends along dimension i, an $(n-1)$-dimensional subarray S of size $[s_1, s_2,..., s_{i-1}, s_{i+1},..., s_{n-1}, s_n]$ is attached to dimension i. Then, h is incremented by one and memorized on H_i. In Fig. 1, the array size is [3, 3] when h is 4. If the array is extended along dimension 1, a one-dimensional subarray of size 3 is attached to dimension 1, and h is incremented to 5 (and is held in H_1 [3]). Each history value can uniquely identify the corresponding extended subarray.

As is well known, element $(i_1, i_2,..., i_{n-1})$ in an $(n-1)$-dimensional fixed-size array of size $[s_1, s_2, ..., s_{n-1}]$ is allocated on memory using an addressing function such as :

$$f(i_1, ..., i_{n-1}) = s_2 s_3 ... s_{n-1} i_1 + s_3 s_4 ... s_{n-1} i_2 + ... + s_{n-1} i_{n-2} + i_{n-1} \qquad (i)$$

We call $<s_2 s_3...s_{n-1}, s_3 s_4...s_{n-1}, ..., s_{n-1}>$ a *coefficient vector*. The vector is computed at array extension and is held in *coefficient table* C_i of the corresponding dimension. Specifically, if $n = 2$, the subarrays are one-dimensional and $f(i_1) = i_1$. Therefore, the coefficient tables can be void if n is less than 3 as in Fig. 1.

Using the above three types of auxiliary tables, history-offset encoding (denoted as *ho-encoding* in the following) of array element $e(i_1, i_2, ..., i_n)$ can be computed as $<h, offset>$, where h is the history value of the subarray in which e is included and *offset* is the offset of e in the subarray computed by (i); e.g., element (3,2) is encoded to $<5,2>$.

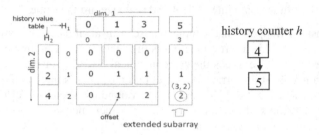

Fig. 1. Two-dimensional extendible array

2.2   History-Pattern Encoding

Fig. 2 illustrates the required data structures for the history-pattern encoding (abbreviated as hp-encoding in the following). scheme. Unlike ho-encoding, when logical extendible array A in the hp-encoding extends its size, a fixed-size subarray equal to the size of the current A in every dimension is attached to the extended dimension. The data structures for A consist of the following two types of tables preserving extension information.

(History table) For each dimension i ($i=1,...,n$), history table H_i is maintained. Each history value h in H_i represents the extension order of A along the i-th dimension. H_i is a one-dimensional array, and each subscript k ($k > 0$) of H_i corresponds to the subscript range from 2^{k-1} to $2^k - 1$ of the i-th dimension. This range is covered by the subarray along the i-th dimension attached to A at the extension when the history counter value is h. For example, as shown in Fig. 2, since $H_1[3]$ is 5, the subscript 3 of H_1 corresponds to the subscript range from 4 to 7 of the first dimension of A.

(Boundary vector table) The *boundary vector table* B is a single one-dimensional array whose subscript is a history value h. It plays an important role for hp-encoding. Each element of B maintains the extended dimension and the boundary vector of the subarray when the history counter value is h. More specifically, the boundary vector represents the past size of A in each dimension when the history counter value is h. For example, the boundary vector in B[3] is $<2, 1>$; therefore, the size of A at a history counter value of 3 is $[2^2, 2^1] = [4, 2]$. Together with the boundary vectors, B also maintains the dimension of A extended at the given history counter value. A includes only the element (0, 0, ..., 0) at its initialization, and the history counter is initialized to 0. B[0] includes 0 as its extended dimension and $<0, 0, ..., 0>$ as its boundary vector

Let h be the current history counter value, and B[h] includes $<b_1, b_2, ..., b_i, ..., b_n>$ as its boundary vector. When A extends along the i-th dimension, B[$h + 1$] includes i as its extended dimension and $<b_1, b_2, ..., b_i + 1, ..., b_n>$ as its boundary vector.

(Logical size and real size) In hp-encoding, A has two size types, i.e., real and logical. Assume that the tuples in n-dimensional dataset M are converted into the set of coordinates. Let s be the largest subscript of dimension k and $b(s)$ be the bit size of s. Then, the real size of dimension k is $s +1$, and the logical size is $2^{b(s)}$. The real size is the cardinality of the k-th attribute; for example, in Fig. 2, the real size is [5, 4], whereas the logical size is [8, 4]. In Fig. 3 below, the real size is [6, 6], and the logical size is [8, 8]. Note that in ho-encoding logical and real size are the same.

(Array extension) Suppose that a tuple whose k-th attribute value emerged for the first time is inserted. This insertion increases the real size of A in dimension k by one. If the increased real size exceeds the current logical size $2^{b(s)}$, A is logically extended along dimension k. That is, current history counter value h is incremented by one, and this value is set to $H_k[b(s+1)]$. Moreover, the boundary vector in B[h] is copied to B[$h+1$] and dimension k of the boundary vector is incremented by one; k is set to the *extended dimension* slot in B[$h+1$], as illustrated in Fig. 2.

Note that h is one-to-one correspondent with its boundary vector in B[h]; this uniquely identifies the past (logical) shape of A when the history counter value is h. To be more precise, for history value $h > 0$, if the boundary vector in B[h] is $<b_1, b_2, ..., b_n>$, the shape of A at h is $[2^{b1}, 2^{b2}, ..., 2^{bn}]$. For example, in Fig. 2, because the boundary vector for the history value 3 is $<2, 1>$, the shape of A when the history counter value was 3 is $[2^2, 2^1]=[4, 2]$. Note that h also uniquely identifies the subarray attached to A at extension when the history counter value was $h-1$. This subarray will be called the *principal subarray* on dimension k at h. For example, in Fig. 2, the *principal subarray* on dimension 2 at $h=4$ is the subarray specified by [0..3, 2..3].

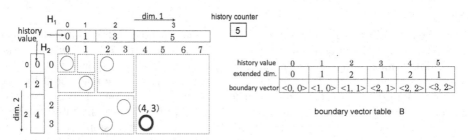

Fig. 2. Data structures for hp-encoding

2.3   Encoding/Decoding

Using the data structures described in Section 2.2, an n-dimensional coordinate $I = (i_1, i_2, ..., i_n)$ can be encoded to the pair $<h, p>$ of *history value* h and *bit pattern* p of the coordinate. The history tables H_i ($i =1,..., n$) and the boundary vector table B are used for the encoding. The history value h for I is determined as max $\{H_k[b(i_k)] \mid k =1,..., n \}$, where $b(i_k)$ is the bit size of the subscript i_k in I. For each history value h, the boundary vector B[h] gives the bit pattern size of each subscript in I. According to this boundary vector, the coordinate bit pattern p can be obtained by concatenating the subscript bit pattern of each dimension by placing in descending order of dimensions on the storage for p from the lower to the higher bits of p. The storage for p can be one machine word length; typically, 64 bits.

For example, consider the array element $(4, 3)$ in Fig.2. According to the above encoding procedure, $H_1[b(4)] > H_2[b(3)]$ because $H_1[b(4)] = H_1[b(100_{(2)})] = H_1[3] = 5$ and $H_2[b(3)] = H_2[b(11_{(2)})] = H_2[2] = 4$. So h is proved to be $H_1[3] = 5$, and element $(4, 3)$ is known to be included in the *principal subarray* (Section 2.2) on dimension 1 at history value 5. Therefore, the boundary vector to be used is $<3, 2>$ in $B[5]$. So the subscript 4 of the element $(4, 3)$ forms the upper 3 bits of p as $100_{(2)}$ and the subscript 3 of the element forms the lower 2 bits of p as $11_{(2)}$. Therefore, p becomes $10011_{(2)} = 19$. Eventually, the element $(4, 3)$ is encoded to $<5, 19>$. Generally, the bit size of the history value h is much smaller than that of pattern p.

Conversely, to decode the encoded pair $<h, p>$ to the original n-dimensional coordinate $I = (i_1, i_2, \ldots, i_n)$, first the boundary vector in $B[h]$ is known. Then the subscript value of each dimension is sliced out from p according to the boundary vector. For example, consider the encoded pair $<h, p> = <5, 19>$. The boundary vector in $B[h]$ is $<3, 2>$, so $p = 10011_{(2)}$ can be divided into $100_{(2)}$ and $11_{(2)}$. Therefore $<5, 19>$ can be decoded to the coordinate $(4, 3)$.

2.4   Hp-Property

From the construction procedure of the boundary vector table B in Section 2.2, the following simple, but important property for our hp-encoding can be known. This property will be called *hp-property* in the following.

[Property 1 (hp-property)] Let $<h, p>$ be an encoded history-pattern of a tuple. h is the total sum of the element values of the boundary vector in $B[h]$ and represents the bit size of the coordinate pattern p for an arbitrary element in the subarray at h.

In our hp-encoding, the favorable property of an extendible array is reflected in the hp-property above. Namely, for the tuples inserted in the subarrays created at the early stage of array extension occupy smaller storage. Consequently, the size of p can be much smaller than in the usual case where each subscript value occupies fixed size storage. It should be noted that the boundaries among the subscript bit patterns in p can be flexibly set to minimize the size of p.

Moreover, the hp-property states that h represents the bit pattern size of p. This simple property together with *shift* and *and/or* register instructions for encoding/decoding makes our encoding scheme to be applied for implementation of large scale multidimensional datasets efficiently with no significant overhead. From this property even if the bit size of p is doubled, h increases only by 1. For example, if the p's current bit size is 255 bits, h is only 1 byte. Therefore our hp-encoding scheme can provide unlimited (logical) history-pattern space size for large and high dimensional dataset with a very small additional storage cost for keeping h.

2.5   Comparison of The Two Encoding Schemes

In this section, we compare hp-encoding with ho-encoding. Let the real size of the extendible array be $[s_1, s_2, \ldots, s_n]$ for both encodings.

(1) Storage Costs for Core Data Structures
For ho-encoding, the core data structures are the history tables and the coefficient tables presented in Section 2.1; for hp-encoding, they are the history tables and the boundary vector table presented in Section 2.2. These data structures guarantee the extensibility of an extendible array.

Here the size of a history table slot is typically assumed to be 2 bytes for ho-encoding and 1 byte for hp-encoding. In ho-encoding, let c be the fixed size in bytes of the coefficients in the coefficient tables. We estimate the storage cost of core data structures for ho-encoding as follows:

(a) History tables: $4*(s_1 + s_2 + ... + s_n)$
(b) Coefficient tables: $c*(n-2)(s_1 + s_2 + ... + s_n)$ $(n > 2)$

For hp-encoding, the storage cost is as follows:

(c) History tables: $\lceil \log_2 s_1 + 1 \rceil + \lceil \log_2 s_2 + 1 \rceil + ... + \lceil \log_2 s_n + 1 \rceil$
(d) Boundary vector table: $n*(\lceil \log_2 s_1 \rceil + \lceil \log_2 s_2 \rceil + ... + \lceil \log_2 s_n \rceil + 1)$

For example, if n is 5 and $s_i = 512$ ($i=1,...,5$), c is computed as 4 bytes, and we have: for ho-encoding (a) is 5120 bytes, (b) is 30720 bytes, and for hp-encoding (c) is 45 bytes, (d) is 225 bytes. The total size of the core data structures for hp-encoding is 0.75% of that for ho-encoding.

(2) Encoding and Decoding Performance
In hp-encoding, encoding/decoding are performed using only *shift* and *and/or* register instructions. Because these instructions do not refer to memory addresses, encoding/decoding can be executed quickly as compared with ho-encoding in which multiplication and division operations are required. In Section 7.3, this will be experimentally confirmed in tuple access and retrieval times.

3   Implementation of History-Pattern Encoding

3.1   Implementation of Core Data Structures

HPMD is an implementation scheme based on the hp-encoding for n-dimensional dataset M. In addition to the core data structures presented in Section 2.2, HPMD includes the following additional data structures:

(1) CVT_i ($1 \leq i \leq n$) is implemented as a B^+ tree. The key value is an attribute value of dimension i; the data value is the corresponding subscript of the extendible array.

(2) C_i ($1 \leq i \leq n$) is a one-dimensional array serving as the *attribute value table*. If attribute value v is mapped by CVT_i to subscript k, the k-th element of C_i keeps v. The element further includes the number of tuples in M, whose attribute value of dimension i is v. This number is used to detect the retrieval completion, which will be described in Section 5.2.

Note that M can be also implemented based on ho-encoding using (1) and (2) above. We call this implementation scheme as HOMD.

Table 1 illustrates an example in which two-dimensional tuples are successively inserted. Fig. 3 shows the constructed HPMD using the inserted tuples from Table 1. By carefully inspecting the given table and Fig. 3, we can trace the change in the related data structures and the produced encoded results generated by each insertion.

3.2   Output File Organization of Encoded Results

In ho-encoding, as discussed in Section 2.1, both the history value and offset of an encoded tuple occupies fixed-size storage, which degrades storage performance. In contrast, in our hp-encoding scheme, we adopt variable length storage according to the pattern size of the encoded results based on the *hp-property* given in Section 2.4.

We can ensure the hp-property by observing h and p in Table 1 and boundary vector table B in Fig. 3. By this property, the history value h can be used as a header of coordinate bit pattern p. It represents p's bit size. Therefore, $<h, p>$ can be treated as a variable length record with size known by h. The hp-property makes it possible to store encoded results in a sequential output file called the *Encoded Tuple File* (ETF), as illustrated in Fig. 4. The encoded results are stored sequentially in the insertion order of the corresponding tuples, similar to a conventional RDBMS. Compared with the size of p, the size of h is sufficiently small, and its size should be fixed. Typically, the size of h is 1 byte and can specify p up to 255 bits.

Table 1. Insertion of the two-dimensional tuples

	inserted tuples	coordinate	history value h	coordinate pattern p		boundary vector
1	<d, p>	(0,0)	0	.		<0,0>
2	<b, p>	(1,0)	1	1.	(1)	<1,0>○
3	<b, q>	(1,1)	2	1.1	(3)	<1,1>○
4	<c, q>	(2,1)	3	10.1	(5)	<2,1>○
5	<f, q>	(3,1)	3	11.1	(7)	<2,1>
6	<e, q>	(4,1)	4	100.1	(9)	<3,1>○
7	<f, t>	(3,2)	5	011.10	(12)	<3,2>○
8	<a, r>	(5,3)	5	101.11	(23)	<3,2>
9	<a, p>	(5,0)	4	101.0	(10)	<3,1>
10	<a, u>	(5,4)	6	101.100	(44)	<3,3>○
11	<b, r>	(1,3)	5	001.11	(7)	<3,2>
12	<e, s>	(4,5)	6	100.101	(37)	<3,3>
13	<d, r>	(0,3)	5	000.11	(3)	<3,2>

Remark 1: the leftmost number represent the insertion orders of the tuples
Remark 2: "." in coordinate bit pattern is a separator between subscripts line feed
Remark 3: ○ denotes that the insertion of the tuple causes the extension of the logical size of the extendible array

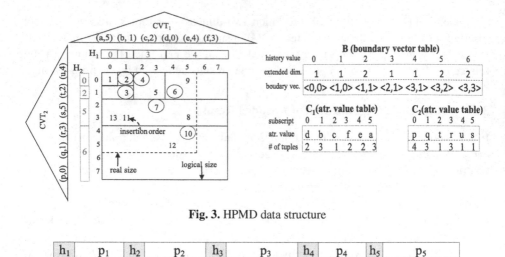

Fig. 3. HPMD data structure

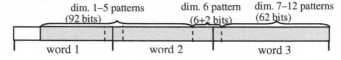

Fig. 4. Storing variable length encoded tuples sequentially in ETF

4   Implementation for Large Scale Datasets

Unfortunately, in hp-encoding, the history-pattern space often exceeds machine word length for high-dimensional and/or large volume datasets. In this section, we provide a scheme to handle such a large history-pattern space with minimal degradation in encoding/decoding speed.

4.1   Extending History-Pattern Space

To handle large-scale tuple datasets using the hp-encoding, the coordinate bit patterns can range over multiple machine words to eliminate the pattern size limitation. For example, Fig. 5 shows the layout for a 162-bit coordinate pattern according to boundary vector <25, 16, 13, 23, 15, 8, 6, 10, 7, 20, 15, 4> of 12 dimensions; this requires three 64-bit words. Compared with a bit pattern within a single word, no storage cost overhead arises with this multiword bit pattern. Furthermore, encoding to and decoding from this multiword bit pattern do not cause significant overhead, since they can be performed by using only *shift, mask, or* register instructions as in a single bit pattern with a little *cut and paste* cost. We omit the details. In contrast, overhead caused by using a multi-precision library in ho-encoding would significantly degrade retrieval performance. Note that the *hp-property* introduced in Section 2.4 is also guaranteed for multiword bit patterns.

Fig. 5. Layout for 162-bit coordinate pattern

4.2   Further Storage Optimizations

Here, we present two optimization strategies for storing encoded results in ETF.

(1) Sharing history value

Due to the *hp-property,* the set of the encoded <*h, p*> pairs can be partitioned into the subsets depending on *h*. The pairs in the same subset have the same history value *h*, so *h* can be shared among these pairs and the bit size of their pattern *p* equals to *h*. Thus the *p*'s of the same *h* are stored in the same node block list as in Fig. 6. Since *h* is one-to-one correspondent with its subarray, the node block list keeps all patterns of the elements in the subarray. If the size of *h* is one byte, and the total number of tuples is *m*, a total of *m*−1 bytes can be saved by this optimization.

(2)   Multi-boundary vector

For the multiword bit patterns in Section 4.1, two types of arrangement of the bit patterns in a node block are considered; (a) by *byte-alignment*, (b) by *word-alignment*.

   In (a), storage cost can be saved, but retrieval cost would increase. But, the situation is just converse in (b). We present the following method to avoid the retrieval overhead inherent in (a) but take advantage of its storage cost savings.

   Assume that the machine word occupies *w* bytes, and *p* bytes are necessary to store a single coordinate bit pattern. Let *l* be the least common multiple of *w* and *p*, and let *bv* be the boundary vector for a single coordinate bit pattern described in Section 4.1. Multi-boundary vector *mbv* is a set of single boundary vectors *bv*s and can be obtained by recalculating and arranging *bv* sequentially *l/p* times. Here we omit the description of its details. Note that *mbv* can be used as if it were a single boundary vector. Using *mbv*, *l/p* single coordinate bit patterns can be stored consecutively in a node block in ETF by byte alignment, while they can be retrieved by word alignment.

   A storage scheme based on the above multi-boundary vector increases the size of the boundary vector table; however, the size is negligibly small compared with ETF size. Consequently, this multi-boundary vector further contributes to generate ETFs compactly without degradation of retrieval performance. we refer to HPMD based on the *multiword bit pattern scheme* in Section 4.1 simply as HPMD, and the HPMD based on the *multi-boundary vector* using node block lists as M-HPMD.

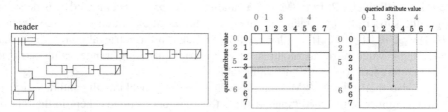

Fig. 6. Node block lists in an ETF   **Fig. 7.** Tuple retrieval in HPMD

5   Tuple Retrieval

5.1   History Value Dependency

We can notice the following important property in both HPMD and M-HPMD.

[**Property 2**] Let h be the history value of the principal subarray PS (see Section 2.2) of the subscript k on dimension i. The array elements with subscript k on dimension i are included only in PS or the subarrays with history values greater than h and extended along the dimension other than dimension i.

This property is shown in Fig. 7 above. The dotted line represents the real size of the extendible array, and the grey colored parts are the candidates of retrieval. We can see that it is not necessary to decode all the elements in ETF, but only the grey colored parts due to the above property. An element in the non-grey subarrays can be checked by its history value, and can be skipped without decoding the pattern part. The total size of the grey parts depends on the history value of the principal subarray of the subscript to be retrieved. This leads to the following property.

[**Property 3 (history value dependency)**]   The subarrays to be decoded for the retrieval of an attribute value v depends on the history value corresponding to v.

5.2   Tuple Retrieval

In HPMD, like in a conventional RDBMS all tuples in ETF should be searched sequentially. Nevertheless, according to Property 2, non-candidate tuples can be skipped without inspecting bit pattern part p by only examining the history value part. In M-HPMD, each tuple is classified in terms of its history value and is stored in the corresponding node block list. Therefore, even history value checking can be avoided in the candidate node block lists.

Let *age* be an integer attribute. For a single value retrieval, such as *age*=20, first the specified attribute value is   searched in CVT_{age} to obtain its subscript value i. Both in HPMD and M-HPMD, the history values for candidates of retrieval are determined according to Property 2. If a candidate $<h, p>$ is encountered, p is decoded to get the subscript of dimension *age*.   If it is i, $<h, p>$ is included in the retrieval results.

For a range value retrieval, such as $10 \leq age < 20$, the set of subscripts covered by the range is obtained by searching the sequence set of CVT_{age}. Based on the obtained subscript set S, the set of the history values for candidates of retrieval are determined according to Property 2. In HPMD, if a candidate $<h, p>$ is encountered, p is decoded to get the subscript i of dimension *age*.   For the attribute value table (see Section 3.1(2)) C_{age}, if $10 \leq C_{age}[i] < 20$, $<h, p>$ is included in the retrieval results. Note that while single value retrieval requires only subscript matching, range value retrieval requires references to the attribute value table.

Note also that in both HPMD and M-HPMD, before checking all candidate tuples, when the number of matched tuples reaches the "number of tuples" kept in the related attribute value table described in Section 3.1, the retrieval can be terminated.

6   Related Work

Substantial research has been conducted on multidimensional indexing schemes based on the mapping strategy in which a multidimensional data point is transformed to a single scalar value. Such mapping strategies include a space-filling curve, such as the

Z-curve [3] or the Hilbert curve [4], which enumerates every point in a multidimensional space. They preserve proximity, which suggests that points near one another in the multidimensional space tend to be near one another in the mapped one-dimensional space. This property of preserving proximity ensures better performance for range key queries against a dynamic multidimensional dataset; however, an important problem with these space-filling curves is that retrieval performance degrades abruptly in high-dimensional data spaces because of the required recursive handling. The UB-tree [2][5][6] maps a spatial multidimensional data point to a single value using a Z-curve; however, a UB-tree has the critical problem that its parameters (e.g., the range of attribute values) need to be properly tuned for effective address mapping [6]. This requirement restricts the usability and performance of the UB-tree.

In contrast to these approaches, in our encoding scheme for an n dimensional tuple, encoding and decoding costs are both $O(n)$, even if n is very large, because these operations are performed using only *shift* and *and/or* machine instructions. Furthermore, the problem of the UB-tree approach is not present in our encoding scheme because of an unlimited extensibility of an extendible array.

The most common scheme for mapping multidimensional data points to scalar values[1] is to use a fixed-size multidimensional array. Much research, such as [7][8][9][10], has been performed using this scheme for the paging environment of secondary storage. The *chunk offset* scheme [10] is a well-known scheme in which multidimensional space is divided into a set of chunks; however, it is not extendible, and a new attribute value cannot be dynamically inserted.

Extendible arrays [11]~[14] provide an efficient solution to this non-extendible problem. In [11], Otoo and Rotem described a method for reducing the size of the auxiliary tables for addressing array elements. [14] proposes the history-offset encoding , by which wider application areas such as in [17] can be developed.

One of the drawbacks inherent in the existing research that uses the addressing function of a multidimensional array is the time cost of decoding for tuple retrieval. The approaches presented in existing research require division operations using the coefficient vector, which are very expensive. Such a drawback is not present with our encoding scheme, and the costs are significantly small. [15] presented the basic idea of the history-pattern encoding.

Another drawback of existing research is that the address computed by the addressing function may exceed machine word length. Some application areas for the ho-encoding scheme are present in the research. [17] provides a labeling scheme of dynamic XML trees. In these applications, however, this address space saturation problem makes it difficult to handle large-scale datasets. Chunking array elements is a means to expanding the address space [10][16], but it only delays saturation of the space. For this problem, in [16], Tsuchida et al. vertically partitioned the tuple set to reduce dimensionality; however, overhead emerges that increased storage costs.

This paper presents an implementation scheme of hp-encoding in order to resolve or alleviate these two drawbacks and provides an efficient implementation for large-scale multidimensional datasets as was confirmed by the analytical and experimental evaluations. As far as we know, there is no research similar to ours.

7   Evaluation Experiments

In this section, performance evaluations are shown for HPMD and M-HPMD in Section 4.2 on the implemented prototype system. These are compared with HOMD and PostgreSQL, which is one of the conventional RDBMS. They all output the tuples sequentially to the output file, and the tuple retrieval is also sequentially performed.

7.1   Evaluation Environment

Construction times, storage sizes, and retrieval times are measured under the following 64 bits computing environment.

CPU: Intel Core i7 (2.67GHz), Main Memory: 12GB, OS: CentOS5.6 (LINUX), PostgreSQL: Version 8.4.4 (64-bit version)

In the measurement of the retrieval times for PostgreSQL, the ¥timing command was used. The command invokes the LINUX system call gettimeofday() and we also used this system call for HOMD, HPMD and M-HPMD. The retrieval time in these implementations includes the time to get the decoded tuples that satisfies the query condition. To suppress the performance deterioration caused by transaction processing in PosgreSQL, the transaction isolation level is set to the lowest level.

7.2   Evaluation Using Large Scale Dataset

The LINEITEM table (Table 2) in TPC-H benchmark data[19] is employed. The size of the input tuple file (*csv* formatted file) generated by TPC-H is about 2.43 GB. The number of tuples is 23,996,604. L_COMMENT column is dropped out. Note that HOMD cannot implement such large table due to the history-offset space overflow.

Table 2. LINEITEM Table

dim.	attribute name	type	cardinality	dim.	attribute name	type	cardinality
1	L_ORDERKEY	int	6000000	9	L_RETURNFLAG	char[1]	3
2	L_PARTKEY	int	800000	10	L_LINESTATUS	char[1]	2
3	L_SUPPKEY	int	40000	11	L_SHIPDATE	char[10]	2526
4	L_LINENUMBER	int	7	12	L_COMMITDATE	char[10]	2466
5	l_QUANTITY	double	50	13	L_RECEIPTDATE	char[10]	2555
6	L_EXTENDEDPRICE	double	1079204	14	L_SHIPINSTRUCT	char[25]	4
7	L_DISCOUNT	double	11	15	L_SHIPMODE	char[10]	7
8	L_TAX	double	9		L_COMMENT	char[44]	15813794

(1) Storage Cost

The total required storage to store a multidimensional dataset includes data structures for encoding/decoding shown in Fig. 3 and ETF to store the encoded tuples.   In HPMD, ETF is a sequential file and in M-HPMD, it is a file of node block lists.   Fig. 8 shows the total required storage sizes for HPMD, M-HPMD, PostgreSQL (denoded by PSQL in the following). In HPMD and M-HPMD the breakdown of the total size is shown. "aux_tables" are the history tables, boundary vector table and attribute value tables in Fig. 3. The maximum history value in the constructed HPMD or M-HPMD was 141 (3 machine words).

As can be seen in Fig. 8, the total sizes for HPMD and M-HPMD are about one-sixth of the size for the PSQL. This indicates that our hp-encoding scheme realizes significant reduction of the storage cost. In M-HPMD the size of ETF is 5% smaller than that in HPMD due to the sharing history value in M-HPMD stated in Section 4.2. It can be noted that while the size for PSQL is about 1.6 times larger than that for the *csv* formatted file, the size for HPMD or M-HPMD is about 30% of the *csv* file size.

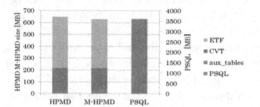

Fig. 8. Storage cost

Table 3. Construction cost (sec)

HPMD	M-HPMD	PSQL
159.95	180.15	224

(2) Construction Time

Table 3 shows the times spent for constructing databases from the *csv* source file. The spent times for HPMD and M-HPMD are about 71% and 80% of that for PSQL. The difference owes to the reduction of output I/O cost; output ETF file size of HPMD and M-HPMD is far less than the output file size of PSQL. It can be observed that the time spent for M-HPMD is 13% larger than that for HPMD. This owes to the time of M-HPMD spent for construction of node block lists.

(3) Retrieval Time

(3-1) Retrieval for single value queries

Fig. 9 shows the retrieval times of single value queries for LINEITEM table. The left side scale is for HPMD and M-HPMD and the right one is for PSQL. Each history value on the horizontal axis represents the leftmost subscript of the principal subarray (see Section 2.2) on the dimension 1 and 6. The retrieval time for the attribute value corresponding to the subscript was measured. We adopt the dimensions since the larger cardinality can better exhibit the properties of our schemes. As was mentioned in Section 5.2, both in HPMD and M-HPMD the retrieval can be terminated without checking all the candidate tuples in ETF by using "num. of tuples" in the attribute value table (See Fig. 3). The measurement was also done in the case all the candidate tuples are checked without using "num. of tuples". We will denote this case as HPMDa and M-HPMDa, and the case using "num. of tuples" as HPMDb and M-HPMDb. The denotations HPMD and M-HPMD will be used for both cases.

In PSQL and HPMDa, the retrieval times are nearly constant irrespective of the queried attribute values, since all the tuples are scanned through. The average times of HPMDa in dim. 1 and dim. 6 are 8.33 and 10.73 times faster than that of PSQL respectively. In contrast, in M-HPMD only the candidate tuples are scanned and decoded. Therefore the history value dependency described in Section 5.1 can be better observed in M-HPMD than in HPMD as in Fig. 9(b). In M-HPMD the time decreases at the maximum history value in both dimension 1 (141) and dimension 6(139). The principal subarrrays corresponding to these history values are located at the end of the logical extendible array. So, the reason of the decrease is that the real size of the logical extendible array in these dimensions is less than that of the logical size, so the tuples in these subarrays do not fill out its logical space.

It can be observed that using "num. of tuples" in the attribute value table of dim. 1 is effective. Since the cardinality of dim. 1 is very large and its attribute values are uniformly distributed, the number of tuples of each attribute value is very small; i.e., less than 10. In HPMD, the attribute values of the same dimension are converted to the dimension subscripts in the ascending order, namely the earlier an attribute value appears, the smaller subscript is assigned to the attribute value. Since the encoded results are stored sequentially in the ETF file, the attribute values covered by the smaller history values are stored earlier in the ETF file, so the number of the tuples satisfying the query quickly attains to "num. of tuples". In M-HPMD, the tuples of the same history value can be directly accessible and in dim. 1 they can be confined in a single node block, so the retrieval times are almost 0. For dim. 6, the cardinality is smaller than that of dim.1, and the attribute values are not so uniformly distributed as those of dim.1, the above advantages for dim. 1 is decreased in both HPMDb and M-HPMDb as can be observed in Fig. 9(b).

For HPMDb and M-HPMb the maximum and minimum ratios of the retrieval times to those of PSQL are shown in Table 4. It can be seen that the ratios are under 11%, and that the maximum retrieval times of M-HPMD are about a half of that of HPMD. This proves the benefit of M-HPMD described in Section 4.2

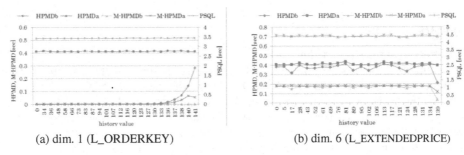

(a) dim. 1 (L_ORDERKEY)   (b) dim. 6 (L_EXTENDEDPRICE)

Fig. 9. Retrieval times of single value

Table 4. Max. and min. ratios of retrieval times

*/PSQL	HPMDb (dim. 1)	M-HPMDb (dim. 1)	HPMDb (dim. 2)	M-HPMDb (dim.2)	HPMDb (dim. 6)	M-HPMDb (dim. 6)	HPMDb (dim. 13)	M-HPMDb (dim. 13)
max	8.24%	0.00%	10.90%	4.97%	9.45%	4.15%	6.68%	3.30%
min	0.00%	0.00%	9.30%	4.30%	4.82%	0.79%	6.60%	3.17%

(3-2) Retrieval for Range Queries

Table 5 shows the ranges of the attribute values for the range queries, on which the retrieval times were measured.   The *selectivities* of the ranges R1, R2 and R3 on each dimension are about 3%, 10%, 20% respectively. Fig. 10 shows the retrieval times for the range queries on dimension 1 and 6. It can be known from Fig.10(a) that the retrieval times for HPMDa is much larger than those of HPMDb. This also dues to the the reason described in (3-1). For HPMDb and M-HPMDb, Table 6 shows the ratios of the retrieval times to those of PostgreSQL on the range queries in Table 5. It should be noted that in both HPMDb and M-HPMDb, dimension subscripts are assigned in ascending order. So, the subscript range corresponding to its attribute value range may often spread over wider than the attribute value range.   This might alleviate the benefits of our HPMD and M-HPMD in some degree; as can be observed in Table 6, the performance on dim. 2 are degraded.

Table 5. Attribute value ranges used in the experiment

	dim. 1	dim. 2	dim. 3	dim. 5
range R1	100,000 ~ 820,000	100,000 ~ 124,000	10,000 ~ 11,200	2~3
range R2	1,000,000 ~ 3400,000	200,000 ~ 280,000	20,000 ~ 24,000	3~9
range R3	4,000,000 ~ 8,800,000	500,000 ~ 660,000	30,000 ~ 38,000	10~19

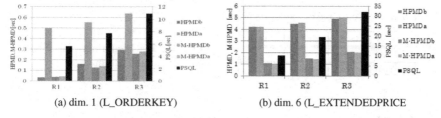

(a) dim. 1 (L_ORDERKEY)   (b) dim. 6 (L_EXTENDEDPRICE)

Fig. 10. Retrieval times of range queries

Table 6. Ratios of retrieval times for range queries

*/PSQL	HPMDb (dim. 1)	M-HPMDb (dim. 1)	HPMDb (dim. 2)	M-HPMDb (dim.2)	HPMDb (dim. 3)	M-HPMDb (dim. 3)	HPMDb (dim. 5)	M-HPMDb (dim. 5)
range R1	1.40%	1.75%	20.64%	11.81%	11.06%	6.98%	5.86%	4.14%
range R2	2.28%	2.82%	12.16%	8.10%	7.29%	5.71%	4.27%	4.03%
range R3	2.76%	3.45%	8.47%	6.42%	4.99%	4.50%	3.87%	4.02%

7.3   Comparison of HOMD and HPMD

ho-encoding is a competitor of our *hp-encoding*. Its HOMD implementation cannot deal with large scale datasets without considerable degradation of retrieval performance. In this section, by using a moderate-scale dataset whose history-offset space is within the machine word size, we briefly compare the performance among HOMD, HPMD and PSQL. The dataset is artificially created as in Table 7.

Tuples in the dataset is uniformly distributed in the history-offset or history-pattern space to evaluate the basic performance of each scheme. In HOMD, the size of history value and offset value are fixed in 2 and 8 bytes respectively, and in HPMD, the history value size is fixed in 1 byte and that of coordinate pattern is variable according to the history value. Both in HOMD and HPMD, the encoded results are output sequentially in the ETF files. Table 8 shows the storage costs for ETF and database construction time. We can observe that the ETF size in HOMD is much larger than that of HPMD. This owes to the advantage of the HPMD implementation, in which the encoded tuples in ETF is variable length records arranged in byte alignment. The construction time in HOMD is about 2.5 times larger than that of HPMD. This owes to the larger time required in encoding tuples.

Fig. 11 shows the retrieval times for single value queries on the 5th dimension. We can see that the retrieval times in HPMDb and M-HPMDb are almost constant irrespective of queried attribute values like in PSQL. This is because that the tuples are uniformly distributed over the attribute domain. On the other hand, the retrieval times for HOMDb are depending on the queried attributes in spite of the uniform tuple distribution due to the attribute value sensibility of HOMD.

Table 7. Used dataset

# of dims	# of tuples	attribute type	cardinality
5	5,000,000	all integer	all 512

Table 8. Storage cost and construction time

	HOMD	HPMD	M-HPMD
ETF (kbytes)	50,000	34,843	30,669
const.time (sec)	20.65	7.04	12.45

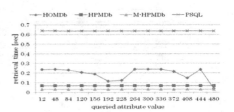

Fig. 11. Retrieval times for single value queries on the 5th dim

8   Conclusion

We have presented a novel encoding/decoding scheme for dynamic multidimensional datasets. The advantage of the scheme lies in the following two points. One is that the scheme provides the minimal encoding/decoding costs avoiding multiplications and divisions inherent in the existing schemes based on multidimensional arrays. The other is that the scheme provides an efficient method to handle a large-scale dataset by alleviating the problem of the address space limitation. These advantages have been confirmed by the experiments.

The implementation scheme presented in this paper does not consider the performance improvement for range queries. Important future work includes the design of the effective strategy for range queries in our framework.

References

1. Zhang, R., Kalnis, P., Ooi, B.C., Tan, K.L.: Generalized multidimensional data mapping and query processing. ACM Transactions on Database Systems, pp. 661–697 (2005)
2. Fenk, R., Markl, R., Bayer, R.: Interval processing with the UB-tree. In: Proc. of IDEAS, pp. 12–22 (2002)
3. Orenstein, J.A., Merrett, T.H.: A class of data structures for associative searching. In: Proc. of PODS, pp. 181–190 (1984)
4. Faloustsos, C., Roseman, S.: Fractals for secondary key retrieval. In: Proc. of PODS, pp. 247–252 (1989)
5. Bayer, R.: The universal B-tree for multidimensional indexing: General concepts. In: Proc. of Worldwide Computing and Its Applications, pp. 198–209 (1997)
6. Ramsak, F., Markl, V., Fenk, R., Zirkel, M., Elhardt, K., Bayer, R.: Integrating the UB-tree into a database system kernel. In: Proc. of VLDB, pp. 263–272 (2000)
7. Sarawagi, S., Stonebraker, M.: Efficient organization of large multidimensional array. In: Proc. of ICDE, pp. 328–336 (1994)
8. Seamons, K.E., Winslett, M.: Physical schemas for large multidimensional arrays in scientific computing applications. In: Proc. of SSDBM, pp. 218–227 (1994)
9. Sarawagi, S., Stonebraker, M.: Efficient organization of large multidimensional arrays. In: Proc. of ICDE, pp. 328–336 (1994)
10. Zhao, Y., Deshpande, P.M., Naughton, J.F.: An array based algorithm for simultaneous multidimensional aggregates. In: Proc. of SIGMOD, pp. 159–170 (1997)
11. Otoo, E.J., Rotem, D.: A storage scheme for multi-dimensional databases using extendible array files. In: Proc. of STDBM, pp. 67–76 (2006)
12. Otoo, E.J., Rotem, D.: Efficient storage allocation of large-scale extendible multidimensional scientific datasets. In: Proc. of SSDBM, pp. 179–183 (2006)
13. Otoo, et. al.: Optimal chunking of large multidimensional arrays for data warehousing. In: Proc. of DOLAP, pp. 25–32 (2007)
14. Hasan, K., Tsuji, T., Higuchi, K.: An Efficient Implementation for MOLAP Basic Data Structure and Its Evaluation. In: Kotagiri, R., Radha Krishna, P., Mohania, M., Nantajeewarawat, E. (eds.) DASFAA 2007. LNCS, vol. 4443, pp. 288–299. Springer, Heidelberg (2007)
15. Tsuji, T., Mizuno, H., Matsumoto, M., Higuchi, K.: A Proposal of a Compact Realization Scheme for Dynamic Multidimensional Datasets. DBSJ Journal 9(3), 1–6 (2009). (In Japanese)
16. Tsuchida, T., Tsuji, T., Higuchi, K.: Implementing Vertical Splitting for Large Scale Multidimensional Datasets and Its Evaluations. In: Cuzzocrea, A., Dayal, U. (eds.) DaWaK 2011. LNCS, vol. 6862, pp. 208–223. Springer, Heidelberg (2011)
17. Tsuji, T., Amaki, K., Nishino, H., Higuchi, K.: History-Offset Implementation Scheme of XML Documents and Its Evaluations. In: Meng, W., Feng, L., Bressan, S., Winiwarter, W., Song, W. (eds.) DASFAA 2013, Part I. LNCS, vol. 7825, pp. 315–330. Springer, Heidelberg (2013)
18. Free Software Foundation, GMP: The GNU Multiple Precision Arithmetic Library (2013). http://gmplib.org
19. Transaction Processing Performance Council: TPC-H (2014). http://www.tpc.org/tpch

Scalagon: An Efficient Skyline Algorithm for All Seasons

Markus Endres[✉], Patrick Roocks, and Werner Kießling

University of Augsburg, Universitätsstr. 6a, 86159 Augsburg, Germany
{endres,roocks,kiessling}@informatik.uni-augsburg.de
http://www.informatik.uni-augsburg.de/en/chairs/dbis/

Abstract. Skyline queries are well-known in the database community and there are many algorithms for the computation of the Pareto frontier. The most prominent algorithms are based on a block-nested-loop style tuple-to-tuple comparison (BNL). Another approach exploits the lattice structure induced by a Skyline query over low-cardinality domains. In this paper, we present *Scalagon*, an algorithm which combines the ideas of the lattice approach and a BNL-style algorithm to evaluate Skylines on arbitrary domains. Since multicore processors are going mainstream, we also present a parallel version of Scalagon. We demonstrate through extensive experimentation on synthetic and real datasets that our algorithm can result in a significant performance advantage over existing techniques.

Keywords: Skyline · High-cardinality · Pre-filter · Optimization

1 Introduction

The Skyline operator [1] has emerged as an important and popular technique for searching the best objects in multi-dimensional datasets. A Skyline query selects those objects from a dataset D that are not dominated by any others. An object p having d attributes (dimensions) dominates an object q, if p is strictly better than q in at least one dimension and not worse than q in all other dimensions, for a defined comparison function. Without loss of generality, we consider subsets of $\mathbb{R}^d$ in which we search for Skylines w.r.t. the natural order $\leq$ in each dimension.

The most cited example on Skyline queries is the search for a hotel that is *cheap* and *close to the beach*. Unfortunately, these two goals are complementary as the hotels near the beach tend to be more expensive. In Figure 1 each hotel is represented as a point in the two-dimensional space of *price* and *distance to the beach*. The *Skyline* consists of all hotels that are not worse than any other hotel in both dimensions. From the Skyline one can now make the final decision, thereby weighing the personal preferences for price and distance.

Since the introduction of the Skyline operator plenty of generic algorithms for Skyline computation have been suggested [2]. Several comparison-based algorithms have been published in the last decade, e.g., the nearest neighbor algorithm [3], SFS (Sort-Filter Skyline) [4], or LESS (Linear Elimination-Sort for

© Springer International Publishing Switzerland 2015
M. Renz et al. (Eds.): DASFAA 2015, Part II, LNCS 9050, pp. 292–308, 2015.
DOI: 10.1007/978-3-319-18123-3_18

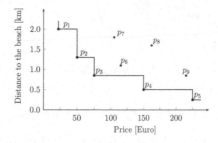

Fig. 1. Skyline example. Select cheap hotels near to the beach.

Skyline) [5], just to name a few. Many of these algorithms have been adapted for parallel Skyline computation, in shared-nothing [6–9], and shared-anything architectures [10–15]. There are also algorithms utilizing an index structure to compute the Skyline, e.g., [16–18].

Another concept are lattice-based algorithms. Instead of direct comparisons of tuples, a lattice structure having the size of the domain is created representing the better-than relationships. Examples for such algorithms are *Lattice Skyline* [19] and *Hexagon* [20]. Principally, such algorithms are limited to *discrete low-cardinality domains*. In [21] this knowledge is used to develop a cost model to guide a balanced pivot point selection in the algorithm *BSkyTree*. However, often there are *continuous high-cardinality* domains involved in a Skyline query, e.g., price information.In these cases Skyline computation using the lattice fails.

Our idea is to combine the advantages of lattice based algorithms and tuple comparison. At first we *scale* the continuous high-cardinality domain down to a *small* domain, and thus a lattice-based algorithm can be efficiently applied. Depending on the granularity of the scaling, many tuples will be put into the same lattice node. This implies that the lattice algorithm on the small domain is just a pre-filtering of the tuples and a comparison-based algorithm is needed afterwards. This idea is realized in the generic *Scalagon* algorithm (The name is composed from the terms *scale*, as the domain is *scaled down*, and *Hexagon*, the lattice-based algorithm from [20]). We show that this two-staged procedure is less expensive than conventional Skyline methods, and that our multi-threaded implementation is a strong competitor to well-known parallel Skyline algorithms.

We restrict our attention to *weakly anti-correlated* data sets as sketched in Figure 1, having a relatively small Skyline size (less than 1% of the tuples). In such settings BNL-style algorithms show a linear behavior and are our strongest competitors, but we will show that Scalagon is superior.

The remainder of this paper is organized as follows: Section 2 contains the formal background used in this paper and recapitulates BNL-style and lattice based algorithms. Based on this background we will present the Scalagon algorithm in Section 3. We conduct an extensive performance evaluation on synthetic and real datasets in Section 4. Section 5 contains our concluding remarks.

2   Skyline Background

2.1   Problem Definition

Assume a set of vectors $D \subseteq \mathbb{R}^d$. We define the so called Pareto ordering for all $x = (x_1, ..., x_d)$, $y = (y_1, ..., y_d) \in D$:

$$x <_\otimes y \Longleftrightarrow \forall j \in \{1, ..., d\} : x_j \leq y_j \ \wedge \ \exists i \in \{1, ..., d\} : x_i < y_i \qquad (1)$$

The *Skyline* of D is defined by the maxima in D according to the ordering $<_\otimes$, or explicitly by the set

$$S(D) = \{t \in D \mid \nexists u \in D : u <_\otimes t\} \ .$$

In this sense we prefer the *minimal values* in each domain. Note that Skylines are not restricted to numerical domains. For any universe Ω and orderings $<_i \in (\Omega \times \Omega)$ $(i \in \{1, ..., d\})$ the Skyline w.r.t. $<_i$ can be computed, if there exist scoring functions $g_i : \Omega \to \mathbb{R}$ for all $i \in \{1, ..., d\}$ such that $x <_i y \Leftrightarrow g_i(x) < g_i(y)$. Then the Skyline of a set $M \subseteq \Omega$ w.r.t. $(<_i)_{i=1,...,d}$ is equivalent to the Skyline of $\{(g(x_1), ..., g(x_d)) \mid x \in M\}$.

2.2   Block Nested Loop Revisited

In general, algorithms of the block-nested-loop class (BNL) [1] linearly scan over the input dataset D. The idea of BNL is to continuously maintain a *window* (or block) of tuples in main memory containing the maximal elements with respect to the data read so far. When a tuple $p \in D$ is read from the input, p is compared to all tuples of the window and, based on this comparison, p is either eliminated, or placed into the window. Three cases can occur: First, p is dominated by a tuple within the window. In this case, p is eliminated and will not be considered in future iterations. Second, p dominates one or more tuples in the window. In this case, these tuples are eliminated; that is, these tuples are removed from the window and will not be considered in future iterations while p is inserted into the window. And third, p is incomparable with all tuples in the window. In this case p is inserted into the window. At the end of the algorithm the window contains the maximal elements, i.e., *the Skyline*. BNL algorithms work particularly well if the Skyline is small [1]. The average case complexity is of the order $\mathcal{O}(n)$, where n counts the number of input tuples. In the worst case, where at least $c \cdot n$ tuples are incomparable for a fixed c, the complexity is $\mathcal{O}(n^2)$ [5].

The major advantage of a BNL-style algorithm is its simplicity and suitability for computing the maxima of arbitrary partial orders. Furthermore, a multitude of optimization techniques have been developed in the last decade. For example, SFS (Sort-Filter Skyline) [4] topologically sorts the dataset, whereas LESS (Linear Elimination-Sort for Skyline) [5] uses dynamic sorting and elimination of tuples from the window. The algorithm *sSkyline* [11] uses a merge method to compute the Skyline, and *BSkyTree* is based on a balanced pivot point selection. There are other approaches like divide-and-conquer [1], or parallel variants of the algorithms above, e.g., *LazyList BNL* [12], *pSkyline* [11], or *APSkyline* [14].

2.3   Lattices for Skylines Revisited

The algorithms *Hexagon* [20] and *Lattice Skyline* [19] exploit the *lattice* induced by a Skyline query over *discrete low-cardinality* domains to compute the best objects. Visualization of such lattices is often done using *Better-Than-Graphs* (*BTG*) (Hasse diagrams), graphs in which edges state dominance [22]. The *nodes* in the BTG represent *equivalence classes*. Each equivalence class contains objects which are mapped to the same feature vector by the scoring function f. All values in the same equivalence class are considered substitutable, hence are *indifferent*. The elements of the dataset D that compose the Skyline set is build up by those nodes in the BTG that have *no path leading to them from another non-empty node*. All other nodes have direct or transitive edges from the Skyline nodes, and therefore are *dominated*. The worst case complexity of such lattice algorithms is linear w.r.t. the number of input tuples and the size of the BTG [19].

For the implementation of such algorithms the lattice is usually represented by an array, where each position stands for one node in the lattice [19]. The array stores the *empty*, *non-empty*, and *dominated* state of a node. For each element $t \in D$ the algorithms compute the unique position in the array and mark this position as *non-empty*. Next, the nodes are visited in a breadth-first order (BFT). Non-empty nodes cause a depth-first traversal (DFT) where the dominance flags are set. Finally those nodes represent the Skyline which are both *non-empty and non-dominated*.

3   The Scalagon Algorithm

3.1   The Idea

The main disadvantage of current lattice based Skyline algorithms (Section 2.3) is their restricted application to discrete low-cardinality domains. First, the BTG must fit into main memory and second the complexity analysis shows that a BTG significantly larger than the number of tuples is slower than BNL-style algorithms [19]. Therefore, the general idea of Scalagon is to *scale the original high-cardinality domain of the input data down to a smaller domain* and apply a lattice algorithm as pre-filter. Using a lattice Skyline method at first step, we have the advantage of a linear runtime complexity and that the lattice based approach is independent from the data distribution, i.e., whether the data is anti-correlated, correlated, or independent distributed [19,20].

Figure 2 shows the entire filtering process on a weakly anti-correlated dataset, where the size of the scaled domain is $\{1, ..., 8\}^2$. At first, using a lattice based approach called *Hexagon product order (HPO)* (based on the Hexagon algorithm in [20]) the maxima on the scaled domain are determined (dark gray tiles) according to the *product order* dominance criteria (a tile is dominated by another tile if it is worse in all components). All objects in the dominated nodes (light gray tiles) are save to be filtered out. The tuples in the dark gray area form a pre-filtered set. On this set a BNL-style algorithm is applied afterwards to finally determine the Skyline w.r.t. the *Pareto order*.

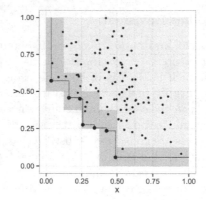

Fig. 2. Filtering of Scalagon and Pareto frontier. The light gray squares are dominated during the HPO phase. The dark gray squares contain the pre-filtered set.

Note that such a method may be prone to outliers: Adding a single tuple like $(10, 10)$ to a scenario as depicted in Figure 2 would introduce a plenty of empty tiles and hence result in a very low number of filtered out tuples (in this case only $(10, 10)$ would be filtered out). We will face this problem in Section 3.4 by excluding the outliers from the scaling and pre-filtering process and adding them directly to the BNL phase.

3.2 Formal Basics

Let $D \subseteq \mathbb{R}^d$ be a d-dimensional dataset and $n = |D|$ the quantity of the input. We use D_i to denote the projection to the i-th component of D, formally

$$D_i := \{x_i \mid (x_1, ..., x_d) \in D\} .$$

For the scaling of the input we assume scale coefficients $s_1, ..., s_d \in \mathbb{N}$ forming the *scale vector* $s = (s_1, ..., s_d)$. The s_i correspond to the cardinality of the scaled domain in the i-th dimension, i.e., the scaled domain is

$$\mathbb{S}(s) = [1..s_1] \times \cdots \times [1..s_d]$$

where $[1..N] := \{1, 2, ..., N\}$ for $N \in \mathbb{N}$. The vectors $b_{\min}$ and $b_{\max}$ form a bounding box of all tuples which shall be scaled. In the simplest case we have $b_{\min} = (\min(D_1), ..., \min(D_d))$ and $b_{\max} = (\max(D_1), ..., \max(D_d))$. In Section 3.4 we discuss tighter limits for the outlier detection.

For the following we define the operators $\{\leq, <\}$ on vectors in the usual way, i.e., $(x_1, ..., x_d) \leq (y_1, ..., y_d) \Leftrightarrow \forall i : x_i \leq y_i$. Formally we realize the scaling of a tuple $x = (x_1, ..., x_d)$ by the mapping

$$f_{s, b_{\min}, b_{\max}} : \mathbb{R}^d \to \mathbb{N}^d \cup \{\text{out}\}$$
$$x \mapsto \begin{cases} (f_1(x_1), ..., f_d(x_d)) & \text{if } b_{\min} \leq x \leq b_{\max} \\ \text{out} & \text{otherwise} \end{cases} \qquad (2)$$

where

$$f_i(x_i) := \left\lfloor s_i \cdot \frac{x_i - b_{\min,i}}{(b_{\max,i} - b_{\min,i}) \cdot (1 + \epsilon)} \right\rfloor + 1 .$$

Therein $\epsilon > 0$ is some very small constant, ensuring that the result of the division is strictly smaller than 1. It follows that $1 \leq f_i(x_i) \leq s_i$ for all x and thus from the definition of f we get that $f(D) \subseteq \mathbb{S}(s) \cup \{\text{out}\}$ for any scale vector s.

3.3   The Algorithm

The Scalagon algorithm can be subdivided in four phases: 1) The Scaling is applied and outliers are isolated. 2) Hexagon Product Order is executed on the scaled dataset which results in a set of non-dominated tiles. 3) All the original tuples which correspond to non-dominated tiles in the scaled dataset are picked out. 4) BNL is applied to the union of these picked tuples and the outliers. In Algorithm 1 we show the pseudo code of the algorithm.

1. After having determined $s, b_{\min}, b_{\max}$ (see Section 3.4 and 3.5), we define

$$S := f_{s,b_{\min},b_{\max}}(D) \backslash \{\text{out}\}, \quad F_0 := \{x \in D \mid f_{s,b_{\min},b_{\max}}(x) = \text{out}\} .$$

2. Hexagon Product Order is applied to determine the set

$$S_{\min} := \{t \in S \mid \nexists u \in S : u < t\} .$$

This set is calculated with the HPO algorithm, which works exactly as the standard Hexagon, but the better-than-relation is the product order instead of the Pareto order. Note that $<$ on vectors corresponds to the product order, i.e. $x < y \iff \forall i \in [1..d] : x_i < y_i$.

3. The set of filtered tuples is calculated, containing all tuples from non-dominated tiles ($S_{\min}$) together with the outliers

$$F := \{x \in D \mid f_{s,b_{\min},b_{\max}}(x) \in S_{\min}\} \cup F_0 .$$

4. Finally, a BNL-style algorithm is applied to find the Pareto optima (Skyline) within F

$$\mathcal{S}(F) := \{t \in F \mid \nexists u \in F : u <_\otimes t\} ,$$

where $<_\otimes$ is the Pareto order from (1). $\mathcal{S}(F)$ is the output of Scalagon.

To show the correctness of Scalagon, we assume the correctness of Hexagon and BNL.

Lemma 1. *The Skyline of the filtered tuples F is equivalent to the Skyline of D, i.e., $\mathcal{S}(D) = \mathcal{S}(F)$.*

Algorithm 1. Scalagon, where f is from Eq. (2)

Input: Skyline order $<_\otimes$, Dataset D, scale vector s, bounds b_{min}, b_{max}
Output: Skyline of D
1: **function** SCALAGON($<_\otimes$,D)
2:                                    ▷ Scaling
3:     $S \leftarrow f_{s,b_{min},b_{max}}(D) \setminus \{\text{out}\}$
4:     $F_0 \leftarrow \{x \in D \mid f_{...}(x) = \text{out}\}$
5:
6:                        ▷ Hexagon product order
7:     $S_{min} \leftarrow \text{HPO}(<, S)$
8:
9:                                ▷ Filter tuples
10:    $F \leftarrow \{x \in D \mid f_{...}(x) \in S_{min}\}$
11:
12:   ▷ BNL
13:    **return** BNL($<_\otimes$, $F \cup F_0$)
14: **end function**

Algorithm 2. Scalagon scaling calculation

Input: D, domain size $(N_i)_{i=1,...,d}$, scale factor α
Output: Scale vector $s = (s_1, ..., s_d)$
1: **function** SCALING($D, (N_i)_{i=1,...,d}, \alpha$)
2:     **for all** $i \in [1..d]$ **do**   ▷ Initialization
3:         $s_i \leftarrow \lfloor (|D|/\alpha)^{1/d} \rfloor$
4:     **end for**
5:     $I \leftarrow \emptyset$
6:                      ▷ Iterative calculation of the s_i
7:     **while** $\exists i \in ([1..d] \setminus I) : N_i < s_i$ **do**
8:         $I \leftarrow \{i \in [1..d] \mid N_i \le s_i\}$
9:         **for all** $i \in I$ **do**
10:            $s_i \leftarrow N_i$
11:        **end for**
12:        **for all** $i \in ([1..d] \setminus I)$ **do**
13:            $s_i \leftarrow \lfloor (|D|/(\alpha \cdot \prod_{j \in I} N_j))^{1/(d-|I|)} \rfloor$
14:        **end for**
15:    **end while**
16:    **return** $(s_1, ..., s_d)$
17: **end function**

Proof. At first we will show $\mathcal{S}(D) \subseteq F$. For sake of readability we omit the scaling parameters, i.e., f is short for $f_{s,b_{min},b_{max}}$.

$$
\begin{aligned}
x \in \mathcal{S}(D) &\iff \nexists u \in D : (u \le x \,\wedge\, \exists i \in \{1, ..., d\} : u_i < x_i) \\
&\Rightarrow \nexists u \in D : u < x \\
&\quad \{f(u) < f(x) \text{ implies } u < x \text{ by isotony of } f\} \\
&\Rightarrow \nexists t \in f(D) : t < f(x) \quad \Rightarrow \quad x \in F
\end{aligned}
$$

Now we have $\mathcal{S}(D) = \{x \in D \mid \nexists u \in D : u <_\otimes x\} = \{x \in F \mid \nexists u \in D : u <_\otimes x\}$ where the second equal sign follows from $\mathcal{S}(D) \subseteq F$ (as above). By definition we have $F \subseteq D$ which implies $D = (D \setminus F) \cup F$. Using this fact we argue for the set condition ($\nexists u \in D : u <_\otimes x$) as follows: For every candidate $u \in D \setminus F$ there exists an element $u' <_\otimes u$ with $u' \in \mathcal{S}(D) \subseteq F$ for which $u <_\otimes x \Rightarrow u' <_\otimes x$ holds because of the transitivity of $<_\otimes$. Hence we can replace $u \in D$ by $u' \in F$ and thus we finally get $\mathcal{S}(D) = \{x \in F \mid \nexists u' \in F : u' <_\otimes x\} = \mathcal{S}(F)$ which shows the claim.   □

3.4   Outlier Detection

The filtering process by tiling and dominating the tiles as illustrated in Figure 2 is only efficient if a large amount of tuples is concentrated in some area. If we do the tiling over the entire domain, Scalagon would be prone to outliers.

Hence we suggest a simple method to detect outliers and set a bounding box where we expect the most tuples. Note that more sophisticated methods for outlier detection or cluster detection are in general too costly for our approach, cf. [23, 24]. As standard BNL has linear complexity in realistic cases [5] the determination of the bounding box and the scale vectors must not cost much.

Figure 3 illustrates our heuristic to detect outliers and how to get the bounding box, which we will describe subsequently in detail:

1. A sample $T = \{t^{(1)}, ..., t^{(N)}\} \subset D$ of $N \ll |D|$ tuples is randomly taken from the input dataset $D \subset \mathbb{R}^d$.

2. For each dimension $k \in [1..d]$ we do:

   (a) A low quantile $q(l)$ and a high quantile $q(h)$ (with $0 < l < h < 1$) of the k-th component of the tuples in T are calculated.

   Assuming that all values in T are distinct we have for $z \in [0, 1]$:

   $$|\{(t_1, ..., t_d) \in T \mid t_k \leq q(z)\}| = z \cdot N$$
   $$|\{(t_1, ..., t_d) \in T \mid t_k > q(z)\}| = (1 - z) \cdot N$$

   (b) We add some summand δ and finally calculate the k-th component of the bounding box with some $0 < \beta < 1$:

   $$\delta := (q(h) - q(l)) \cdot \beta$$
   $$b_{\min,k} := q(l) - \delta$$
   $$b_{\max,k} := q(h) + \delta$$

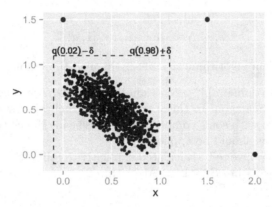

Fig. 3. Outlier detection: The Scalagon filtering is restricted to a bounding box. For example we used $l = 0.02$, $h = 0.98$ and $\beta = 0.2$.

3.5 Scaling

Subsequently we want to determine an optimal scaling factor. For the entire runtime of the Scalagon algorithm we have two different cost producers. 1) The operations on the tiles from Hexagon Product Order, cf. Section 2.3. 2) The comparisons caused by a BNL-style algorithm. For anticorrelated realistic datasets we have $k \cdot |D|$ comparisons, where $k \ll |D|$ as long as $|\mathcal{S}(D)| \ll |D|$.

With a fine-grained scaling the filtering efficiency raises, i.e., the number of comparisons in the final BNL phase decreases. At the same time the HPO phase calculating the non-dominated tiles becomes more costly. Hence the decisive criterion is the choice of the scaling coefficients s_i compared to the dataset size $|D|$. Therefore we introduce the *scale factor* α, defined by the ratio of the dataset size $|D|$ and the size of the scaled domain:

$$\alpha = \frac{|D|}{|\mathbb{S}(s)|}$$

For the first, this factor α is assumed to be known. We dedicate Section 4.2 to the determination of an optimal α. This factor allows to continuously change between a BNL-style and a lattice based algorithm. For $\alpha \to 0$ we are dealing with a lattice based algorithm where $f(x)$ is unique for every tuple $x \in D$. For $\alpha \to \infty$ we retrieve $|\mathbb{S}(s)| = 1$ meaning that there is just one equivalence class. Hence there is no filtering and Scalagon acts like a BNL-style algorithm.

Let us now assume that the domain size $(N_1, ..., N_d)$ is given. Typically N_i is either small (low-cardinality values like gender, nation, ...) or $N_i = \infty$ (domains like salary, price, ...), meaning that we are faced with a continuous domain.

Obviously it does not make sense to choose a scale vector with $s_i > N_i$. This would introduce just empty tiles in the lattice structure of the HPO-phase. Hence the problem of determining the scaling coefficients for a given α reads as follows:

$$\alpha \approx \frac{|D|}{\prod_{i=1}^{d} s_i} \tag{S1}$$

$$s_i \leq N_i \quad \text{for all } i \in [1..d] \tag{S2}$$

To solve this problem algorithmically we distinguish the following cases:

1. If the size of the domain $\prod_{i=1}^{d} N_i$ is smaller than $|D|/\alpha$ it does not make any sense to apply Scalagon. In this case, Scalagon behaves as a purely lattice based algorithm with some computational overhead. Hence the usual lattice based algorithms should be applied.
2. Otherwise the s_i have to be calculated explicitly to fulfill (S2). To this end we describe the algorithm SCALING (Algorithm 2) to determine the s_i accordingly.

The idea to iteratively determine the s_i is as follows: First we initialize s_i by

$$s_1 := ... := s_d := \left\lfloor \left(\frac{|D|}{\alpha} \right)^{1/d} \right\rfloor .$$

This fulfills (S1). If the domain size $(N_1, ..., N_d)$ is sufficiently large (or continuous), (S2) is also fulfilled and we are done.

Otherwise for those i where $s_i > N_i$ we take the original cardinality $s_i := N_i$. Afterwards the other s_i are recalculated such that (S1) is fulfilled. This step is iterated until a solution for (S1) and (S2) is found.

In the pseudo code of the algorithm (cp. Algorithm 2) we use $I \subseteq [1..d]$ as an index set to mark those s_i which are already set to N_i. We define the complement $\bar{I} := [1..d] \backslash I$. To determine the s_i in the iteration step we calculate

$$\alpha = \frac{|D|}{|\mathbb{S}(s)|} = \frac{|D|}{\prod_{i=1}^{d} s_i} = \frac{|D|}{\prod_{i \in I} N_i \cdot \prod_{i \in \bar{I}} s_i} .$$

Requiring identical s_i values for $i \in \bar{I}$ leads us to

$$s_i = \left\lfloor \left(\frac{|D|}{\alpha \cdot \prod_{i \in I} N_i} \right)^{1/|\bar{I}|} \right\rfloor \quad \text{for all } i \in \bar{I} .$$

This corresponds to line 13 in Algorithm 2.

Note that the while-condition (line 7) together with the assignment to I (line 8) ensures that the set I is strictly growing w.r.t. the inclusion order $\subsetneq$ in every iteration step. As soon as $I = [1..d]$ is reached we have $\bar{I} = \emptyset$. Thus the while-condition is trivially false and the algorithm terminates.

In the continuous case we have a (hyper-)quadratic scaling $s_1 = \ldots = s_d$. One may criticize this as simplistic, and this is only reasonable if the input data is distributed more or less uniformly over the bounding box from the outlier detection step. But to the best of our knowledge all more sophisticated solutions (like constructing an equi-depth histogram for each dimension and then place the values in the corresponding bucket) would introduce too much costs. An expensive pre-filtering easily leads to methods which cannot compete to the simple BNL approach.

Subsequently we will present a concrete numerical example on the iterative scaling procedure.

Example 1. Assume a scaling factor $\alpha = 50$, $|D| = 10^6$, and domain cardinalities $N_1 = 2$, $N_2 = 50$, $N_3 = 10^5$, i.e., $D \subseteq [1..2] \times [1..50] \times [1..10^5]$. Applying the SCALING algorithm leads to values depicted in Table 1. After the second iteration the algorithm terminates with $s_1 = 2$, $s_2 = 50$, $s_3 = 200$.

Table 1. A run of the Scalagon scaling calculation

Iteration	I	s_1	s_2	s_3
0	$\emptyset$	27	27	27
1	$\{1\}$	2	100	100
2	$\{1,2\}$	2	50	200

3.6 Complexity Analysis

Concerning the complexity of Scalagon, the first phase is the Hexagon product order (HPO) algorithm. This causes linear costs w.r.t. the input size $n = |D|$ and the size of the lattice, which is proportional to the size of the scaled domain $\mathbb{S}(s)$, i.e., $|\mathbb{S}(s)| = s_1 \cdot \ldots \cdot s_d$, cf. [19]. Hence we have for the HPO phase the complexity $C_{t,\text{HPO}}(n, s) = \Theta(n + s_1 \cdot \ldots \cdot s_d)$.

For the filtering phase we can assume that it takes constant time to check $f(t) \in S_{\min}$, hence this does not add additional complexity.

Finally we consider the BNL phase. In the worst case the BNL-style algorithm takes $\Theta(n^2)$ time, which can be realized for a set of incomparable objects, e.g., a correlation of $c = -1$. This means that the complexity in phase 4 of Scalagon is bounded by $C_{t,\text{BNL}}(n) = \mathcal{O}(n^2)$ and the overall theoretical *time complexity* is

$$C_t(n, s) = \mathcal{O}(n^2 + s_1 \cdot \ldots \cdot s_n) \ .$$

The space complexity of the HPO algorithm corresponds to the size of lattice, $C_{s,\text{HPO}}(s) = \Theta(s_1 \cdot \ldots \cdot s_d)$. Regarding the filtering phase and the BNL phase the space complexity is bounded by $\mathcal{O}(|F|)$, even if all the filtered tuples are put into main memory. In the worst case no tuples are filtered, and we have a space complexity of $\mathcal{O}(n)$. In summary we are dealing with a *space complexity* of

$$C_s(n, s) = \mathcal{O}(s_1 \cdot \ldots \cdot s_d + n) \ .$$

Note that in average and realistic cases we have linear costs in the BNL phase [5] and the actual performance mainly depends on choosing α.

More sophisticated analysis of the complexity would necessitate some assumptions about the dataset and the distribution. Scalagon is an "all seasons algorithm" which has the only assumption that most of data (without the outliers) is concentrated in some area. This occurs in most realistic datasets and is a quite "weak" assumption.

3.7   Memory Requirements and α

One disadvantage of *Hexagon* [20] and *Lattice Skyline* [19] is that the lattice must fit into main memory. With our Scalagon approach we can *scale down the original domain* such that the scaled lattice fits into memory, independent from the available memory size.

In contrast to the usual lattice Skyline algorithms described in Section 2.3, we only need the *dominance* state of every tile and not the *empty/non-empty* states in the HPO phase. This means that the required main memory capacity of HPO corresponds to a bit array in the same size as the scaled domain, i.e., $\prod_{i=1}^{d} s_i$. This array stores if the area is dominated (light gray tiles in Figure 2) or is relevant for the BNL phase (white and dark grey tiles).

Assume the memory is restricted to M bytes. Then we have to insure that the scaled domain $\mathbb{S}(s) = [1..s_1] \times \ldots \times [1..s_d]$ fulfills $mem\_BTG(s_1, .., s_d) \leq M$:

$$\frac{1}{8} \prod_{i=1}^{d} s_i \leq M \ .$$

Since $\alpha = |D| / \prod_{i=1}^{d} s_i$ we have $\alpha \geq |D|/(8 \cdot M)$.

Note that an externalization of our algorithm is straight forward: determine the right value for α such that the BTG fits into the available main memory M and apply an external BNL-style algorithm in phase 4, cp. Section 3.3. Therefore, Scalagon is not restricted to main memory.

4   Experiments

4.1   Framework

The algorithms used in our benchmarks have been implemented in Java 7.0 using only built-in techniques for locking, compare-and-swap operations, and thread management for the parallel algorithms. The experiments were performed on a single node running Debian Linux 7.1 equipped with two Intel Xeon 2.53 GHz quad-core processors using hyper-threading, that means a total of 16 cores.

For our synthetic datasets we used the data generator commonly used in Skyline research [1]. We focused on weakly anti-correlated data distributions, because this is the most challenging task for Skyline computation. For the experiments on real-world data we used the well-known *NBA* and *Household* datasets[1]. The *NBA* dataset is a small 5-dimensional dataset containing 17264 tuples, where each entry records performance statistics for a NBA player [25]. *Household* is a larger 6-dimensional dataset having 127931 points and a domain of $[1..10^4]^6$.

4.2   Scalagon and the α Factor

A scale-factor $\alpha \rightarrow 0$ implies a similar algorithm like Hexagon, as every tuple has its own tile in the scaled setting. Analogously $\alpha \rightarrow \infty$ leads to one tile for all tuples and results in a BNL-style algorithm with some useless computational overhead. That means, choosing the scaling factor α is a crucial point in the Scalagon algorithm. We experimentally search for an α value to achieve maximum performance. A similar idea was already used in the implementation of Quicksort by [26]. Thereby a *CutOff*-parameter is introduced which combines Quicksort and insertion sort (the latter is more efficient for small arrays). Its value is determined experimentally and reduces the runtime significantly.

Test Setting 1. In Figure 4 we show the run time of Scalagon w.r.t. different values for α. For every sample we did five runs, each sample contains $n = 10^7$ tuples in the domain $[0, 1]^2$ with a correlation value of -0.7, i.e. a typical weak anti-correlation. We tested the values $\alpha \in \{10^{-2}, 10^{-1}, ..., 10^6\}$. We did this experiment in the *statistical computing software* R, where the package *rPref* offers the BNL and Scalagon algorithm. The script is available at the web [27].

The experiment shows that $\alpha \in [1, 1000]$ is approximately optimal and within this interval the run time is nearly identical. Outside this interval the run time increases and for $\alpha < 0.1$ and $\alpha > 10^5$ the performance of Scalagon tends to be worse than BNL. Therefore, choosing α seems to be easy. In all test settings we could find a clear performance maximum plateau within the interval $[1, 1000]$.

---

[1] These datasets were crawled from *www.nba.com* and *www.ipums.org*.

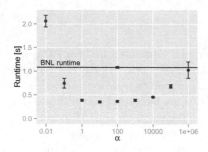

Algorithm	α	$|F|$	C	T_{hpo}	T_{sSky}	T_{all}
Std-BNL			21053410			12.26
LESS			20106892			11.16
sSkyline			11554728			8.95
BSkyTree			7564			**7.92**
Hexagon			–			104.95
Scalagon	10	223390	6106593	0.93	4.39	5.32
	20	277089	6378523	0.40	4.75	5.15
	50	332737	7518080	0.30	5.51	5.81
	100	387930	7963012	0.27	6.20	6.47
	200	425117	8245772	0.25	5.73	5.98

Fig. 4. Test setting 1. Runtime w.r.t. α compared with BNL.

Fig. 5. Test setting 2. Filtering of Scalagon ($|F|$) and number of comparisons (C). All times ($T_{(...)}$) in seconds.

Test Setting 2. In Figure 5 we substantiate the results of Test Setting 1 using the domain $[1..100]^2 \times [1..10^4]$ and varying α to measure the influence of the scaling factor to the performance. We used $|D| = 10^6$ anti-correlated input objects. We compared Scalagon to standard **Std-BNL** [1], **LESS** [5], **sSkyline**[11], **BSkyTree** [21], and **Hexagon** [20]. In **Scalagon** we used *sSkyline* for phase 4, because it is a simple Skyline algorithm without any pre-computation. The Skyline set has a size of 476 objects. The *overall runtime* T_{all} (in **seconds**) of BSkyTree performs better than Hexagon, although Hexagon has a linear runtime complexity. This is due to the fact that the depth-first dominance test in the large BTG takes a lot of time. The HPO phase of Scalagon eliminates about two third of the input objects ($|F|$ is the number of tuples after the HPO pre-filter process). Therefore the number of *tuple comparisons* C in phase 4 is much less then in the other algorithms. Note that in the BSkyTree algorithm the computation of the pivot element is the most time consuming part, but implies only 7564 tuple comparisons. Scalagon with $\alpha = 20$ clearly outperforms all other algorithms. Thereby, T_{hpo} is the runtime of HPO (phases 1–3) and T_{sSky} is the runtime of sSkyline (phase 4). The scaled domain for $\alpha = 20$ is $\mathbb{S} = [1..36]^3$ which leads to a BTG size of 46656 nodes for the HPO stage.

4.3 Parallel Scalagon

In this section we provide performance benchmarks for our parallel implementation of Scalagon. Thereby we used the setting described in Section 4.1. We have implemented a parallel Hexagon product order algorithm, which is a modification of the algorithm ARL-S described in [15]. Combining the parallel Hexagon product order method with a parallel BNL-style Skyline algorithm for the second phase leads to the parallel Scalagon algorithm **pScalagon**.

As competitors we considered parallel BNL (**pBNL**) using a Lazy List [12], **pSkyline** [11], and **APSkyline** using the equi-volume angle-based partitioning strategy [14]. We skipped the algorithm D&C [12] and HPL-S and ARL-S from [15], because they are outperformed by another algorithm or cannot be applied

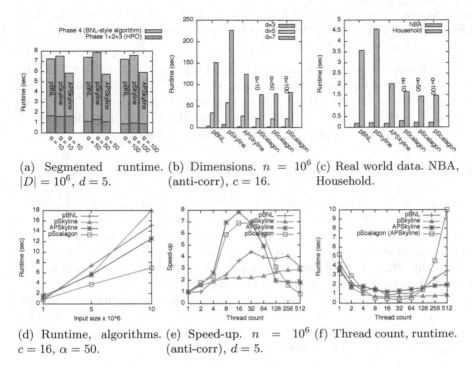

(a) Segmented runtime. $|D| = 10^6$, $d = 5$.

(b) Dimensions. $n = 10^6$ (anti-corr), $c = 16$.

(c) Real world data. NBA, Household.

(d) Runtime, algorithms. $c = 16$, $\alpha = 50$.

(e) Speed-up. $n = 10^6$ (anti-corr), $d = 5$.

(f) Thread count, runtime.

Fig. 6. Parallel Skyline algorithms

due to high-cardinality domains. Note that all our results are in line with the results presented in [12] and [14].

Test Setting 3 – Segmented Runtime. In Figure 6(a) we show the segmented runtime for pScalagon using different parallel BNL-style algorithms for phase 4. Using $\alpha = 10$, the HPO is a little bit slower than for $\alpha = 100$ because of a larger BTG and a more expensive DFT. For larger α the BNL-style algorithms have to do more work.

Test Setting 4 – Dimensions. In Figure 6(b) we used different domains $[1..1000]^d$, with $d \in \{3, 5, 7\}$ dimensions, which is a realistic case in Skyline queries. For $d = 3$ all runtimes are quite similar, whereas for higher dimensions pScalagon (with APSkyline) outperforms its competitors. Note that the size of the Skyline set normally increases with the dimensionality of the dataset on anti-correlated data [28], making Skyline processing for algorithms relying on tuple comparison more demanding. This experiment verifies the advantage of our hybrid algorithm using a data independent lattice approach as pre-filter.

Test Setting 5 – Real-World Data. In Figure 6(c) we report the comparison results for real world data. We expected that pScalagon is worse than its competitors considering the NBA dataset, because of the hybrid approach which produces some overhead on small datasets. However, for the Household dataset pScalagon outperforms all other algorithms.

Test Setting 6 – Runtime. In this test we compared pScalagon (with APSkyline) to the state-of-the-art parallel Skyline algorithms, cp. Figure 6(d). The domain is $[1..5] \times [1..10^3] \times [1..10^5]$. For the given $\alpha = 50$ this leads to the scaled domain $\mathbb{S} = [1..5] \times [1..63] \times [1..63]$ in the case of $n = 10^6$. pScalagon clearly outperforms its competitors, in particular for large datasets.

Test Setting 7 – Thread Count. In Figure 6(e) we measured the speed-up on 5 dimensions: $[1..2]^2 \times [1..5] \times [1..10^4]^2$. We observed that APSkyline and pScalagon have a good speed-up up to 8 threads. From the ninth thread on, the performance only marginally increases and beyond 16 threads it gradually decreases. This can be explained with decreasing cache locality and increasing communication costs as our test system uses two quad-core processors with Hyper-Threading. Starting with the ninth core, the 2nd processor must constantly communicate with the first. The same effect is mentioned in [12].

For evaluating the runtime performance of the parallel algorithms in absolute numbers, we also measured the computation time. The results can be found in Figure 6(f). The bad performance of pScalagon using more than 256 threads can be explained by a very high locking of the lattice nodes.

5 Conclusion and Outlook

Scalagon can be seen as the smooth integration of the lattice Skyline idea, which was originally developed for discrete domains, into a general Skyline algorithm for continuous domains and mixed discrete/continuous domains. Scalagon does not depend on a plenty of configuration factors or hard to determine pre-sorting functions like some other algorithms, but can be tuned by the factor α. With the only assumption that the most of the tuples are more or less uniformly distributed, Scalagon is well suited for a large variety of input data. Our experiments showed the superior characteristics and performance of Scalagon in different settings.

For future work we want to use GPUs to optimize the pre-filter phase of Scalagon. Since there reside many comparable objects inside a node of the lattice, it would be possible to run a BNL-style algorithm inside a tile to eliminate dominated objects. But this could only be done using massive parallel hardware architectures like GPUs.

Acknowledgments. We want to thank Selke et al. [12] for providing us with the source code of the parallel BNL, sSkyline, pSkyline, and pDC algorithms. The implementation of APSkyline is based on the source code made available by Liknes et al. [14]. This work has been partially funded by the Bavarian Ministry of Economic Affairs, Infrastructure, Transport and Technology, grant numbers IUK-1109-0003//IUK398/002 and IUK-1307-0004//IUK434/003.

References

1. Börzsönyi, S., Kossmann, D., Stocker, K.: The Skyline Operator. In: ICDE 2001 Proceedings of the 17th International Conference on Data Engineering, pp. 421–430. IEEE Computer Society, Washington, DC (2001)
2. Chomicki, J., Ciaccia, P., Meneghetti, N.: Skyline Queries, Front and Back. SIGMOD Rec. **42**(3), 6–18 (2013)
3. Kossmann, D., Ramsak, F., Rost, S.: Shooting stars in the sky: an online algorithm for skyline queries. In: VLDB 2002 Proceedings of the 28th International Conference on Very Large Data Bases, pp. 275–286. VLDB Endowment (2002)
4. Chomicki, J., Godfrey, P., Gryz, J., Liang, D.: Skyline with presorting. In: ICDE 2003 Proceedings of the 19th International Conference on Data Engineering, pp. 717–816 (2003)
5. Godfrey, P., Shipley, R., Gryz, J.: Algorithms and Analyses for Maximal Vector Computation. The VLDB Journal **16**(1), 5–28 (2007)
6. Balke, W.-T., Güntzer, U., Zheng, J.X.: Efficient distributed skylining for web information systems. In: Bertino, E., Christodoulakis, S., Plexousakis, D., Christophides, V., Koubarakis, M., Böhm, K. (eds.) EDBT 2004. LNCS, vol. 2992, pp. 256–273. Springer, Heidelberg (2004)
7. Wu, P., Zhang, C., Feng, Y., Zhao, B.Y., Agrawal, D.P., El Abbadi, A.: Parallelizing skyline queries for scalable distribution. In: Ioannidis, Y., Scholl, M.H., Schmidt, J.W., Matthes, F., Hatzopoulos, M., Böhm, K., Kemper, A., Grust, T., Böhm, C. (eds.) EDBT 2006. LNCS, vol. 3896, pp. 112–130. Springer, Heidelberg (2006)
8. Cosgaya-Lozano, A., Rau-Chaplin, A., Zeh, N.: Parallel computation of skyline queries. In: HPCS 2007 Proceedings of the 21st International Symposium on High Performance Computing Systems and Applications, p. 12. IEEE Computer Society, Washington, DC (2007)
9. Rocha-Junior, J.B., Vlachou, A., Doulkeridis, C., Nørvåg, K.: AGiDS: a grid-based strategy for distributed skyline query processing. In: Hameurlain, A., Tjoa, A.M. (eds.) Globe 2009. LNCS, vol. 5697, pp. 12–23. Springer, Heidelberg (2009)
10. Gao, Y., Chen, G.-C., Chen, L., Chen, C.: Parallelizing progressive computation for skyline queries in multi-disk environment. In: Bressan, S., Küng, J., Wagner, R. (eds.) DEXA 2006. LNCS, vol. 4080, pp. 697–706. Springer, Heidelberg (2006)
11. Park, S., Kim, T., Park, J., Kim, J., Im, H.: Parallel Skyline Computation on Multicore Architectures. In: ICDE 2009 Proceedings of the 2009 IEEE International Conference on Data Engineering, pp. 760–771. IEEE Computer Society, Washington, DC (2009)
12. Selke, J., Lofi, C., Balke, W.-T.: Highly scalable multiprocessing algorithms for preference-based database retrieval. In: Kitagawa, H., Ishikawa, Y., Li, Q., Watanabe, C. (eds.) DASFAA 2010. LNCS, vol. 5982, pp. 246–260. Springer, Heidelberg (2010)
13. Afrati, F.N., Koutris, P., Suciu, D., Ullman, J.D.: Parallel skyline queries. In: ICDT 2012 Proceedings of the 15th International Conference on Database Theory, pp. 274–284. ACM, New York (2012)
14. Liknes, S., Vlachou, A., Doulkeridis, C., Nørvåg, K.: APSkyline: improved skyline computation for multicore architectures. In: Bhowmick, S.S., Dyreson, C.E., Jensen, C.S., Lee, M.L., Muliantara, A., Thalheim, B. (eds.) DASFAA 2014, Part I. LNCS, vol. 8421, pp. 312–326. Springer, Heidelberg (2014)
15. Endres, M., Kießling, W.: High parallel skyline computation over low-cardinality domains. In: Manolopoulos, Y., Trajcevski, G., Kon-Popovska, M. (eds.) ADBIS 2014. LNCS, vol. 8716, pp. 97–111. Springer, Heidelberg (2014)

16. Tan, K.-L., Eng, P.-K., Ooi, B.C.: Efficient progressive skyline computation. In: VLDB 2001 Proceedings of the 27th International Conference on Very Large Data Bases, pp. 301–310. Morgan Kaufmann Publishers Inc, San Francisco (2001)

17. Papadias, D., Tao, Y., Fu, G., Seeger, B.: An optimal and progressive algorithm for skyline queries. In: SIGMOD 2003 Proceedings of the 2003 ACM SIGMOD International Conference on Management of Data, pp. 467–478. ACM, New York (2003)

18. Lee, K., Zheng, B., Li, H., Lee, W.-C.: Approaching the skyline in Z order. In: VLDB 2007 Proceedings of the 33rd International Conference on Very Large Data Bases, pp. 279–290. VLDB Endowment (2007)

19. Morse, M., Patel, J.M., Jagadish, H.V.: Efficient skyline computation over low-cardinality domains. In: VLDB 2007 Proceedings of the 33rd International Conference on Very Large Data Bases, pp. 267–278. VLDB Endowment (2007)

20. Preisinger, T., Kießling, W.: The hexagon algorithm for evaluating pareto preference queries. In: MPref 2007 Proceedings of the 3rd Multidisciplinary Workshop on Advances in Preference Handling (in conjunction with VLDB 2007) (2007)

21. Lee, J., Hwang, S.-W.: BSkyTree: scalable skyline computation using a balanced pivot selection. In: EDBT 2010 Proceedings of the 13th International Conference on Extending Database Technology, pp. 195–206. ACM, New York (2010)

22. Davey, B.A., Priestley, H.A.: Introduction to Lattices and Order, 2nd edn. Cambridge University Press, Cambridge (2002)

23. Aggarwal, C.C., Yu, P.S.: Outlier detection for high dimensional data. In: SIGMOD 2001 Proceedings of the 2001 ACM SIGMOD International Conference on Management of Data, vol. 30, pp. 37–46. ACM, New York, May 2001

24. Aggarwal, C.C.: Outlier Analysis. Springer, New York (2013)

25. Chan, C.-Y., Jagadish, H.V., Tan, K.-L., Tung, A.K.H., Zhang, Z.: On high dimensional skylines. In: Ioannidis, Y., Scholl, M.H., Schmidt, J.W., Matthes, F., Hatzopoulos, M., Böhm, K., Kemper, A., Grust, T., Böhm, C. (eds.) EDBT 2006. LNCS, vol. 3896, pp. 478–495. Springer, Heidelberg (2006)

26. Bentley, J.L.: Programming Pearls. Addison-Wesley (2000)

27. Roocks, P.: R script for α determination (2014). http://www.informatik. uni-augsburg.de/en/chairs/dbis/db/staff/roocks/publications/rpref_alpha.zip

28. Shang, H., Kitsuregawa, M.: Skyline operator on anti-correlated distributions. In: VLDB 2013 Proceedings of the 39rd International Conference on Very Large Data Bases, vol. 6, pp. 649–660 (2013)

Towards Order-Preserving
SubMatrix Search and Indexing

Tao Jiang$^{(\boxtimes)}$, Zhanhuai Li, Qun Chen, Kaiwen Li, Zhong Wang, and Wei Pan

School of Computer Science and Technology,
Northwestern Polytechnical University, 710072 Xi'an, China
{jiangtao,likaiwen,zhongwang}@mail.nwpu.edu.cn,
{lizhh,chenbenben,panwei1002}@nwpu.edu.cn
http://wowbigdata.net.cn/

Abstract. *Order-Preserving SubMatrix* (OPSM) has been proved to be important in modelling biologically meaningful subspace cluster, capturing the general tendency of gene expressions across a subset of conditions. Given an OPSM query based on row or column keywords, it is desirable to retrieve OPSMs quickly from a large gene expression dataset or OPSM data via *indices*. However, the time of OPSM mining from gene expression dataset is long and the volume of OPSM data is huge. In this paper, we investigate the issues of indexing two datasets above and first present a naive solution *pfTree* by applying prefix-Tree. Due to it is not efficient to search the tree, we give an optimization indexing method *pIndex*. Different from *pfTree*, *pIndex* employs row and column header tables to traverse related branches in a *bottom-up* manner. Further, two pruning rules based on *number* and *order* of keywords are introduced. To reduce the number of column keyword candidates on fuzzy queries, we introduce a First Item of keywords roTation method *FIT*, which reduces it from $n!$ to n. We conduct extensive experiments with real datasets on a single machine, Hadoop and Hama, and the experimental results show the efficiency and scalability of the proposed techniques.

Keywords: OPSM · Gene expression data · *pIndex* · *FIT* · Queries

1 Introduction

DNA microarray enables simultaneously monitoring of the expression level of tens of thousands of genes over hundreds of experiments. Gene expression data, fixed on DNA microarrays, can be viewed as an $n \times m$ matrix with n genes (rows) and m experiments (columns), in which each entry denotes the expression level of a given gene under a given experiment. Traditional clustering methods

This work was supported in part by National Basic Research Program 973 of China (No. 2012CB316203), Natural Science Foundation of China (Nos. 61033007, 61272121, 61332006, 61472321), National High Technology Research and Development Program 863 of China (No. 2013AA01A215), Graduate Starting Seed Fund of Northwestern Polytechnical University (No. Z2012128).

© Springer International Publishing Switzerland 2015
M. Renz et al. (Eds.): DASFAA 2015, Part II, LNCS 9050, pp. 309–326, 2015.
DOI: 10.1007/978-3-319-18123-3_19

do not work well for gene expression data, due to a fact that most genes are tightly coexpression only under a subset of experiments, and are not necessarily expression at the same or similar expression level. Thus, it makes pattern-based subspace clustering as the popular tool to find meaningful clusters. Recently, *Order-Preserving SubMatrix* (OPSM) [2], a special model of pattern-based clustering, has been accepted as a biologically meaningful cluster model. In essence, an OPSM is a subset of rows and columns in a data matrix where all the rows induce the same liner ordering of the columns, e.g., rows g_3 and g_6 have an increasing expression level on columns 2, 7, 5, and 1. And OPSM cluster model focuses on the relative order of columns rather than the actual values. As costs of gene expression analysis continue to decrease, large volume of gene expression datasets and OPSM mining results are accumulated, but not efficiently used by biologists, who would like to find supporting rows or columns based on keyword queries. This problem *OPSM query is to retrieve one or some supporting rows or columns based on column or row keywords from a given data matrix*, which plays an important role in inferring gene coregulated networks.

Most of the previous studies, such as [15,10,6,21,17,8,9], address the problem of OPSM mining, few work is studied for OPSM query. OPSM problem is first proposed by Ben-Dor [2]. Then, Liu and Trapp et al. [15,17] try to develop efficient mining methods. Gao et al. [10] give *KiWi* framework to find small twig clusters. Chui [6], Zhang [21], and Fang et al. [7,8,9] present noise-tolerant models. OPSM mining tools, such as *GPX* [11] and *BicAT* [1], have a common feature that it uses an indirect way to give queried results. And indirect way is not efficient, thus we want to present a direct-way query tool. The most similar work with OPSM query is presented in work [11], Jiang et al. [11] give an interactive method to facilitate OPSM search, which can drill down and roll up.

OPSM query is similar to string matching problem [13,3], the common work of both problems is to find whether a pattern string is in the given string. Two famous methods of string matching are *KMP* [13] and *BM* [3], which are efficient for string matching without gaps, but cannot work well for OPSM query which allows having gaps in the given string. Thus, these methods cannot be directly used by us. Clearly, it is necessary to build indices in order to help processing OPSM queries. Prefix tree [14] and suffix tree [19,16,18] are two common models, due to the former one allows two strings share the common prefix string, which reduces the spaces, we choose it as the basic model.

Designing a direct-way OPSM query tool is a challenging work due to the reasons below. First, the numbers and sizes of datasets are huge. As the cost of gene expression analysis decreases, the numbers and sizes of gene expression datasets have been growing at an ever-increasing rate. Further, large volume of OPSM datasets are accumulated. Second, how to devise a common tool for two different datasets above. As we all know, OPSM mining time from gene expression data is larger than the search time from OPSM data, but the amount of OPSM dataset is much larger than that of gene expression data. Last but not the least, the index size should be small enough to save in memory, index update should be efficient, and the queries on the index should be fast and scalable.

To address the challenges above, we first present a naive solution *pfTree* by applying prefix-Tree, which can reduce some duplicate data. Due to it is not efficient to search the tree, an optimization method *pIndex* is given. Meanwhile, two indices can incorporate two kinds of data, i.e., it delays OPSM mining until query or directly uses mining results. In this way, we take advantage of pros of two different datasets. Further, *pIndex* utilizes row and column header tables for index update and OPSM queries. To improve the performances of queries, it gives pruning methods to reduce the scans of useless branches. To reduce the number of column keyword candidates on fuzzy queries, we introduce a First Item of keywords roTation method *FIT*, which reduces it from $n!$ to n.

We have applied *pfTree* and *pIndex* to gene expression and OPSM datasets. And we conduct extensive experiments, and experimental results demonstrate that both indices are compact in size, the proposed techniques are very efficient for index update and OPSM queries. Further, we implement the methods on a single machine, Hadoop and our modified Hama platform [12], and it is also efficient and scalable. The main contributions of this paper are as follows:

- We propose a naive prefix tree based indexing method *pfTree* and an optimization mechanism *pIndex* associated with row and column header tables.
- Index update (insert or delete) and query methods are presented. *FIT* and some pruning methods improve the query performances.
- We evaluate *pfTree* and *pIndex* on a single machine, Hadoop and our modified Hama platform [12], and the study confirms that *pIndex* has better performance in terms of processing cost and scalability.

The rest of paper is organized as follows: Section 2 gives preliminary concepts and presents the basic framework for Order-Preserving SubMatrix indexing and search. Section 3 illustrates the index building and updating method and how to construct header tables. Exact and fuzzy queries using header table based search paradigm are discussed, *FIT* strategy and pruning methods are proposed in Section 4. We report empirical studies, and give related work in Section 5 and 6, respectively. Section 7 concludes this study.

2 Preliminaries

In this section, we introduce preliminary concepts and outline an indexing framework to address *Order-Preserving SubMatrix* query problem.

In this paper, we use the following notations listed in *Table 1*.

	Table 1. Notations			**Table 2.** An OPSM Dataset	
Notation	Description	Notation	Description	Row No.	Column No.
G	gene set	C	condition set	1,2,5	VI,III,I,VIII,XVI
g	subset of G	c	subset of C	3,6,9	VI,III,I,II,XIII
g_i	a gene	c_i	a condition	7,10,11	VI,II,III
$D(G,C)$	a matrix	e_{ij}	entry of D	4,8,12	III,II,XVI
τ	c threshold	δ	g threshold	4,6	VI,III,I,VIII,XVI

Definition 1. (ORDER-PRESERVING SUBMATRIX (OPSM)). *Given a dataset* $(n \times m \text{ matrix}) D(G, C)$, *an OPSM is a pair* (g, c), *where g is a subset of the n rows and c is a permutation of a subset of the m columns which satisfies the condition: for each row in g, the data values e are monotonically increasing or decreasing with respect to the permutation of the indexes of columns, i.e.,* $e_{i1} \prec e_{i2} \prec ... \prec e_{ij} \prec ... \prec e_{ik}$ $(e_{i1} \succ e_{i2} \succ ... \succ e_{ij} \succ ... \succ e_{ik})$, *where* $(i1, ..., ij, ..., ik)$ *is the permutation of the indexes of columns* $(1, ..., j, ..., k)$.

Definition 2. (EXACT QUERY ON GENES (EQ_g)). *Given an OPSM datasets and a subset of genes* $g = (g_i, ..., g_j, ..., g_k)$, *exact query on g returns the subsets of conditions* $c = (c_x, ..., c_y, ..., c_z)$ *above length threshold* τ *in OPMSs that contain all the items in g.*

Definition 3. (EXACT QUERY ON CONDITIONS (EQ_c)). *Given an OPSM datasets and a subset of conditions* $c = (c_x, ..., c_y, ..., c_z)$, *exact query on c returns the subsets of genes* $g = (g_i, ..., g_j, ..., g_k)$ *above size threshold* δ *in OPMSs that contain all the items in c and keep the order of c.*

Definition 4. (FUZZY QUERY ON GENES (FQ_g)). *Given an OPSM datasets and a subset of genes* $g = (g_i, ..., g_j, ..., g_k)$, *fuzzy query on g returns the subsets of conditions* $c = (c_x, ..., c_y, ..., c_z)$ *above length threshold* τ *in OPMSs that contain a subset of the items in g above size threshold* δ.

Definition 5. (FUZZY QUERY ON CONDITIONS (FQ_c)). *Given an OPSM datasets and a subset of conditions* $c = (c_x, ..., c_y, ..., c_z)$, *fuzzy query on c returns the subsets of genes* $g = (g_i, ..., g_j, ..., g_k)$ *above size threshold* δ *in OPSMs that contain a subset of c above length threshold* τ *and need not to keep order of c.*

The processing of OPSM queries can be divided into three major steps:

- *Index construction and update*: which is the fundamental part. It uses prefix tree to incorporate the two kinds of datasets. If several OPSMs have a common prefix, they share the prefix in the tree and each suffix is added after the prefix. Index update includes *index inserting* and *index deleting*. To make it efficient is the critical task.
- *Header table design*: which is the assistant data structure. It consists of two parts: (1) *Row header table*, which helps *pIndex* deletion and OPSM queries on *row keywords*. (2) *Column header table*, which aids *pIndex* deletion and OPSM queries on *column keywords*.
- *Query processing*: which consists of two substeps: (1) *Search*, which traces the tree with row or column keywords in the way of *bottom-up*, and gets column branches or row sets respectively. (2) *Filter*, which does the intersection and checks whether the intermediate results are above the user-defined threshold.

3   pIndex

To design a compact and efficient index for OPSM queries, let's first examine an example *Example 1*. The row No. (gene name) and column No. (experimental condition) of OPSMs are listed in the first and second columns of *Table 2*.

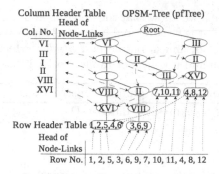

Fig. 1. pIndex (pfTree, column and row header tables)

In the following, we use OPSM dataset as an example, due to each row in gene expression data can be seen as an OPSM. A compact index can be designed based on the observations below:

- There are a number of overlaps in the OPSMs. If each overlap only be stored one time, it may avoid some unnecessary work, and thus save some spaces.
- If multiple OPSMs share the same order of the same columns, they can be merged into one with the row No. record merging.
- If two OPSMs share a common prefix, the shared parts can be merged using one prefix structure and two branches are added after the prefix.

With these observations, we give an example to explain how to construct a prefix tree of OPSMs, called OPSM-Tree or *pfTree*, a naive indexing method.

Example 1. Table 2 shows a sample OPSM dataset. This dataset will be used as our running example. The procedure of pfTree construction is plotted in Fig. 1.

In *Example 1*, first, one may create the root of a tree, labelled with "null". Scan the OPSM dataset, the scan of the first OPSM leads to the construction of the first branch of the tree: <VI, III, I, VIII, XVI>. Notice that we keep the item order in an OPSM, since it is the permutation of expression values with column No. And add a leaf node (1, 2, 5). For the second OPSM, since its item list <VI, III, I, II, XIII> shares a common prefix <VI, III, I> with the existing path <VI, III, I, VIII, XVI>. One new node (II) is created and linked as a child of (I), and another new node (XIII) is created and linked as the child of (II). The leaf node records row No. <3, 6, 9>. For the third OPSM, since its column No. list <VI, II, III> shares only the node (VI) with VI-prefix subtree, and a new node (II) is created and linked as a child of (VI), another node (III) is created and linked as a child of (II), and the leaf node records row No. <7, 10, 11>. The scan of the forth OPSM leads to the construction of the second branch of the tree <III, II, XVI>, and the leaf node records row No. <4, 8, 12>. For the last OPSM, since its item list <VI, III, I, VIII, XVI> is identical to the first one, it only adds the row No. <4, 6> in the leaf node (1, 2, 5). *pfTree* build method is introduced in lines 6-7 of *Algorithm 1*.

Through experiments, we know that the deleting and queries based on *pfTree* is not efficient, although it is fast to build index and insert index. To advance the performance, we give an optimization method *pIndex*, which uses two auxiliary data structures, column and row header tables, to facilitate tree traversal.

Algorithm 1. (PINDEX-BUILD) pIndex Construction

Input: OPSM or gene expression dataset D
Output: *pIndex*

```
1  treeRoot ← null;
2  while ((opsm ← D.nextLine()) ≠ null) do
3  │    nameList ← opsm.g; arrayInt ← opsm.c; curNode ← treeRoot;
4  │    for (it : arrayInt) do
5  │    │    linkFlag ← false;
6  │    │    if (false = currentNode.hasThisChild(it)) then
7  │    │    │    curNode.addChild(it); linkFlag ← true;
8  │    │    curNode ← curNode.getChild(it); curNode.frequencyIncrease(|g|);
9  │    │    if (null = columnHeadTable.get(it)) then
10 │    │    │    columnHeadTable.put(it, curNode); curNode.setBkLink(null);
11 │    │    else
12 │    │    │    itNode ← columnHeadTable.get(it);
13 │    │    │    while (itNode.getFwdLink() ≠ null) do
14 │    │    │    │    itNode ← itNode.getFwdLink();
15 │    │    │    if (linkFlag) then
16 │    │    │    │    itNode.setFwdLink(curNode); curNode.setBkLink(itNode);
17 │    curNode.setFinal(true); curNode.setName(nameList);
18 │    for (name : nameList) do
19 │    │    if (null ≠ rowHeadTable.get(name)) then
20 │    │    │    nodeSet.addAll(rowHeadTable.get(name));
21 │    │    nodeSet.add(curNode); rowHeadTable.put(name, nodeSet);
22 return pIndex;
```

For *pIndex* deletion and queries on conditions c, a column header table is built in which the order is conducted based on the occurrences of items from left to right and from top to bottom, and each item points to its occurrence in the tree via a column head of node-link. Nodes with the same column No. are linked in sequence via such bidirection node-links. After scanning all OPSMs, OPSM-tree with the column node-links is built, and column header table is shown in the left side of *Fig. 1*. The method is illustrated in lines 9-16 of *Algorithm 1*.

For *pIndex* deletion and queries on genes g, a row header table is built in which the order is conducted based on the occurrences of row Nos from left to right, and the tree nodes which have the same row No. will be saved in one hash set. For convenience of explanation, nodes with the same row No. are linked in sequence via a row head of node-link in *Fig. 1*. After scanning all OPSMs, the tree with the associated row header table is built, and row header table is shown in the bottom of *Fig. 1*. The method is plotted in lines 18-21 of *Algorithm 1*.

Algorithm 2. (PINDEX-DEL-ROW) pIndex Deletion by Rows

   Input: gene names g
   Output: *pIndex*
1 **for** $(name : g)$ **do**
2   nodeSet $\leftarrow$ rowHeadTable.get(name);
3   **for** $(node : nodeSet)$ **do**
4   **if** $(1 < |nodeSet|)$ **then** rowHeadTable.get(name).remove(node);
5   **else if** $(1 = |nodeSet|)$ **then** rowHeadTable.remove(name);
6   **else if** $(0 = |nodeSet|)$ **then** break;
7   **if** $(1 < |node.getName()|)$ **then** node.deleteName(name);
8   **else if** $(1 = |node.getName()|)$ **then** delNodesAndCols(node);

9 **return** *pIndex*;

Algorithm 3. (DEL-NODECOL) Delete Nodes And Columns

   Input: nodes *nodes*
   Output: *pIndex*
1 **while** $(treeRoot \neq node)$ **do**
2   node.frequencyDecrease();
3   node.getParent().getChildren().remove(node.getItem());
4   tmp $\leftarrow$ node; node $\leftarrow$ node.getParent(); tmp $\leftarrow$ setParent(null);
5   **if** $(null \neq tmp.getBkLink()$ & $null \neq tmp.getFwLink())$ **then**
6   new $\leftarrow$ tmp.getFwdLink(); new.setBkLink(tmp.getBkLink());
7   tmp.getBkLink().setFwLink(new);
8   **else if** $(null \neq tmp.getBkLink()$ & $null = tmp.getFwLink())$ **then**
9   tmp.getBkLink().setFwLink(null); tmp.setBkLink(null);
10   **else if** $(null = tmp.getBkLink()$ & $null = tmp.getFwLink())$ **then**
11   columnHeadTable.remove(tmp.getItem());
12   **else if** $(null = tmp.getBkLink()$ & $null \neq tmp.getFwLink())$ **then**
13   columnHeadTable.remove(tmp.getItem());
14   tmp.getFwLink().setBkLink(null);
15   columnHeadTable.put(tmp.getItem(), tmp.getFwLink());
16   **if** $(1 < node.getFrequency)$ **then** break;

17 **return** *pIndex*;

To facilitate *pIndex* update, we give *pIndex* insertion and deletion methods. Due to *pIndex* insertion is similar with *pIndex* construction, we only introduce *pIndex* deletion method, which consists of two way, *by-row* and *by-column*.

For *pIndex deletion by rows*, we give the method in *Algorithm 2*. First, it gets the keywords of rows (gene names). For each name, it consists of three substeps operations, i.e., *deleting from row header table*, *deleting tree nodes*, and *deleting from column header table.* For the first step, it fetches leaf nodes from row header table (lines 1-2). Then, for the leaf nodes, it first checks the number. If it is 0, it ends the processing (line 6). If it is 1, it removes the key-value record of the name (line 5). If it is more than 1, it only removes the value (node) from the key-value record (line 4). For the other steps, it invokes *Algorithm 3*.

Algorithm 4. (PINDEX-DEL-COL) pIndex Deletion by Columns

Input: columns c
Output: $pIndex$

1 key $\leftarrow$ c.get($|c|$ - 1); keyNode $\leftarrow$ columnHeadTable.get(key);
2 **while** ($null \neq keyNode$) **do**
3   itNode $\leftarrow$ keyNode.getParent(); count $\leftarrow$ $|c|$ - 2;
4   **while** ($0 \leq count$) **do**
5   **if** ($c.get(count) = itNode.getItem()$) **then**   count $\leftarrow$ count - 1;
6   **if** ($treeRoot = itNode$) **then**   break;
7   **else**   itNode $\leftarrow$ itNode.getParent();
8   **if** ($0 \leq count$) **then**   keyNodes.clear(); keyNodes.add(keyNode);
9   keyNode $\leftarrow$ keyNode.getFwdLink();
10   **for** ($node : keyNodes$) **do**   leafNodes $\leftarrow$ findLeafNodes(node);
11   **for** ($node : leafNodes$) **do**
12   **for** ($name : node.getName()$) **do**
13   **if** ($null = rowHeadTable.get(name)$) **then**   break;
14   **else if** ($1 = |rowHeadTable.get(name)|$) **then**
15   rowHeadTable.remove(name);
16   **else if** ($1 < |rowHeadTable.get(name)|$) **then**
17   rowHeadTable.get(name).remove(node);
18   delNodesAndCols(node); (*Algorithm 3*)
19 **return** $pIndex$;

Now, we give the detailed information about *Algorithm 3*. First, it checks whether the node is the root of the tree. If it is, it ends (line 1). Otherwise, it reduces the frequency of the node by one (line 2). Due to it is the leaf node, and shared by only one row, we first delete it from the child set of its parent, then we set its parent is null (lines 3-4). There are four cases to delete the current node from column header table. (1) If it links forward and back nodes, we make the forward node and back node to point each other (lines 5-7). (2) If it only has back node, we set the forward link of back node and back link of current node be null (lines 8-9). (3) If it does not have both forward and back nodes, we remove the key-value of current node from the column header table (lines 10-11). (4) If it only has the forward node, we set the back node of forward node be null, and put the forward node into the column header table instead of current node (lines 12-15). Finally, if the frequency of the parent of current node is more than 1, we end the processing (line 16).

To delete *pIndex* by columns, we introduce *Algorithm 4*. Due to it uses the *bottom-up* way to trace the tree, we first get the last keyword (column No.) and the head node in the column header table (line 1). Based on the node-links having the same item (column No.), if the branch contains the reversed keywords, we fetch the bottom source node in the node-link. If not, it goes to the next node on the node-link (lines 3-9). After it gets all the bottom source nodes, we get the leaf nodes of each nodes (line 10). Further, we delete the branches and the row node-links in the *bottom-up* way (lines 11-17). And it invokes *Algorithm 3* to delete related column node-links (line 18). We use a recursive way to find the leaf nodes, which is illustrated in *Algorithm 5*.

Algorithm 5. Find Leaf Nodes

Input: tree node *node*
Output: leaf nodes *leafNodes*

1 **if** $(node.isLeaf())$ **then** *leafNodes*.add(node);
2 **if** $(node.hasChild())$ **then**
3 $\quad$ children $\leftarrow$ node.getChildren();
4 $\quad$ **while** $(child : children)$ **do** findLeafNodes(child);
5 **return** *leafNodes*;

---

4   OPSM Queries

In the section, we explore OPSM queries on *pIndex* with two header tables.

Algorithm 6. (EQ_g) Exact Query on Genes

Input: Gene name keywords g, Length threshold τ
Output: HashMap<g, set of conditions> *result*

1 **for** $(name : g)$ **do**
2 $\quad$ **if** $(null = rowHeadTable.get(name))$ **then return** *null*;
3 $\quad$ nodeList $\leftarrow$ rowHeadTable.get(name);
4 $\quad$ **for** $(node : nodeList)$ **do**
5 $\quad\quad$ colList $\leftarrow$ branch of node;
6 $\quad\quad$ **if** $(colList.length \geq \tau)$ **then** fstLists.add(colList);
7 $\quad$ hashMap.put(name, fstLists); fstLists.clear();
8 **if** $(1 = |g|)$ **then return** *hashMap*;
9 fstLists $\leftarrow$ hashMap.get(g_0); flag $\leftarrow$ false;
10 **for** $(i \leftarrow 1\ to\ |g| - 1)$ **do**
11 $\quad$ **if** $(flag)$ **then** fstLists $\leftarrow$ resLists; resLists.clear();
12 $\quad$ flag $\leftarrow$ true; secLists $\leftarrow$ hashMap.get(g_i);
13 $\quad$ **for** $(out : fstLists, in : secLists)$ **do**
14 $\quad\quad$ lcs $\leftarrow$ findLongestCommonSubsequence(out, in);
15 $\quad\quad$ **if** $(lcs.length \geq \tau)$ **then** out $\leftarrow$ lcs; resLists.add(lcs);
16 **if** $(resLists.size() > 0)$ **then** result.add(g, resLists);
17 **return** *result*;

For exact queries on genes EQ_g, we first locate the row numbers, then trace back the branches of all leaves of row numbers, and further find the longest common subsequences ($LCSs$) of all branches, finally output the column numbers above length threshold τ in $LCSs$. The details are illustrated in *Algorithm 6*.

Example 2. (Exact Query on Genes EQ_g) We use Table 2 and Fig. 1 as example. Given some gene keywords 2, 3, 9, and length threshold 3, search the subset of conditions contained by these genes. First, find the heads of node-links of 2, 3, 9 from the row header table, then get the branch $<VI,III,I,VIII,XVI>$ that contain gene 2 and the branch $<VI,III,I,II,VIII>$ that contain gene 3 and 9, further find the LCS of the two branches above, and the result is $<VI,III,I,VIII>$ which is above length threshold 3.

From *Example 2*, we get the pruning rule of query result, *Rule 1*.
Rule 1. (Keyword No. based Pruning) For the keywords of column and row in exact queries, the branches should contain all the elements of c in query based on conditions, and the leaves should contain all the elements of g in query based on genes. If not, prune the branch or leaf.

For exact queries on conditions EQ_c, we first locate column keywords using column header table, from located node to tree root. If keywords are in the same branches, we test the orders. If the order is the same as input order, we return the genes above size threshold δ as the result. Otherwise, we test until there are no column keywords to locate. The details are illustrated in *Algorithm 7*.

Algorithm 7. (EQ_c) Exact Query on Conditions

Input: Condition keywords c, Size threshold δ
Output: HashMap$<c$, set of genes$>$ *result*

```
1  key ← c.get(|c| - 1); keyNode ← columnHeadTable.get(key);
2  if (null = keyNode) then return null;
3  while (null ≠ keyNode) do
4      itNode ← keyNode.getParent(); count ← |c| - 2;
5      while (count ≥ 0) do
6          if (c.get(count) = itNode.getItem()) then count ← count - 1;
7          if (itNode = treeRoot) then break;
8          else itNode ← itNode.getParent();
9      if (count < 0) then nodes.add(keyNode);
10     keyNode ← keyNode.getFwdLink();
11 for (inNode : nodes) do nameSet ← getNamesInLeaves(inNode);
12 if (nameSet.size() ≥ δ) then result.put(c, nameSet);
13 return result;
```

Example 3. (Exact Query on Conditions EQ_c) We use Table 2 and Fig. 1 as example. Given some condition keywords VI, III, I, and size threshold 3, search the subset of genes contained by these conditions. First, find the heads of node-links of I, III, VI (reverse order of input keywords) from column header table, then get the branch $<VI,III,I,VIII,XVI>$ and the branch $<VI,III,I,II,VIII>$ that *contain I, then check whether the two branches contain III, VI and keep the order of I, III, VI, and they meet the conditions above, further, get the gene set $<1,2,4,5,6,3,6,9>$ which is above size threshold 3.*

From *Example 3*, we get the pruning rule of query result, *Rule 2*.
Rule 2. (Order based Pruning) For the keywords of column subset c in Exact Queries, the branches contain all the elements in c should also keep the order of elements in c. If not, prune the branch.

Algorithm 8. (FQ_g) Fuzzy Query on Genes

Input: Gene name keywords g, Length threshold τ, Size threshold δ
Output: HashMap<subset of g, set of conditions> *result*
1 **for** $(i \leftarrow \delta \ to \ |g|)$ **do** querySetLists $\leftarrow$ combination of i items in g;
2 **for** $(querySet : querySetLists)$ **do** result $\leftarrow EQ_g$(querySet, τ);
3 **return** *result*;

Algorithm 9. (FQ_c) Fuzzy Query on Conditions

Input: Condition keywords c, Size threshold δ, Length threshold τ
Output: HashMap<subset of c, set of genes> *result*
1 **for** $(i \leftarrow 0 \ to \ |c| - 1)$ **do**
2   key $\leftarrow$ columnHeadTable.get(c_i); **if** $(null = key)$ **then** **return** *null*;
3   **while** $(null \neq key)$ **do**
4   list.add(c_i); node $\leftarrow$ key.getParent(); no $\leftarrow |c|$ - 2;
5   **while** $(no \geq (|c| - \tau))$ **do**
6   **for** $(it : c)$ **do**
7   **if** $(it = node.getItem())$ **then** no- -; list.add(it); break;
8   **if** $(node = treeRoot)$ **then** break;
9   **else** node $\leftarrow$ node.getParent();
10   **if** $(no < (|c| - \tau))$ **then**
11   **while** $(node \neq treeRoot)$ **do**
12   **for** $(it : c)$ **do**
13   **if** $(it = node.getItem())$ **then** no- -; list.add(it); break;
14   node $\leftarrow$ node.getParent();
15   nodes.add(key);
16   key $\leftarrow$ key.getFwdLink();
17   **for** $(inNode : nodes)$ **do** nameSet $\leftarrow$ getNamesInLeaves(inNode);
18   **if** $(result.hasKey(list))$ **then** nameSet.addAll(result.get(list));
19   result.put(list, nameSet); nameSet.clear(); list.clear();
20   **for** $(res : result)$ **do**
21   **if** $(|res.value| < \delta)$ **then** result.remove(res);
22 **return** *result*;

For fuzzy queries on genes FQ_g, we first compute the combination of row keywords above size threshold δ, then locate the row numbers in each combination, further fetch the branches of row numbers and find the longest common subsequences above length threshold τ as the results, which is the same as exact queries on genes EQ_g. The details are illustrated in *Algorithm 8*.

Example 4. (Fuzzy Query on genes) We use Table 2 and Fig. 1 as example. Given some gene keywords 2, 3, 9, size threshold 2, and length threshold 3, search the subset of conditions contained by these genes. First, we compute the combinations of gene keywords above size threshold 2, and get the combination <2,3>, <2,9>, <3,9> and <2,3,9>. Then, we fetch the branch < VI,III,I,VIII,XVI> that contain gene 2 and the branch < VI,III,I,II,VIII> that contain gene 3 and 9, further find the longest common subsequences of combinations above, and get results of <2,3>, <2,9> and <2,3,9> are < VI,III,I,VIII>, and result of <3,9> is < VI,III,I,II,VIII>. And all the results are above length threshold 3.

For fuzzy queries on conditions FQ_c, we first rotate the first element of keywords. Then, locate column keywords with column header table, from located node to tree root. If the number of keywords in the same branches is above length threshold τ, we get gene names in the leaves. And we test whether the number of gene names above size threshold δ, if it is true, add the keyword set as key and gene name set as value into final results. Otherwise, we test until each keyword as first element one time (FIT). The details are plotted in *Algorithm 9*.

Example 5. (Fuzzy Query on Conditions) We use Table 2 and Fig. 1 as example. Given some condition keywords VI, III, I, length threshold of keywords 2, and size threshold of results 3, search the subset of genes contained by these conditions. First, use <I> as the first element, and get branch <VI,III,I> which is above length threshold 2, then we fetch the gene name set <1,2,5,4,6,3,9> which is above size threshold 3. Further use <III> as the first element, and get branch <VI,III> and < VI,II,III> which are above length threshold 2, then we fetch the gene name set <1,2,5,4,6,3,9,7,10,11>. When using <VI> as the first element, there are no results. Now, we return keyword sets and gene sets above as results.

5   Experimental Evaluation

In this section, we report our experiments that validate the effectiveness and efficiency of *pIndex*. Due to this is the first work of OPSM query, we only compare *pIndex* with *pfTree*, which is a naive approach proposed in *Section 3* and does not use any auxiliary structure. Experiments demonstrate that:

- The index size of both methods, *pIndex* and *pfTree*, is the same, and the compact ratio is close to 0.98 when the number of conditions is smaller.
- On single machine, although *pfTree* performs well on index build and index insert, *pIndex* outperforms *pfTree* by 1 to 2 orders of magnitude in various cases on index delete, EQ_g, EQ_c, FQ_g and FQ_c.

- *pIndex* is also implemented on Hadoop and our modified Hama platform [12], and it has better scalability.

We use two kinds of datasets in our experiments: real datasets [4] and a series of synthetic datasets. Most of our experiments have been performed on the real datasets since it is the source of real demand. All our experiments are performed on 1.87GHz, 16GB memory, Inspur servers running Ubuntu 12.04. Both *pIndex* and *pfTree* are implemented in Java and complied with Eclipse 4.3. And the versions of Hadoop and Hama are 0.20.2 and 0.4.0, respectively. Limited by space, we only report the results on real datasets here.

Table 3. Details of the Gene Expression Datasets

Dataset	File Name	Rows	Columns	Dataset	File Name	Rows	Columns
D_1	adenoma	12488	6	D_4	krasla	12422	50
D_2	a549	22283	11	D_5	bostonlungstatus	12625	94
D_3	5q_GCT_file	22278	24	D_6	bostonlungsubclasses	12625	202

5.1   Evaluation on Single Machine

We first test the index size of *pfTree* and *pIndex*. As mentioned before, both *pfTree* and *pIndex* use prefix tree to index, although *pIndex* also utilizes header tables, it only occupies small memory space. Thus, the index size of both methods are nearly the same. Fig. 2.(a) depicts the compact ratio of index with number of rows varied from $1k$ to $12k$ on 4 different columns (6, 11, 24 and 50). The curves clearly show that the smaller columns, the larger compact ratio. They also illustrate a salient property: the compact ratio is stable with different rows. When row number of dataset increases, compact ratio does not change much.

Having verified the index size of *pfTree* and *pIndex*, we now check their behaviours. Fig. 2.(b) and (c) presents the index construction time of two methods on varying rows and columns, respectively. As shown in the figures, *pfTree* outperforms *pIndex* in every row and column case. The reason is that *pIndex* spends additional time on building header tables. Although performance of *pIndex* on index build is worse than that of *pfTree*, *pIndex* does better on other tests. Similar with index build, when inserting 10 to 2000 rows and 202 fixed columns ($|C|$=202), and inserting 6 to 202 columns and 100 fixed rows, performance of *pfTree* is also better than that of *pIndex*, which is shown in Fig. 2.(d) and (e).

The scalability of performance on index delete is presented in Fig. 2.(f) and (g). We select $10k$ rows of D_6 as the test dataset. If there are no specific notifications, row and column numbers of dataset to build index are $|G|$=10k and $|C|$=202. First, we test run time of deleting five set genes. As seen from Fig. 2.(f), the run time of *pfTree* increases from $754ms$ to $2449ms$, while that of *pIndex* increases from $20ms$ to $256ms$. The increasing trend of *pIndex* is much slower than that of *pfTree*. Similarly in Fig. 2.(g), when we delete five set columns, the run time of *pfTree* decreases from $1154ms$ to $1007ms$, while that of *pIndex* decreases from $535ms$ to $99ms$. And the decreasing trend of *pIndex* is also faster.

Next, we check the scalability on EQ_g and FQ_g. When we vary row keywords from 2 to 6, on dataset D_6, as shown in Fig. 2.(h), run times of both methods are nearly two horizontal lines, but run time of *pfTree* is above 30 times larger than that of *pIndex*. When we test on six different datasets, as shown in Fig. 2.(i), run time of *pfTree* is nearly 35 times larger than that of *pIndex*. Fig. 2.(j) and (k) illustrate the performance of FQ_g on different row keywords and datasets. The run times of *pfTree* increases dramatically, while that of *pIndex* is nearly a horizontal line. Overall, *pIndex* outperforms *pfTree* by 70 to 360 times and by 8 to 130 times in two cases. The tests demonstrate the efficiency of row header table and the scalability of *pIndex* on EQ_g and FQ_g.

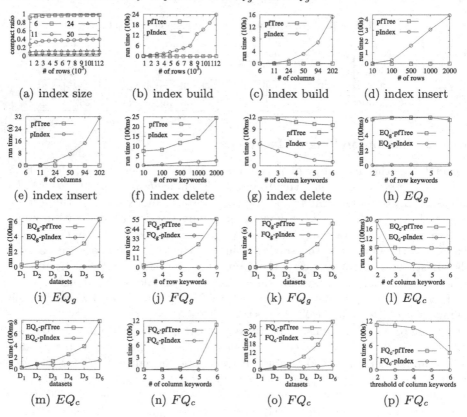

(a) index size   (b) index build   (c) index build   (d) index insert

(e) index insert   (f) index delete   (g) index delete   (h) EQ_g

(i) EQ_g (j) FQ_g (k) FQ_g (l) EQ_c

(m) EQ_c (n) FQ_c (o) FQ_c (p) FQ_c

Fig. 2. Performance Evaluation on a Single Machine

Last, we give the tests on EQ_c and FQ_c. As shown in Fig. 2.(l), when we vary column keywords form 2 to 6, on dataset D_6, run times of *pIndex* outperforms *pfTree* nearly in every case, except 2 column keywords. The underlying reason is that *pfTree* goes through less nodes in the tree when input smaller keywords. When we test on six different datasets, as shown in Fig. 2.(m), *pIndex* outperforms *pfTree* by 1 to 9 times. Now, we test the scalability on FQ_c. When we vary the column keywords form 2 to 6, on dataset D_6, as shown in Fig. 2.(n), run time of *pfTree* increases dramatically, while that of *pIndex* is nearly a horizontal line. When we test on six different datasets, as shown in Fig. 2.(o), the run time of *pfTree* increases from $120ms$ to $33589ms$, while that of *pIndex* increases from $176ms$ to $3265ms$. The increasing trend of *pIndex* is much slower than that of *pfTree*. When we vary the threshold of column keywords, and query 6 column keywords on dataset D_6, as shown in Fig. 2.(p), *pIndex* outperforms *pfTree* by 461 to 759 times. The tests demonstrate the efficiency of column header table and *FIT* method, and the scalability of *pIndex* on EQ_c and FQ_c.

5.2   Evaluation on Single Machine, Hadoop and Hama

In the following experiments, we use *pIndex* as example to compare its performance on SM, Hadoop and Hama (2 nodes). Since we have had detail tests of *pIndex* on SM, we only give its behaviours on index build, EQ_c and FQ_c.

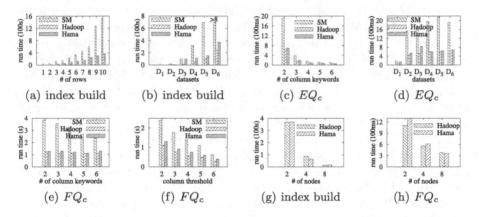

(a) index build   (b) index build   (c) EQ_c (d) EQ_c

(e) FQ_c (f) FQ_c (g) index build   (h) FQ_c

Fig. 3. Performance Evaluation on SM, Hadoop and Hama

Fig. 3.(a) and (b) present the index build time of *pIndex* on varying rows and columns, respectively. As shown in the figures, Hadoop and Hama outperform SM by 2 to 6 times in each row and column case, and Hadoop has much better behaviours than Hama obviously in Fig. 3.(b). The reason is the skew in data partition on Hama, while Hadoop can use small files to solve the problem.

The EQ_c performance on different column keywords and datasets is shown in Fig. 3.(c) and (d). When we vary the column keywords from 2 to 6, on dataset D_6, in Fig. 3.(c), run times on Hadoop and Hama (2 nodes) outperform by more than 2 times that on SM nearly in each case. When we test on six different datasets, in Fig. 2.(d), Hadoop and Hama also outperform SM by 2 to 3 times.

The scalability of FQ_c on three platforms is presented in Fig. 3.(e) and (f). When we vary the column keywords from 2 to 6, on dataset D_6, as shown in Fig. 2.(e), Hadoop and Hama have nearly the same behaviour, and both platform show 2 to 3 times better performance than SM. When we vary the threshold of column keywords, and query 6 column keywords on dataset D_6, as shown in Fig. 2.(f), similarly with the former one, Hadoop and Hama have nearly the same behaviour, and both platform show 2 to 3 times better performance than SM.

Last, we test the scalability of *pIndex* on different number of nodes. In Fig. 3.(g), when varying node number, index build time on 4(8) nodes improves more than 4 times on that on 2(4) nodes, and Hadoop and Hama give nearly the same behaviour. Similarly, in Fig. 3.(h), run time of FQ_c based on 6 column keywords on 4(8) nodes improves more than 2 times on that on 2(4) nodes. Through the experiments, we know *pIndex* has good scalability. Although *pIndex* build spends more time than *pfTree*, we can use parallel platform to address.

6 Related Work

In this section, we review previous work on OPSM mining, since OPSM queries are based on OPSM mining. Besides the mining techniques, we also review the work of OPSM queries, since it is the core work of this paper.

The earliest work on pattern mining [20] is the cluster model, which is the general case of OPSM, proposed in [5]. The model simultaneously clusters both genes and conditions, and overcomes some problems of existing cluster methods. The concept of OPSM is first proposed by Ben-Dor et al. in [2]. Since it is NP-hard, they propose a probabilistic model to mine an OPSM from a random matrix, but it cannot Guarantee finding all OPSMs. To address the problem, Liu et al. [15] introduce a deterministic algorithm. They develop an auxiliary data structure called *OPC-Tree*, which can search the full order space and thus can find all OPSMs. Gao et al. [10] observe biologists are in particular interest to reveal small groups of genes, thus they propose the *KiWi* framework which substantially reduces the search space and problem scale. However due to the noisy nature of real data, existing methods fail to discover some significant ones, Zhang et al. [21] present a noise-tolerant model called approximate order preserving cluster (*AOPC*). Fang et al. [7] propose a relaxed OPSM model called *ROPSM*, and employ *OPSM-Growth* method that includes column-centric and row-centric strategies to mine *ROPSM* patterns. Later, they give a new relaxed OPSM model in [8] by consider the linearity relaxation, which is called Bucket OPSM (*BOPSM*) model.

For OPSM query problem, there are few work on this topic. The most similar works with the topic of OPSM query are presented in these works [15,11], the exact topic is OPSM mining, but Liu et al. [15] introduce an auxiliary data structure which makes the volume of OPSMs smaller, and Jiang et al. [11] present an interactive OPSM mining method, which can drill down and roll up. And it facilitates the OPSM search from the massive results. To the best our knowledge, our work is the first one to design one tool of OPSM query for biologists, and inherits the advantages of the previous work.

7 Conclusions

This paper presents an index associated two header tables, called *pIndex*, for *Order-Preserving Submatrix* query, and *pIndex* shows good behaviours on exact search, fuzzy query, and index deletion. Although it does not work so well on index build and insert, experimental results suggest that we can use distributed parallel platform to address. For the future work, we will explore online sharing the query results for follow-up OPSM queries, rather than find the query results on index from scratch each time.

References

1. Barkow, S., Bleuler, S., Prelić, A., Zimmermann, P., Zitzler, E.: Bicat: a biclustering analysis toolbox. Bioinformatics **22**(10), 1282–1283 (2006)
2. Ben-Dor, A., Chor, B., et al.: Discovering local structure in gene expression data: the order-preserving submatrix problem. In: RECOMB, pp. 49–57 (2002)
3. Boyer, R.S., Moore, J.S.: A fast string searching algorithm. Communications of the ACM **20**(10), 762–772 (1977)
4. BroadInstitute: Datasets.rar and 5q_gct_file.gct. http://www.broadinstitute.org/cgi-bin/cancer/datasets.cgi
5. Cheng, Y., Church, G.M.: Biclustering of expression data. In: Bourne, P.E., Gribskov, M., et al. (eds.) ISMB, pp. 93–103. AAAI (2000)
6. Chui, C.K., Kao, B., et al.: Mining order-preserving submatrices from data with repeated measurements. In: ICDM, pp. 133–142. IEEE Computer Society (2008)
7. Fang, Q., Ng, W., Feng, J.: Discovering significant relaxed order-preserving submatrices. In: KDD, pp. 433–442. ACM (2010)
8. Fang, Q., Ng, W., Feng, J., Li, Y.: Mining bucket order-preserving submatrices in gene expression data. IEEE Trans. Knowl. Data Eng. **24**(12), 2218–2231 (2012)
9. Fang, Q., Ng, W., Feng, J., Li, Y.: Mining order-preserving submatrices from probabilistic matrices. ACM Transactions on Database Systems (TODS) **39**(1), 6 (2014)
10. Gao, B.J., Griffith, O.L., Ester, M., Jones, S.J.M.: Discovering significant opsm subspace clusters in massive gene expression data. In: Eliassi-Rad, T., Ungar, L.H., Craven, M., Gunopulos, D. (eds.) KDD, pp. 922–928. ACM (2006)
11. Jiang, D., Pei, J., Zhang, A.: Gpx: interactive mining of gene expression data. In: Nascimento, M.A., Özsu, M.T., Kossmann, D., Miller, R.J., Blakeley, J.A., Schiefer, K.B. (eds.) VLDB, pp. 1249–1252. Morgan Kaufmann (2004)
12. Jiang, T., Li, Z., Chen, Q., Wang, Z., Pan, W., Wang, Z.: Parallel partitioning and mining gene expression data with butterfly network. In: Decker, H., Lhotská, L., Link, S., Basl, J., Tjoa, A.M. (eds.) DEXA 2013, Part I. LNCS, vol. 8055, pp. 129–144. Springer, Heidelberg (2013)
13. Knuth, D.E., Morris Jr., J.H., Pratt, V.R.: Fast pattern matching in strings. SIAM Journal on Computing **6**(2), 323–350 (1977)
14. KNUTTn, D.: The art of computer programming, vol 3: Sorting and searching (1973)
15. Liu, J., Wang, W.: Op-cluster: Clustering by tendency in high dimensional space. In: ICDM. pp. 187–194. IEEE Computer Society (2003)
16. McCreight, E.M.: A space-economical suffix tree construction algorithm. Journal of the ACM (JACM) **23**(2), 262–272 (1976)
17. Trapp, A.C., Prokopyev, O.A.: Solving the order-preserving submatrix problem via integer programming. INFORMS Journal on Computing **22**(3), 387–400 (2010)
18. Ukkonen, E.: On-line construction of suffix trees. Algorithmica **14**(3), 249–260 (1995)

19. Weiner, P.: Linear pattern matching algorithms. In: IEEE Conference Record of 14th Annual Symposium on Switching and Automata Theory, 1973. SWAT2008, pp. 1–11. IEEE (1973)

20. Yang, J., Wang, W., Wang, H., Yu, P.S.: δ-clusters: capturing subspace correlation in a large data set. In: Agrawal, R., Dittrich, K.R. (eds.) ICDE, pp. 517–528. IEEE Computer Society (2002)

21. Zhang, M., Wang, W., Liu, J.: Mining approximate order preserving clusters in the presence of noise. In: Alonso, G., Blakeley, J.A., Chen, A.L.P. (eds.) ICDE, pp. 160–168. IEEE (2008)

Database Storage and Index II

Large-Scale Multi-party Counting
Set Intersection Using a Space
Efficient Global Synopsis

Dimitrios Karapiperis[1]([✉]), Dinusha Vatsalan[2],
Vassilios S. Verykios[1], and Peter Christen[2]

[1] School of Science and Technology, Hellenic Open University, Patras, Greece
{dkarapiperis,verykios}@eap.gr
[2] Research School of Computer Science, The Australian National University,
Canberra, ACT 0200, Australia
{dinusha.vatsalan,peter.christen}@anu.edu.au

Abstract. Privacy-preserving set intersection (PPSI) of very large data
sets is increasingly being required in many real application areas including
health-care, national security, and law enforcement. Various techniques
have been developed to address this problem, where the majority of them
rely on computationally expensive cryptographic techniques. Moreover,
conventional data structures cannot be used efficiently for providing count
estimates of the elements of the intersection of very large data sets. We
consider the problem of efficient PPSI by integrating sets from multiple
(three or more) sources in order to create a global synopsis which is the
result of the intersection of efficient data structures, known as Count-Min
sketches. This global synopsis furthermore provides count estimates of
the intersected elements. We propose two protocols for the creation of
this global synopsis which are based on homomorphic computations, a
secure distributed summation scheme, and a symmetric noise addition
technique. Experiments conducted on large synthetic and real data sets
show the efficiency and accuracy of our protocols, while at the same time
privacy under the Honest-but-Curious model is preserved.

1 Introduction

Computing set operations, such as intersection, union, equi-join and disjointness, efficiently and privately among different parties is an important task in
privacy-preserving data mining [3,9]. In this paper, we study the problem of
privacy-preserving set intersection (PPSI), which is also known as private data
matching [15], of multi-sets of an arbitrary large number of distinct elements
held by three or more parties. Growing privacy concerns and government laws
preclude the exchange of sensitive and private values stored in databases across
different organizations for calculating the intersection of those private values.

This research was partially funded by the Australian Research Council under Discovery Project DP130101801.

M. Renz et al. (Eds.): DASFAA 2015, Part II, LNCS 9050, pp. 329–345, 2015.
DOI: 10.1007/978-3-319-18123-3_20

This has led to an active research area, known as PPSI, in the field of privacy-preserving data mining [3] and specifically in private data matching [19,35].

PPSI is useful in many real-world applications, ranging from health-care, crime detection, national security, to finance and business. An example motivating application would be a health surveillance system, where by monitoring drug consumption at pharmacies or hospitals located at different places, alerts could be issued whenever consumption of certain drugs exceeds a threshold at all or some of these hospitals. A crime detection or national security application could be the monitoring of the number of times certain on-line services are accessed, by applying the intersection operation on requests made to these on-line services from different Internet Service Providers (ISPs). These examples illustrate that often large sets of sensitive elements held by different parties (or organizations) need to be intersected so that a set of common elements, accompanied by their counts of occurrences, can be identified. However, privacy and confidentiality concerns, as well as other business regulations, commonly prevent the sharing and exchange of such private values across several parties.

There have been several solutions proposed in the literature addressing the problem of PPSI. Most of them are either based on Secure Multi-Party computation (SMC) [27] techniques that are computationally expensive and thus are not scalable to large sizes of sets and larger number of parties, or they only perform intersection of two sets (from two parties). Moreover, in the applications described above, we are not only interested in learning the set of intersection of elements but also their number of occurrences. This problem cannot be efficiently solved by conventional data structures, such as hash tables or vectors, due to the large number of distinct elements that need to be monitored.

In this paper, we propose the creation of a privacy-preserving global synopsis by integrating data from many sources and attaining the common elements. More specifically, we tackle the challenge of identifying the counts of these elements from a potentially very large multi-set as they occur. Each party independently summarizes its elements in a local synopsis, which is implemented by a Count-Min sketch [12], and then these local synopses are intersected in order to create the global synopsis. This global synopsis (a) provides collective count estimates for the common elements attained and (b) hides the contribution of each party to these estimates. We propose and evaluate two protocols for the creation of this global synopsis:

1. the first protocol, which relies on homomorphic operations, exhibits accurate and reliable results but adds high communication cost. The number of homomorphic operations has a logarithmic relation to the total number of occurrences of elements in the sets to be intersected.
2. the second protocol relies completely on simple secure computations, exhibiting high performance and simultaneously highly accurate results.

The remainder of this paper is structured as follows. We next review related work in Section 2. In Section 3, we formulate the problem to be addressed, and in Section 4 we describe the building blocks used for creating a privacy-preserving global synopsis. Then, we present our protocols for the creation of the global synopsis in detail in Section 5, while in Section 6 we empirically evaluate our protocols using both synthetic and real data sets. Finally, in Section 7 we summarize our work and discuss directions for future work.

2 Related Work

Various techniques have been developed addressing the PPSI problem over the past decades. Most of the solutions proposed so far rely on general SMC-based cryptographic techniques. General two-party secure computation was introduced by Yao [36] and extended to multi-parties by Goldreich et al. [18].

Agrawal et al. [4] developed two-party protocols based on SMC commutative encryption schemes for three set operations: intersection, intersection size, and equi-join. The protocols allow for information integration with minimal data sharing. However, they are expensive in terms of computation and communication complexity. Freedman et al. [15] proposed two-party PPSI protocols based on homomorphic encryption and balanced hashing for both the semi-honest and the malicious adversary models. In their work, the sets are represented as roots of polynomials. This work was extended by Kissner et al. [24], who utilize the power of polynomial representation of multi-sets for PPSI.

Hazay and Lindell [20] adopted a pseudo-random-function-based solution for the two-party PPSI problem, which can be used either for one malicious and one semi-honest party or for two covert parties [5]. Dachman-Soled et al. [13] addressed the problem of PPSI for two malicious parties using homomorphic encryption and polynomial functions. In addition, several approaches have been proposed on variants of the PPSI problem, such as privacy-preserving union [16], privacy-preserving equality test [30], or privacy-preserving disjointness [23]. These works also employ SMC-based privacy techniques, which makes the solutions not efficient and scalable to large sets held by multiple parties.

In order to overcome the drawback of high computational overhead with SMC-based techniques, privacy-preserving set operations, which rely on efficient privacy techniques such as Bloom filter-based encoding, have recently being investigated in the areas of privacy-preserving record linkage [35] and privacy-preserving data mining [3]. Lai et al. [32] proposed an efficient PPSI on multiple sets using Bloom filters. The parties distributively compute a conjuncted Bloom filter by applying the logical AND operation on their partitions and then each party checks its elements with this conjuncted Bloom filter in order to determine if they are in the intersection set. A similar approach using Counting Bloom filters was proposed by Many et al. [28]. Dong et al. [14] introduced an efficient PPSI protocol between two sets using Garbled Bloom filters (GBfs) and an oblivious transfer (OT) protocol. First, the participating parties encode their sets as GBFs and then run the OT protocol in order to obtain the intersection set. A private equi-join approach on multiple databases was presented by Kantarcioglu et al. [22], where a secure equi-join is applied on k-anonymized databases.

Roughan and Zhang [33] proposed an efficient private set union solution by using Count-Min sketches [12] in order to collect Internet-wide statistics.

Charikar et al. [8] proposed the Count-Sketch data structure, which was adjusted for self-join size estimation by Cormode and Garofalakis [11]. However, space requirements of Count sketches are far higher than those of Count-Min sketches [12], making them less suitable for large-scale applications.

3   Problem Formulation

Let us suppose a distributed environment, which consists of m parties, each denoted by p_i, where $i = 1, \ldots, m$, and a global authority G, which plays the role of a central public semi-trusted regulatory agency. Each p_i should monitor the number of occurrences of an element e_j (like an IP address, a drug, or the registration plate of a car), where $j = 1, \ldots, n$, and n is a large number possible in the tens or even hundreds of millions. By doing so, a local summary S is built by each p_i, which includes the total number of occurrences of each e_j for a certain time period, or for a specified total number of e_js monitored, denoted by N, where each distinct e_j might appear several times. Therefore, N is equal to $\sum_{j=1}^{n} V(e_j)$, where $V(\cdot)$ returns the exact number of occurrences of an e_j. An update operation is required when an e_j should be monitored by a p_i, which should increase the number of occurrences of this e_j in the local summary. A query operation is also needed, which should return the current number of occurrences of an e_j. Authority G, by exploiting the local summaries, should answer collective queries regarding the number of occurrences of each e_j above a specified threshold θ, as monitored globally by every p_i. In essence, these e_js constitute the intersection set $\mathcal{S}$, defined formally as $\mathcal{S} = \{e_j | e_j \in p_1 \wedge \ldots \wedge e_j \in p_m\}$.

Such a query could be "How many times ($> \theta$) has a certain web site been accessed by all p_is?". All summaries should be collected by G on a frequent basis, so that G can produce an almost real-time global summary. Moreover, each p_i should maintain its summary in main memory to allow fast updating, when an e_j is monitored. Therefore, the size of the data structure used to realize each summary should be as small as possible, although the number of the e_js can be large. The reluctance of some p_is to disseminate their summaries due to privacy concerns is an additional problem. For instance, a hospital might be a reluctant p_i not willing to share any medical information, such as the drugs consumed which are the e_js in this case, in order to protect the privacy of its patients.

The solution of building each summary, by utilizing a vector, where each position represents a distinct element and holds the number of its occurrences, is prohibitive due to the size of the summary which will grow linearly with the number of elements represented. Also, by using a vector, the size and the update time for each element is $\mathcal{O}(n)$. By utilizing a hash table, we can achieve $\mathcal{O}(1)$ update time but the size required remains the same as by using a vector. For this reason, each p_i creates a local synopsis, denoted by S_i, which is a specialized sketching data structure [12] that consumes a fixed amount of space in main memory regardless of the number of e_js represented, at the cost of an allowable configurable error. After all the S_is are created, a global synopsis, denoted by

GS, should be generated by performing the intersection operation among the S_is. The GS produced should provide collective count estimates for all the elements of the intersection set S. Privacy of each p_i should be protected, so that it is not possible to infer the contribution of each p_i to these collective count estimates. Formally, given a collective count estimate for an e_j ($e_j \in S$) one cannot make any inferences or estimates regarding each $V_{p_i}(e_j)$, where $V_{p_i}(\cdot)$ denotes the exact number of occurrences for an e_j at a certain party p_i.

4 Background

In this section, we give a brief outline of the building blocks utilized in order to create the privacy-preserving global synopsis.

4.1 Creating a Local Synopsis

An efficient way for creating a local synopsis is by using a Count-Min sketch [12] (sketch). The main feature of a sketch is that it utilizes space that is sublinear with the number of e_js represented by it. A sketch is an array that consists of D rows and W cells in each row, initialized to 0. Both D and W are specified later. In order to update an e_j in a sketch, D hash operations are performed by randomly chosen, pairwise independent hash functions of the form $h_d(e_j) = [(a_d e_j + b_d) \mod P] \mod W$, where $d = 1, \ldots, D$, P is a large prime number (e.g., $2^{31} - 1$) greater than n, and each a_d, b_d are randomly chosen integers from $(0, P)$. Thus, there are D hash results, each corresponding to a cell in each row, where its value is incremented by 1, namely $S[d][h_d(e_j)] = S[d][h_d(e_j)] + 1$, for each of the $d \in D$ rows. The query operation $query(S, e_j)$ returns the count estimate of an e_j by hashing this e_j using the same hash functions as in the update operation and then picking the minimum hash value.

An interesting property is the linearity of sketches; the sketch produced by adding cell-wise two sketches (both built by using the same hash functions) is the union of these two sketches. This property makes sketches particularly useful because collective count estimates can be provided in distributed environments. By using sketches, we can detect frequent elements, such as IPs flooding in a network or drugs consumption above a certain frequency of appearance, denoted by ϕ. These frequent elements, the so-called heavy hitters [12], may indicate a certain anomaly, which may require an immediate course of proactive actions. For example, over-consumption of certain drugs by all pharmacies state-wide may highlight an outbreak of an infectious disease. By setting $D = \lceil \ln(N/\delta) \rceil$ and $W = \lceil 1/\epsilon \rceil$ [12], heavy hitters which exhibit a number of occurrences more than a specified threshold[1] θ, are identified, where $\theta = \lceil \phi N \rceil$ with $0 < \phi < 1$. Simultaneously, any e_j occurring less than $\lceil (\phi - \epsilon)N \rceil$ times, where $\epsilon \ll \phi$ e.g., $\epsilon = \phi/10$, is ignored, with confidence $1 - \delta$.

---

[1] In the literature, the term threshold can also be found as *support*.

Algorithm 1. Secure distributed summation scheme.

Input: $x_1, \ldots, x_m, r, F$
Output: s ▷ the summation returned
  1: $q_1 \leftarrow (x_1 + r) \mod F$ ▷ p_1 produces q_1
  2: **for** $(i = 2, \ldots, m)$ **do**
  3:     $q_i \leftarrow (q_{i-1} + x_i) \mod F$ ▷ p_i produces q_i by using q_{i-1}
  4: **end for**
  5: $s \leftarrow (q_m - r) \mod F$ ▷ p_1 produces s which is the summation

---

4.2   Homomorphic Computations for Preserving Privacy

A reliable Secure Multi-Party (SMC) technique of performing a joint computation among several parties is the partially homomorphic Paillier cryptosystem [31]. A joint computation could be the addition of some values, where these values should remain secret due to privacy concerns. Successive encryption of the same value generates different cipher texts with high probability. A trusted authority is required in order to issue a public/private key pair, needed for the encryption and decryption operations respectively. Given two values (messages), x_1 and x_2, encryption is performed by using the public key and the produced cipher texts are denoted by $\widetilde{x}_1$ and $\widetilde{x}_2$ respectively. Given the cipher texts, we can perform either homomorphic addition $(\widetilde{x}_1 \oplus \widetilde{x}_2)$ or multiplication with a constant c $(c \odot \widetilde{x}_1)$. The cipher texts can be decrypted by the trusted authority by using its private key.

4.3   Secure Distributed Summation

A simple summation scheme, as introduced in [9], can be applied by the p_is in order to perform a joint summation. This scheme is much more efficient than the homomorphic approach described above. By using this scheme, each p_i masks the corresponding values in its S_i, such that if p_{i+1} obtains S_i it cannot reproduce the actual values of S_i. We illustrate the scheme by using a simple example. Each p_i has monitored e_1 x_i times. Then, the p_is should sum up jointly each x_i such that none of the x_is are disclosed to any p_i. First, p_1 generates a random number r, which lies in the interval $(0, F)$, where F is a large integer greater than the summation calculated. The steps of the scheme are illustrated in Algorithm 1. The number of parties m must be more than two, because otherwise x_2 will be revealed to p_1.

5   Protocols for Creating a Privacy-Preserving Intersection Global Synopsis

In the simple non privacy-preserving scenario, G performs the intersection operation, among the S_is, following a naive protocol where each p_i submits to G its corresponding S_i and then G merges them. This approach overwhelms G with the S_is, imposing high computational overhead and network traffic. Additionally, since the hash functions are common for both updating an S_i and querying GS, G can easily perform an iterative query process to an S_i, for each possible e_j, and consequently find out sensitive information regarding a single p_i. In order to mitigate these concerns, the GS can also be generated by the p_is jointly, without the participation of G, as we will describe later.

The basic idea of the proposed protocols is the secure generation of the global synopsis GS', which holds the common e_js without the count estimates. More specifically, if a cell in a S_i contains a value which is below a specified threshold θ, then the GS' should indicate this and the corresponding cell in the GS should be set to 0. Thus, if we query GS for an e_j which is hashed to a cell of a row in GS holding 0, this will result in returning 0 as the global estimate. This happens because during the query operation of the GS, we choose the minimum value of the D estimates. By doing so, we can use GS' to easily exclude those cells in GS which correspond to elements not included in the intersection set.

The communication complexity of both protocols is linear in the number of parties, namely $\mathcal{O}(m)$. We make the assumption that the participating parties (p_is and G) follow the Honest-but-Curious (HBC) model [19,27], in that they follow the protocol steps while being curious to learn about other party's data. Furthermore, we assume that there is no collusion among them. Our protocols are secure in that the contribution of each p_i to the count estimates is hidden from both the rest of the p_is and from G. We utilize the secure distributed summation scheme combined with homomorphic operations in order to both deliver accurate results and to protect the privacy of the p_is in an efficient manner.

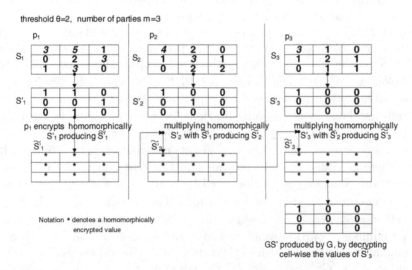

Fig. 1. Each $\widetilde{S}'_{i-1}$ is multiplied cell-wise by each S'_i. In the end, G decrypts $\widetilde{S}'_m$ and attains the common elements ($i = 2, \ldots, m$).

Table 1. The steps of the homomorphic protocol for creating a privacy-preserving intersection global synopsis

step 1	G issues the pair of public/private keys needed for the homomorphic operations and sends out the public key to each p_i.
step 2	Each p_i by using its S_i, produces S_i', by replacing the values above θ with 1 and those below θ with 0.
step 3a	p_1 encrypts homomorphically S_1', producing $\widetilde{S}_1'$, which is submitted to p_2.
step 3b	p_1 produces R (one random value per cell) and then Q_1, which are $D \times W$ arrays and correspond to r and q_1 respectively depicted in Algorithm 1, needed for the secure cell-wise summation of each S_i. Q_1 is submitted to p_2.
step 4a	Each p_i $(i = 2, \ldots, m)$ performs cell-wise homomorphic multiplications between $\widetilde{S}_{i-1}'$ and S_i', producing $\widetilde{S}_i'$, which is submitted to p_{i+1}.
step 4b	Each p_i $(i = 2, \ldots, m)$, by using Q_{i-1}, produces Q_i $(Q_i[d][w] = (Q_{i-1}[d][w] + S_i[d][w]) \mod F)$, which is submitted to p_{i+1}.
step 5a	p_m produces $\widetilde{S}_m'$ and then submits it to G.
step 5b	p_m produces Q_m and then submits it to p_1.
step 6	p_1 produces GU, by subtracting R from Q_m, which is submitted to G.
step 7a	G decrypts $\widetilde{S}_m'$ in order to produce GS'.
step 7b	G multiplies cell-wise GU with GS' in order to obtain GS.

5.1   The Homomorphic Protocol

The homomorphic protocol (HP) is two-fold; it identifies the common e_js and calculates the union of the S_is, by exploiting the linearity of sketches as explained in Sect. 4.1. Any cells which contain values above θ in all S_is are securely identified by performing homomorphic operations using encrypted data, as if we are using the initial plain values. First, each p_i produces a new synopsis, denoted by S_i', by replacing the cells of the S_i which hold values above threshold θ with 1 and below it with 0. Then, p_1 encrypts S_1' homomorphically, producing $\widetilde{S}_1'$, which is submitted to p_2. Following the protocol, each p_i $(i = 2, \ldots, m)$ performs cell-wise homomorphic multiplications between $\widetilde{S}_{i-1}'$ and S_i', producing $\widetilde{S}_i'$, which is submitted to p_{i+1}, as shown in Fig. 1. This is necessary because if a cell in any S_i' contains 0, then GS' in this cell should contain 0, regardless whether there are other S_i's that contain 1 in this particular cell. By doing so, we identify the common e_js, as monitored by every p_i in a secure manner since each p_i cannot infer anything by inspecting $\widetilde{S}_{i-1}'$. If instead of multiplying we added cell-wise $\widetilde{S}_{i-1}'$ to $\widetilde{S}_i'$, we would create a GS' where each cell would contain the exact number of parties, which exceed θ. In this case, we can apply the rule of the minimum number of parties, where an $S_i'[d][w]$ should exceed θ in order to be included in the GS'. Synopsis GS' though cannot be used to provide count estimates because the values of its cells are equal to either 1 or 0. For this reason, we also obtain GU, by using the secure distributed summation scheme, which is the result of the union operation among the S_is. Finally, by multiplying cell-wise

---

Algorithm 2. Multiplying $\widetilde{S}'_{i-1}$ with S'_i, where $i = 2, \ldots, m$ (step 4a of the HP).

Input: $\widetilde{S}'_{i-1}$, S'_i
Output: $\widetilde{S}'_i$
1: **for** $(d = 1, \ldots, D)$ **do**
2:     **for** $(w = 1, \ldots, W)$ **do**
3:         $\widetilde{S}'_i[d][w] \leftarrow \widetilde{S}'_{i-1}[d][w] \odot S'_i[d][w]$
4:     **end for**
5: **end for**

Algorithm 3. p_i produces Q_i by using Q_{i-1}, where $i = 2, \ldots, m$ (steps 4b and 2b of the HP and NBP, respectively.)

Input: S_i, Q_{i-1}, F
Output: Q_i
1: **for** $(d = 1, \ldots, D)$ **do**
2:     **for** $(w = 1, \ldots, W)$ **do**
3:         $Q_i[d][w] \leftarrow (Q_{i-1}[d][w] + S_i[d][w]) \mod F$
4:     **end for**
5: **end for**

GU with GS', we obtain GS, which can be used to provide count estimates of the common e_js, realizing the result of the intersection operation among the underlying S_is. This protocol, as illustrated in Table 1, prevents G, or any p_i, from inferring any information from the intermediate synopses circulated. Algorithm 2 illustrates step 4a of the protocol where each p_i, by performing homomorphic multiplications between $\widetilde{S}'_{i-1}$ and S'_i, produces $\widetilde{S}'_i$ which is submitted to p_{i+1}. At each party, the number of homomorphic operations is $\mathcal{O}(D \times W)$ where D has a logarithmic dependency on N and W depends on ϵ regardless of N or n.

In Algorithm 3, it is shown how each p_i performs the secure distributed summation scheme cell-wise, by exploiting the linearity of the S_is (step 4b of the protocol). The main computational overhead of this protocol is the encryption of the S_1 in step 3a. Also, each $\widetilde{S}'_i$ adds high communication cost due to its size, which is proportional to the size of S'_i (or S_i), multiplied by a constant factor (e.g., 2), which depends on the implementation of the Paillier cryptosystem [31].

5.2   The Noise-Based Protocol

In the Noise-Based Protocol (NBP), instead of producing each $\widetilde{S}_i$ in order to generate the GS', we apply cell-wise the secure distributed summation scheme (see Algorithm 1). As shown in Algorithm 4, if an $S_i[d][w]$ contains a value above θ, then $Q'_i[d][w]$ becomes $Q'_{i-1}[d][w]$ plus 1. Otherwise, p_i assigns to $Q'_i[d][w]$ the value of $Q'_{i-1}[d][w]$ plus some symmetric noise (eg., a random value drawn from a Laplace or a Gaussian distribution, where the location and scale parameters are set to 0 and 1, respectively).

Algorithm 4. p_i produces Q'_i by using Q'_{i-1}, where $i = 2, \ldots, m$ (step 2a of the NBP).

Input: S_i, Q'_{i-1}, F, θ
Output: Q'_i
1: **for** $(d = 1, \ldots, D)$ **do**
2:     **for** $(w = 1, \ldots, W)$ **do**
3:         **if** $S_i[d][w] > \theta$ **then**
4:             $Q'_i[d][w] \leftarrow (Q'_{i-1}[d][w] + 1) \mod F$
5:         **else**
6:             $Q'_i[d][w] \leftarrow (Q'_{i-1}[d][w] + Lap(0,1)) \mod F$ ▷ adding Laplace noise
7:         **end if**
8:     **end for**
9: **end for**

We add symmetric noise in order to sanitize $Q'_i[d][w]$ with respect to m. By doing so, p_1 is prevented from making any inferences, such as the exact number of parties which exceed θ, by inspecting the cells of the Q'_m after subtracting R'. By revealing the value of m for a cell, which is the case where all parties exceed θ in that cell, neither p_1 nor G learns anything that can breach the privacy of any party. Fig. 2 illustrates how three parties create the GS' by following the steps of this protocol. Finally, G receives two global synopses, namely GU and GS'. Synopsis GU is the result of the union operation among the S_is while GS' holds for each cell either m or a sanitized value, which is the result of the noise applied in the case where at least one cell in the same coordinates of any S_i contains an unacceptable value (below θ). Authority G, by checking cell-wise GS', identifies accurately which cells should be discarded from GU and sets them to 0. This protocol is illustrated in Table 2.

6 Evaluation

We evaluate our protocols in terms of the accuracy of the count estimates, the execution time, the space required, the precision, and the recall of the results, by using both synthetic and real data sets. For measuring accuracy, we specify the completeness measure C as:

$$C = 1 - \frac{\sum_{\forall e_j \in \mathcal{S}} \mid query(GS, e_j) - V(e_j) \mid}{\sum_{\forall e_j \in \mathcal{S}} V(e_j)}, \tag{1}$$

where $V(\cdot)$ returns the exact global number of occurrences of an e_j. The completeness measure shows the overall accuracy of the estimates, as compared with the exact global number of occurrences of the e_js. A value for completeness near 1 denotes high accuracy for the estimates provided by the corresponding GS. Recall is the number of the correct elements found as a percentage of the number of the truly correct elements. On the other hand, precision is the number of the correct elements found as a percentage of the entire output. For our experiments,

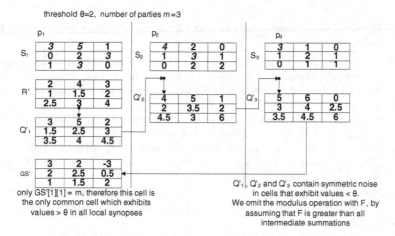

Fig. 2. In each cell of each Q'_i, we add either 1 or symmetric noise, depending on the corresponding value in each S_i $(i = 2, \ldots, m)$

we have chosen as the application domain the monitoring and the identification of common web resources appearing at five local parties. All experiments were conducted on a Pentium Dual Core at 2GHz with 4GB RAM. The software components are developed using the Java programming language version 1.7, and are available from the authors.

We compare our protocols with the intersection operation included in the Sepia library presented in [7] and [28], where Counting Bloom filters (CBfs) are used, as initially introduced in [10]. A CBf is an integer array, where each e_j is hashed K times, by using the HMAC-MD5 hash function [26]. For each cell that an e_j is hashed to, we increment this cell's value by 1. In order to derive the count estimate of an e_j, we hash it and then take the minimum of the values retrieved. More specifically, we use the weighted intersection (I-SEPIA) of the Sepia library, where an e_j should be represented by the global CBf ($gCBf$) only if it has been monitored by every p_i. Each p_i submits to G two CBfs: A flag-based CBf that actually states if an e_j has been monitored by this p_i, and another CBf which holds the number of occurrences of each e_j. The flag-based CBfs are cell-wise multiplied producing $gCBf_1$, which includes the common e_js that should be represented by the $gCBf$. The CBfs holding the number of occurrences are cell-wise added producing $gCBf_2$, which holds the global summations. A straightforward cell-wise multiplication between these two $gCBf$s produces the final $gCBf$. The number of hash functions K and the size L of each CBf depend on the specified false-positive probability, denoted by FPR. This rate can be considered as the acceptable error rate, when a CBf provides a count estimate for an e_j which it has never been hashed to it. The specified error rate is achieved by setting $K = \ln(2)L/n$ and $L = cn$, where c is a small constant in order to minimize the formula $(0.6185)^{L/n}$. An analysis of the derivation of these optimal values is given in [6].

Table 2. The steps of the noise-based protocol for creating a privacy-preserving intersection global synopsis

---

step 1a p_1 produces R' and Q'_1, which are $D \times W$ arrays and correspond to r and q_1 respectively depicted in Algorithm 1, but each cell of Q'_1 instead of holding the real values of S_1 holds 1 if $S_1[d][w] > \theta$ and some symmetric noise if not (plus the random values of R').

step 1b p_1 produces R and then Q_1, both needed for the secure cell-wise summation of each S_i. Both Q_1 and Q'_1 are submitted to p_2.

step 2a Each p_i $(i = 2, \ldots, m)$, by checking cell-wise each S_i, adds $Q'_{i-1}[d][w] + 1$ to $Q'_i[d][w]$, only if $S_i[d][w]$ is above θ. Otherwise, $Q'_i[d][w]$ becomes $Q'_{i-1}[d][w]$ plus some symmetric noise.

step 2b Each p_i $(i = 2, \ldots, m)$, by using Q_{i-1}, produces Q_i ($Q_i[d][w] = (Q_{i-1}[d][w] + S_i[d][w])$ mod F). Both Q_i and Q'_i are submitted to p_{i+1}.

step 3a p_m produces Q'_m.

step 3b p_m produces Q_m and then submits it to p_1 along with Q'_m.

step 4a p_1 produces GS' by subtracting R' from Q'_m.

step 4b p_1 produces GU, by subtracting R from Q_m. Both GU and GS' are submitted to G.

step 5 G checks cell-wise GS' and if $GS'[d][w] \neq m$, then $GU[d][w] = 0$. Finally, $GS = GU$.

6.1   Evaluation Using Synthetic Data Sets

Each party uses a synthetic data set where we generate $N = 10^9$ occurrences from an alphabet of $n = 10^6$ distinct elements. The parameters used for building the synopses are depicted in Table 3.

Table 3. The parameters used for building the synopses

ϕ	ϵ	D	W	δ
0.1	0.01	26	100	0.01
0.01	0.001	26	1000	0.01
0.001	0.0001	26	10,000	0.01

Elements generated for each party follow the Zipf distribution, since there are numerous studies reporting that web resource popularity obeys power-law long-tailed distributions [2,17,21,25]. We set the skew parameter z to 1 (Fig. 3a) and to 2 (Fig. 3b) for higher skew. As shown in these figures, by setting $z = 2$, the differences between the exact values and the corresponding estimates are almost eliminated, since the accuracy of Count-Min sketches is increased by using highly skewed data distributions [34].

For I-SEPIA, we set $FPR = 0.1$, which yields nearly 5×10^6 cells for the corresponding $CBFs$ and the $gCBF$. In Fig. 4a, we illustrate the completeness rates, where our protocols outperform I-SEPIA. By setting higher skew ($z = 2$),

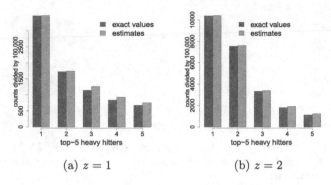

(a) $z = 1$ (b) $z = 2$

Fig. 3. Absolute estimates and exact values, by generating e_js, following the Zipf distribution

the completeness rate of our protocols is even higher than the one of I-SEPIA. In Fig. 4b, we show the completeness rates for both protocols, where we observe that these rates are constantly above 0.9. Also, by setting skew to 2, the completeness rates, as expected, exhibit higher rates exceeding 0.95. We also observe that, as ϕ is set to lower values, the completeness rates fall, exhibiting the lowest value by setting $\phi = 0.001$ ($\theta = 0.001 \times 10^9 = 10^6$ and $\epsilon = 0.0001$). This happens because exact counts of the e_js near $\theta = 10^6$ become more and more uniform and this uniformity of values, as reported in [34], results in reducing the accuracy of the Count-Min sketches. Recall rates for both I-SEPIA and our protocols are constantly at 1.0. In terms of precision rates (Figs. 4c and 4d), our protocols perform slightly lower than I-SEPIA. Especially, by setting $\phi = 0.001$ with skew $z = 1$, precision falls below 0.95, which means that there is an amount of elements returned where their exact number of occurrences is below θ (false positives). When we increase skew ($z = 2$), precision rates increase accordingly reaching almost 1.0. As expected, the number of occurrences for some false positives returned by our protocols lie within the interval ($\lceil (\phi - \epsilon)N \rceil, \theta$).

The higher precision rates achieved by I-SEPIA are outweighed by the cost of exceptionally large space required, whereas our protocols utilize orders of magnitude smaller data structures, as shown in Fig. 4e. The space utilized by our protocols depends on the specified parameters ϵ, δ, and on N. By setting lower values for ϕ, in cases where we need a lower threshold, we consequently decrease ϵ, which eventually results in using more space, as it is clearly shown in Fig. 4e by setting ϕ to 0.001. We illustrate the space requirements by assuming the RAM model [29], where a plain integer is represented by a machine word while in HP a homomorphically encrypted integer is represented by two machine words respectively. Obviously, the space utilized affects execution time as well, since the number of operations required is proportional to the number of cells of each synopsis (S_i', S_i, Q_i and Q_i'), which depends on ϵ and consequently on ϕ, as illustrated in Fig. 4f. The extra time required for HP is due to the initial encryption of the cells of S_1' in step 3a, while the homomorphic multiplications in step 4a add negligible overhead.

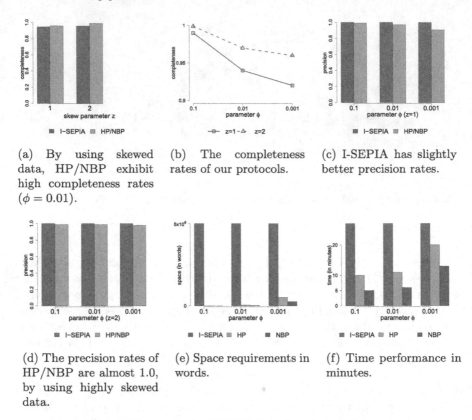

(a) By using skewed data, HP/NBP exhibit high completeness rates ($\phi = 0.01$).

(b) The completeness rates of our protocols.

(c) I-SEPIA has slightly better precision rates.

(d) The precision rates of HP/NBP are almost 1.0, by using highly skewed data.

(e) Space requirements in words.

(f) Time performance in minutes.

Fig. 4. Comparing to I-SEPIA by using skewed data sets

6.2 Evaluation Using Real Data Sets

We received from OTS SA [1], which is a big Greek IT company, an anonymized list of the top $1,000$-ranked web sites, as determined by the visits paid by their employees in January 2014. In Fig. 5a, the distribution of the occurrences of the e_js, where an e_j is a hit to a web site, is illustrated which indicates a stretched-exponential distribution with the stretching parameter set between 0.8 and 0.9. The total number of hits $N = 10^8$ and the number of distinct e_js $n = 10^3$. We set the frequency ϕ to 0.1, 0.01 and 0.005. If we set ϕ to 0.001 the corresponding threshold would be considerably reduced, namely it would be equal to 10^5, and the synopses would return too many e_js as heavy hitters. We distributed randomly the e_js to five parties except for the heavy hitters, for each value of frequency ϕ, which appear universally in order to allow for evaluating the performance of both our protocols and I-SEPIA. In terms of completeness, our protocols, as expected due to the skewness of the data set, exhibit slightly better rates, as depicted in Fig. 5b. The recall rates are consistently 1.0 while the precision rates are almost the same for both our protocols and I-SEPIA, as shown in Fig. 5c.

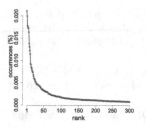 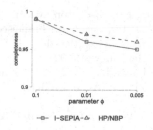 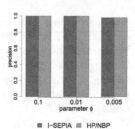

(a) Distribution of the occurrences of the e_js according to their rank.

(b) The completeness rates of our protocols are constantly above 0.95.

(c) The precision rates for both our protocols and I-SEPIA are almost the same, very close to 1.0.

Fig. 5. Intersecting web logs provided by OTS SA

7   Conclusions

We have proposed two protocols for efficient and privacy-preserving set intersection (PPSI) by using Count-Min sketches. The aim of these protocols is to allow an authority to create a global synopsis by performing private intersection of the sets that are individually summarized as local synopses by three or more parties. This global synopsis also provides count estimates of the intersected elements. Our protocols use a combination of privacy techniques, namely homomorphic computations, a secure distributed summation scheme, and noise addition. An empirical study conducted on large synthetic and real data sets validates the efficiency and accuracy of our protocols as compared to an existing PPSI protocol. In the future, we aim to (a) improve further the accuracy and scalability of our protocols, (b) apply time widows for monitoring the e_js, and (c) study other private set operations and techniques using efficient data structures.

References

1. OTS SA (2014). http://www.ots.gr/
2. Adamic, L., Huberman, B.: Zipf's law and the internet. Glottonmetrics **11**, 143–150 (2002)
3. Aggarwal, C., Yu, P.: A general survey of privacy-preserving data mining models and algorithms. Adv. Datab. Sys. **34**, 11–52 (2008)
4. Agrawal, R., Evfimievski, A., Srikant, R.: Information sharing across private databases. In: SIGMOD, San Diego, California, USA, pp. 86–97 (2003)
5. Aumann, Y., Lindell, Y.: Security against covert adversaries: Efficient protocols for realistic adversaries. J. of Cryptol. **23**(2), 281–343 (2010)
6. Broder, A., Mitzenmacher, M.: Network applications of Bloom filters: A survey. Internet Math. **1**(4), 485–509 (2002)
7. Burkhart, M., Dimitropoulos, X.: Privacy-preserving distributed network troubleshooting - bridging the gap between theory and practice. ACM Trans. Inf. Sys. Sec. **14**(4) (2011)

8. Charikar, Moses, Chen, Kevin, Farach-Colton, Martín: Finding frequent items in data streams. In: Widmayer, Peter, Triguero, Francisco, Morales, R., Hennessy, Matthew, Eidenbenz, Stephan, Conejo, Ricardo (eds.) ICALP 2002. LNCS, vol. 2380, pp. 693–703. Springer, Heidelberg (2002)
9. Clifton, C., Kantarcioglou, M., Vaidya, J., Lin, X., Zhu, M.Y.: Tools for privacy preserving distributed data mining. ACM SIGKDD Explor. Newsl. 4(2), 28–34 (2002)
10. Cohen, S., Matias, Y.: Spectral Bloom filters. In: SIGMOD, San Diego, California, pp. 241–252 (2003)
11. Cormode, G., Garofalakis, M.: Sketching streams through the net distributed approximate query tracking. In: VLDB, Trondheim, Norway, pp. 13–24 (2005)
12. Cormode, G., Muthukrishnan, S.: An improved data stream summary: the Count-Min sketch and its applications. J. of Algor. 55(1), 58–75 (2005)
13. Dachman-Soled, D., Malkin, T., Raykova, M., Yung, M.: Efficient robust private set intersection. Appl. Cryptog. 2(4), 289–303 (2012)
14. Dong, C., Chen, L., Wan, Z.: When private set intersection meets big data: an efficient and scalable protocol. In: SIGSAC, Berlin, Germany, pp. 789–800 (2013)
15. Freedman, Michael J., Nissim, Kobbi, Pinkas, Benny: Efficient private matching and set intersection. In: Cachin, Christian, Camenisch, Jan L. (eds.) EUROCRYPT 2004. LNCS, vol. 3027, pp. 1–19. Springer, Heidelberg (2004)
16. Frikken, K.: Privacy-preserving set union. Appl. Cryptog. Network Sec. 4521, 237–252 (2007)
17. Glassman, S.: A caching relay for the world wide web. Comput. Netw. ISDN Syst. 27(2), 165–173 (1994)
18. Goldreich, O., Micali, S., Wigderson, A.: How to play ANY mental game. In: STOC, New York, USA, pp. 218–229 (1987)
19. Hall, Rob, Fienberg, Stephen E.: Privacy-preserving record linkage. In: Domingo-Ferrer, Josep, Magkos, Emmanouil (eds.) PSD 2010. LNCS, vol. 6344, pp. 269–283. Springer, Heidelberg (2010)
20. Hazay, Carmit, Lindell, Yehuda: Efficient protocols for set intersection and pattern matching with security against malicious and covert adversaries. In: Canetti, Ran (ed.) TCC 2008. LNCS, vol. 4948, pp. 155–175. Springer, Heidelberg (2008)
21. Jauhari, M., Saxena, A., Gautam, J.: Zipf's law and the number of hits on the world wide web. Annals of Lib. and Inf. Studies 54, 81–84 (2007)
22. Kantarcioglu, Murat, Jiang, Wei, Malin, Bradley: A privacy-preserving framework for integrating person-specific databases. In: Domingo-Ferrer, Josep, Saygın, Yücel (eds.) PSD 2008. LNCS, vol. 5262, pp. 298–314. Springer, Heidelberg (2008)
23. Kiayias, A., Mitrofanova, A.: Testing disjointness of private datasets. In: Patrick, Andrew S., Yung, M. (eds.) FC 2005. LNCS 3570, vol. 3570, pp. 109–124. Springer, Heidelberg (2005)
24. Kissner, Lea, Song, Dawn: Privacy-preserving set operations. In: Shoup, Victor (ed.) CRYPTO 2005. LNCS, vol. 3621, pp. 241–257. Springer, Heidelberg (2005)
25. Krashakov, S., Teslyuk, A., Shchur, L.: On the universality of rank distributions of website popularity. Comp. Netw. 50(11), 1769–1780 (2006)
26. Krawczyk, H., Bellare, M., Canetti, R.: HMAC: keyed-hashing for message authentication, Internet RFC 2104 (1997). http://tools.ietf.org/html/rfc2104
27. Lindell, Y., Pinkas, B.: Secure multiparty computation for privacy-preserving data mining. J. Priv. Conf. 1(1) (2009)
28. Many, D., Burkhart, M., Dimitropoulos, X.: Fast private set operations with sepia. Tech. Rep. no. 345, ETH Zurich (2012)

29. Motwani, R., Raghavan, P.: Randomized Algorithms. Cambridge University Press (1995)
30. Naor, M., Pinkas, B.: Oblivious transfer and polynomial evaluation. In: STOC, Atlanta, Georgia, USA, pp. 245–254 (1999)
31. Paillier, Pascal: Public-key cryptosystems based on composite degree residuosity classes. In: Stern, Jacques (ed.) EUROCRYPT 1999. LNCS, vol. 1592, p. 223. Springer, Heidelberg (1999)
32. Pierre, K., Lai, S., Yiu, K., Chow, C., Chong, L., Hui, C.: An efficient Bloom filter based solution for multiparty private matching. In: SAM (2006)
33. Roughan, M., Zhang, Y.: Secure distributed data-mining and its application to large-scale network measurements. SIGCOMM Comput. Commun. Rev. **36**(1), 7–14 (2006)
34. Rusu, F., Dobra, A.: Statistical analysis of sketch estimators. In: SIGMOD, Beijing, China, pp. 187–198 (2007)
35. Vatsalan, D., Christen, P., Verykios, V.S.: A taxonomy of privacy-preserving record linkage techniques. J. Inf. Sys. **38**(6), 946–969 (2013)
36. Yao, A.: How to generate and exchange secrets. In: SFCS, Toronto, Canada, pp. 162–167 (1986)

Improved Weighted Bloom Filter and Space Lower Bound Analysis of Algorithms for Approximated Membership Querying

Xiujun Wang[1,4][✉], Yusheng Ji[2], Zhe Dang[3], Xiao Zheng[1], and Baohua Zhao[4]

[1] Anhui University of Technology, Ma'anshan, China
wxj@mail.ustc.edu.cn
[2] National Institute of Informatics, Tokyo, Japan
kei@nii.ac.jp
[3] Washington State University, Pullman, USA
zdang@eecs.wsu.edu
[4] University of Science and Technology of China, Hefei, China

Abstract. The elements in a large universe U have different membership likelihoods and query frequencies in many applications. Thus, the number of hash functions assigned to each element of U in the traditional Bloom filter can be further optimized to minimize the average false positive rate. We propose an improved weighted Bloom filter (IWBF) that assigns an optimal number of hash functions to each element and has a less average false positive rate compared to the weighted Bloom filter. We show a tight space lower bound for any approximated membership querying algorithm that represents a small subset S of U and answers membership queries with predefined false positive rates, when the query frequencies and membership likelihoods of the elements in U are known. We also provide an approximate space lower bound for approximated membership querying algorithms that have an average false positive rate, and show that the number of bits used in IWBF is within a factor of 1.44 of the approximate space lower bound.

1 Introduction

The traditional Bloom filter [4] is an elegant memory-efficient randomized data structure for representing the elements in a small subset S of a large universe U ($S \subset U$), and its use is in approximated membership queries. The elements that occur in S are called members. Given a small subset S of a large universe U ($|U| \gg |S|$) and a false positive rate ϵ, the traditional Bloom filter uses a memory space of $|S| \log_2(e) \log_2(1/\epsilon)$ bits and $k = \log_2(1/\epsilon)$ bit probes for each query. It is a well-known technique for approximated membership querying.

Many alternative set representations exist that support membership query with the false positive ϵ shown above, e.g., perfect hashing [7], cuckoo hashing [25], and d-left counting Bloom filter [6]. However, the representations require complicated rehash mechanisms to resolve hash collisions and, thus, are less versatile than the traditional Bloom filters. In fact, the Bloom filter continues

M. Renz et al. (Eds.): DASFAA 2015, Part II, LNCS 9050, pp. 346–362, 2015.
DOI: 10.1007/978-3-319-18123-3_21

to attract interest because of its space efficiency and ease of implementation. There has been much research on Bloom filter and its variants for approximately detecting duplicates [8,15,16,17,18,19,23,24,26].

The traditional Bloom filter makes a hidden assumption that all elements in a large universe U have equal membership likelihood (the probability that element e in the universe U occurs in S) and equal query frequency (the probability that the querying client makes a request about whether e occurs in S or not). However, it is commonly observed that elements from a large universe U do not necessarily have the same membership likelihood, i.e., some elements occur more frequently in S than the others [8]. The elements also don't necessarily have the same query frequency. For example, in URL crawling, a search engine periodically crawls the web to enlarge its web-page databases. For a URL directed to a web page, the search engine must query its URL database to find out whether the page has been fetched or not. The URL of a web page at a hot web site has a larger probability to be queried by a search engine than URLs of other web pages.

This paper follows the definition of the expectation of the false positive probability in [8] (the expectation of false positive probability, shown in the equation (4) in [8] , is called the average false positive rate in this paper). Notice that some other definitions [26] can be treated as a special case as the one in [8]. Ref. [8] uses random variables to model the information about query frequencies and membership likelihoods of elements in U. Then Ref. [8] defines an average false positive rate as a mathematical expectation of the false positive probability $E[P_{FP}]$ (see the equation (4) in [8]), and try to minimize $E[P_{FP}]$ by controlling the number of hash functions assigned to each element of U. However, the numbers of hash function so chosen are not optimal, in the following sense: we call the numbers of hash functions optimal if these numbers produce the global minimal value of the average false positive rate. In Section 3, we propose IWBF, which assigns an optimal number of hash functions to each element in U when the query frequencies and membership likelihoods of the elements in U are known, and identify the reason for WBF's failure to obtain the optimal numbers of hash functions.

We have not found in literatures a space lower bound for any approximated membership querying algorithm that has an average false positive rate θ (the algorithm represents a subset S and answers membership queries with an average false positive rate but no false negative errors), when the query frequencies and the membership likelihoods of elements in U are known.

Our Contributions

(1) We propose an improved weighted Bloom filter (IWBF), which has an optimal number of hash functions assigned to each element $e \in U$. IWBF is proven to have no false negative errors and a lower average false positive rate than that of the weighted Bloom filter (WBF) in [8], given the same memory space.

(2) Based on information cost in communication complexity theory [20], we show a tight space lower bound for any approximated membership querying algorithm that represents a small subset S of a large universe U in a memory-efficient way and answers membership queries with predefined false positive rates for the elements in U.

(3) We derive an approximate space lower bound for any approximated membership querying algorithm that represents a small subset S of U and answers membership queries with an average false positive rate but no false negative errors. Then we show that the number of bits used in an IWBF is within a factor of $\log_2(e) \approx 1.44$ of the approximate space lower bound. We also show that the lower bound of space $|S| \log_2(1/\epsilon)$, based on the pigeonhole principle in [7,9], is a special case of the approximate space lower bound obtained in this paper.

2   Membership Likelihood and Query Frequency Definition

For ease of illustration, we give each element a number identifier and assume $U = \{1, 2, ..., N\}$ (the universe contains N elements). We adopt the same model as [8] to represent the membership likelihoods and query frequencies as follows.

An indicator random variable X_i is used for each element $i \in U$ to represent whether i occurs in S:

$$X_i = 1 \text{ when } i \in S; \quad X_i = 0 \text{ when } i \notin S. \tag{1}$$

$X_1, .., X_N$ are independent random variables. A random variable Y represents a query element chosen from U by the querying client in one query request:

$$Y = i, \text{ if } i \text{ is queried in the request.} \tag{2}$$

An indicator random variable Y_i is used for each element $i \in U$: $Y_i = 1$ when $Y = i$; $Y_i = 0$ otherwise. The membership likelihood and query frequency of each element $i \in U$ are represented by $P(X_i = 1) = x_i \in (0, 1)$ and $P(Y_i = 1) = y_i \in (0, 1)$ respectively. n denote the expected number of elements in S:

$$n = E[|S|] = E[\sum\nolimits_{i \in U} X_i] = \sum\nolimits_{i \in U} x_i. \tag{3}$$

3   Improved Weighted Bloom Filter

3.1   Weighted Bloom Filter [8]

A weighted Bloom filter (WBF) [8] uses an m-bit array and assigns k_i uniformly random and independent hash functions $h_1, .., h_{k_i}$ ($h_j(i) \in \{1, .., m\}, j = 1, .., k_i$) to each $i \in U$. It is used to represent a subset S of n elements (called members) from a large universe U ($S \subset U$), then answer the membership queries as follows (the values of $k_i, i \in U$ in the two steps are shown later):

- firstly, for each element in $i \in S$, the k_i bits at positions $h_j(i), j = 1, .., k_i$ of the m-bit array are set to $'1'$;
- secondly, for each query element $q \in U$, the answer to $q \in S$? is $'yes'$ if all of the bits in positions $h_1(q), .., h_{k_q}(q)$ are $'1'$, and $'no'$ otherwise.

Table 1. Notations

N	the number of elements in a large universe U
n	the expected number of elements in S ($S \subset U$), $n = \sum_{i \in U} x_i$
m	the number of memory bits used in WBF or IWBF
x_i	the membership likelihood of the element $i \in U$
r_i	the normalized query frequency of the element $i \in U$, shown in (5)
$R_+ = (0, +\infty)$	the set of positive real numbers
$\overline{k^\circ} = (k_1^\circ, .., k_N^\circ)$	numbers of hash functions assigned to elements in U in WBF
$\overline{k^*} = (k_1^*, .., k_N^*)$	the global minimizer of the average false positive rate $FP(\overline{k})$ over R_+^N

When assigning $k_i \in R_+$ hash functions for each $i \in U$, obviously, each non-member element may have a different false positive probability in a WBF. Ref. [8] defines the average false positive rate denoted by $E[P_{FP}]$ as the expectation of the weighted sum of these false positive probabilities (see the equation (4) in [8]):

$$E[P_{FP}] = E[\sum_{i \in U} r_i (1 - e^{-\sum_{j \in U} X_j k_j / m})^{k_i}] = T(\overline{k}), \qquad (4)$$

where r_i denote the normalized query frequency of $i \in U$:

$$r_i = (1 - X_i) Y_i / (\sum_{j \in U} (1 - X_j) Y_j). \qquad (5)$$

Clearly $E[P_{FP}]$ is a function of $\overline{k} \in R_+^N = (0, +\infty)^N$, denoted by $T(\overline{k})$ in (4).
Let K denote the expected number of hash functions used by elements in S:

$$K = E[\sum_{i \in U} X_i k_i] = \sum_{i \in U} x_i k_i. \qquad (6)$$

Considering that the average false positive rate shown in (4) is hard to minimize directly, Ref. [8] uses the assumption that $\sum_{i \in U} X_i k_i$ is sharply concentrated around its expected value $K = \sum_{i \in U} x_i k_i$, then rewrites (4) as follows (shown in equation (20) in [8]):

$$E[P_{FP}] \approx \sum_{i \in U} E[r_i](1 - e^{-K/m})^{k_i} = FP(\overline{k}). \qquad (7)$$

$\sum_{i \in U} E[r_i](1 - e^{-K/m})^{k_i}$ is a function of $\overline{k} \in R_+^N$, denoted by $FP(\overline{k})$ in (7).
By Assuming that $k_i, i \in U$ can be any real number in R_+ and setting $\partial FP(\overline{k})/\partial k_i, i \in U$ to 0 in the proof of Theorem 3.1 of [8], Ref. [8] finds that the minimizer of $FP(\overline{k})$ is (see the equation (5) or (47) in [8]):

$$k_i^\circ = (m/n) \ln 2 + (\ln E[r_i] - \sum_{j \in U} (x_j/n) \ln E[r_j]) / \ln 2, i \in U. \qquad (8)$$

Actually, WBF assigns k_i° hash functions for $i \in U$. More specifically, given a WBF with m-bit array, for each $i \in S$, the k_i° bits at positions $h_1(i), h_2(i), .., h_{k_i^\circ}(i)$ of the m-bit array are set to $'1'$ when storing a subset S, then for each query element $q \in U$, the answer to $q \in S$? is $'yes'$ if all of the bits in positions $h_1(q), .., h_{k_q^\circ}(q)$ are $'1'$, and $'no'$ otherwise. The average false positive rate of WBF is taken to be $FP(\overline{k^\circ})$ ($\overline{k^\circ} = (k_1^\circ, .., k_N^\circ)$), see the equation (7) in [8].
However, there are two problems in WBF. **(1)**: $\overline{k^\circ}$ is not the global minimizer of $FP(\overline{k})$ in R_+^N. Because $FP(\overline{k})$ is not a convex function of $\overline{k}$ in R_+^N,

then $\overline{k}^\circ$, which satisfy $\partial FP(\overline{k})/\partial \overline{k}^\circ = 0$ (see equation (23) in [8]), are not necessarily a global minimizer of $FP(\overline{k})$. **(2)**: The actual average false positive rate
of a WBF, when assigning k_i° hash function to $i \in U$, is determined by $T(\overline{k}^\circ)$
($\forall \overline{k} \in R_+^N, T(\overline{k})$ is defined in (4)). Ref. [8] doesn't provide any proof to show the
difference between $T(\overline{k}^\circ)$ and $FP(\overline{k}^\circ)$ is small that can be ignored.

3.2   Improved Weighted Bloom Filter Design

Let $\overline{k}^* = (k_1^*, .., k_N^*)$ denote the global minimizer of $FP(\overline{k})$ in (7) over the set
R_+^N and let $K^* = \sum_{i \in U} x_i k_i^*$ denote the optimal value of the expected number
of hash functions. The basic idea of an improved Weighted Bloom filter (IWBF)
is as same as WBF, but IWBF assigns an optimal number of hash functions
k_i^* (shown in (10)) to each element $i \in U$, thus has a much less average false
positive rate as compared with WBF, i.e., $FP(\overline{k}^*)$ (shown in (11)) is less than
$FP(\overline{k}^\circ)$. The idea of finding $\overline{k}^*$ is as follows:

- firstly divide the feasible region R_+^N of $\overline{k}$ into sets $A_K = \{\overline{k} = (k_1, k_2, ..., k_N)|$
  $\overline{k} \in R_+^N, \sum_{i \in U} k_i x_i = K\}$, where K takes value in R_+, and show that $\forall K \in$
  R_+, $FP(\overline{k})$ is a convex function in A_K;
- secondly, for a set A_K, we find the minimizer $\overline{f(K)} = (f_1(K), f_2(K), .., f_N(K))$
  that produces the minimal value $FP(\overline{f(K)})$ of $FP(\overline{k})$ over A_K;
- lastly, we find the global minimizer K^* of $FP(\overline{f(K)})$ for $K \in R_+$. Since
  $FP(\overline{f(K)})$ shown in (22) is a function of K, then clearly $\overline{k}^* = \overline{f(K^*)}$.

Considering that K is fixed for $\overline{k} \in A_K$ and A_K is a convex set, so each component function $(1 - e^{-K/m})^{k_i}, i \in U$ is convex in A_K, then $FP(\overline{k})$ in (7) (the
nonnegative linear combination of these functions) is also convex in A_K. But
$FP(\overline{k})$ is not convex in R_+^N, and this is the reason why [8] fails to get the global
minimizer of $FP(\overline{k})$ in R_+^N.

In the above, we only consider the minimization of $FP(\overline{k})$, then get $\overline{k}^*$ as
the global minimizer of $FP(\overline{k})$. However, the actual false positive rate of IWBF,
when assigning k_i^* hash functions for $i \in U$, is determined by $T(\overline{k}^*)$. We shall
give a sufficient condition (shown in (12)) to guarantee that difference between
$T(\overline{k}^*)$ and $FP(\overline{k}^*)$ is small enough that can be ignored, i.e., $FP(\overline{k}^*)$ can safely be
taken to be the actual average false positive rate of IWBF, under this condition.
Thus, in Theorem 1, we first show how to find the values of K^* and $\overline{k}^*$, then a
sufficient condition (12) is provided to guarantee that $T(\overline{k}^*) \approx FP(\overline{k}^*)$.

Lemma 1. *Let $X_1, .., X_N$ be independent random variables, where X_i is defined
in (1), and $P(X_i = 1) = x_i, i \in U$. The expected number of hash functions used
by elements in S is $K = \sum_{i \in U} x_i k_i$ in (6) with $\overline{k} = (k_1, .., k_N) \in R_+^N$, and the
expected number of elements in S is $n = \sum_{i \in U} x_i$ in (3). Furthermore we define
$v = \sum_{i \in U} x_i k_i^2$ and $k_{max} = \max\{k_1, .., k_N\}$. Then we have*

$$P(|\sum_{i \in U} X_i k_i - K| \geq n^{-1/3} m) \leq e^{\frac{-m^2}{2vn^{2/3}}} + e^{\frac{-m^2}{2(vn^{2/3} + \frac{1}{3}k_{max} m n^{1/3})}}. \qquad (9)$$

Proof. Let $\lambda = n^{-1/3}m$ (λ is shown in the inequalities (3.1) and (3.2) in [11]), then by Theorem 3.3 in [11] we have the conclusion.

Theorem 1. *We are given the expected size n of S in (3), and an m-bit array in an improved weighted Bloom filter (IWBF). Then $K^* = m \ln 2$, and*

$$k_i^* = (m/n)\ln 2 + \log_2 E[r_i]/x_i - \sum\nolimits_{j \in U} (x_j/n)\log_2(E[r_j]/x_j) \qquad (10)$$

Accordingly,

$$FP(\overline{k^*}) = 2^{-(m/n)\ln 2} \cdot n \cdot \prod\nolimits_{j \in U} (E[r_j]/x_j)^{x_j/n}. \qquad (11)$$

Moreover, suppose that:

$$\max_{i,j \in U} |\log_2(E[r_i]/x_i) - \log_2(E[r_j/x_j])| < (m/n)\ln 2, \qquad (12)$$

Then $E[P_{FP}] = T(\overline{k^*})$ the actual false positive rate of IWBF, when assigning k_i^* hash functions for $i \in U$, satisfies:

$$T(\overline{k^*}) - FP(\overline{k^*}) < 2e^{-\frac{n^{1/3}}{3}}. \qquad (13)$$

Proof. The main steps for proving (10) and (11) are: at first, we find the minimal value $FP(\overline{f(K)})$ of $FP(\overline{k})$ for each set A_K; secondly, we find $\overline{k^*}$ by analyzing the function $FP(\overline{f(K)})$ in R_+. The main idea for proving (13) is: $T(\overline{k^*})$ can be decomposed into two parts, then we show that the first part is less than $2e^{-n^{1/3}/3}$ and the second is approximately equal to $FP(\overline{k^*})$, when the sufficient condition in (12) is true. The detailed proof is given in Appendix.

3.3   False Negative Rate and False Positive Rate Analysis

From Algorithm 1, it is easy to see that the IWBF has no false negative error.

In this section, $FP(\overline{k^\circ})$ and $FP(\overline{k^*})$ are called the idealized average false positive rate of WBF and IWBF respectively. Clearly, since $\overline{k^*}$ is the global minimizer of $FP(\overline{k})$ over the set R_+^N, we have $FP(\overline{k^*}) \leq FP(\overline{k^\circ})$. In the following, at first, we show the numerical values of the ratio between $FP(\overline{k^\circ})$ and $FP(\overline{k^*})$ over Zipf's Distribution. Secondly, we give an example, which shows that the actual average false positive rate of IWBF, determined by $T(\overline{k^*})$, is less than $FP(\overline{k^*})$ the idealized average false positive rate WBF.

Numerical comparison of the idealized average false positive rates of the WBF and IWBF over Zipf's Distribution:

Let $U = \{1, 2, .., N\}$. In a Zipf's distribution with parameter α, for element i, its membership likelihood is defined as: $x_i = nc(1/i^\alpha)$, where c is a positive constant to ensure that: $\sum_{i \in U} x_i/n = 1$, and α controls the skewness of the distribution.

Given that the expected size of S is n, the size of U is N, and the IWBF and WBF each have an m-bit array, we define R: the ratio between $FP(\overline{k^*})$ of IWBF and $FP(\overline{k^\circ})$ of WBF. It is easy to see that: $R = \frac{FP(\overline{k^*})}{FP(\overline{k^\circ})} = (n/N)\prod_{i \in U}(x_i)^{-x_i/n}$ ≤ 1 (m is eliminated). Since R is only related to the values of n, N and membership likelihoods $x_i, i \in U$, thus, in the following, we shall show the values of

Algorithm 1. An IWBF for approximated membership querying

Input: $S,M,k_i(i \in U),h_1,..h_g,Q$
S: a small subset of U, whose elements taken from distributions of $X_i, i \in U$; M: m-bit array whose bits are all initialized to $'0'$; k_i: the number of hash functions assigned to an element i is set to the nearest positive integer from k_i^*; $h_1,..,h_g$: uniform and independent hash functions, $g = \max_{i \in U}(k_i)$; Q: a sequence of query elements generated by the distribution of Y, $Q = q_1,..,q_j,..$ $q_j \in U$;
Output: a sequence of yes/no answers corresponding to each query element in Q;
Representing part: //*store S into M*//
For each $e \in S$
    set the k_e bits $M[h_j(e)], j = 1,..,k_e$ to $'1'$;
Querying part:
For each $q \in Q$
    probe the k_q bits $M[h_j(q)], j = 1,..,k_q$ in the bit array M;
    If all k_q bits $M[h_j(q)], j = 1,..,k_q$ are $'1'$ **Then** $q \in S$, Output $'yes'$ for q;//*q is assumed to be a member of S*//
        Else $q \notin S$, Output $'no'$ for q; //*q is not a member of S*//
//* The implementation issues of this algorithm are discussed in Section 3.4*//

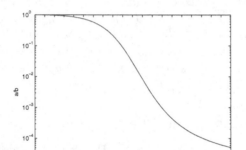

Fig. 1. Ratio of the idealized average false positive rate of IWBF (a) to that of WBF (b) for Zipf's Distributions

R by setting $n = 10^2, N = 10^5, x_i = nc(1/i^\alpha), i \in U$ and $\alpha \in (0,1)$. Since the value of R is irrelevant to the value m, the value of m is not set (thus the values of $FP(\overline{k^*})$ and $FP(\overline{k^\circ})$ are not shown in this comparison). Figure 1 depicts the R values corresponding to different values of α. When α is large, the idealized average false positive rate of the IWBF becomes much lower than that of the WBF. For example, when $\alpha = 1$, $R = 0.0289$, the idealized average false positive rate of IWBF is only about 2.89% of that of WBF.

Comparison of the actual average false positive rate of IWBF with the idealized average false positive rate of the WBF on an example
We give an example to show that the actual false positive rate is less than the idealized average false positive rate of WBF.

For a universe U of $N = 2 \times 10^7$ elements, we assign to each element a number identifier and assume $U = \{1, 2, .., 2 \times 10^7\}, U_1 = \{1, 2, .., 10^7\}, U_2 = \{10^7 + 1, .., 2 \times 10^7\}$. For $i \in U_1$, let $x_i = 1.6 \times 10^{-3}$ and $E[r_i] = 5 \times 10^{-8}$; for $i \in U_2$, let $x_i = 0.4 \times 10^{-3}$ and $E[r_i] = 5 \times 10^{-8}$. The expected size of a subset S of U is $n = \sum_{i \in U} x_i = 1.6 \times 10^4 + 0.4 \times 10^4 = 2 \times 10^4$. Suppose that both IWBF and WBF have an array of $m = 4 \times 10^4$ bits. $m = 4 \times 10^4$ is not large, since it is only two times as large as $n = 2 \times 10^4$ the expected size of a subset S.

Based on Theorem 1, the number of hash functions assigned to $i \in U$ in IWBF is $k_i^* = 0.9862, i \in U_1$ and $k_i^* = 2.9863, i \in U_2$. Rounding to integers, IWBF shall use one hash function for $i \in U_1$ ($k_i^* = 1, i \in U_1$), and three hash functions for $i \in U_2$ ($k_i^* = 3, i \in U_2$), then $K^* = \sum_{i \in U_1} 1.6 \times 10^{-3} \times 1 + \sum_{i \in U_2} 0.4 \times 10^{-3} \times 3 = 2.8 \times 10^4$. Based on (7), the idealized average false positive rate of IWBF is $FP(\overline{k^*}) = \sum_{i \in U_1} 5 \times 10^{-8}(1 - e^{-0.7})^1 + \sum_{i \in U_2} 5 \times 10^{-8}(1 - e^{-0.7})^3 = 0.3155$ (since k_i^* is rounded to an integer, we can't use (11) as the idealized average false positive rate of IWBF). Clearly, $v = \sum_{i \in U} x_i(k_i^*)^2 = 5.2 \times 10^4$ and $k_{max} = \max\{k_1, .., k_N\} = 3$, then, by (9) in lemma 1, we can get:

$$P(|\sum_{i \in U} X_i k_i^* - K^*|/m \geq n^{-1/3}) = P(|\sum_{i \in U} X_i k_i^*/m - 0.7| \geq 0.0368) < 2.4 \times 10^{-9}.$$

The above inequality shows that the random variable $\sum_{i \in U} X_i k_i^*/m$ is sharply concentrated around its expected value $K^*/m = 0.7$. Then we have $T(\overline{k^*}) \approx FP(\overline{k^*}) = 0.3155$, i.e., $FP(\overline{k^*}) = 0.3155$ is a good approximation for the actual false positive rate of IWBF, determined by $T(\overline{k^*})$, in this example.

However, for this example, by Theorem 3.1 in [8], $k_i^\circ = 1.3863, i \in U$ (assuming k_i° can be real number) then the idealized average false positive rate of WBF is $FP(\overline{k^\circ}) = 0.3825$ (by the equation (7) in [8]). Furthermore, rounding $k_i^\circ = 1.3863$ to 1, WBF shall use one hash function for $i \in U$. Based on (7), the idealized average false positive rate of WBF is $FP(\overline{k^\circ}) = \sum_{i \in U_1} 5 \times 10^{-8}(1 - e^{-0.5})^1 + \sum_{i \in U_2} 5 \times 10^{-8}(1 - e^{-0.5})^1 = 0.3935$. Similarly, rounding $k_i^\circ = 1.3863$ to 2, WBF uses two hash functions for $i \in U$, and the idealized average false positive rate of WBF is 0.3996. Thus, for this example, we can see that the actual false positive rate of IWBF is less than the idealized false positive rate WBF, no matter $k_i^\circ, i \in U$ is a integer or real number in WBF.

3.4   Discussion on the Practical Implementation of IWBF

In Algorithm 1, there are two practical problems when we implement IWBF: how to estimate the values of membership likelihoods and query frequencies of elements in U in real systems; how to determine the value of k_q for $q \in U$ efficiently during querying. We shall discuss the problems in the following.

The techniques [3,12,13,14,21] used for estimating the values of membership likelihoods and query frequencies when implementing WBF for real system are well discussed in [8]. Obviously, these techniques can be applied in IWBF directly. In some real system, due to the stability of the distribution that generates the member elements and query elements, our estimates of these values can be more accurate as more elements are observed.

Since we assume the sufficient condition (12) in IWBF, it is easy to see $\forall i \in U, k_i^* \in (0, (2m \ln 2)/n), k_i = [k_i^*] \in \{1, 2, .., \lceil (2m \ln 2)/n \rceil\}$ in Algorithm 1. m/n is usually set to be a small constant in practice. Since $\forall i \in U, k_i \in \{1, .., \lceil (2m \ln 2)/n \rceil\}$, U can be divided into $B = \lceil (2m \ln 2)/n \rceil$ categories (B is a small constant), then we can extract a number of common features form the elements in one category by some well-known pattern mining algorithms. More specifically, we can build B buckets, each bucket is associated with a unique integer in $\{1, .., B\}$ and some simply checkable features that can determine whether a element is contained in this bucket or not. For those elements that do not have any checkable features contained in the B buckets, we may assign them with an average number $[B/2]$. The B buckets will provide each element $i \in U$ with k_i by checking the attributes of i. Thus we can determine k_q for each $q \in U$ efficiently during querying as well as storing S. When the assumption in (12) is not true in practice, we may simply put the element i that satisfies $k_i^* > B$ to the B-th category (when $k_i^* > B$, $k_i = B$).

4   Space Lower Bound Analysis

The road map of the analysis of the space lower bound is:

- we first give background knowledge of information theory and communication complexity;
- then in the two-party one-way communication model [1,2,20], we shall analyze the information cost lower bound of the best ε-error protocol for the membership function in (14), where ε represent the false positive rates;
- finally, based on the information cost lower bound, we prove a tight space lower bound for any approximated membership querying algorithm that represents a small subset S of a large universe U and answers membership queries with predefined false positive rates ε for the elements in U; we also show an approximate space lower bound for any approximated membership querying algorithm that has a given average false positive rate θ.

4.1   Background of Communication Complexity

In the two-party randomized communication complexity model [2,20], two computationally powerful probabilistic players, Alice and Bob, are required to compute a function $F : \mathscr{X} \times \mathscr{Y} \to \mathscr{Z}$ together, where $\mathscr{X}$ and $\mathscr{Y}$ are the respective input sets to Alice and Bob; $\mathscr{Z}$ is the output set. They exchange messages according to a shared protocol Π. Given fixed inputs $x \in \mathscr{X}$ for Alice and $y \in \mathscr{Y}$ for Bob, the random variable $\Pi(x, y)$ represents the message transcript obtained when Alice and Bob follow the protocol Π on the input pair x and y (the randomness is over the coins of Alice and Bob). For the definitions of a δ-error protocol for a function F and the δ-error randomized communication complexity of F in the two-party one-way communication model, please see [2,20].

Information cost is the amount of information that the message transcript $\Pi(X, Y)$ carries about the inputs of the two parties in a protocol Π. For the definition of the information cost of a protocol Π, please see [2,20]. In the following, I denotes the mutual information, and H the information entropy.

4.2 Information Cost Analysis for the Membership Function

In the following analysis, we use the two-party one-way communication model [2,20], which means that Alice sends exactly one message to Bob in the running of a protocol Π, then Bob computes the function value based on the message transcript (sent by Alice) and his input. Given that the universe $U = \{1, .., N\}$, we assume the random vector $X = (X_1, .., X_N)$ is the input to Alice, the random variable Y is the input to Bob ($X_i, i \in U$ and Y are defined in (1) and (2) respectively). Then, we use the random variable $\Pi(X)$ to represent the message transcript sent to Bob by Alice for a protocol Π in the two-party one-way communication model ($\Pi(X)$ depends only on the random variable X the input to Alice).

We now define the membership function f:

$$f(X, Y) = 1 \text{ if } X_Y = 1; f(X, Y) = 0 \text{ if } X_Y = 0; \tag{14}$$

Obviously, $f(X, Y = i) = X_i$. We generalize the error probability δ in the definition of a δ-error protocol for a function F in [2,20] to a vector of error probabilities $\varepsilon = (\varepsilon_1, .., \varepsilon_N)$ and define an ε-error protocol for the membership function f as follows:

Definition 1. (an ε-error protocol for the membership function f). Let $\varepsilon = (\varepsilon_1, .., \varepsilon_N)$, $\varepsilon_i \in [0, 1], i \in N = \{1, .., N\}$. In the two-party one-way communication model, a protocol Π is called an ε-error protocol for the membership function f (14) if there exists a function G such that for each $Y = i \in U$, $P(G(\Pi(X), Y = i) = 1 | f(X, Y = i) = 0) \leq \varepsilon_i$.

Similar to the definition of the δ-error randomized communication complexity of a function F in [2,20], we denote the cost of the best ε-error protocol for f by $R_\varepsilon(f)$, and call it the ε-error randomized communication complexity of f.

In the following, we will give a lower bound for $R_\varepsilon(f)$, by explicitly calculating the information cost lower bound of the best ε-error protocol for f (X and Y are the input to Alice and Bob respectively). For ease of illustration, we define $h(\omega) = \omega \log_2(\omega^{-1}) + (1 - \omega) \log_2((1 - \omega)^{-1})$.

Theorem 2. In the two-party one-way communication model, assumes that the input X and Y of Alice and Bob are independent, and $X_i, i \in U$ are also independent of other $X_j, j \neq i$. Then, the ε-error randomized communication complexity $R_\varepsilon(f)$ for the membership function f in (14) satisfies the following:

$$R_\varepsilon(f) \geq \sum_{i \in U} \{h(x_i) - [\varepsilon_i(1 - x_i) + x_i] h(\frac{\varepsilon_i(1 - x_i)}{\varepsilon_i(1 - x_i) + x_i})\} = Lower Bound_1 \tag{15}$$

Proof. Let Π denote the best ε-error protocol for the membership function in (14) in the two-party one-way communication model. Let $|\Pi|$ denote the length of the longest message transcript produced by the protocol Π over all possible values of X in $\{0,1\}^N$; then we have $R_\varepsilon(f) = |\Pi| \geq H(\Pi(X)) \geq I(X,Y;\Pi(X))$.

It is obvious that the random variable $\Pi(X)$ only depends on X and $\Pi(X)$ is independent of Y in the two-party one-way communication model. We also note that $X_i, i \in U$ are also independent of other $X_j, j \neq i$. Then we get: $I(X,Y;\Pi(X)) \geq \sum_{i \in U} \{h(x_i) - H(X_i|\Pi(X))\}$.

Table 2. Joint distribution of X_i and G_i

G_i \ X_i	0	1	marginal
0	$(1-\varepsilon_i)(1-x_i)$	$\varepsilon_i(1-x_i)$	$1 - x_i$
1	0	x_i	x_i
marginal	$(1-\varepsilon_i)(1-x_i)$	$\varepsilon_i(1-x_i)+x_i$	

Since Π is an ε-error protocol for f in (14), Definition 1 indicates that there is a function G satisfies: $P(G(\Pi(X), Y = i) = 1|X_i = 0) \leq \varepsilon_i$. We denote $G(\Pi(X), Y = i)$ by G_i for $i \in U$.

Since G is a function, the value of $G(\Pi(X), Y = i)$ is fixed when the value of $\Pi(X)$ is known. Then by the basic properties of entropy, we have $H(X_i|\Pi(X)) = H(X_i|\Pi(X), Y = i) = H(X_i|\Pi(X), Y = i, G_i) \leq H(X_i|G_i)$. It's easy to see that $H(X_i|G_i)$ in the above formula is maximized, if $P(G_i = 1|X_i = 0) = \varepsilon_i$. Thus, we get the joint distribution of G_i and X_i for each $i \in U$ in Table 2. On the basis of this table, we can get $H(X_i|G_i) = (\varepsilon_i(1-x_i)+x_i)h(\frac{\varepsilon_i(1-x_i)}{\varepsilon_i(1-x_i)+x_i})$. By summing over all $i \in U$, we get (15).

4.3 Tight Space Lower Bound for Approximated Membership Querying Algorithms

In this paper, we assume that an approximated membership querying algorithm is composed of two parts: the representing part, which represents a small subset S of U in a memory-efficient sketch structure; and the querying part, which answers membership queries with predefined false positive rates. We furthermore assume that the input to the representing part is independent from the input to the querying part. These assumptions are quite general, and almost all approximated membership querying algorithm based on memory sketch structure (cuckoo hashing, Bloom filters and its variant, and others data structures in [5,6,8,10,15,19,22,23,24,26]) can be decomposed into these two parts.

Thus, if the representing part of an approximated membership querying algorithm Γ plays the role of Alice and the querying part plays the role of Bob, we can easily transform an approximated membership querying algorithm into an ε-error protocol Π^* for membership function f (defined in (14)) in the two-party one-way communication model. The cost of Π^* is the number of bits used in the sketch structure. It follows that we can give the space lower bound for any approximated membership querying algorithm based on the information cost lower bounds $LowerBound_1$ in (15).

Let us assume that the inputs of an algorithm Γ_ε are the same as the inputs to Alice and Bob in an ε-error protocol, which are $X = (X_1, .., X_N)$ and Y ($X_i, i \in U$ and Y are defined in (1) and (2)). Let $\varepsilon = (\varepsilon_1, .., \varepsilon_N)$.

Definition 2. *(ε-error approximated membership querying algorithm). Given false positive rate ε_i for each query element $i \in U$, an ε-error approximated membership querying algorithm Γ_ε is an algorithm that represents an input set $S \subset U$ into a memory-efficient structure, and answers membership queries with false positive rate (or probability) ε_i for each element $i \in U$. The elements in S, called members, are generated subject to the membership likelihoods $(P(X_i = 1) = x_i$, see (1)) of the elements in U.*

To be more specific, for a query element $i \in U$, if $i \notin S$, the probability that Γ_ε falsely classifies i as a member of S is ε_i.

Theorem 3. *When the inputs X and Y are independent, and $X_i, i \in U$ are also independent of other $X_j, j \neq i$, then $LowerBound_1$ in (15) is a space lower bound for any Γ_ε.*

Proof. We can easily transform an ε-error approximated membership querying algorithm Γ_ε into an ε-error protocol Π for the membership function f defined in (14). Clearly, they solve the same function. This transformation can be done by changing the memory space used by the ε-error algorithm Γ_ε into the message transcript of an ε-error protocol, which is similar to the blackboard model in Proposition 1 of [20]. The inputs of Alice and Bob in the ε-error protocol are X and Y respectively. Thus, $R_\varepsilon(f)$ is a space lower bound of any ε-error approximated membership querying algorithm Γ_ε.

Then the information cost lower bound $LowerBound_1$ (shown in (15)) of $R_\varepsilon(f)$ is also a space lower bound of any ε-error approximated membership querying algorithm Γ_ε. We have the conclusion.

Next, we will prove that $LowerBound_1$ in (15) is a tight space lower bound for any ε-error algorithm.

Corollary 1. *Suppose that inputs X and Y are independent, and $X_i, i \in U$ are also independent of other $X_j, j \neq i$. Then $LowerBound_1$ in (15) is a tight space lower bound for any ε-error approximated membership querying algorithm Γ_ε.*

Proof. If we take $\varepsilon_i = 1$, then $LowerBound_1$ in (15) is simply equal to 0. We can design an $\varepsilon = \{1\}^N$-error approximated membership querying algorithm Γ^* as follows:
For any input S, Γ^* doesn't use any bits of memory. For each query element $Y = i \in U$, Γ^* always outputs that i is an member of S. Thus, the false positive rate for $i \in U$ is $P(\Gamma^*$ outputs i is an member of $S|X_i = 0) = 1$, Thus we have the conclusion.

4.4   The Approximate Space Lower Bound for Approximated Membership Querying Algorithms

Definition 3. *(average false positive rate of an ε-error approximated membership querying algorithm Γ_ε). Let $\varepsilon = (\varepsilon_1, .., \varepsilon_N)$. We define a weighted sum $\sum_{i \in U} E[r_i \varepsilon_i] = \sum_{i \in U} E[r_i] \varepsilon_i$ as the average false positive rate of an ε-error approximated membership querying algorithm Γ_ε, where r_i is defined in (5).*

Theorem 4. *Let the expected size of S and the size of U be n in (3) and N, respectively. Suppose that inputs X and Y are independent, and $X_i, i \in U$ are also independent of other $X_j, j \neq i$. Then any approximated membership querying algorithm that has an average false positive rate θ but no false negative errors must use a number of bits larger than the following approximately:*

$$\sum_{i \in U} x_i \log_2((E[r_i]n)/(x_i \theta)) \tag{16}$$

Proof. We give the intuition of the proof. Since $LowerBound_1$ shown in (15) gives space lower bound for Γ_ε with various false positive rates $\varepsilon = (\varepsilon_1, .., \varepsilon_N) \in [0,1]^N$, we can find the specific values of $\varepsilon_i, i \in U$ that minimize $LowerBound_1$ under the constraint that $\sum_{i \in U} E[r_i] \varepsilon_i = \theta$. This leads to the space lower bound (16). We outline the main steps in the following.

Firstly, since $x_i \ll 1$ usually, then $LowerBound_1$ is approximately equal to $\sum_{i \in U} x_i \log_2 \frac{1}{\varepsilon_i}$. Then we minimize $\sum_{i \in U} x_i \log_2 \frac{1}{\varepsilon_i}$ under the equality constraint that $\sum_{i \in U} E[r_i] \varepsilon_i = \theta$. We set up a function: $L(\varepsilon_1, .., \varepsilon_N, \lambda) = \sum_{i \in U} x_i \log_2 \frac{1}{\varepsilon_i} + \lambda(\sum_{i \in U} E[r_i] \varepsilon_i - \theta)$, where λ is a Lagrange multiplier. Taking the derivative with respect to $\varepsilon_i, i \in U$ of $L(\varepsilon_1, .., \varepsilon_N, \lambda)$, and setting it to zero yields:

$$\varepsilon_i = x_i/(\lambda E[r_i] \ln 2) \tag{17}$$

Through plugging (17) into the equality constraint $\sum_{i \in U} E[r_i] \varepsilon_i = \theta$ and using that fact that the expected size of S is n (shown in (3)), we have $\lambda = n/(\theta \ln 2)$. Plugging $\lambda = n/(\theta \ln 2)$ into (17), we get $\varepsilon_i = (\theta x_i)/(n E[r_i])$.

Considering that $\sum_{i \in U} x_i \log_2 \frac{1}{\varepsilon_i}$ is a convex function of $\varepsilon_i, i \in U$, thus $\sum_{i \in U} x_i \log_2 \frac{1}{\varepsilon_i}$ is minimized when $\varepsilon_i = (\theta x_i)/(n E[r_i])$, and we get (16).

Following Theorem 1, it is easy to see that the memory space used by an IWBF is within a factor of $\log_2 e \approx 1.44$ of the space lower bound in (16).

4.5   Generalization of the Space Lower Bound based on Pigeonhole Principle

We show the space lower bound analysis based on the pigeonhole principle describe in [7,9] is a special case of the approximate lower bound (16) in Theorem 4.

In the analysis of the lower bound in [7,9], given N is the size of a universe U and n the size of S, it is assumed that all element $i, i \in U$ have equal membership likelihood and query probability. Thus $P(X_i = 1) = x_i = n/N$, $P(Y = i) = y_i = 1/N$ and $E[r_i] = 1/N$. We can plug these into (16), and get $n \log_2(1/\theta)$, which is the space lower bound proposed in [7].

5   Conclusions

We have proposed an improved weighted Bloom filter (IWBF) and proved it to have a lower average false positive rate than that of the weighted Bloom filter given the same memory space. In the two-party one-way communication model, based on information cost in communication complexity theory, we have shown a tight space lower bound for any approximated membership querying algorithm that has predefined false positive rates for elements in a large universe U, when the query frequencies and membership likelihoods of the elements in U are known. We also provided an approximate space lower bound for any approximated membership querying algorithm that has an average false positive rate. We then showed that the number of bits used in an IWBF is within a factor of 1.44 of the approximate space lower bound.

Appendix: Proof of Theorem 1

Proof. Given a set A_K, we shall find the minimizer $\overline{f(K)} = (f_1(K), .., f_N(K))$ that produces the minimum value $FP(\overline{f(K)})$ of $FP(\overline{k})$ in A_K. Obviously, this is a minimization problem with the equality constraint $\sum_{i \in U} x_i k_i = K$. We set up a function $L(\overline{k}, \lambda) = FP(\overline{k}) + \lambda(\sum_{i \in U} x_i k_i - K)$, where λ is a Lagrange multiplier, $\overline{k} = (k_1, .. k_N)$, and $FP(\overline{k})$ is as in (7). Taking the derivative with respect to k_i for $L(\overline{k}, \lambda)$ and setting it to zero, $\partial L(\overline{k}, \lambda)/\partial k_i = 0$, we have:

$$(1 - e^{-K/m})^{k_i} = (-\lambda x_i)/(E[r_i] \ln(1 - e^{-K/m})). \qquad (18)$$

Raising both sides by the power x_i, we get $(1 - e^{\frac{-K}{m}})^{x_i k_i} = [\frac{-\lambda x_i}{E[r_i] \ln(1 - e^{\frac{-K}{m}})}]^{x_i}$, and thus

$$\prod_{j \in U}(1 - e^{\frac{-K}{m}})^{x_j k_j} = \prod_{j \in U}[\frac{-\lambda x_j}{E[r_j] \ln(1 - e^{\frac{-K}{m}})}]^{x_j}. \qquad (19)$$

By $\sum_{j \in U} x_j k_j = K$ and the $\sum_{j \in U} x_j = n$, (19) can be reduced to:

$$\lambda = -\ln(1 - e^{-K/m})(1 - e^{-K/m})^{K/n} \prod_{j \in U}[E[r_j]/x_j]^{x_j/n}. \qquad (20)$$

Plugging (20) into (18) and eliminating λ, we can find the minimizer $\overline{f(K)} = (f_1(K), f_2(K), .., f_N(K))$ for $FP(\overline{k})$ over set A_K as follows: $\forall i \in U$,

$$f_i(K) = K/n + \sum_{j \in U} \frac{x_j}{n \ln(1 - e^{-K/m})} \ln \frac{x_i E[r_j]}{x_j E[r_i]}. \qquad (21)$$

We plug (21) into $FP(\overline{k})$ in (7) $FP(\overline{f(K)})$ can be written as

$$FP(\overline{f(K)}) = n e^{-(m/n) \ln(e^{-K/m}) \ln(1 - e^{-K/m})} \prod_{j \in U}[E[r_j]/x_j]^{x_j/n}. \qquad (22)$$

Since $FP(\overline{k})$ is convex in set A_K, it is obvious that $\forall \overline{k} \in A_K$, $FP(\overline{k}) \geq FP(\overline{f(K)})$. Noting $\ln(x) \ln(1 - x)$ is maximized when $x = 0.5$, it is easy to see that $FP(\overline{f(K)})$ in (22) as a function of K has a global minimizer

$K^* = m \ln 2$, thus $\forall K' \in R_+, FP(\overline{f(K')}) \geq FP(\overline{f(K^*)})$. Furthermore, $\forall \overline{k} \in R_+^N$, we have $\overline{k} \in A_{K'}$: $\overline{k}$ must be contained in a set $A_{K'}, K' \in R_+$. Thus we have $FP(\overline{k}) \geq FP(\overline{f(K')}) \geq FP(\overline{f(K^*)})$. Then we know $\overline{f(K^*)}$ is the global minimizer of $FP(\overline{k})$ in R_+^N ($k_i^* = f_i(K^*), i \in U$). Plugging $K^* = m \ln 2$ into (21) and (22), we obtain (10) and (11).

In the following, we will prove $T(\overline{k^*}) - FP(\overline{k^*}) < 2e^{-n^{1/3}/3}$ based on the sufficient condition shown in (12).

Let $Z = \sum_{j \in U} X_j k_j^*$, According to (4), we have $T(\overline{k^*}) = E[\sum_{i \in U} r_i (1 - e^{\frac{-Z}{m}})^{k_i^*}]$, which can be decomposed as follows:

$$T(\overline{k^*}) = E\Big[\sum_{i \in U} r_i(1 - e^{\frac{-Z}{m}})^{k_i^*} \Big| |Z - K^*| \geq m/n^{\frac{1}{3}}\Big] P\Big(|Z - K^*| \geq m/n^{\frac{1}{3}}\Big)$$
$$+ E\Big[\sum_{i \in U} r_i(1 - e^{\frac{-Z}{m}})^{k_i^*} \Big| |Z - K^*| < m/n^{\frac{1}{3}}\Big] P\Big(|Z - K^*| < m/n^{\frac{1}{3}}\Big). \quad (23)$$

Noting that the membership likelihoods $x_i \in (0,1)$, and $\sum_{i \in U} x_i/n = 1$ as shown in (3), then based on the assumption in (12), it is easy to see that $k_i^* \in (0, \frac{2m}{n} \ln 2), \forall i \in U$. Thus $k_{max} = \max\{k_1^*, .., k_N^*\} < (2m/n) \ln 2$. From $K^* = m \ln 2$, it is easy to see $v = \sum_{i \in U} x_i^2 k_i^* \leq k_{max} \sum_{i \in U} x_i k_i^* < m^2/n$. By $v < m^2/n$ and $k_{max} < (2m/n) \ln 2$, we know $\frac{m^2}{2vn^{2/3}} > \frac{n^{1/3}}{3}$ and $\frac{m^2}{2(vn^{2/3} + \frac{1}{3}k_{max}mn^{1/3})} > \frac{n^{1/3}}{3}$. Then, plugging these two inequalities into (9) of lemma 1, we can get

$$P(|\sum_{i \in U} X_i k_i^* - K^*| \geq m/n^{1/3}) < 2e^{-\frac{n^{1/3}}{3}}. \quad (24)$$

Obviously, $\forall i \in U$, since $k_i^* > 0$, it is easy to see $(1 - e^{-\sum_{j \in U} X_j k_j^*/m})^{k_i^*} < 1$, thus we have the following (r_i is defined in (5)):

$$E\Big[\sum_{i \in U} r_i(1 - e^{\frac{-\sum_{j \in U} X_j k_j^*}{m}})^{k_i^*} \Big| |\sum_{i \in U} X_i k_i^* - K^*| \geq \frac{m}{n^{\frac{1}{3}}}\Big] < E[\sum_{i \in U} r_i] = 1. \quad (25)$$

Conditioning on $|\sum_{i \in U} X_i k_i^* - K^*| < m/n^{1/3}$ and $K^* = m \ln 2$, we get

$$e^{-K^*/n} e^{-1/n^{1/3}} < e^{\frac{-\sum_{j \in U} X_j k_j^*}{m}} < e^{-K^*/n} e^{1/n^{1/3}}.$$

Considering that n (the expected size of a subset S) is usually assumed to be large in the setting for approximated membership querying, we can get

$$E\Big[\sum_{i \in U} r_i(1 - e^{\frac{-\sum_{j \in U} X_j k_j^*}{m}})^{k_i^*} \Big| |\sum_{i \in U} X_i k_i^* - K^*| < m/n^{\frac{1}{3}}\Big] \approx FP(\overline{k^*}). \quad (26)$$

Based on (23), (24), (25) and (26), we can see that $T(\overline{k^*}) < 2e^{-\frac{n^{1/3}}{3}} + FP(\overline{k^*})$, which is the inequality shown in (13). Then $T(\overline{k^*}) \approx FP(\overline{k^*})$.

Acknowledgments. This work is supported by the Natural Science Foundation of China (No.61402008 and No.61402009), Natural Science Foundation of Anhui Province (No.1408085QF128), and Major Program of the Natural Science Foundation of the Anhui Higher Education Institutions of China (No.KJ2014ZD05).

References

1. Ablayev, F.: Lower bounds for one-way probabilistic communication complexity and their application to space complexity. Theoretical Computer Science **157**, 139–159 (1996)
2. Bar-Yossef, Z., Jayram, T., Kumar, R., Sivakumar, D.: An information statistics approach to data stream and communication complexity, vol. 68, pp. 702–732. Academic Press Inc. (2004)
3. Berinde, R., Indyk, P., Cormode, G., Strauss, M.J.: Space-optimal heavy hitters with strong error bounds. ACM Transactions on Database Systems (TODS) **35**(4), 26 (2010)
4. Bloom, B.H.: Space/time trade-offs in hash coding with allowable errors. Communications of the ACM **13**(7), 422–426 (1970)
5. Bonomi, F., Mitzenmacher, M., Panigrah, R., Singh, S., Varghese, G.: Beyond bloom filters: from approximate membership checks to approximate state machines. In: ACM SIGCOMM Computer Communication Review, vol.36, pp. 315–326. ACM (2006)
6. Bonomi, F., Mitzenmacher, M., Panigrahy, R., Singh, S., Varghese, G.: An improved construction for counting bloom filters. In: Azar, Y., Erlebach, T. (eds.) ESA 2006. LNCS, vol. 4168, pp. 684–695. Springer, Heidelberg (2006)
7. Broder, A., Mitzenmacher, M.: Network applications of bloom filters: A survey. Internet Mathematics **1**(4), 485–509 (2004)
8. Bruck, J., Gao, J., Jiang, A.: Weighted bloom filter. In: 2006 IEEE International Symposium on Information Theory, pp. 2304–2308. IEEE (2006). Extented version in http://www.paradise.caltech.edu/papers/etr072.pdf
9. Carter, L., Floyd, R., Gill, J., Markowsky, G., Wegman, M.: Exact and approximate membership testers. In: Proceedings of the Tenth Annual ACM Symposium on Theory of Computing, pp. 59–65. ACM (1978)
10. Chakrabarti, K., Chaudhuri, S., Ganti, V., Xin, D.: An efficient filter for approximate membership checking. In: Proceedings of the 2008 ACM SIGMOD International Conference on Management of Data, pp. 805–818. ACM (2008)
11. Chung, F., Lu, L.: Concentration inequalities and martingale inequalities: a survey. Internet Mathematics **3**(1), 79–127 (2006)
12. Cormode, G., Hadjieleftheriou, M.: Methods for finding frequent items in data streams. The VLDB Journal **19**(1), 3–20 (2010)
13. Cormode, G., Muthukrishnan, S.: What's hot and what's not: tracking most frequent items dynamically. ACM Transactions on Database Systems (TODS) **30**(1), 249–278 (2005)
14. Demaine, E.D., López-Ortiz, A., Munro, J.I.: Frequency estimation of internet packet streams with limited space. In: Möhring, R.H., Raman, R. (eds.) ESA 2002. LNCS, vol. 2461, pp. 348–360. Springer, Heidelberg (2002)
15. Deng, F., Rafiei, D.: Approximately detecting duplicates for streaming data using stable bloom filters. In: Proceedings of the 2006 ACM SIGMOD International Conference on Management of Data, pp. 25–36. ACM (2006)
16. Guo, D., Li, M.: Set reconciliation via counting bloom filters. IEEE Transactions on Knowledge and Data Engineering **25**(10), 2367–2380 (2013)
17. Guo, D., Liu, Y., Li, X., Yang, P.: False negative problem of counting bloom filter. IEEE Transactions on Knowledge and Data Engineering **22**(5), 651–664 (2010)
18. Guo, D., Wu, J., Chen, H., Yuan, Y., Luo, X.: The dynamic bloom filters. IEEE Transactions on Knowledge and Data Engineering **22**(1), 120–133 (2010)

19. Hua, Y., Xiao, B., Veeravalli, B., Feng, D.: Locality-sensitive bloom filter for approximate membership query. IEEE Transactions on Computers **61**(6), 817–830 (2012)
20. Jayram, T.: Information complexity: a tutorial. In: Proceedings of the Twenty-Ninth ACM SIGMOD-SIGACT-SIGART Symposium on Principles of Database Systems, pp. 159–168. ACM (2010)
21. Karp, R.M., Shenker, S., Papadimitriou, C.H.: A simple algorithm for finding frequent elements in streams and bags. ACM Transactions on Database Systems (TODS) **28**(1), 51–55 (2003)
22. Kirsch, A., Mitzenmacher, M.: Less hashing, same performance: building a better bloom filter. In: Azar, Y., Erlebach, T. (eds.) ESA 2006. LNCS, vol. 4168, pp. 456–467. Springer, Heidelberg (2006)
23. Liu, Y., Chen, W., Guan, Y.: Near-optimal approximate membership query over time-decaying windows. In: 2013 Proceedings IEEE, INFOCOM, pp. 1447–1455. IEEE (2013)
24. Metwally, A., Agrawal, D., El Abbadi, A.: Duplicate detection in click streams. In: Proceedings of the 14th International Conference on World Wide Web, pp. 12–21. ACM (2005)
25. Pagh, R., Rodler, F.F.: Cuckoo hashing. Journal of Algorithms **51**(2), 122–144 (2004)
26. Zhong, M., Lu, P., Shen, K., Seiferas, J.: Optimizing data popularity conscious bloom filters. In: Proceedings of the Twenty-Seventh ACM Symposium on Principles of Distributed Computing, pp. 355–364. ACM (2008)

Tree Contraction for Compressed Suffix Arrays on Modern Processors

Takeshi Yamamuro[1]([⊠]), Makoto Onizuka[2], and Toshimori Honjo[1]

[1] NTT Corp., Tokyo, Japan
{yamamuro.takeshi,honjo.toshimori}@lab.ntt.co.jp
[2] Graduate School of Information Science and Technology,
Osaka University, Osaka, Japan
onizuka@ist.osaka-u.ac.jp

Abstract. We propose a novel processor-aware compaction technique for pattern matching that is widely-used in databases, information retrieval, and text mining. As the amount of data increases, it is getting important to efficiently store data on memory. A compressed suffix array (CSA) is a compact data structure for in-memory pattern matching. However, CSA suffers from tremendous processor penalties, such as a flood of instructions and cache/TLB misses due to the lack of processor-aware design. To mitigate these penalties, we propose a novel compaction technique for CSA, called suffix trie contraction (STC). The frequently accessed suffixes of CSA are transformed to a trie (e.g., a suffix trie), and then inter-connected nodes in the trie are repeatedly 'contracted' to a single node, which enables lightweight sequential scans in a processor-friendly way. In detail, STC consists of two contraction techniques: fixed-length path contraction (FPC) and sub-tree contraction (SC). FPC is applied to the parts with a few branches in the trie, and SC is applied to the parts with many branches. Our experiment results indicate that FPC outperforms naive CSA by two orders of magnitude for short pattern queries and by three times for long pattern queries. As the number of branches inside the trie increases, SC gradually becomes superior to CSA and FPC for short pattern queries. Finally, the latency and throughput of STC are 7 times and 72 times better than those of CSA for the TREC test data set at the expense of additional 7.1 % space overhead.

Keywords: Pattern matching · Tree traversal · SIMD · Compression

1 Introduction

Pattern matching is a well-known and important task in many domains such as databases, information retrieval, and text mining. As the amount of data increases, it is getting important to efficiently store data on memory. A compressed suffix array (CSA) is a compact data structure for in-memory pattern matching and is a compressed form of a suffix array (SA). SA is composed of two data structures: suffixes S and a suffix array SA. When an input of characters

© Springer International Publishing Switzerland 2015
M. Renz et al. (Eds.): DASFAA 2015, Part II, LNCS 9050, pp. 363–378, 2015.
DOI: 10.1007/978-3-319-18123-3_22

is $T[0...N-1]=t_0t_1...t_{N-1}$, $S[i]$ is defined as $T[i...N-1]$. Here, $T[i]$ is a symbol of characters: $T[i] \in \Sigma$ ($0 < i < N$ - 1 and Σ is a set of characters). SA is a sequence of array indices referring to S and it is alphabetically sorted. SA enables fast pattern matching; given a pattern query $P[0...M-1]=p_0p_1...p_{M-1}$ ($P[i] \in \Sigma$, where $0 < i < M$ - 1), the search algorithm returns the array indices of the suffixes whose prefixes are equal to P. Since SA is alphabetically sorted, the algorithm is executed by using binary searches on SA [5].

CSA exploits succinct data structures in order to compress SA and was extensively studied in the 2000s [8]. Algorithms for succinct data structures use *rank* and *select* operations [15]; given $B[0...N-1]=b_0b_1...b_{N-1}$ and $B[i]=\{b_i|b_i \in \{0,1\}\}$, $rank_1(B,i)$ returns the total count of 1s in $B[0...i]$ and $select_1(B,i)$ returns the (i+1)-th position of 1s. Although CSA has an advantage in terms of space complexity, two major performance issues arise for pattern matching on modern processors as follows.

- CSA calls *rank* and *select* operations a large number of times. These calls heavily put pressure on modern processors; they cause a flood of instructions and random access on memory.
- The papers [16][7][2] reported that binary searches are inefficient on modern processors in terms of cache/TLB[1] misses and branch penalties.

To illustrate these inefficiencies of CSA on processors, we made a preliminary experiment. The relative performance penalties of CSA over SA are shown in Fig.1: the number of executed instructions, data references, and L1/L2 cache misses. The setting of this experiment was the same as those of the experiments described in Section 4. These details are further described in Section 4.1. This result indicates that CSA is inferior to SA by three orders of magnitude in terms of the number of instructions and data references. All the penalties gradually grow as the pattern length increases.

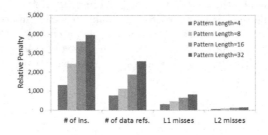

Fig. 1. Relative penalties of CSA compared to those of SA by varying the pattern length between 4 and 256. A 1GiB portion of gov2 in the TREC Terabyte Track was used as a test data set.

Our goal is to design a processor-friendly technique for pattern matching to be more efficient without losing the compression advantage of CSA. We propose

---

[1] Translation Look-aside Buffer, it represents a cache mechanism that holds often-used translation entries from logical addresses to physical ones in memory.

a novel compaction technique, called suffix trie contraction (STC). Our idea is that the frequently accessed suffixes of CSA are transformed to a trie (e.g., a suffix trie), and then inter-connected nodes in the trie are repeatedly '*contracted*' to a single node in physically consecutive memory addresses. This contraction enables pattern matching to exclude the inefficient *rank/select* operations and the binary searches of CSA by using lightweight sequential scans in a processor-friendly way. This is efficient because frequently accessed suffixes are co-located on memory and the co-location leads to improving cache efficiency. STC consists of two contraction techniques depending on the trie structure: fixed-length path contraction (FPC) and sub-tree contraction (SC). FPC selects a fixed-length path with a few branches in the trie from the root to a leaf, and then contracts it to a single node. As for the parts with many branches in the trie, SC contracts a sub-tree instead of a path. FPC and SC are repeatedly applied to sub-trees, and then the trie is re-constructed to be efficient on processors. STC switches FPC and SC optimally depending on the number of branches in the trie.

Our contributions are as follows;

- We propose a novel contraction technique, called STC. The frequently accessed suffixes of CSA are transformed to a trie, and then STC repeatedly contracts nodes in the trie, which enables lightweight sequential scans in a processor-friendly way.
- STC consists of FPC and SC and switches the two contraction strategy optimally depending on the number of branches in the trie; FPC contracts a fixed-length path and SC contracts a sub-tree instead of a path.
- We implemented these techniques on our prototype and verified their effectiveness; FPC outperforms naive CSA by two orders of magnitude for short pattern queries and by three times for long pattern queries. As the number of branches increases, SC gradually becomes superior to CSA and FPC for short pattern queries. Finally, the latency and throughput of STC are 7 times faster and 72 times higher than those of CSA for the TREC data set at the expense of additional 7.1 % space overhead.

This paper is organized as follows; we present the overview of our proposals and the details in Section 2 and Section 3, respectively. We describe how we implemented our techniques and verify the effectiveness in Section 4. Finally, Section 5 discusses related work and Section 6 concludes our findings.

2   Design Overview

A design overview of our proposed technique is shown in Fig.2. The frequently accessed suffixes of CSA are transformed to a processor-optimized trie, and the trie is used to search the suffixes that have the same prefixes in common with given pattern queries. For example, the bold path $\alpha_1 b_2 \alpha_3 b_3 \alpha_4$ in Fig.2 represents nodes and edges in the trie, and the leaf refers to the suffixes that have $\alpha_1 b_2 \alpha_3 b_3 \alpha_4$ as a prefix in common. That is, the bold path is regarded as short-cut to the corresponding suffixes of CSA. We assume that an input query is

$\alpha_1 b_2 \alpha_3 b_3 \alpha_4$ x (x is an arbitrary character). The processor-optimized trie is used to find a set of suffixes that have $\alpha_1 b_2 \alpha_3 b_3 \alpha_4$, and then $\alpha_1 b_2 \alpha_3 b_3 \alpha_4$ x is finally searched from the set in a similar way of CSA.

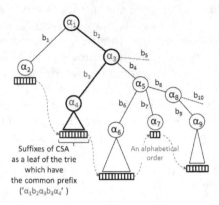

Fig. 2. Overview of our proposed technique. The frequently accessed suffixes of CSA are transformed to a processor-optimized trie (the higher layer of nodes and edges in the figure), and the trie is used to search the suffixes that have the same prefixes in common with given pattern queries. That is, the trie is regarded as short-cut to the corresponding suffixes of CSA.

Our proposed suffix tree contraction (STC) builds a processor-optimized trie as follows; first, S (a string set of suffixes) is build from T in a similar way of the original SA. The frequently accessed suffixes of SA are transformed to a trie (e.g., a suffix trie) where nodes and edges are labeled with symbols. Every single node can have multiple symbols (possibly empty $\emptyset$). Every edge has a single symbol that is called a 'branching character'. Given a string set S, the longest common prefix α in S is labeled to the root, and the next divergent symbol represent a branching character. On the basis of the assumption that one symbol among the branching symbols is b, S_b is obtained by removing α and b from the prefix of S: $S_b = \{\beta | \alpha b \beta \in S\}$. S_b is assigned as the symbol of the node that is connected to the root by an edge labeled b. STC selects inter-connected nodes in the trie, and then 'contracts' them to a single node in physically-consecutive memory addresses. This contraction is repeatedly processed in the subsequent sub-trees so that pattern matching can exclude the inefficient $rank/select$ operations and the binary searches of CSA by using sequential scans in a processor-friendly way. Finally, a pair of the trie and CSA are used for efficient pattern matching.

STC consists of two contraction techniques: fixed-length path contraction (FPC) and sub-tree contraction (SC). FPC selects a fixed-length path ($\alpha_1 b_2 \alpha_3 b_3 \alpha_4$ in Fig.2) from the root to a leaf, and contracts it to a single node. Similarly, SC contracts a sub-tree ($\alpha_1 b_1 \alpha_2$, $\alpha_1 b_2 \alpha_3 b_3 \alpha_4$, and $\alpha_1 b_2 \alpha_3 b_4 \alpha_5$ in Fig.2) instead of a path. FPC is applied to paths with a few branches in the trie, and SC is applied to the parts with many branches. The node selection in STC and

SC needs to choose frequently accessed nodes which refer to the most frequently accessed suffixes because getting these nodes together leads to improving cache efficiency. Therefore, FPC and SC select nodes depending on the number of symbols under the nodes in the trie (the size of the solid triangles in Fig.2) because we expect that a path or a sub-tree with more symbols tend to have a higher probability of reference. The algorithms of the node selection and the traversal of the optimized trie are further described in following Section 3.1 and 4.1.

3   Tree Contraction

3.1   Fixed-Length Path Contraction

FPC is a simple contraction technique to improve access locality for cache efficiency. We modify a previously proposed technique, named path decomposition [1]; it selects a single path from the root to one leaf, and then contracts 'all' nodes on the selected path to a single node. In contrast, our technique contracts the predefined constant fixed-length path from the root. There are two reasons for this modification; first, the referential probability of symbols farther away from the root become lower relatively. Co-locating frequently accessed nodes with infrequently accessed ones may make performance worsen because of cache inefficiency. Second, it is difficult for processors to handle variable-length data as compared to fixed-length data. Recent studies proposed state-of-the-art techniques to process fixed-length data by using SIMD (e.g., Single Instruction, Multiple Data) instructions for cache efficiency and branch-free structures [16][7][2]. Our technique also uses such instructions in order to boost symbol comparisons for pattern matching. This contraction is repeatedly applied to subsequent sub-trees the specified number of times. As a result, the contracted nodes are expected to be accessed together because the most of highly-referenced nodes are located in physically consecutive memory addresses. An overview of FPC is shown in Fig.3, in which $\alpha_1 b_2 \alpha_3 b_4 \alpha_5 b_5 \alpha_6$ is contracted to β_1 in physically consecutive memory addresses.

Node Selection. Algorithm 1 shows a pseudo code to select P_{len}-length paths with the largest weight from input suffixes S. Notice that symbols in Σ strictly correspond to integers from 1 to $|\Sigma|$ with the alphabetical orders preserved. Algorithm 1 selects a path symbol-by-symbol with P_{len} loops by using a heap tree (lines 7-21). First, it looks for a path $path + s$ in S by using a backward search (lines 12-13). The backward search returns the corresponding range of SA indices referring to the suffixes that have the same prefixes in common with $path + s$ [9]. If it exists, it calculates $weight$ and pushes s with $weight$ together to a heap tree that has a node with the maximum weight as the root (lines 14-15). Finally, it picks up a symbol p with the maximum $weight$ from the heap tree and appends p to the tail of $path$ (lines 18-19). The process is repeated up to P_{len} and a P_{len}-length path is selected. Algorithm 1 is repeatedly executed E_{num} times to contract paths with the highest weight.

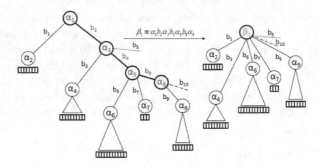

Fig. 3. Overview of the FPC behavior. FPC selects a path from the root towards a leaf by the predefined constant fixed-length of nodes, and then the path $(\alpha_1 b_2 \alpha_3 b_4 \alpha_5 b_5 \alpha_8)$ is contracted to the single node (β_1). This contraction is repeatedly applied to the subsequent sub-trees depicted by triangles with solid lines.

Algorithm 1 Pseudo code to select a P_{len}-length path by FPC

```
 1: /* IN: Plen is a predefined constant fixed-length of nodes */
 2: /* IN: S is a set of suffixes */
 3: /* IN: SA is a suffix array */
 4: /* IN: N is the size of SA */
 5: /* IN: path is an empty string */
 6: /* OUT: A single path selected by FPC */
 7: for i ← 1 to Plen do
 8:    range = (0, N-1);
 9:    for s ← 1 to |Σ| do
10:       /* range is an index range of SA, which has path + s */
11:       /* as a common prefix of the suffixes in S */
12:       backward_search(path, s, range, SA);
13:       if path + s exists in S then
14:          Calculate weight for path + s;
15:          Push s with weight to a heap tree;
16:       end if
17:    end for
18:    Pop s from the root of a heap tree;
19:    Append s to the tail of path;
20:    Reset the heap tree;
21: end for
```

Node Traversal. Fig.4 illustrates how paths selected by Algorithm 1 are traversed. We assume that $L_{path1} = \alpha_1 b_2 \alpha_3 b_4 \alpha_5 b_8 \alpha_8$ and L_{path2}, which is a contracted node having α_4 as the root, are selected by FPC in Fig.3. Two additional data structures are needed so as to traverse nodes from L_{path1} to L_{path2}: N_{small} and N_{large}. They represent the corresponding pointers of L_{path1} to next subtrees (the solid triangles in Fig.3). For example, we assume that an input query is $\alpha_1 b_2 \alpha_3 b_3 \alpha_4$ in Fig.4. Symbols in the query are compared with L_{path1}; they match substring $\alpha_1 b_2 \alpha_3$ of each other, and then the next corresponding symbols are different (b_3 and b_4). b_3 is 'smaller' than b_4 and p_{b_3} in N_{small} is selected as a pointer to L_{path2} below α_4. Symbol comparisons in L_{path1} can be executed with a vectorized SIMD instruction, or *pcmpestri* of Intel SSE4.2. They have the ability to compare 16-byte symbols in L_{path1} simultaneously and output the comparison results as a byte array. This array can be directly mapped to

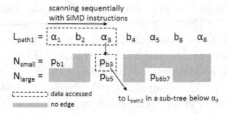

Fig. 4. Overview of traversal between paths (L_{path1} and L_{path2}) contracted from Algorithm 1. An input query is assumed to be $\alpha_1 b_2 \alpha_3 b_3 \alpha_4$. The method of the traversal applies a SIMDized sequential scan to L_{path} for comparing symbols, checking a corresponding pointer (p_{b3}) in N_{small}, and jumping to L_{path2} below α_4.

pointers in N_{small} and N_{large} by using a predefined look-up table. The performance results of both scalar and vectorized implementations are compared in Section 4.

3.2   Sub-tree Contraction

SC is a technique to process paths having many branching symbols in a trie. It selects not a single path but inter-connected multiple paths (e.g., a sub-tree) from the root, and then contracts them to a single node. An overview of SC and three paths (α_1 to α_4, α_1 to α_6, and α_1 to α_7) contracted to a single node γ_1 are shown in Fig.5. The contracted node γ_1 dispatches a query for pattern matching to such descendant edge whose symbol matches to the prefix of the query (e.g., $b_{\gamma11}$ and $b_{\gamma12}$).

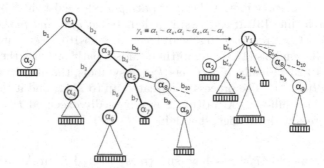

Fig. 5. Overview of the SC behavior. SC chooses a sub-tree (α_1 to α_4, α_1 to α_6, and α_1 to α_7) from the root, and then contracts it to a single node γ_1.

The contracted node γ_1 is made up of N-byte integers that are generated from the three selected paths as shown in Fig.6. The order of the three paths needs to be preserved as integer forms in γ_1 for comparing paths correctly. The paths from α_1 to α_4, from α_1 to α_6, and from α_1 to α_7 lie in an ascending alphabetical order and this order should be preserved in d_1 from the first path, d_2 from the second path, and d_3 from the last path. N-byte integers are obtained from paths by packing the symbols in each path to a single integer by bit-shifting; we assume that paths are expressed as *abc* and *abd*. In the *ASCII* code, *abc* is 97, 98 ,99, and *abd* is 97, 98, 100. *abc* is replaced by 6,382,179, or $(97 \ll 16) \mid (98 \ll 8) \mid 99$, and *abd* is replaced by 6,382,180, or $(97 \ll 16) \mid (98 \ll 8) \mid 100$. This transposition maintains the alphabetical order[2]. These integers can be sequentially scanned in a processor-friendly way so as to search next sub-trees.

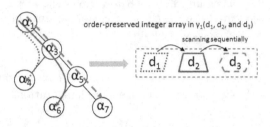

Fig. 6. The detail of the node γ_1 in Fig.5. The multiple paths selected by SC are transformed to the array of N-byte integers. Notice that the order of these paths needs to be preserved as integer forms in γ_1 for comparing paths correctly.

Node Selection. Algorithm 2 shows a pseudo code to select P_{num} paths depending on the weights. While it is similar to Algorithm 1, the major difference is that it selects not a single path, but multiple paths (*paths*). For example, Fig.5 shows that a sub-tree consists of three paths: the paths from α_1 to α_4, from α_1 to α_6, and from α_1 to α_7. Algorithm 2 selects P_{num} paths that have the same prefixes in common; first, it checks if path *path* + *s* exists in S by using the backward search (line 13). If it exists, it calculates *weight* and pushes (*path* + *s*, *range*) with *weight* together to a heap tree (lines 15-16). It executes this procedure to all the symbols $(1...|\Sigma|)$, and then finds a path that has the maximum *weight* (lines 10-23). Finally, if *path* has T_{depth} symbols, the algorithm adds the path to *paths* (line 24). The process is repeated up to P_{num} and it selects P_{num} paths that form a sub-tree. Algorithm 2 is repeatedly executed E_{num} times to contract sub-trees with the highest weight.

Node Traversal. Fig.7 shows how sub-trees selected from Algorithm 2 are traversed. Two data structures are needed in order to traverse sub-trees: L_{tree} and N_{pos}. L_{tree} has order-preserved N-byte integers and N_{pos} represents the corresponding pointers of L_{tree} to next sub-tree. We assume that an input query is transposed to integer d_Q. If d_Q follows $d_2 < d_Q < d_3$, $p_{b_{\gamma_{13}}}$ is selected as a pointer to a next sub-tree.

---

[2] This transposition needs more than 3-byte integers, e.g., *int* in C. The more symbols the paths have, the larger bytes of integers it must use at the expense of extra costs.

Algorithm 2 Pseudo code to select a sub-tree by SC

```
 1: /* IN: P_{num} is the number of paths that the output sub-tree contains */
 2: /* IN: T_{depth} is the number of symbols that the sub-tree has */
 3: /* IN: S is a set of suffixes */
 4: /* IN: SA is a suffix array */
 5: /* IN: N is the size of SA */
 6: /* IN: path is an empty string */
 7: /* IN: paths is an empty set of strings */
 8: /* OUT: A sub-tree selected by SC */
 9: for i ← 1 to P_{num} do
10:    loop
11:        range = (0, N-1);
12:        for s ← 1 to |Σ| do
13:            backward_search(path, s, range, SA);
14:            if path + s exists in S then
15:                Calculate weight with path + s;
16:                Push (path + s, range) for weight to a heap tree;
17:            end if
18:        end for
19:        Pop (cand, range) from the heap tree;
20:        if cand has T_{depth} symbols then
21:            break;
22:        end if
23:    end loop
24:    paths = paths ∩ cand;
25:    Pops (cand, range) from the root of the heap tree;
26:    path = cand;
27: end for
```

3.3 STC: Hybrid Strategy of FPC and SC

FPC and SC have some pros and cons; FPC is advantageous for a few branches in a trie and SC is robust over many branches. STC switches the contraction strategy optimally depending on the candidate set of nodes that represents a path in FPC or a sub-tree in SC. STC determines whether FPC or SC is the better strategy; it calculates the referential probability of a path and a sub-tree and compare these probability to determine which is better, FPC or SC. As described in Section 2, the referential probability is calculated from the number of symbols under the nodes in the trie because we expect that a path or a sub-tree with more symbols tend to have a higher probability of reference.

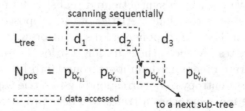

Fig. 7. Overview of traversal between sub-trees selected from Algorithm 2. It applies SIMDized sequential scans to order-preserved N-byte integers (d_1, d_2, and d_3) in L_{tree}, checks a corresponding pointer ($p_{b_{\gamma 13}}$) in N_{pos}, and jumps to a next sub-tree.

4   Evaluation

4.1   Environment Settings

Baseline Performance Values. We used a naive CSA implementation as a baseline in following experiments. CSA has the many variants of implementations and we assume the most basic implementation [8]; S and SA are transformed to ψ and a BWT (Burrows-Wheeler Transform)-applied text T by using context information. This transformation makes these data easily compressible and they are compressed by using following compression methods; ψ is split to the chunks that have integers sorted in an ascending order and each chunk is compressed with randomly-accessible gamma codes[3]. BWT-applied T is compressed with LZEnd [3] which is a variant of the LZ77 algorithms. LZEnd enables random access to a compressed text directly by using succinct data structures. Notice that our proposed technique is orthogonal to CSA implementations because the processor-optimized trie built by STC is an auxiliary structure for CSA and can be applied for arbitrary CSA implementations. Table 1 lists the baseline performance values of our CSA implementation as compared to those of a SA library, or libdivsufsort v2.0.1[4]. A 1GiB portion of gov2 in the TREC Terabyte Track was used as a test data set. SA defeats CSA by two orders of magnitude in all the length of queries because CSA has high penalties on processors (Fig.1). In contrast, CSA is beneficial in terms of space; the index size of SA is 5.0GiB compared to about 1.3GiB of CSA.

Table 1. Baseline performance values (μs) of our CSA implementation. The performance deteriorates in line with pattern length because it has high processor penalties as shown in Fig.1.

Pattern Length	4	8	16	32
CSA BASELINE	665.41	1092.66	1669.75	2352.73
SA	6.81	6.75	5.98	5.87

Other Settings. We used both synthetic and realistic data sets in our experiments so as to investigate the characteristics of the proposed contraction techniques: FPC, SC, and STC. An exponential distribution was used to generate a sequence of synthetic test symbols; the equation is $-\alpha\log(1\text{-}rand())$ where $rand()$ is a function to generate randomized numbers from 0.0 to 1.0. When α is low, the equation generates low-entropy data corresponding to a trie with fewer branches. The higher α is, the more branches a trie has. These proposed techniques were also applied to the realistic data set, or gov2 in the TREC Terabyte Track and queries in the TREC 2009 Million Query Track. Our performance experiments

---

[3] https://github.com/pfi/dag_vector
[4] http://code.google.com/p/libdivsufsort

were executed on a server with a 6-cores Xeon X5670 processor and 16GiB of memory. Caches incorporated in the processor were 32KiB, 256KiB, and 12MiB as L1, L2, and LL (Last Level) caches, respectively. The codes were written in C++ and complied by GNU Compiler Collection v4.7.1 with an option -O2.

4.2   Experimental Results

Fig.8, Fig.9, and Fig.10 plot the overview results of FPC, SC, and CSA. We used randomly-generated 1GiB test data sets with different α parameters; α was set to 0.75 in Fig.8, 1.25 in Fig.9, and 2.0 in Fig.10, respectively. FPC used $P_{len}=16$ and the scalar implementation as described in Section 3. SC used $P_{num}=15$ and $T_{depth}=8$. The E_{num} value of both techniques was set to 10,000. Additional space from 1% to 6% of the CSA index size is required for the processor-optimized trie built by FPC and SC. Fig.8 shows that FPC outperforms CSA by two orders of magnitude for short pattern queries and by three times for long pattern queries, respectively. Moreover, FPC defeats SC in all the patterns with $\alpha=0.75$. As α increases, that is, as the number of branches increases, the performance of FPC gradually approaches that of CSA in Fig.9 and Fig.10. SC is effective in a trie with many branches and defeats CSA and FPC for short pattern queries by four orders of magnitude in Fig.10. Contractly, SC has few advantages over CSA for long pattern queries. Note that both FPC and SC outperform CSA for short pattern queries from 4 to 8. This is because the frequently accessed entries of CSA are located in physically consecutive memory areas, and thus, pattern matching is mostly executed inside the processor caches. As the pattern length becomes longer, the matched pattern becomes overflowed and the performance declined as shown in the figures. The following sections describe parameter-calibrated tests; first, the relationship between space overhead and performance is investigated by varying E_{num} values in FPC and SC. The performance of FPC and SC are evaluated using the parameters described in Section 3.

FPC Analysis. This section describes our investigation of FPC in the same condition as indicated in Fig.8. Improvement indicators are plotted in Fig.11 by varying E_{num} values from 10,000 to 40,000. The indicator represents a relative latency value over CSA and is calculated as: $\frac{CSA\ Latency}{FPC\ Latency}$. Fig.11 shows that the improvement logarithmically increases from 14.3x to 20.7x with the increase of the E_{num} value and there is a slight performance gain between 30,000 and 40,000. This experiment indicates that the E_{num} value has a trade-off in terms of space and performance gains.

Fig.12 plots the calibrated tests of P_{len}. The case of $P_{len}=16$ is superior to the other cases. As the pattern length becomes longer, the performance gradually declines. This result follows our expectation as explained in Section 3; nodes farther away from the root have relatively low referential probability and the transformation to contract a long path to a single node possibly makes performance values worsen. As a result, contracting a short fixed-length path is more efficient than the previously proposed path decomposition that contracts all paths from the root to a leaf.

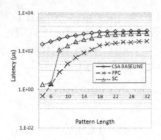 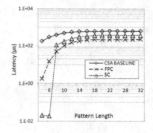 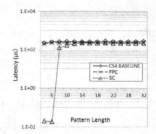

Fig. 8. Latency comparisons for CSA, FPC, and SC with $\alpha=0.75$ and $E_{num}=10,000$. The additional space of FPC and SC is 5.5% and 0.78% of the total index space of CSA, respectively.

Fig. 9. Latency comparisons for CSA, FPC, and SC with $\alpha=1.25$ and $E_{num}=10,000$. The additional space of FPC and SC is 4.4% and 1.3% of the total index space of CSA, respectively.

Fig. 10. Latency comparisons for CSA, FPC, and SC with $\alpha=2.00$ and $E_{num}=10,000$. The additional space of FPC and SC is 3.7% and 1.7% of the total index space of CSA, respectively.

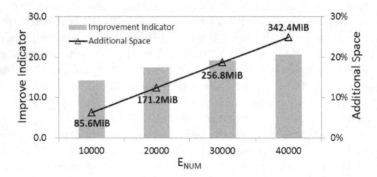

Fig. 11. FPC improvement indicators with $P_{len}=16$. Bar charts indicate that they increase logarithmically with the increase of the E_{num} value.

Fig.13 plots the throughput performance of the vectorized FPC implementation with *pcmpestri*. Note that the vectorization has little effect on latency and they are slightly better than those of the scalar implementation. In contrast, the throughput of the vectorized implementation is superior to that of the scalar one. The performance gaps narrow as the pattern length gets longer and α increases. Therefore, the vectorization is effective when pattern matching on a trie with fewer branches is expected to be processed inside processor caches. $P_{len}=16$ was used for the vectorized FPC implementations in the following experiments, unless otherwise stated.

SC Analysis. This section is the same structure as the previous one; first, it discusses improvement indicators by varying E_{num} values. P_{num} and T_{depth} were calibrated in a following experiment. The improvement indicators of SC are shown in Fig.14. The results are similar to those of Fig.11; the improvements

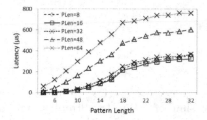

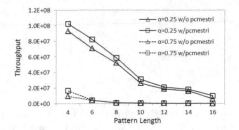

Fig. 12. FPC parameter-calibrated tests for P_{len}. The case of $P_{len}=16$ is the fastest among the others. As the length increases, the performance gradually declines.

Fig. 13. Performance results for the FPC vectorized implementation with *pcmpestri*. The vectorization is found to have a good effect when patterns are short and α is small.

logarithmically increase from 2.7x to 3.1x and there are few differences in the larger E_{num} values.

Fig.15 plots benchmark results by varying P_{num} and T_{depth}. The case with $P_{num}=15$ and $T_{depth}=12$ has the best results among the others. Notice that the calibration of P_{num} leads to the slight differences in the performance and $P_{num}=15$ is better in terms of space overhead. On the basis of these results, we used $P_{num}=15$ and $P_{depth}=12$ in the following experiments for the better trade-off of space costs and performance unless otherwise stated.

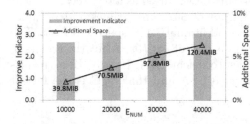

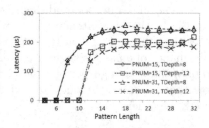

Fig. 14. SC improvement indicators with $P_{num}=15$ and $T_{depth}=8$. The indicators increase logarithmically in line with Fig.11.

Fig. 15. Benchmark results of SC by varying P_{num} and T_{depth}. The results improve as T_{depth} increases, whereas P_{num} has little effect on the performance.

STC Analysis. Fig.16, Fig.17, and Fig.18 plot how performance changes by shifting the contraction strategy of STC from FPC to SC. In Fig.16, the pattern length was fixed at 4 and α varied from 0.25 to 2. We observe that the strategy shift correctly occurs except when α is 0.75. When looking at this in detail, the weight at 0.75 is similar between FPC and SC, and the weight of SC is slightly higher than that of FPC. This is because we expect that a path with more symbols has a higher probability of references as described in the design overview of Section 3. This assumption causes the prediction error, and then the number of symbols is not the best indicator to represent the weight of STC. The

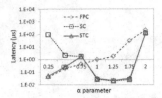
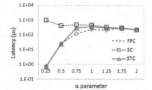
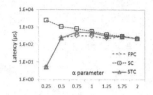

Fig. 16. Strategy shift of STC with pattern length of 4

Fig. 17. Strategy shift of STC with pattern length of 16

Fig. 18. Strategy shift of STC with pattern length of 32

same incorrect behavior occurs in Fig.17 and Fig.18. A following experiment uses a different indicator for the wight to evaluate this assumption.

Realistic Benchmark. Finally, the benchmark results of the TREC test data are plotted in Fig.19. The parameters of FPC and SC were based on the observations made in Sections 4.2.1 and 4.2.2, and the E_{num} value was set to 10,000 in terms of space overhead and performance gains. The total index size of CSA is about 1.30GiB and the additional space ratios of FPC, SC, and STC are 5.3%, 5.2%, and 7.1%, respectively. There are two different assumptions underlying this benchmark; first, FPC and SC select nodes depending on the number of symbols under the nodes in a trie as described in the design overview of Section 3 (labeled as *'uniform'*). Second, we predict the future referential probability of suffixes by using the queries (labeled as *'biased'*). A 20% portion of the queries was used as a learning set, and the other portion was used as a test set. It is clear from Fig.19 that the latency and throughput of STC are 7 times and 72 times better than those of CSA at the expense of additional 7.1 % space overhead. The performance of *biased* is slightly better than that of *uniform*, but the results of *uniform* have enough performance values to defeat CSA. Finally, these results show that all our techniques are superior to CSA and that STC is the fastest among the others.

5 Related Work

This section describes two related topics: succinct data structures and processor-optimized techniques. The theory of succinct data structures was mainly studied in the 1980s. Since practical implementations were proposed in the early 2000s, most of them have been used in pattern matching: compressed suffix arrays (CSA), backward searches (BS) [9], FM-indexes (FMI) [10], and LZ77 self-indexes (LZ77SI) [4]. BS was proposed to directly use ψ for pattern matching instead of transforming ψ to SA because the transformation causes critical performance penalties. In contrast, FMI only exploits *rank* operations for pattern matching by using BWT. LZ77SI is also an efficient technique for pattern matching based on LZEnd which is a succinct representation of LZ77. The comprehensive survey of them can be found in [8]. Our technique is orthogonal to these structures and easily applied into frequently accessed entries in them.

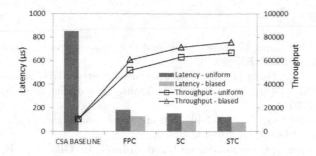

Fig. 19. Realistic benchmarks of CSA, FPC, SC, and STC

Modern high-end processors are very fast and incorporate many cores and special instructions, e.g., SIMD ones. However, Intel corp. has reported that performance gaps between processor-optimized codes and naive ones could be created by an average of 23x and at most 53x [11]. Actually, many technical issues need to be concerned to boost implemented program codes: instruction-level parallelism, cache/TLB miss penalties, conditional penalties, and memory-bandwidth limits. Research communities proposed many techniques designed to prevent these anomalies from search algorithms; early techniques [12][13] calculate optimal node size on B+Tree to minimize cache/TLB misses and another technique uses query buffering so as to save memory bandwidth [14]. Recent approaches [16][2][7] exploit SIMD instructions to realize branch-free structures and the vectorized comparisons for tree traversal. As processors evolve, the hardware awareness becomes considerable so as to improve existing algorithms.

6   Conclusions

We presented the novel contraction technique, called STC, to achieve fast pattern matching for CSA without losing the compression advantage of succinct data structures. Our experiments show that FPC outperforms naive CSA on a trie with a few branches. As the number of branches increases, SC becomes superior to CSA and FPC for short pattern queries. STC is a hybrid technique of FPC and SC, and exhibits the best performance among the other techniques for the TREC test data set at the little expense of space overhead. We believe that the amount of data increases drastically and it is getting necessary to carry out the efficient in-memory pattern matching in many domains such as databases, information retrieval, and text mining.

References

1. Grossi, R., Ottaviano, G.: Fast compressed tries through path decompositions. In: Proceedings of ALENEX (2012)
2. Kim, C., et al.: Designing Fast Architecture-sensitive Tree Search on Modern Multi-core/Many-core Processors. ACM Transaction on Database Systems **36**(4), 22:1–22:34 (2011)
3. Kreft, S., Navarro, G.: LZ77-like Compression with fast random access. In: Proceedings of DCC, pp. 239–248 (2010)
4. Kreft, S., Navarro, G.: Self-indexing based on LZ77. In: Giancarlo, R., Manzini, G. (eds.) CPM 2011. LNCS, vol. 6661, pp. 41–54. Springer, Heidelberg (2011)
5. Manber, U., Myers, G.: Suffix arrays: a new method for on-line string searches. In: Proceedings of SODA, pp. 319–327 (1990)
6. Manzini, G.: An Analysis of the Burrows Wheeler Transform. J. ACM **48**(3), 407–430 (2001)
7. Yamamuro, T., et al. Vast-tree: a vector-advanced and compressed structure for massive data tree traversal. In: Proceedings of EDBT, pp. 396–407 (2012)
8. Navarro, G., Mäkinen, V.: Compressed Full-Text Indexes. ACM Computing Surveys (CSUR) **39**(1) (2007)
9. Mäkinen, V., Navarro, G., Sadakane, K.: Advantages of backward searching: efficient secondary memory and distributed implementation of compressed suffix arrays. In: Fleischer, R., Trippen, G. (eds.) ISAAC 2004. LNCS, vol. 3341, pp. 681–692. Springer, Heidelberg (2004)
10. Ferragina, P., Manzini, G.: Opportunistic data structures with applications. In: Proceedings of FOCS (2000)
11. Kim, C., et al.: Closing the Ninja Performance Gap through Traditional Programming and Compiler Technology. Technical report, Intel Lab. (2012)
12. Hankins, R.A., Patel, J. M.: Effect of node size on the performance of cache-conscious B+trees. In: Proceedings of SIGMETRICS, pp. 283–294 (2003)
13. Chen, S., Gibbons, P.B., Mowry, T.C.: Improving index performance through prefetching. In: Proceedings of SIGMOD, pp. 235–246 (2001)
14. Zhou, J., Ross, K.A.: Buffering accesses to memory-resident index structures. In: Proceedings of VLDB, pp. 405–416 (2003)
15. Okanohara, D., Sadakane, K.: Practical entropy-compressed rank/select dictionary. In: Proceedings of ALENEX, pp. 60–70 (2006)
16. Schlegel, B., Gemulla, R., Lehner, W.: K-ary search on modern processors. In: Proceedings of DaMoN, pp. 52–60 (2009)

Scalable Top-k Spatial Image Search on Road Networks

Pengpeng Zhao[1,2]($\boxtimes$), Xiaopeng Kuang[1,2], Victor S. Sheng[3], Jiajie Xu[1,2],
Jian Wu[1,2], and Zhiming Cui[1,2]

[1] School of Computer Science and Technology, Soochow University, Suzhou, China
ppzhao@suda.edu.cn
[2] Collaborative Innovation Center of Novel Software Technology
and Industrialization, Suzhou 215006, People's Republic of China
[3] Computer Science Department, University of Central Arkansas, Conway, USA
ssheng@uca.edu

Abstract. A top-k spatial image search on road networks returns k images based on both their spatial proximity as well as the relevancy of image contents. Existing solutions for the top-k text query are not suitable to this problem since they are not sufficiently scalable to cope with hundreds of query keywords and cannot support very large road networks. In this paper, we model the problem as a top-k aggregation problem. We first propose a new separate index approach that is based on the visual vocabulary tree image index and the G-tree road network index and then propose a query processing method called an external combined algorithm(CA) method. Our experimental results demonstrate that our approach outperforms the state-of-the-art hybrid method more than one order of magnitude improvement.

Keywords: Top-k spatial image query · Separate index · Road networks

1 Introduction

A picture paints a thousand words and it is difficult to describe the content of the query image clearly in just a few words. Traditional text-based image retrieval techniques that depend on manual annotations cannot meet the increasing needs. Content-based image search is very useful to user and has a wide range of applications, which has attracted more and more attention in the computer vision and multimedia community [5, 23, 27, 33–35].

Modern-era mobile phones and tablets have evolved into powerful image and video processing devices, equipped with high-resolution cameras, color displays, and hardware accelerated graphics. They are also equipped with location sensors (GPS receivers, compasses, and gyroscopes), and connected to broadband wireless networks, allowing fast information transmission. All these enable a new class of applications to use a camera phone to initiate search queries about objects in visual proximity with a natural human-computer interaction. Such applications

© Springer International Publishing Switzerland 2015
M. Renz et al. (Eds.): DASFAA 2015, Part II, LNCS 9050, pp. 379–396, 2015.
DOI: 10.1007/978-3-319-18123-3_23

can be used, e.g., for identifying products, comparison shopping, and finding information about movies, CDs, real estates, print media or artworks. The first deployments of such systems include Google Goggles [12], Nokia Point and Find [22], Kooaba [17], Ricoh iCandy [10,13,15] and Amazon Snaptell [28].

With the prevalence of mobile Internet and GPS-enabled devices, users can conveniently take a photo, mark its geo-location and annotate it with tags on Flickr [1] so that the photo is associated with spatial information. Consequently, massive amount of images that are geo-tagged are being generated at an unprecedented scale. A recent study [9] shows that there are over 95 million geotagged photos on Flickr with a daily growth rate of around 500,000 new geo-tagged photos. Furthermore, we can generate geo-tags automatically for non-geo-tagged images according to the associated address text information, which is called reverse geo-tagging. For example, images in business directories such as Google Places for Business and Yahoo! Local, as well as rating and review services such as TripAdvisor and Dianping. These data are featured with both image visual contents and geo-spatial contents, and we refer to them as geo-image objects. The fusion of geo-locations and images enables queries that take into account both spatial proximity and image relevancy. For example, users are usually only interested in information (such as products, restaurants or scenic spots) related to certain locations, e.g., "Where is a similar style of shoes sold nearby", "Where are similar attractions nearby", and "Where are similar snacks nearby". We refer to these queries, which consist of both images and spatial conditions, as spatial image queries.

To the best of our knowledge, this paper proposes a new kind of top-k spatial image queries that takes into account both geographic proximity and image content relevancy for the image query associated with geo-location. We call this type of queries a top-k spatial image query. It consists of a spatial component (the location of the photo taken) and an image component. The answer to such a top-k image query is a list of k ranked images, according to a ranking function that combines their distances to the query location and their relevance of their content descriptions to the query image. we provide two examples of the two top-k spatial image query examples are as follows:

Example 1: Mobile product search. *As shown in Fig.1, a customer finds a pair of very favourite fancy shoes in a shopping store but without a right size. She wants to see whether there are some similar styles of shoes in nearby stores. She submits a top-k spatial image query by taking a photo with a mobile phone and gets k similar shoes on sale in nearby businesses.*

Example 2: Tourist attraction search. *As shown in Fig.2, a traveler wants to take some photos of nearby seaside beach around a tourist attraction. Based on her location and the scene she wants to capture (seaside and beach), a top-k spatial image query on the geo-tagged image datasets uploaded by other visitors can provide some useful suggestions.*

In this paper, we focused on top-k spatial image search on road networks because the people's trajectories are usually constrained by road networks. The

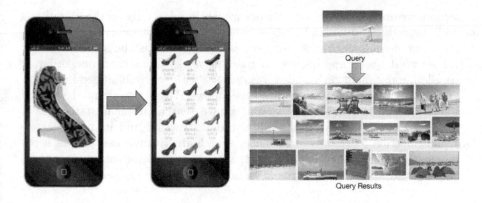

Fig. 1. An example of mobile product **Fig. 2.** An example of tourist attraction
search   search

scale of these geo-tagged image databases and the demand for real-time response
make it very critical to develop efficient query processing mechanisms. Although
top-k spatial keyword query models, query processing techniques, and indexing
mechanisms have been well studied for years[8,25,29,31,32](surveyed in [4,6]),
which is an important tool in exploring useful information from a spatial textual
database. We cannot use spatial keyword indices directly to solve spatial image
queries efficiently because there are many differences between texts and images.
A query image typically has hundreds of feature points, and could be quantized
into hundreds of visual words. A keyword query generally only has 2-5 keywords
[8,25,32]. Existing solutions are not sufficiently scalable to cope with hundreds
of query keywords. Recently, Zhang et al. proposed a new Rank-aware combined
algorithm (RCA) method [31], which can handle more number of query keywords.
However, RCA focuses on the Euclidean space. It is not suitable to the problem
on the road networks space. Existing solution of spatial keyword search on road
networks [14,18,24,30], using a hybrid index technology, still cannot support very
large road networks (e.g. the entire USA road network). The main limitation is
either high memory consumption or heavy search overhead.

In this paper, we model the problem as a top-k aggregation problem and
propose a new separate index approach that supports both top-k pruning and
efficient merging. We separately use a vocabulary tree based inverted index as
an image index and a G-tree [36] for organizing the spatial information of images
on road networks. A visual vocabulary tree has been widely used for large scale
content-based image retrieval in recent years and performed well empirically in
many image retrieval tasks. G-tree is selected because it is an efficient tree-based
index for k-nearest neighbor (kNN) search on large scale road networks. Based
on the two separate index, we propose a query processing algorithm External
CA, which extends CA algorithm proposed by Fagin et al.[11] with more effective
pruning. In summary, the contributions of this paper are listed as follows.

First, we formulate the top-k spatial image query on road networks problem
and identify its applications.

Second, we use a separate index structure which relies only on the ubiquitous and well-understood index and propose a new and efficient query approach.

Third, we demonstrate that our approach outperforms the state-of-the-art spatial keyword query approaches on road networks more than one order of magnitude improvement on a real-world dataset with one million geo-images.

The rest of this paper is organized as follows. We review related work in Section 2 and present the top-k spatial image search problem statement in Section 3. Section 4 presents a separated index structure and its maintenance algorithms. Section 5 presents our efficient search algorithm based on this separated index. Section 6 experimentally compares our approach to other algorithms with an extensive performance study. Section 7 makes the final conclusion.

2 Related Work

To the best of our knowledge, there is no existing work on the problem of spatial image search on road networks. Below, we introduce three important categories of related work.

Content-Based Image Search. In recent years, with the introduction of local invariant features [3,20], the Bag-of-Words(BoW) model representation [27] has been significantly scaled up by the use of large hierarchical vocabulary trees [21]. With vocabulary trees, which typically contain millions of leaf nodes (representing visual words), images are represented by a very sparse BoW vectors. Further, by indexing images with the inverted file structure, the scalability of image search is achieved.

The high dimensional indexing mechanisms for large scale image search can mainly be divided into three categories: tree-based index, hashing-based index, and visual words based inverted index. Locality Sensitive Hashing (LSH) [2] is known as an effective technique to index dense features, but it is not a good one to index sparse features. KD-tree [19] can only be applied when the dimensionality of the feature space is about a dozen. Visual vocabulary tree based inverted index has been widely used for large scale content-based image retrieval in recent years[21].

Spatial Keyword Search. Spatial keyword search has been actively studied in recent years due to its great potential in both industry applications and research problems (See [4,6] for a comprehensive survey). A type of popular spatial keyword query is the boolean spatial keyword search which aims to retrieve all objects whose text descriptions contain a given set of keywords and whose locations are near to the query location [6,7]. Another type of popular query is the top-k spatial keyword search which aims to find a set of geo-tagged objects based on the spatial proximity and the query keywords similarity. Many efficient indexing techniques have been proposed such as IR[8], S2I[25], I3[32], RCA[31]. Although this is similar to the top-k spatial image query. However, the image of a query contains hundreds of query keywords, and existing hybrid index solutions cannot handle this scale number of keywords.

Spatial Keyword Search on Road Network. Most of the existing spatial keyword search articles focus on the Euclidian space. Spatial keyword search on road networks has attracted increasing research efforts in recent years [14, 18,24,30]. Li et al. [18] addressed a range constrained spatial keyword query on road networks that returns nodes whose textual descriptions are relevant to the query keywords within a specified area. Zhang et al. [30] studied the problem of diversified spatial keyword search on road networks which considers both the relevance and the spatial diversity of the results. Guo et al. [14] investigated continuous top-k spatial keyword query on road networks and proposed an incremental manner method for moving queries. The top-k spatial-keyword query [24] was proposed to find the top-k ranked objects, measured as a combination of their road network distances and the relevances of their text descriptions. It is similar to our problem. However, its hybrid technology cannot scale to large road networks.

3 Problem Definition

In this section, we formally define the problem of top k spatial image query on road networks.

Road Networks. We use a weighted graph to describe a road network, which is denoted as $G = (V, E, W)$, where V is a set of vertices, the edge set is denoted as E and W is a set of weights denoting the cost on the corresponding edge, like travel time or distance. A vertex $\nu \in V$ indicates that ν is an intersection or an end point of edge in the graph. $(\nu, \nu') \in E$ denotes an edge, and $w_{\nu,\nu'}$ is a weight of the edge (ν, ν') represents the length (network distance) $|\nu, \nu'|$ of the road segment. For simplicity, we assume undirected graphs. However, directed graphs is also support by our approach. In this case, $(\nu, \nu') \neq (\nu', \nu)$ and the distance $|\nu, \nu'|$ may be different from $|\nu', \nu|$.

Geo-Image. Let D be a spatial image database, which consists of N images. Each spatial image I in D is associated with a spatial location and defined as a pair($I.loc$, $I.content$), where $I.loc$ is a geographical space descriptor composed of latitude and longitude, and $I.content$ is a image content descriptor. Image $I.content$ encoded into a bag-of-words (BoW) model [27] with millions of visual words.

Problem Statement. Intuitively, a top-k spatial image query on road networks retrieves at most k images in database D for a given query Q such that their locations are near to the location specified in Q on road networks and their image content descriptions are relevant to the contents in Q.

Formally, given a query $Q = (loc, content)$ where $Q.loc$ is a location descriptor and $Q.content$ is a vector encoded into a bag-of-words (BoW) model, the k

images returned are ranked according to a ranking function $F(D(Q.loc, I.loc),$ $S(Q.content, I.content))$, where $D(Q.loc, I.loc)$ is the road network distance between Q and I and $S(Q.content, I.content)$ is the image similarity of I according to the content of the query image. The paper's proposals are applicable to a wide range of ranking functions, namely all functions that are monotone with respect to distance proximity $D(Q.loc, I.loc)$ and image relevancy $S(Q.content, I.content))$. We follow existing work and use linear interpolation[8]. Specifically, we derive a ranking function as a linear interpolation of normalized factors for ranking an image I with regard to a query Q.

$$F(Q, I) = \alpha D(Q.loc, I.loc) + (1 - \alpha)S(Q.content, I.content) \tag{1}$$

where $\alpha \in (0, 1)$ is a parameter used to balance the spatial proximity and the image relevance, which allows users to set their preferences between the image relevancy and the location proximity when conducting a query.

Without loss of generality, we assume that both the query location and the geo-tagged images are at vertices. This assumption can be bypassed finding the nearest edge and the nearest vertex of a given location.

Road Network proximity give the importance of the location of a spatial image to the query location. The road network proximity measure can be defined as

$$D(Q.loc, I.loc) = d(Q.loc, I.loc)/d_{max} \tag{2}$$

where $d(Q.loc, I.loc)$ is the shortest path between $Q.loc$ and $I.loc$ and d_{max} is the largest network distance between any object and any location in the road network. d_{max} can be obtained by traversing the network from each object until the entire network is expanded and keeping the maximum distance. $D(Q.loc, I.loc)$ is in the range of [0,1].

Image Similarity. Most content-based image search approaches rely on the popular Bag-of-Visual-Words (BoW) model [27]. Generally, an image is represented by a set of local features, which are encoded into a bag-of-words vector with millions of visual words. Those local features are extracted from image keypoints detected with Difference of Gaussian (DoG) [20] detector. Then around a keypoint, a local patch is described into a local feature, such as SIFT [20] and SURF [3]. Local features are usually of high dimension. Matching raw local descriptors is expensive to compute because each image may contains hundreds of local features. An efficient solution is to achieve a compact representation, a vocabulary tree is defined and then local features can be quantized to visual words.

The vocabulary tree can be constructed off-line by unsupervised clustering algorithm (e.g. hierarchical k-means (HKM) [21]) and typically contains millions of leaf nodes (representing visual words). An image is then represented as a bag of visual words, and these are entered into an index for later querying and retrieval. Nister and Stewenius [21] have demonstrated very inspiring retrieval performance using a large hierarchical vocabulary tree. The use of a tree structure dramatically reduces the computation time required to quantize a feature descriptor into one of millions of words.

Image represented into BoW model with millions of visual words is a sparse vector. Therefore, the inverted file structure, which has been successfully applied in textual information retrieval, is leveraged to index large-scale image database. We formulate the image retrieval as a voting problem. Each visual word in the query image votes on its matched images. The query and each image in the corpus is represented as a sparse vector of term (visual word) occurrences and search proceeds by calculating the similarity between the query vector and each image vector, using an L2 distance. We use the well-known TF-IDF [26] weighting scheme to compute the relevance score.

Specifically, given a query Q and an image I, the ranking function for this query is to give a relevance score S based on the normalized difference between the query and database vectors as follows:

$$S(Q.content, I.content) = \|\frac{q_i}{\|q\|} - \frac{d_i}{\|d\|}\| \tag{3}$$

4 The Rank-Aware Separate Index Method

Inspired by work [31], we model the spatial image query problem as a top-k aggregation problem [11]. By transforming the problem as a top-k aggregation problem, we can design an efficient index and query solution that relies on separated index such as the widely used the visual vocabulary tree image index and the road network index.

Top-k Aggregate Query. Consider a set of grouping attributes $G = g_1, ..., g_r$, and an aggregate function F that is evaluated on each object. A top-k aggregate query returns the k objects, based on G, with the highest F values.

Given an image query Q with m vision vocabularies, each geo-image I in a database D can be modeled as a tuple with $m + 1$ elements, i.e., $(\omega_1, \omega_2, \ldots, \omega_m, \omega_{m+1})$, where $\omega_i (1 \le i \le m)$ is the weight of the corresponding visual vocabulary, and ω_{m+1} is a road network spatial distance between the database image I and the query location. By giving the aggregation function, the problem of the top-k spatial image query on road networks is now transformed as a top-k aggregation problem.

Example 3. Consider a geo-image database system where several visual features are extracted from each image. Example features include color histograms, edge orientation, texture and geo-location, as shown in Fig. 3. Features are stored in separate relations indexed using high-dimensional indexes that support similarity queries. Suppose that a user is interested in the top 5 images, which are the most similar to a given query image based on a set of visual features and the spatial distance.

4.1 The Concept of Top-k Aggregation Algorithms

In this section, we present an overview of top-k aggregation algorithms. Ilyas and Beskales [16] provided a comprehensive survey over related algorithms.

Color Features

Edge Features

Texture Features

Location

Top-k Aggregation Results

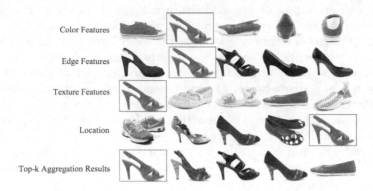

Fig. 3. Multi-feature Queries in an Image Database

For convenience, we describe top-k aggregation algorithms in the context of spatial image search on road networks as follows. Given a query with m visual vocabularies and a spatial location, we assume that $m+1$ sorted lists are already existing, i.e., $L_1, L_2, \ldots, L_m, L_{m+1}$. Each list L_i, $i \in [1, m]$, is sorted in a descending order according to the image visual vocabulary relevance TF-IDF score of the ith query visual vocabulary; and L_{m+1} is sorted in a descending order by the image road networks distance to the query location. Thus, each sorted list item is a pair of image ID and a score (image content similarity or road networks distance).

The first top-k aggregate query algorithm TA is proposed by Fagin [11], which consists of two main steps.

1. Conduct a sorted access in parallel to each of the $m+1$ sorted lists L_i. As a new object o is seen under the sorted access in some list, conduct random access to the other lists L_i to find its score in L_i. Compute the aggregated score of the object o using Equation 1. If this score is among the k highest scores seen so far, then remember object o and its score $f(o)$ (ties are broken arbitrarily, so that only k objects and their scores are remembered at any time).

2. For each list L_i, let b_i be the score of the last object seen under the sorted access. Define the threshold value T to be the aggregated score of b_i using Equation 1. As soon as k objects have been seen with scores equal to or higher than T, halt.

Although the TA algorithm uses a threshold criterion that is provably optimal on every possible instance of the data, it can potentially incur very high processing overheads arising from the random accesses for the object attributes. Subsequently, many variations and extensions of the TA method have been developed in the last few years [16]. This paper builds based on the particular variant the Combined Algorithm (CA) [11], which is a combination of NRA(No Random Access) and TA and takes into account the cost of random access and supports top-k pruning by using a group of objects as a unit when sorted lists are given. CA retrieves k objects with the highest combined scores while accessing sorted lists in parallel until top-k objects whose scores are higher than the threshold value are obtained. CA maintains the threshold value by applying the aggregated score of the last object accessed from each list into the scoring function. If an object o is accessed from only one list, CA randomly accesses o in the other lists to obtain the component score.

4.2 Rank-Aware Separate Indexes

Our approach uses separate indexes, which include a road network index for searching by location and a visual vocabulary tree index for searching by image.

The Road Network Index. G-Tree is an efficient and elegant tree-based index for kNN search on road networks [36]. We utilize it to efficiently and dynamically compute the nearest neighbor images on road networks as a sorted spatial list. G-Tree is constructed by partitioning the road network recursively, until we get a small enough sub-network for efficient search. Each node of G-tree is a sub-network and each leaf node of G-Tree is a set of nodes on the road network. A leaf node is considered on the border of its parent if it has a direct edge to other outside leaf nodes. The shortest path between two border leaf nodes in each partition is calculated off-line. For the nodes of G-Tree with the same parent, the shortest path distance between the border nodes of these G-Tree nodes was kept. To search a shortest path between two leaf nodes p_1,p_2 with parent P_1 and P_2, it utilizes the off-line calculated shortest paths between the border nodes of P_1 and P_2 and the shortest path of p_1/p_2 to the border of P_1/P_2. The search complexity is positively correlated to the height of the G-tree. Because of space limitation, the details of G-tree search can be found in its original paper [36].

The Image Index. With the introduction of local invariant features [3,20], the Bag-of-Words(BoW) model representation [27] has been significantly scaled up by the use of large hierarchical visual vocabulary trees [21]. Given an image, its local features, such as SIFT [20], are first extracted, and then quantized into the closest visual words ('codebooks') using the visual vocabulary tree, which are pre-learned on a training dataset. Finally, a high-dimensional vector is generated using the BoW descriptor to represent the image. Therefore, the vector is so sparse that inverted index files are well suited to implement the indexing and searching efficiently. The BoW descriptor is indexed by an inverted file which has an entry for each visual word followed by a list of all the images in which the visual word occurs. Visual words are weighted by the TF-IDF, where the IDF reflects their discriminative abilities in database images and the TF indicates their importance in a query image.

5   A Query Processing Algorithm

The location of a query image is dynamic and the sorted spatial list is query dependent. A spatial list cannot be pre-computed statically but need to be sorted at runtime using the query location to compute the road network distance. We will introduce how to efficiently and dynamically construct the sorted spatial list in the following.

5.1 Dynamically Constructing Sorted Spatial List

We denote the sorted spatial list for the distance scores on road networks as L_s. We dynamically construct the sorted spatial list L_s by incrementally accessing the tree-based road network index, G-tree, using a best-first kNN search algorithm [36].

Our top-k aggregation algorithm external CA sequentially accesses the sorted spatial list L_s by a fixed number of objects at every iteration. So we do not need to build the whole sorted spatial list at one time. We apply the progressive best-first kNN search algorithm in the unit of η objects instead of entire objects. For incremental best-first kNN on G-tree, we maintain a priority queue Q that sorts the objects and G-tree nodes by their network distance. We obtain the current η objects R_s from the front queue of $R_{candidates}$ ($R_{candidates}$ is a global variable) and maintain the remaining objects in $R_{candidates}$ for the next iteration. If the current size of $R_{candidates}$ is less then η, we access the next group of candidates from the priority queue Q and insert into $R_{candidates}$ until the size of $R_{candidates}$ is larger than or equal to η except when the priority queue is empty. The detail of progressive constructing L_s and retrieving the current η objects algorithm is shown in Algorithm 1.

5.2 External CA Algorithm

In this section, we discuss our external CA algorithm in details. The CA algorithm uses a parameter η to control the depth of sequential access. In each iteration, η objects in each list are sequentially accessed. η is set to be the ratio of the cost of a random access to the cost a sequential access. For each accessed image img, let $Bound(img)$ denote an upper bound of the aggregated score of img. An image img is defined to be viable if $Bound(img)$ is larger than the kth best score that has been computed so far. At the end of each iteration, the viable image with the maximum $Bound(img)$ value is selected for random access to determine its aggregated score. The algorithm terminates when at least k distinct images have been seen and there are no viable images. Our approach adopts a different random access strategy and termination criteria from the the traditional CA algorithm. The CA algorithm selects the viable image with the maximum upper bound score for random access and terminates if this upper bound score is no greater than the score of the kth best image seen so far (denoted as T_k). In contrast, our external CA algorithm does not maintain $Bound(img)$ to store the upper bound score for each viable image. Moreover, our random access is applied to the viable images in the min-heap storing top-k results so that threshold T_k can be increased as much as possible towards an early termination.

Based on our sequential access approach, we can also compute a tighter upper bound for an image aggregated score (denoted by $Bound_k$). Specifically, after the ith iteration, $Bound_k$ can be calculated by

$$Bound_k = \alpha \cdot Bound_s(i) + (1 - \alpha) \cdot m \cdot Bound_t(i) \tag{4}$$

Algorithm 1. exploreCurrentSpatialList

---

1 **Input**: $q.loc$
2 **Output**: the current η objects from L_s
3 $R_s:=\{\}$;
4 **while** $|R_{candidates}| < \eta$ **do**
5      $e := Q.\text{Dequeue}()$; //Q is a priority queue
6   **if** e *is an object* **then**
7   insert e into $R_{candidates}$;
8   **else**
9   **if** e *is a leaf node* **then**
10   **if** $q.loc \in e$ **then**
11   MINDIST-INSIDE-LEAF($q.loc,e$); /*minimum distance between two nodes in one leaf*/
12   **else**
13   MINDIST-OUTSIDE-LEAF($q.loc,e$); /*minimum distance between two nodes in two leafs*/
14   **for** *each* $v \in L(e)$ **do**
15   /*$L(e)$ is an occurrence list composed of objects IDs appear in the leaf node*/
16                  $Q.\text{Enqueue}(\langle v,\text{SPDist}(q.loc,e) \rangle)$ /*SPDist return minimum distance between two tree nodes*/
17   **else if** e *is a non-leaf node* **then**
18   **for** *each child node* $c \in L(e)$ **do**
19   **if** $q.loc$ *is in* c **then**
20                      $Q.\text{Enqueue}(\langle c,\text{SPDist}(q.loc,c)=0 \rangle)$; /*$q.loc$ belongs to c so that the distance $|q.loc, c|$ is 0.*/
21   **else**
22   MINDIST-OUTSIDE-NONLEAF($q.loc,c$);
23                      $Q.\text{Enqueue}(\langle c,\text{SPDist}(q.loc,c) \rangle)$;
24 Store η objects of $R_{candidates}$ into R_s;
25 $R_{candidates} := R_{candidates} - R_s$;
26 **return** R_s ;

---

where $Bound_s$ represents the spatial upper bound, $Bound_t$ represents the textual upper bound and m is the number of query words. If $Bound_k \leq T_k$, we stop the sequential access on the sorted lists as it is guaranteed that no unseen image could have an aggregated score higher than T_k. However, the algorithm cannot be terminated at this point because there could be some viable images not in the top-k heap but with an upper bound score larger than T_k. For these images, we need to conduct random access to get their full aggregated score and update the top-k heap if we find a better result. In this way, we can guarantee that no correct result is missed.

Our external CA algorithm processing top-k spatial image queries on road networks is shown in Algorithm 2. The input parameter L_t is the collection of textual lists sorted by TF-IDF values of visual vocabularies, G_s is the G-tree road network index, m is the number of query visual words and k means return k results. A top-k heap is initialized in line 1 and the pointer p_t is initialized for sequential access the textual lists in line 5. In each iteration, we perform sequential access in the textual lists by calling a function *exploreTextList* (line 7) and scans in the spatial lists sequentially and progressively by calling the procedure *exploreCurrentSpatialList*, i.e., Algorithm 1 (lines 8). After the sequential access, we perform random access on the viable images in the top-k heap (lines 10-11 in Algorithm 2). Finally, we update $Bound_k$ according to Equation 4. If $Bound_k \leq T_k$, we stop the sequential access. For each viable image not in top-k, we perform random access to obtain its complete score and update the top-k results if it is a better result.

Algorithm 2. External CA Algorithm(L_t, G_s, m, k)

---

1 initialize a min-heap *topk* with k dummy images with score 0
2 $T_k = 0$ //T_k is the mininum score in *topk*
3 $Bound_k = 1$ //$Bound_k$ is the upper bound score for all the unseen images
4 **for** $i = 1; i \leq m; i++$ **do**
5 $\quad$ $p_t[i] = 0;$
6 **while** $T_k < Bound_k$ **do**
7 $\quad$ exploreCurrentTextList();
8 $\quad$ exploreCurrentSpatialList();
9 $\quad$ **for** *each viable image ImgId in topk* **do**
10 $\quad\quad$ randomAccess($ImgId$);
11 $\quad\quad$ update($topk, T_k$);   /*update T_k*/
12 $\quad$ $Bound_k = \alpha \cdot Bound_s(i) + (1 - \alpha) \cdot m \cdot Bound_t(i);$ /*update $Bound_k$*/
13 **for** $ImgId \in W$ **do**
14 $\quad$ **if** $ImgId \notin topk$ *and ImgId is viable* **then**
15 $\quad\quad$ randomAccess($ImgId$);
16 $\quad\quad$ update($topk, T_k$);   /*update T_k*/
17 return *topk*;

18 **exploreCurrentTextList()**
19 **for** $i = 1; i \leq m; i++$ **do**
20 $\quad$ **for** $j = 0; j \leq \eta; j++$ **do**
21 $\quad\quad$ $doc = L_t[i][p_t[i]];$
22 $\quad\quad$ seqAccess($ImgId, i, \alpha \cdot ImgId.score$);
23 $\quad\quad$ $p_t[i] = p_t[i] + 1;$
24 $\quad$ return

---

6   Performance Evaluation

In this section, we conduct a comprehensive experiments to evaluate the efficiency and scalability of the proposed approach in the paper. We compare our external CA approach with the state-of-the-art top-k spatial keyword query on road networks approach Overlay [24]. Overlay combines the state-of-the-art approaches for road networks index with the spatio-textual index, and employs an overlay network on the top of the road network index to prune regions that cannot contribute with relevant objects improving the query processing performance. All methods are implemented in Java. We conduct the experiments on a PC with 8G memory and I5-3470CPU@3.20GH, running the Windows 7 Operating System.

6.1 Dataset

We first evaluate the scalability and performance of our system on an image dataset of over one million images crawled from the photo-sharing site, Flickr [1], using Oxford landmarks as queries. For the scalability and performance evaluation, we randomly sampled five sub datasets whose sizes vary from 200,000 to 1000,000 from the image dataset. We further evaluate the performance of proposed approaches on four different size real road networks datasets, i.e., the road networks of San Francisco (SF)[1], Florida (FLA), central USA (CTR), and USA (USA)[2]. We first extracted all the local features of geo-images using SIFT [20], and then quantized into the visual words with a pre-learned vocabulary tree. The number of local features of each geo-image is from 1 to 300. The average number of features is about 110.

The statistics of the geo-image datasets are listed in Table 1. The characteristics of four road networks datasets are shown in Table 2. And the settings of the parameters in the experiments are shown in Table 3.

Table 1. Geo-image datasets

Datasets	Number of Images	Number of Distinct Visual Words	Average Number of Visual Words	Disk Storage size
200K	200,000	612792	112.4	196M
400K	400,000	614767	112.9	390M
600K	600,000	541571	111.8	580M
800K	800,000	616403	112.1	775M
1M	1,000,000	616785	131.7	971M

Table 2. Characteristics of the road networks datasets

Attributes	SF	FLA	CTR	USA
Total size	11MB	111MB	1.45GB	2.53GB
Total number of vertices	174956	1070376	14081815	23947347
Total number of edges	223001	2712797	34292496	58333344

---

[1] http://www.cs.utah.edu/~lifeifei/SpatialDataset.htm
[2] http://www.dis.uniroma1.it/challenge9/download.shtml

Table 3. Parameters evaluated in the experiments. The default values are presented in bold.

Parameter	Values
Top-k	10 50 **100** 150 200
Number of keywords	10 50 **100** 150 200
α	0.9 **0.7** 0.5 0.3 0.1
Size of image Datasets	**200000** 400000 600000 800000 1000000
Road Networks	**SF** FLA CTR USA

6.2   Experimental Results

In this section, we evaluate the query processing performance of our proposed external CA and the comparison approach Overlay on different size of geo-image datasets and on different size of road networks.

Varying the Number of Results (k). In this experiment, we increase the number of results k from 10 to 200. Each query processes on the 200k image dataset. It contains 100 visual query words, and the spatial ratio α is set to 0.7. The experimental results are shown in Fig.4. From Fig.4, we can see that our external CA performs much better than Overlay. The response time of our external CA always keeps at a very low level under the different number of results k. Furthermore, its response time doesn't increase with the increment of the number of results k. The response time of Overlay is much higher. With the increment of the number of results, its response time increases linearly when $k \leq$ 100. This is because Overlay doesn't stop finding more candidates by expanding regions until either the query is terminated prematurely or all the candidates are accessed. With the increment of the number of results, Overlay should access more candidates to meet the query requirements. Our external CA just loads the related inverted text lists according to the visual query words and finds closer edges or vertices quickly through querying on G-Tree, which just costs a little I/Os and the memory used is smaller than that used in Overlay as well.

Varying the Number of Query Keywords. To investigate the response time of our external CA algorithm under queries with different number of visual words. In this experiment, we increase the number of visual query words from 10 to 200. The experimental results are shown in Fig. 5. From Fig. 5, we can see that the response time of our external CA algorithm is much lower than that of Overlay. With the increment of the number of query words, the response time of both algorithms only has a slight increment. As we know, a geo-image is considered to be relevant if it contains at least one visual query word. Thus, the number of candidates grows dramatically as the number of query words increases. The larger the number of visual words in the query, the larger the number of objects that may be relevant for the query. However, we don't see this significant increment in both algorithms. Overall, our external CA algorithm is around one order of magnitude better in response time enhanced.

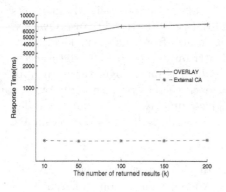

Fig. 4. Response time under varying the number of returned results

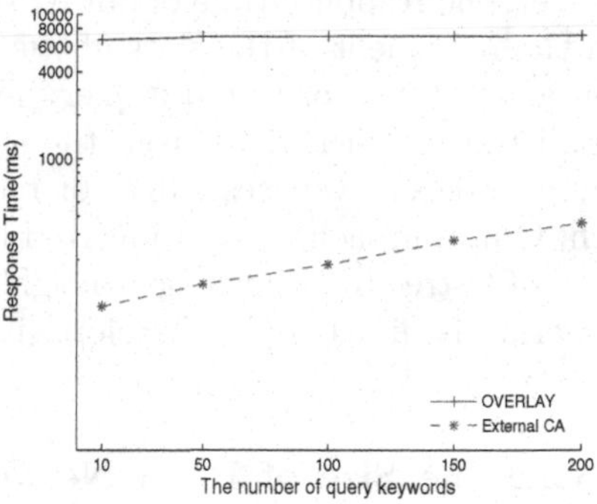

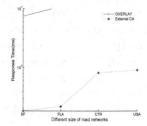

Fig. 5. Response time under varying the number of keywords

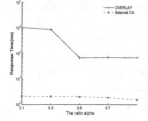

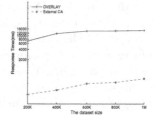

Fig. 6. Response time under varying the ratio α between image content and spatial

Fig. 7. Response time under varying the size of the image datasets.

Fig. 8. Response time under varying the size of road networks.

Varying the Query Preference Ratio α. In this experiment, we evaluate the impact of the ratio α between image content and spatial. The experimental results are shown in Fig. 6. A small value of α gives more preference to the textual description of the objects, while a high value of α gives more preference to the network proximity. From Fig.6, we can see that the response time of Overlay increases significantly as α is less than 0.5. However, our external CA algorithm always has a much lower response delay. This is because Overlay would access more candidates and terminate the query later. However, our external CA does not affect by this. It loads the closer objects quickly step by step.

Varying Image Dataset Size. In this experiment, we study the response time under different sizes of image datasets. The experimental results are shown in Fig. 7. From Fig. 7, we can see that the response time of Overlay is significantly higher than our external CA algorithm. Furthermore, the response time of Overlay significantly increase with the increment of the sizes of the image datasets.

However, the response time of our external CA algorithm only slightly increases with the increment of the sizes of the image datasets. This is because Overlay, given a set of keywords and a query location, starts from the query location to expand the search regions from the spatial attribute only. Its I/O data access includes adjacent vertices, edges or regions, which is much lager than our approach. Our approach keeps a inverted list for each visual words and employs the ability of G-tree to find the near neighbor objects quickly. In addition, the index of Overlay is disk-based. Therefore, it does not scale well when the geo-image datasets scale large.

Varying the Size of Road Networks. In this experiment, we study the response time of Overlay and our external CA algorithm on four different sizes of road networks, i.e., SF, FLA, CTR and USA. The experimental results are shown in Fig. 8. According to the road network size (refer to Table 2), we present the four road networks in the increment order in the horizontal axis of Fig. 8. From Fig. 8, we can notice that we only have the response time on the SF road network for Overlay. This is because Overlay does not scale well for large scale road networks due to its hybrid index structure and the query response time is over 10 second on FLA road networks. We have the response time on all four road networks. Fig. 8 shows that our external CA performs not bad even if running on lager road networks such as CTR and USA. When running on the USA road network, the response time is still within one second. That is, our external CA approach can be applied to different scales of road networks.

7   Conclusions

In this paper, we introduced top-k spatial image queries on road networks. Given a spatial location and a query image, a top-k spatial image query on road networks returns the k best images ranked in terms of both content similarity to the query image and the shortest road network distance to the query location. We formulated the spatial image query problem as a top-k aggregation query problem and presented a separate index structure that is based on the visual vocabulary tree image index and G-tree road networks index. Then, we proposed an external CA algorithm that works well on inverted lists sorted by visual words relevance and G-tree road networks index. Finally, we conduct experiments on the Flickr dataset with up to one million geo-images and on four road networks. We evaluate the performance of our proposed approach from different aspects. Our experimental results show that our proposed approach is superior over the state-of-the-art hybrid solutions.

Acknowledgments. This work was partially supported by Chinese NSFC project (61003054, 61170020, 61402311, 61440053), the US National Science Foundation (IIS-1115417), and Collaborative Innovation Center of Novel Software Technology and Industrialization.

References

1. Flickr. http://www.flickr.com/
2. Andoni, A., Indyk, P.: Near-optimal hashing algorithms for approximate nearest neighbor in high dimensions. In: 47th Annual IEEE Symposium on Foundations of Computer Science, FOCS 2006, pp. 459–468. IEEE (2006)
3. Bay, H., Tuytelaars, T., Van Gool, L.: SURF: speeded up robust features. In: Leonardis, A., Bischof, H., Pinz, A. (eds.) ECCV 2006, Part I. LNCS, vol. 3951, pp. 404–417. Springer, Heidelberg (2006)
4. Cao, X., Chen, L., Cong, G., Jensen, C.S., Qu, Q., Skovsgaard, A., Wu, D., Yiu, M.L.: Spatial keyword querying. In: Atzeni, P., Cheung, D., Ram, S. (eds.) ER 2012 Main Conference 2012. LNCS, vol. 7532, pp. 16–29. Springer, Heidelberg (2012)
5. Cao, Y., Wang, C., Li, Z., Zhang, L., Zhang, L.: Spatial-bag-of-features. In: 2010 IEEE Conference on Computer Vision and Pattern Recognition (CVPR), pp. 3352–3359. IEEE (2010)
6. Chen, L., Cong, G., Jensen, C.S., Wu, D.: Spatial keyword query processing: an experimental evaluation. Proceedings of the VLDB Endowment 6(3), 217–228 (2013)
7. Christoforaki, M., He, J., Dimopoulos, C., Markowetz, A., Suel, T.: Text vs. space: efficient geo-search query processing. In: Proceedings of the 20th ACM International Conference on Information and Knowledge Management, pp. 423–432. ACM (2011)
8. Cong, G., Jensen, C.S., Wu, D.: Efficient retrieval of the top-k most relevant spatial web objects. Proceedings of the VLDB Endowment 2(1), 337–348 (2009)
9. Doherty, A.R., Smeaton, A.F.: Automatically augmenting lifelog events using pervasively generated content from millions of people. Sensors 10(3), 1423–1446 (2010)
10. Erol, B., Antúnez, E., Hull, J.J.: Hotpaper: multimedia interaction with paper using mobile phones. In: Proceedings of the 16th ACM International Conference on Multimedia, pp. 399–408. ACM (2008)
11. Fagin, R., Lotem, A., Naor, M.: Optimal aggregation algorithms for middleware. Journal of Computer and System Sciences 66(4), 614–656 (2003)
12. Google: Goggles. http://www.google.com/mobile/goggles/
13. Graham, J., Hull, J.J.: Icandy: a tangible user interface for itunes. In: CHI 2008 Extended Abstracts on Human Factors in Computing Systems, pp. 2343–2348. ACM (2008)
14. Guo, L., Shao, J., Aung, H.H., Tan, K.L.: Efficient continuous top-k spatial keyword queries on road networks. GeoInformatica, 1–32 (2014)
15. Hull, J.J., Erol, B., Graham, J., Ke, Q., Kishi, H., Moraleda, J., Van Olst, D.G.: Paper-based augmented reality. In: 17th International Conference on Artificial Reality and Telexistence, pp. 205–209. IEEE (2007)
16. Ilyas, I.F., Beskales, G., Soliman, M.A.: A survey of top-k query processing techniques in relational database systems. ACM Computing Surveys (CSUR) 40(4), 11 (2008)
17. Kooaba: http://www.kooaba.com
18. Li, W., Guan, J., Zhou, S.: Efficiently evaluating range-constrained spatial keyword query on road networks. In: Han, W.-S., Lee, M.L., Muliantara, A., Sanjaya, N.A., Thalheim, B., Zhou, S. (eds.) DASFAA 2014. LNCS, vol. 8505, pp. 283–295. Springer, Heidelberg (2014)
19. Liu, T., Moore, A.W., Yang, K., Gray, A.G.: An investigation of practical approximate nearest neighbor algorithms. In: Advances in Neural Information Processing Systems, pp. 825–832 (2004)

20. Lowe, D.G.: Distinctive image features from scale-invariant keypoints. International Journal of Computer Vision **60**(2), 91–110 (2004)
21. Nister, D., Stewenius, H.: 2006 IEEE Computer Society Conference on Scalable recognition with a vocabulary tree. In: Computer Vision and Pattern Recognition, vol. 2, pp. 2161–2168. IEEE (2006)
22. Nokia: Point and find. http://www.pointandfind.nokia.com
23. Philbin, J., Chum, O., Isard, M., Sivic, J., Zisserman, A.: Object retrieval with large vocabularies and fast spatial matching. In: IEEE Conference on Computer Vision and Pattern Recognition, CVPR 2007, pp. 1–8. IEEE (2007)
24. Rocha-Junior, J.B., Nørvåg, K.: Top-k spatial keyword queries on road networks. In: Proceedings of the 15th International Conference on Extending Database Technology, pp. 168–179. ACM (2012)
25. Rocha-Junior, J.B., Gkorgkas, O., Jonassen, S., Nørvåg, K.: Efficient processing of top-k spatial keyword queries. In: Pfoser, D., Tao, Y., Mouratidis, K., Nascimento, M.A., Mokbel, M., Shekhar, S., Huang, Y. (eds.) SSTD 2011. LNCS, vol. 6849, pp. 205–222. Springer, Heidelberg (2011)
26. Salton, G., Buckley, C.: Term-weighting approaches in automatic text retrieval. Inf. Process. Manage. **24**(5), 513–523 (1988)
27. Sivic, J., Zisserman, A.: Video google: a text retrieval approach to object matching in videos. In: 2003 Proceedings of the Ninth IEEE International Conference on Computer Vision, pp. 1470–1477. IEEE (2003)
28. SnapTell: http://www.snaptell.com
29. Zhang, C., Zhang, Y., Zhang, W., Lin, X.: Inverted linear quadtree: efficient top k spatial keyword search. In: 2013 IEEE 29th International Conference on Data Engineering (ICDE), pp. 901–912. IEEE (2013)
30. Zhang, C., Zhang, Y., Zhang, W., Lin, X., Cheema, M.A., Wang, X.: Diversified spatial keyword search on road networks. In: EDBT, pp. 367–378 (2014)
31. Zhang, D., Chan, C.Y., Tan, K.L.: Processing spatial keyword query as a top-k aggregation query. In: Proceedings of the 37th International ACM SIGIR Conference on Research & Development in Information Retrieval, pp. 355–364. ACM (2014)
32. Zhang, D., Tan, K.L., Tung, A.K.: Scalable top-k spatial keyword search. In: Proceedings of the 16th International Conference on Extending Database Technology, pp. 359–370. ACM
33. Zhang, S., Huang, Q., Hua, G., Jiang, S., Gao, W., Tian, Q.: Building contextual visual vocabulary for large-scale image applications. In: Proceedings of the International Conference on Multimedia, pp. 501–510. ACM (2010)
34. Zhang, S., Tian, Q., Hua, G., Huang, Q., Gao, W.: Generating descriptive visual words and visual phrases for large-scale image applications. IEEE Transactions on Image Processing **20**(9), 2664–2677 (2011)
35. Zhang, S., Tian, Q., Hua, G., Huang, Q., Li, S.: Descriptive visual words and visual phrases for image applications. In: Proceedings of the 17th ACM International Conference on Multimedia, pp. 75–84. ACM (2009)
36. Zhong, R., Li, G., Tan, K.L., Zhou, L.: G-tree: an efficient index for knn search on road networks. In: Proceedings of the 22nd ACM International Conference on Information & Knowledge Management, pp. 39–48. ACM (2013)

Social Networks II

An Efficient Method to Find the Optimal Social Trust Path in Contextual Social Graphs

Guanfeng Liu[1,2](✉), Lei Zhao[1,2], Kai Zheng[3], An Liu[1,2], Jiajie Xu[1,2], Zhixu Li[1,2], and Athman Bouguettaya[4]

[1] School of Computer Science, Soochow University, Suzhou 215006, China
{gfliu,zhaol,anliu,jjxu,zxli}@suda.edu.c
[2] Collaborative Innovation Center of Novel Software Technology
and Industrialization, Jiangsu, China
[3] School of Information Technology and Electrical Engineering,
The University of Queensland, Brisbane 4072, Australia
kevinz@itee.edu.au
[4] School of Computer Science and Information Technology,
RMIT University, Melbourne 3001, Australia
athman.bouguettaya@rmit.edu.au

Abstract. Online Social Networks (OSN) have been used as platforms for many emerging applications, where trust is a critical factor for participants' decision making. In order to evaluate the trustworthiness between two unknown participants, we need to perform trust inference along the social trust paths formed by the interactions among the intermediate participants. However, there are usually a large number of social trust paths between two participants. Thus, a challenging problem is how to effectively and efficiently find the optimal social trust path that can yield the most trustworthy evaluation result based on the requirements of participants. In this paper, the core problem of finding the optimal social trust path with multiple constraints of social contexts is modelled as the classical NP-Complete Multi-Constrained Optimal Path (MCOP) selection problem. To make this problem practically solvable, we propose an efficient and effective approximation algorithm, called T-MONTE-K, by combining Monte Carlo method and our optimised search strategies. Lastly we conduct extensive experiments based on a real-world OSN dataset and the results demonstrate that the proposed T-MONTE-K algorithm can outperform state-of-the-art MONTE_K algorithm significantly.

1 Introduction

1.1 Background

In recent years, social networking sites have been used as platforms for a variety of activities. For example, there is an increasing interest of employers to use OSNs as part of the recruitment process for academic positions (globalacademyjobs.com). In addition, by connecting with OSNs (e.g., Facebook and

© Springer International Publishing Switzerland 2015
M. Renz et al. (Eds.): DASFAA 2015, Part II, LNCS 9050, pp. 399–417, 2015.
DOI: 10.1007/978-3-319-18123-3_24

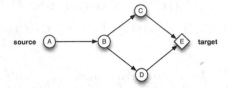

Fig. 1. A social graph

Twitter) at some e-commerce websites like ThisNext (thisnext.com) and eBay (ebay.com), a buyer can recommend the products available on these e-commerce websites to his/her friends who are also in the OSNs. In these activities, trust is one of the most important factors that influence participants' decision making. However, most users in OSNs do not have previous direct interactions, which calls for approaches and mechanisms for evaluating the trustworthiness between participants who are unknown to each other.

An OSN is usually modelled as a graph, where each vertex represents a participant, and each edge corresponds to a real-world or online interaction between two participants. For instance, an edge may indicate they are colleagues in real-world, or they have online follower-followee relationship in micro-blogging sites (e.g., $A \rightarrow B$ in Fig. 1). Each participant usually has relationship with many others. Thus given two non-adjacent vertex, there may be multiple paths from the source participant (e.g., A) to the target participant (e.g., E) (For example, paths $A \rightarrow B \rightarrow C \rightarrow E$ and $A \rightarrow B \rightarrow D \rightarrow E$ in Fig. 1). This path linking a source and a target can be used to evaluate the trustworthiness of the target participant, which is called a *social trust path* [1]. In addition to the vertexes and edges, an OSN contains various contextual information, such as social positions of participants, social relationships and social trust between participants [2], which also have significant influence on trust evaluation [3–6]. The process of the trust evaluation for the target participants based on the social trust paths and social contexts is called *trust propagation* [2,7].

1.2   The Problem and Challenges

In a large social graph, there could be a large number of social trust paths between a source and the target [8]. Enumerating all the social trust paths and evaluating their trustworthiness are computationally infeasible [9]. Alternatively, we can look for an optimal social path yielding the most trustworthy propagation result from multiple paths by taking the social contexts and the preferences of a source participant in trust evaluation.

Example 1: In Fig. 1, suppose C has no knowledge of IT, and C does not know E well personally. $C \rightarrow E$ is formed as C bought a used car from E. But D is an IT expert and D is familiar with E as D supervised E in an IT project. When given the trust query of *looking for a programmer* by A, based on the trust theory in Social Psychology [10,11], D's recommendation of E is more credible than C. So the trust evaluation result based on the social trust path $A \rightarrow B \rightarrow D \rightarrow E$ is more reliable than the one based on $A \rightarrow B \rightarrow C \rightarrow E$.

So given a trust query under a certain context, a social trust path has *higher quality* if the path contains more important intermediate participants, who have closer relationships. The problem of optimal social trust path is to find the path with the best quality and satisfying the preferences of a source participant in path finding. This problem has proved to be NP-Complete in our prior work [2].

In the literature, Lin *et al.* [12] proposed an optimal social path finding method. In their model, the shortest path between the source participant and the target one is selected as the optimal one. In another work [1], the path with the maximal propagated trust value is selected as the most trustworthy path. These methods do not consider the social contextual information, like social position and social relationship, which has significant influence on trust propagation [3, 13]. Ie trust propagation to illustrate the preference of the source. However, this has not been considered in previous work. With considering the constraints of social contexts, the optimal social trust path finding becomes a challenging NP-Complete problem [2]. In order to solve this challenging problem, we have proposed several social trust path finding models where the social information, like social trust, social relationship and social position have been taken into account [2,14–17]. However, ONS is in a real-time environment [18,19], where the social interactions between participants and the social information, like social position, social trust can change in real-time. To deliver a solution, some of the existing methods need to access all the vertexes and edges. Therefore, when facing a large social graph, they have low efficiency. Moreover, other methods are based on the probabilistic methods. Although they can deliver a solution very fast without accessing all the vertexes and edges of a social graph, they are suffering low effectiveness due to the disadvantages of the heuristic search strategies.

1.3 Contribution

This paper builds upon and extends the approach described in [2], where we first propose a new concept, called *Quality of Social Trust Path*, which is essential to illustrate the ability of a social trust path to guarantee a certain trust level during trust evaluation. We then propose an efficient approximation algorithm, T-MONTE-K, for the classical Multi-Constrained Optimal Path (MCOP) selection problem based on Monte Carlo method [20] and new optimization strategies. Our algorithm achieves the time complexity of $O(mu)$, where m is the number of simulations, and u is the maximal outdegree of nodes in social networks. Finally, We conduct experiments on a real-world dataset of OSNs, Epinions (epinions.com). The experimental results illustrate that T-MONTE-K can find better social trust path than the existing methods more efficiently.

2 Related Work

In this section, we introduce the existing works about social trust path finding with and without social contexts.

2.1   Social Trust Path Finding without Social Contexts

SmallBlue [12] is an OSN created for IBM staff. In this system, if a source would like to find a target (e.g., a C++ programmer), it considers up to 16 social paths between them with the path length of no more than 6 hops, among which, the shortest one is taken as the optimal path. Hang *et al.* [1] proposed a social trust path finding method in online social networks, where trust between participants is considered in the path selection. In their model, the aggregated belief value (trust value) of a social trust path is computed by multiplying the trust value between any two intermediate nodes in the path. Among all the social trust paths, the one with the highest aggregated belief (i.e., the maximum of aggregated trust value) is selected as the optimal path that yields the most trustworthy result of trust propagation between a source participant and the target participant. Wang *et al.* [21] proposed a social trust path finding method where a source participant can specify a threshold. Their method first aggregated trust values given to each of the recommenders (i.e., the intermediate nodes) in the network between a source participant and the target participant. If the aggregated trust value of a recommender is greater than the threshold specified by the source participant, the recommender is kept in the trust network. Otherwise, the recommender (the node) is deleted from the trust network. After node deletion, the rest of the social trust paths are kept for trust evaluation.

However, the above existing methods select the optimal social trust path(s) from a large volume of paths based on different selection criteria, which indeed reduces the computation complexity of the trust evaluation between two unknown participants. However, in the above methods, the social information including *social relation* and *social position* of participants are not taken into account in path finding. In addition, a source participant can have different purposes in evaluating the trustworthiness of the target participants (e.g., recruitment or buying products). The different trust evaluation criteria in different applications should be reflected by specifying certain constraints of the above social information for social trust path finding. Thus, although trust information is taken into consideration in some of the existing trust path finding methods, they cannot be expected to find the trustworthy trust paths without considering social information and complex trust criteria specification.

2.2   Social Trust Path Finding with Social Contexts

To address these issues, we have proposed an optimal trust path finding model by taking the social contexts and the constraints of these contexts into account in our priori work [2]. In the work, the social contexts in a social trust path are aggregated and constrained for the path finding. As the constrained optimal path finding is an NP-Complete problem [2], we have proposed several heuristic algorithms [14–16] by adopting the Dijkstra's algorithm. These algorithms can guarantee to find a social trust path if there exists a feasible one in a given social graph. But they do not have satisfactory performance as they need to access all the vertexes and edges several times. In addition, we have proposed an approximation algorithms [2] based on the Monte Carlo method However, this algorithm suffers low effectiveness as it greedily looks for the social trust path with the highest utility.

3 Preliminary

In this section, we first review the contextual social graph, and propose a novel concept *Quality of Social Trust Path* (QoSTP). We then present the model of the optimal social trust path finding with end-to-end QoSTP constraints.

3.1 Contextual Social Graph

Contextual Social Graph (CSG) is a labeled directed graph $G = (V, E, LV, LE)$ consists of (1) a set nodes V; (2) a set of edges $E \in V \times V$, where $(v, w) \in E$ denotes a directed edge from node v to w; and (3) a function LV defined on V such that for each node v in V, $LV(v)$ is a label for v. Intuitively, the node labels may present e.g., social roles. Moreover, LE defined on E such that for each link v, w in E, $LE(v, w)$ is a label for (v, w), like social relationships and social trust. The labelled social contexts have significant influence on trust evaluation and trust path finding. Below we introduce the social impact factors that comes from the social contexts and affects the social trust path selection.

Social Impact Factors

1. **Social Trust:** In our model, let $T_{A,B}^{D_i} \in [0,1]$ denote the trust value that A assigns to B in domain i. If $T_{A,B}^{D_i} = 0$, it indicates that A completely distrusts B in the domain, while $T_{A,B}^{D_i} = 1$ indicates that A completely believes B's future action can lead to the expected outcome in that domain.
2. **Social Intimacy Degree:** Let $r_{A,B} \in [0,1]$ denote the *Social Intimacy Degree* between A and B in online social networks. $r_{A,B} = 0$ indicates that A and B have no intimate social relationship while $r_{A,B} = 1$ indicates they have the most intimate social relationship.
3. **Role Impact Factor:** Let $\rho_A^{D_i} \in [0,1]$ denote the *Role Impact Factor* of A, illustrating the impact of participant A in domain i, which is determined by the expertise of A. $\rho_A^{D_i} = 1$ indicates that A is a domain expert and has the greatest impact in the domain while $\rho_A^{D_i} = 0$ indicates that A has no knowledge and has the least impact in that domain.

Although it is difficult to build up comprehensive social intimacy degrees and role impact factors in all domains, it is feasible to build them up in some specific social communities by using data mining techniques [22–24], which is out of the scope of this paper. Fig. 2 depicts a Contextual Social Graph (CSG), where we can see that each of B, C, D and E is associated with a role impact factor in domain i, and each edge is associated with the social trust in domain i and social intimacy degree.

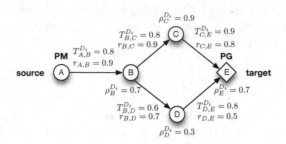

Fig. 2. A contextual social graph

3.2   Optimal Social Trust Path Queries in CSG

Definition 1: *Quality of Social Trust Path* (QoSTP) is the ability to guarantee a certain level of trustworthiness in trust propagation along a social trust path, taking trust (T), social intimacy degree (r), and role impact factor (ρ) as attributes.

In service composition, service consumers can set multiple end-to-end constraints for the attributes of QoS (e.g., cost, delay and availability) to satisfy their requirements of services. Different requirements have different constraints (e.g., total cost<\$20, delay<5s and availability>70%). In our model, a source participant can set multiple end-to-end constraints for QoSTP attributes (i.e., T, r and ρ) as the requirements of trust propagation in social trust paths. $QoC_{AM}^{T^{D_i}}, QoC_{AM}^{r}$ and $QoC_{AM}^{\rho^{D_i}}$ are denoted as the QoSTP constraints of T, r and ρ in domain i respectively.

Based on the theories in *Social Psychology* [25], we adopt the multiplication method to aggregate T and r values of a path, and adopt the average method to aggregate the ρ values of the vertices in a path. The details of the aggregation method has been discussed in [2].

Utility Function. In our model, we define a feasible utility (denoted as $\mathcal{F}$) as the measurement of the trustworthiness of social trust paths, which takes the QoSPT attributes T, r and ρ as arguments.

$$\mathcal{F}_{p(a_1,...,a_n)} = \omega_T * T_{p(a_1,...,a_n)}^{D_i} + \omega_r * r_{p(a_1,...,a_n)} + \omega_\rho * \rho_{p(a_1,...,a_n)}^{D_i} \tag{1}$$

where ω_T, ω_r and ω_ρ are the weights of T, r and ρ respectively; $0 < \omega_T, \omega_r, \omega_\rho < 1$ and $\omega_T + \omega_r + \omega_\rho = 1$.

A source participant can specify different weights for different QoSTP attributes in path finding. For example, if a source participant believes the social position of participants is more important in the domain of employment, he/she can specify a relative high value for ω_{CIF}. In contrast, if he/she regards the social relationship is more important, he/she can specify a relatively high value for ω_r.

An Optimal Social Trust Path Finding Query. Given a group of QoSTP constraints $QoC^{D_i}_{v_s,v_t}$ in domain i and a group of weights ω_φ $\varphi \in \{T, r, \rho\}$ the Multiple QoSTP Constrained Optimal Social Trust Path Finding (MQCOTP) Problem is to find a path p from a source v_s to a target v_t such that:

1. $T^{D_i}_{p_{v_s,v_t}} \geq QoC^{T^{D_i}}_{v_s,v_t}$, $r_{p_{v_s,v_t}} \geq QoC^r_{v_s,v_t}$ and $\rho^{D_i}_{p_{v_s,v_t}} \geq QoC^{\rho^{D_i}}_{v_s,v_t}$
2. $\mathcal{F}_{p_{v_s,v_t}}$ is maximised over all feasible trust paths satisfying the above condition 1.

Then MQCOTP can be modelled as the classical Multi-Constrained Optimal Path (MCOP) selection problem which is NP-Complete [26].

4   Social Trust Path Finding

In this section, we propose an efficient and effectiveness approximation algorithm, T-MONTE-K, based on the Monte Carlo method [20] and our optimization strategies.

4.1   Monte Carlo Method

The Monte Carlo method [20] is one of the techniques for solving NP-complete problems [20,27]. Generally, the Monte Carlo method consists of four steps: (1) defining a domain of inputs, (2) generating inputs randomly, (3) performing a computation on each input, and (4) aggregating the results into the final result.

4.2   T-MONTE-K

In this section, we propose an new efficient and effective approximation algorithm T-MONTE-K. It adopts Twice the Monte Carlo method to search a network from v_t (the target) to v_s (the source), and from v_s to v_t respectively. During this process, T-MONTE-K selects up to K candidates at each of the search step. Next, we will introduce the details of T-MONTE-K.

In social trust path finding, if a path satisfies multiple QoSTP constraints, it means that each aggregated QoSTP attribute (i.e., T, r or ρ) of that path should be larger than the corresponding QoSTP constraint. Therefore, we propose an *objective function* in Eq. (2) to investigate whether the aggregated QoSTP attributes of a path can satisfy the QoSTP constraints. From Eq. (2), we can see that if any aggregated QoSTP attribute of a social trust path does not satisfy the corresponding QoSTP constraint, then $\delta(p) > 1$. Otherwise $\delta(p) \leq 1$.

$$\delta(p) \triangleq max\{(\frac{1 - T^{D_i}_p}{1 - QoC^{T^{D_i}}_p}), (\frac{1 - r_p}{1 - QoC^r_p}), (\frac{1 - \rho^{D_i}_p}{1 - QoC^{\rho^{D_i}}_p})\} \tag{2}$$

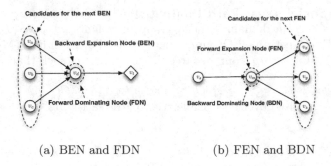

(a) BEN and FDN   (b) FEN and BDN

Fig. 3. BEN and FEN

Backward_Search: In the *Backward_Search* procedure from v_t to v_s, T-MONTE-K calculates δ values of the social trust paths from v_t to the neighbors of the current *Backward Expansion Node* (BEN) (e.g., node v_d in Fig. 3(a)). Then based on the following *Strategy 1*, T-MONTE-K selects up to K neighbors with the K minimal δ values as the candidates (e.g., nodes v_a, v_b and v_c in Fig. 3(a)). Then one of them is selected as the next BEN based on Eq. (3) .

In this searching process, if a node v_{k_i} is selected as the next BEN, the aggregated $T^{D_i}_{p_{v_s} \to v_{k_i}}$ $r_{p_{v_s} \to v_{k_i}}$ and $\rho^{D_i}_{p_{v_s} \to v_{k_i}}$ are recorded at v_{k_i} based on the following optimization *Strategy 2*. According to the following *Theorem 1*, the path identified by *Backward_Search* procedure can investigate whether there exists a feasible path in the sub-network.

Theorem 1: The path identified by the *Backward_Search* procedure with the minimal δ converge to a feasible solution if one exists in a sub-network.

Proof: Let p_s be a path from v_t to v_s with the minimal δ at v_t delivered by the *Backward_Search* procedure, and p_* be a feasible solution. Then, based on the *Strategy 1*, $\delta(p_s) \leq \delta(p_*)$. Assume p_s is not a feasible solution, then $\exists \varphi \in \{T, r, \rho\}$ that $\varphi_{p_s} < Q^{\varphi}_{v_s, v_t}$. Hence, $\delta(p_s) > 1$. Since p_* is a feasible solution, then $\delta(p_*) \leq 1$ and $\delta(p_s) > \delta(p_*)$. This contradicts $\delta(p_s) \leq \delta(p_*)$. Therefore, p_s is a feasible solution. □

With the increase of the Backward Simulation Times (BST), the solution delivered by the *Backward_Search* procedure can always convergency to the path with the minimal δ. If $\delta_{min} > 1$, it indicates there is no feasible solution in the sub-network. If $\delta_{min} \leq 1$, it indicates there exists at least one feasible solution and the identified path is a feasible solution.

Strategy 1: K-path with the Minimal δ. According to Eq. (2), the lower the δ value of a path, the higher the probability for that path to be a feasible solution. Thus, given a partially identified social trust path from v_t to v_d ($v_d \neq v_s$, see Fig. 3(a)), we calculate the δ values of the paths from v_t to each neighboring node of v_d and record up to K neighboring nodes (e.g., nodes v_a, v_b and v_c in Fig. 3(a)) that yield up to K minimum δ values as candidates for the next BEN. One of them will be selected as the next BEN based on the probability computed by Eq. (3).

$$\mathbf{Pro}^B(v_{k_i}) = \frac{\delta(p^B_{v_{k_i} \to v_t})}{\sum_{i=1}^{K} \delta(p^B_{v_{k_i} \to v_t})} \tag{3}$$

where $\mathbf{Pro}^B(v_{k_i})$ is the probability of v_{k_i} to be selected as the next BEN.

As this strategy selects no more than K neighbors at each step in social trust path finding, it can reduce the search space and deliver high efficiency.

Strategy 2: Optimization at A Backward Dominating Node (BDN).
During the *Backward_Search*, if the *outdegree* of a node is greater than one (e.g., node v_m in Fig. 3(b)) in the social network, then the node, e.g., v_m, is regarded as a *Backward Dominating Node (BDN)*. To obtain a near-optimal solution, T-MONTE-K performs multiple simulations. In the first simulation, if a social trust path from v_t to v_m (denoted as path $p^B_{m_1}$) is identified, we store the $\delta(p^B_{m_1})$ and the aggregated value of each QoSTP attribute of $p^B_{m_1}$ at v_m. In all subsequent simulations, if a different social path from v_t to v_m (denoted as path $p^B_{m_i}$, where $i > 1$) is identified, the optimization is performed in the following situations.

Situation 1: If $v_m = v_s$ and $\delta(p^B_{m_i}) > \delta(p^B_{m_1})$, it indicates $p^B_{m_i}$ is worse than $p^B_{m_1}$. Thus we replace the values of $p^B_{m_i}$ (i.e., T, r, ρ, and δ) with the ones stored at v_m.

Situation 2: If $v_m = v_s$ and $\delta(p^B_{m_i}) < \delta(p^B_{m_1})$, it indicates $p^B_{m_i}$ is better than $p^B_{m_1}$. Thus, we store the aggregated QoSTP attribute values of $p^B_{m_i}$ at v_m.

Situation 3: If $v_m \neq v_s$, $\delta(p^B_{m_i}) > \delta(p^B_{m_1})$ and each of the aggregated QoSTP value of $p^B_{m_i}$ is less than the corresponding value of $p^B_{m_1}$, it indicates $p^B_{m_i}$ is worse than $p^B_{m_1}$. Thus we replace the values of $p^B_{m_i}$ with the ones stored at v_m.

Situation 4: If $v_m \neq v_s$, $\delta(p^B_{m_i}) < \delta(p^B_{m_1})$, and each of the aggregated QoSTP value of $p^B_{m_i}$ is less than the corresponding value of $p^B_{m_1}$, it indicates $p^B_{m_i}$ is better than $p^B_{m_1}$. Thus, we store the aggregated QoSTP attribute values of $p^B_{m_i}$ at v_m.

Following *Strategy 2*, BDN v_m records T, r, ρ, $\mathcal{F}$ and δ values of the locally optimal social trust path from v_s to v_m. The optimization at v_m can guarantee that the delivered solution from v_s to v_m is locally optimal.

Forward_Search: If there exists a feasible solution identified by the *Backward_Search* procedure, a forward search is executed from v_s to v_t. This process uses the information provided by the above *Backward_Search* to identify whether there is another path p_t which is better than the above returned best path p_s (i.e., $\mathcal{F}(p_t) > \mathcal{F}(p_s)$). In this procedure, at each of the neighboring nodes of the the the current *Forward Expansion Node* (FEN), T-MONTE-K calculates the aggregated QoSTP attribute values of the path from v_s to an intermediated node v_m (denoted as path p^F_m). Let p^B_m denote the path from v_m to v_t identified by the *Backward_Search* procedure, then a *foreseen path* from v_s to v_t via v_m (denoted as $p_{fm} = p^F_m + p^B_m$) can be identified. Let h denote the number of hops of path p_{fm}. The aggregated QoS attribute values of p_{fm} can be calculated as $T^{D_i}_{p_{fm}} = T^{D_i}_{p^F_m} * T^{D_i}_{p^B_m}$, $r_{p_{fm}} = r_{p^F_m} * r_{p^B_m}$ and $\rho^{D_i}_{p_{fm}} = (\rho^{D_i}_{p^F_m} + \rho^{D_i}_{p^B_m})/(h-1)$. Then, based on the following *Strategy 3*, T-MONTE-K selects up to K candidates that 1) the foreseen path from v_s to v_t each of the K next FENs is feasible, and 2) they have the K maximal $\mathcal{F}$ values. Then one of them is selected as the next FEN based on the probability computed by Eq. 4. Finally, T-MONTE-K calculates $\mathcal{F}$ value of the path from v_s to the new selected FEN, and update the aggregated QoSTP attributes values based on the following *Strategy 4*.

$$\mathbf{Pro}^F(v_{k_i}) = \frac{\delta(p^F_{v_s \to v_{k_i}})}{\sum_{i=1}^{K} \delta(p^F_{v_s \to v_{k_i}})} \tag{4}$$

where $\mathbf{Pro}^F(v_{k_i})$ is the probability of v_{k_i} to be selected as the next FEN.

Strategy 3: K-path with the Maximal $\mathcal{F}$. As introduced in *Section 4.3*, the goal of the optimal social trust path finding is to find the path with the best utility and meets the QoSTP constraints. Thus, given a partially identified social trust path from v_s to v_m ($v_m \neq v_t$), if p_{fm} is feasible, we calculate the path utilities of v_s to each neighboring node of v_m and record up to K neighboring nodes (e.g., node v_x, v_y and v_z in Fig. 3(a)) that yield up to K maximal path utilities as candidates for the next FEN selection.

As this strategy selects no more than K neighbors at each step in social trust path finding, it can reduce the search space and deliver high efficiency.

Strategy 4: Optimization at A Forward Dominating Nodes (FDN). If the indegree of a node, like v_d in Fig. 3(a) ($v_d \neq v_s$) is greater than one in the social network, then node v_d is regarded as a *Forward Dominating Node* (FDN). To obtain a near-optimal solution, T-MONTE-K performs multiple simulations. In the first simulation, if a social trust path from v_s to v_d (denoted as path $p^F_{d_1}$) is selected, we store the utility $\mathcal{F}$ and the aggregated value of each QoSTP attribute of $p^F_{d_1}$ at v_d. In all subsequent simulations, if a different social path from v_s to $p^F_{d_1}$ (denoted as path $p^F_{d_i}$, where $i > 1$) is selected, the optimization is performed in the following situations.

Proof: Assume the optimal solution in the sub-network is denoted as p_o, and the path identified by the Forward_Search procedure is denoted as p_t. If $\mathcal{F}(p_o) > \mathcal{F}(p_t)$, then $\exists v_i \in p_o$ and $\exists v_j \in p_t$ ($v_i \neq v_t, v_j \neq v_t$), $\mathcal{F}(p_{v_s \to v_i}) = 0$ and $\mathcal{F}(p_{v_s \to v_j}) = 1$. As $T^{D_i}_{p_{v_s \to v_i}} = 0$, $r_{p_{v_s \to v_i}} = 0$ and $\rho^{D_i}_{p_{v_s \to v_i}} = 0$. Then $T^{D_i}_{p_o} = 0$ and $r_{p_o} = 0$, and thus cannot satisfy the constraints $QoC^\varphi \in (0,1)$, ($\varphi \in \{T, r, \rho\}$). Then p_o is an infeasible solution, which contradicts p_o is an optimal solution. Therefore, $\mathcal{F}(p_o) = \mathcal{F}(p_t)$. So, *Theorem 4* is correct.   □

The process of T-MONTE-K is as follows.

Situation 1: If $v_d = v_t$ and $\mathcal{F}(p^F_{d_i}) < \mathcal{F}(p^F_{d_1})$, it indicates $p^F_{d_i}$ is worse than $p^F_{d_1}$. Thus we replace the values of $p^F_{d_i}$ (i.e., T, r, ρ, $\mathcal{F}$ and δ values) with the ones stored at v_d .

Situation 2: If $v_d = v_t$ and $\mathcal{F}(p^F_{d_i}) > \mathcal{F}(p^F_{d_1})$, it indicates $p^F_{d_i}$ is better than $p^F_{d_1}$. Thus, we store the aggregated QoSTP attributes values of $p^F_{d_i}$ at v_d.

Situation 3: If $v_d \neq v_t$, $\mathcal{F}(p^F_{d_i}) < \mathcal{F}(p^F_{d_1})$ and $\delta(p^B_{d_i}) > \delta(p^B_{d_1})$, it indicates $p^F_{d_i}$ is worse than $p^F_{d_1}$. Thus we replace the aggregated QoSTP attributes values of $p^F_{d_i}$ with the ones stored at v_d .

Situation 4: If $v_d \neq v_t$, $\mathcal{F}(p_{d_i}^F) > \mathcal{F}(p_{d_1}^F)$ and $\delta(p_{d_i}^F) < \delta(p_{d_1}^F)$, it indicates $p_{d_i}^F$ is better than $p_{d_i}^F$. Thus, we store the aggregated QoSTP attributes values of $p_{d_i}^F$ at v_d.

Following *Strategy 4*, the FDN v_d records T, r, ρ, $\mathcal{F}$ and δ values of the locally optimal social trust path from v_s to v_d. The optimization at v_d can guarantee that the delivered solution from v_s to v_d is locally optimal.

The following *Theorem 2* illustrates that the social trust path p_t identified by the *Forward_Search* procedure can not be worse than the feasible social trust path p_s identified by the *Backward_Search* procedure. Namely, $\mathcal{F}(p_t) \geq \mathcal{F}(p_s)$.

Theorem 2: With the social trust path p_s identified by the *Backward_Search* procedure and the social trust path p_t identified by the *Forward_Search* procedure in T-MONTE-K, if p_s is a feasible solution, then p_t is feasible and $\mathcal{F}(p_t) \geq \mathcal{F}(p_s)$.

Proof: Assume that path p_s consists of $n+2$ nodes $v_s, v_1, ..., v_n, v_t$. In the *Forward_Search* procedure, T-MONTE-K searches the neighboring nodes of v_s and chooses v_1 from these nodes when a foreseen path from v_s to v_t via v_1 is feasible and the current path from v_s to v_1 has the maximal $\mathcal{F}$. This step is repeated at all the nodes between v_1 and v_n until a social trust path p_t is identified. If at each search step, only one node (i.e., $v_1, ..., v_n$) has a feasible foreseen path, then p_t is the only feasible solution in the sub-network between v_s and v_t. According to *Theorem 1* and *Strategy 4*, then $p_t = p_s$. Thus, $\mathcal{F}(p_t) = \mathcal{F}(p_s)$. Otherwise, based on *Strategy 4*, IFF p_t and p_s do not have any joint nodes except for v_t, $\mathcal{F}(p_t) < \mathcal{F}(p_s)$. However, in a sub-network from v_s to v_t, in addition to v_t, v_s is the joint node for both p_t and p_s. Therefore, $\mathcal{F}(p_t) > \mathcal{F}(p_s)$. So, *Theorem 2* is correct.   □

Based on *Theorem 2*, If there is only one feasible solution existing in the trust network, both *Backward_Search* and *Forward_Search* procedures can mine that path and the path is the optimal one.

The following *Theorem 3* illustrates that the social trust path identified by the *Forward_Search* procedure converges to the optimal solution.

Theorem 3: If K is not less than the maximal outdegree of a sub-network, the solution p_t identified by the *Forward_Search* procedure converges to the optimal solution with the increase of the Forward Simulation Times (FST).

Proof: Assume the optimal solution in the sub-network is denoted as p_o, and the path identified by the Forward_Search procedure is denoted as p_t. If $\mathcal{F}(p_o) > \mathcal{F}(p_t)$, then $\exists v_i \in p_o$ and $\exists v_j \in p_t$ $(v_i \neq v_t, v_j \neq v_t)$, $\mathcal{F}(p_{v_s \to v_i}) = 0$ and $\mathcal{F}(p_{v_s \to v_j}) = 1$. As $T_{p_{v_s \to v_i}}^{D_i} = 0$, $r_{p_{v_s \to v_i}} = 0$ and $\rho_{p_{v_s \to v_i}} = 0$. Then $T_{p_o}^{D_i} = 0$ and $r_{p_o} = 0$, and thus cannot satisfy the constraints $QoC^\varphi \in (0,1)$, $(\varphi \in \{T, r, \rho\})$. Then p_o is an infeasible solution, which contradicts p_o is an optimal solution. Therefore, $\mathcal{F}(p_o) = \mathcal{F}(p_t)$. So, *Theorem 3* is correct.   □

The process of T-MONTE-K is as follows.

Initialization: Mark the status of all nodes in the network as unvisited. Add v_s into set BEN_{set} and v_t into the FEN_{set} that stores the BENs and FENs identified by the *Backward_Search* procedure and the *Forward_Search* procedure respectively.

Backward_Search:

Step 1: Get an unvisited node v_t from BEN_{set} and mark v_t as visited. Select up to K neighboring nodes v_{Bj} of v_t based on *Strategy 1*.

Step 2: For each v_{Bj}, calculate the probability of v_{Bj} to be selected as the next BEN based on Eq. (3). Then, select one of them (denoted as $v_{Bk}, 1 \leqslant k \leqslant j$) based on $\{\mathbf{Pro}^B(p_{v_{Bj}})\}$.

Step 3: Calculate the aggregated QoSTP values (i.e., T, r and ρ) of $p_{v_{Bk} \to v_t}^B$, and store (or replace) the corresponding aggregated value at v_{Bk} based on *Strategy 2*.

Step 4: If not reach the simulation times, go to *Step 1*. Otherwise, if $v_{Bk} = v_s$ and $\delta(p_{v_{Bk} \to v_t}^B) \leq 1$, start *Forward_Search* procedure. Else if $v_{Bk} = v_s$ and $\delta(p_{v_{Bk} \to v_t}^B) > 1$, return infeasible path in the sub-network.

Forward_Search:

Step 5: Get an unvisited node v_s from FEN_{set} and mark v_s as visited. Select up to K neighboring nodes v_{Fj} of v_s based on *Strategy 1*.

Step 6: For each v_{Fj}, calculate the probability of v_{Fj} to be selected as the next BEN based on Eq. (3). Then, Select one of them (denoted as $v_{Fk}, 1 \leqslant k \leqslant j$) based on $\{\mathbf{Pro}^F(p_{v_{Fk}})\}$.

Step 7: Calculate the path utility $\mathcal{F}(p_{v_s \to v_{Fk}}^F)$, and store (or replace) the path utility at v_{k_j} based on *Strategy 2*.

Step 8: If not reach the simulation times, go to *Step 6*. Otherwise, return the identified optimal social trust path $p_{v_s \to v_t}^F$ and its utility.

Based on the properties of Monte Carlo method and social networks [2], the time complexity of T-MONTE-K can reach $O(mu)$ that is the same as MONTE_K [2], where m is the number of simulations; and u is the maximal outdegree of nodes in social networks. According to the *power-law* characteristic [28], only a few nodes have a large outdegree in social networks. For example, in *Enron* email corpus, an social network forming by sending and receiving emails, 94.7% nodes have an outdegree less than 15. Moreover, the average outdegree is 3.4 and the maximum is 1567. Therefore, in T-MONTE-K, each node can keep a small search space without pruning a large number of neighboring nodes (i.e., candidates) of a node in K-path selection, which results in high efficiency and a higher probability of finding the optimal solution. By considering both the path utility and path feasibility in the algorithm design, T-MONTE-K can deliver better solutions than our proposed the most promising algorithm, MONTE_K [2] with higher efficiency.

5   Experiments

Currently, there is no complete contextual trust-oriented social network structure which contains all the social contexts [16], i.e., T, r and ρ. Thus, we cannot find any existing dataset that completely fits the experiments. On the other hand, since the main purpose of our algorithm is to find the optimal social trust path in OSNs, in order to study the performance of our proposed algorithm on path finding, we need a dataset which contains social network structures. Then, we select a real-world dataset of OSN to conduct experiments. I.E., *Epinions* dataset (trustlet.org), where each link is formed by the trust relations specified by a truster to a trustee. In addition, the *Epinions* dataset has also been proved to possess the properties of social networks [29], and has been widely used in the studies of trust in OSNs [30,31]. Thus, we select the *Epinions* dataset available at TrustLet (trustlet.org) with 88,180 nodes (participants) and 71,7667 links for our experiments.

5.1   Experimental Settings

Below are the settings of our experiments.

1. We firstly extract a large-scale sub-network from each of the datasets by randomly select a pair of source and target from them respectively. This sub-network contains 1746 nodes and 12,220 links.
2. As we discussed in *Section 3,* these social contextual impact factor values (i.e., T, r and ρ) can be mined from the existing social networks, and some solutions have already been proposed to obtain accurate social impact factor values. But mining the social contextual impact factor values is another very challenging problem, which is out of the scope of our work. Moreover, in the real cases, the value of these impact factors can vary from low to high values. So, there are no fixed patterns for the value of social contextual impact factors. Without loss of generality, we randomly set the values of these impact factors by using the function $rand()$ in *Matlab*.
3. Although both of the heuristic algorithms, H_OSTP and MFPB-HOSTP are effective in social trust path finding, they are not suitable for real-time OSN environments. Moreover, in that real-time environment, MONTE_K is so far the most promising algorithm. Therefore, we compare the performance of our T-MONTE-K with that of MONTE_K in the experiments.
4. Considering the small-world characteristic of OSN, we set the maximal search hops of all the algorithms to 6. In addition, a set of relative low QoSTP constraints (i.e., $QoC^T = 0.005$, $QoC^r = 0.005$, $QoC^\rho = 0.05$ to ensure the high possibility of having one feasible solution in an OSN. Otherwise, no solution might be delivered by both of the algorithms and thus cannot investigate the performance of them.

Both T-MONTE-K and MONTE_K are implemented using Matlab R2013a running on an Dual core PC with Intel Xeon E5645 2.40GHz CPU, 3GB RAM, Windows 7 operating system and MySql 5.6.14 database.

5.2   Experimental Results

The path utility and execution time of T-MONTE-K and MONTE_K are averaged based on three independent executions. The results are plotted in Fig. 4 to Fig. 10.

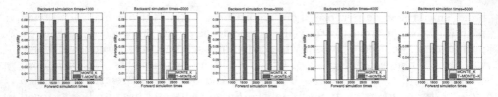

Fig. 4. Average path utility of the two algorithms with different BST

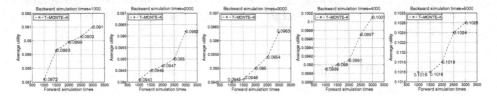

Fig. 5. Average path utility of T-MONTE-K with different BST

Average Path Utility. In order to investigate the performance of the algorithms in the quality of all the identified social trust paths during experiments, we first evaluate the average path utilities with different BST (Backward Simulation Times) and K respectively.

Fig. 4 plots the average path utility with BST varying from 1,000 to 5,000. From the figure, we can see that in all cases, T-MONTE-K can deliver higher average path utility than MONTE_K. This is because that T-MONTE-K considers the QoSTP constraints in the *Backward_Search* procedure. This increases the probability of finding a feasible path with high path utility in the *Forward_Search* procedure. In addition, under a certain BST, the average path utility delivered by T-MONTE-K increases with the increase of FST (Forward Simulation Times), but it does not alway increase for MONTE_K. This is because that based on *Strategy 4*, T-MONTE-K adopts optimizations at FDNs, and thus the more the FST, the higher the probability to deliver a path with higher utility; but MONTE K may spend many FST to search those infeasible path as it greedily searches the path with the maximal path utility. Furthermore, from Fig. 5, we can also see that under a certain FST, the average path utility delivered by T-MONTE-K increases with the increase of BST (e.g., the average path utility is 0.0911 when FST=3500 and BST=1000, and that is 0.0962 when FST=3500 and BST=2000). This is because that based on *Strategy 2*, T-MONTE-K adopts optimizations at BDNs, and thus the more the BST, the more the nodes that are accessed by the *Backward_Search*. This can provide more information for the *Forward_Search* procedure to find the social paths with higher utilities.

Fig. 6 plots the average path utility delivered by T-MONTE-K and MONTE_K respectively with K vary from 5 to 25. From the figure, we can see that in all cases T-MONTE-K can deliver higher average path utility than MONTE_K. This property is similar as the one depicted in Fig. 4, which further justifies the conclusion that with

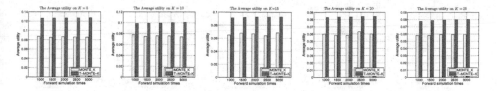

Fig. 6. Average path utility of the two algorithms with different K

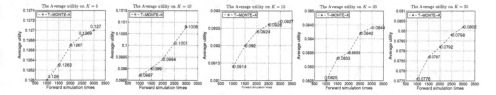

Fig. 7. Average path utility of T-MONTE-K with different K

the assistant of the information provided by the *Backward_Search*, T-MONTE-K can increase the probability of finding a feasible path with high path utility.

In addition, under a certain BST, the average path utility delivered by T-MONTE-K is increased with the increase of FST, but it does not alway increase for MONTE_K.. This property is similar as the one depicted in Fig. 6, which further justifies the conclusion that in T-MONTE-K, the more the FST, the higher the probability to deliver a path with higher utility. Furthermore, from Fig. 7, we can see that under a certain FST, the average path utility decreases with the increase of K. This is because that the increase of the search space cannot ensure all the new included social trust paths are better than the identified paths based on the existing search space.

Based on the statistics, on average the path utility delivered is 0.0962 by T-MONTE-K and is 0.0701 by MONET_K. T-MONTE-K can deliver 37.2% more of the average path utility than MONTE_K.

The Maximum Path Utility. In order to investigate the performance of the algorithms in the optimal social path finding, we evaluate the maximal path utility delivered by T-MONTE-K and MONTE_K with different BST and K respectively.

Fig. 8 plots the maximal path utility with different BST varying from 1,000 to 5,000. From the figure, we can see that (1) with the increase of FST, the maximal path utility delivered by the both algorithms increase, and (2) in all the cases, our T-MONTE-K can always deliver better utility than MONTE_K. This is because that (i) based on *Strategy 4* and the property of MONTE_K, the more the FST, the higher the probability of delivering a better solution, and (ii) T-MONTE-K considers both QoSTP constraints and path utility. Thus, it has higher probability to identify those feasible paths with high utilities. Therefore, T-MONTE-K can deliver better solutions than MONTE_K.

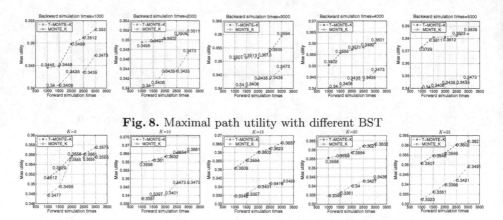

Fig. 8. Maximal path utility with different BST

Fig. 9. Maximal path utility with different K

Fig. 9 plots the maximal path utility with different K varying from 5 to 25. From the figure, we can see that under a certain K value, the maximal path utility delivered by both of the algorithms are increased with the increase of FST. In addition, T-MONTE-K can deliver better solution in all cases, which further justifies the above conclusion that T-MONTE-K can deliver better solution with taking both QoSTP constraints and path utility into account. Furthermore, from the figure, we can see that under a certain FST, there is no fixed relation between K value and the maximal path utility. Namely, even adopting large K value (e.g., the maximal path utility is 0.3615 when $K = 20$ and $FST = 3500$), T-MONTE-K may deliver a worse solution than the one adopting a small K value (e.g., the maximal path utility is 0.3657 when $K = 15$ and $FST = 3500$). This is because that the K value is corresponding to a local optimal solution, a larger K can not ensure more better global solutions to be included in the path finding.

Based on the statistics, on average the maximal path utility delivered by T-MONTE-K is 0.3586 and by MONET_K is 0.2747. T-MONTE-K can deliver 30.5% more for the path utility than MONTE_K in their identified optimal social trust paths.

The Execution Time. In order to investigate the efficiency of T-MONTE-K, we evaluate the average execution time of the two algorithms.

Fig.10 plots the average execution time of T-MONTE-K and MONTE_K under different BST and FST. From the figure we can see that our T-MONTE-K consumes more execution time than MONTE_K in all cases because in addition to the *Forward_Search* from v_s to v_t, T-MONTE-K needs to perform *Backward_Search* procedure. Therefore it spends more execution time than MONTE_K. In addition, under the same FST, the execution time of T-MONTE-K is increased with the increase of BST. This is because that the more the BST, the more the information provided by the *Backward_Search* procedure, then the more the execution time of the *Forward_Search* procedure in social trust path finding by using this information.

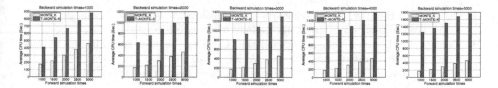

Fig. 10. Execution time of T-MONTE-K and MONTE_K

Based on the statistics. on average T-MONTE-K consumes 3.4 times of the execution time of MONTE_K while delivering better solutions in sub-networks. This is much better than our previously proposed heuristic algorithms H-OSTP [14] (more than 6 times of the execution time of MONTE_K) and MFPB-HOSTP [16] (more than 7.8 times of the execution time of MONTE_K). Since T-MONTE-K has the same polynomial time complexity (i.e, $O(mu)$) as MONTE_K, it has very good efficiency in social trust path finding.

6   Conclusion

In this paper, we have proposed T-MONTE-K, an effective and efficient approximation algorithm, which considers both the QoSTP constraints and the path utility, thus avoiding lots of unnecessary probing of infeasible paths. The results of extensive experiments conducted based on a real-world dataset demonstrate that T-MONTE-K can outperform the existing state-of-the-art algorithm MONTE_K.

In our future work, we will conduct a more exhaustive experiment using a larger dataset to investigate the scalability of the approach.

Acknowledgments. This work was partially supported by Natural Science Foundation of China (Grant Nos. 61303019, 61232006, 61402312, 61402313, 61440053, 61003044), Australian Research Council DP (DP140103171), Doctoral Fund of Ministry of Education of China (20133201120012), and Collaborative Innovation Center of Novel Software Technology and Industrialization, Jiangsu, China.

References

1. Hang, C., Wang, Y., Singh, M.: Operators for propagating trust and their evaluation in social networks. In: AAMAS 2009, pp. 1025–1032 (2009)
2. Liu, G., Wang, Y., Orgun, M.A.: Optimal social trust path selection in complex social networks. In: AAAI 2010, pp. 1397–1398 (2010)
3. Adler, P.S.: Market, hierarchy, and trust: The knowledge economy and the future of capitalism. Organization Science **12**(12), 215–234 (2001)
4. Ashri, R., Ramchurn, S., Sabater, J., Luck, M., Jennings, N.: Trust evaluation through relationship analysis. In: AAMAS 2005, pp. 1005–1011
5. Brass, D.J.: A Socal Network Prespective On Industral/organizational psychology. Industrial/Organizational Handbook (2009)
6. Dalton, M.: Men Who Manage. Wiley, New York (1959)

7. Golbeck, J., Hendler, J.: Inferring trust relationships in web-based social networks. ACM Transactions on Internet Technology **6**(4), 497–529 (2006)
8. Kunegis, J., Lommatzsch, A., Bauckhang, C.: The slashdot zoo: Mining a social network with negative edges. In: WWW 2009, pp. 741–750
9. Baase, S., Gelder, A.: Computer Algorithms Introduction to Design and Analysis. Addision Wesley
10. Christianson, B., Harbison, W.S.: Why isn't trust transitivie? In: Lomas, M. (ed.) Security Protocols 1996. LNCS, vol. 1189, pp. 171–176. Springer, Heidelberg (1997)
11. Mansell, R., Collins, B.: Trust and Crime in Information Societies. Edward Elgar Publishing (2005)
12. Lin, C., Cao, N., Liu, S., Papadimitriou, S., Sun, J., Yan, X.: Smallblue: Social network analysis for expertise search and collective intelligence. In: ICDE 2009, pp. 1483–1486 (2009)
13. Miller, R., Perlman, D., Brehm, S.: Intimate Relationships, 4th edn. McGraw-Hill College (2007)
14. Liu, G., Wang, Y., Orgun, M., Lim, E.P.: A heuristic algorithm for trust-oriented service provider selection in complex social networks. In: SCC, pp. 130–137 (2010)
15. Liu, G., Wang, Y., Orgun, M.A.: Finding k optimal social trust paths for the selection of trustworthy service providers in complex social networks. In: ICWS 2011, pp. 41–48
16. Liu, G., Wang, Y., Orgun, M.A., Lim, E.P.: Finding the optimal social trust path for the selection of trustworthy service providers in complex social networks. IEEE Transactions on Services Computing (TSC) (2011)
17. Liu, G., Wang, Y., Wong, D.: Multiple qot constrained social trust path selection in complex social networks. In: TrustCom 2012
18. Sakaki, T., Okazaki, M., Matsuo, Y.: Tweet analysis for real-time event detection and earthquake reporting system development. IEEE Transactions on Knowledge and Data Engineering **25**(4), 919–931 (2013)
19. Shin, Y., Lim, J., Park, J.: Joint optimization of index freshness and coverage in real-time search engines. IEEE Transactions on Knowledge and Data Engineering **24**(12), 2203–2217 (2012)
20. Gentle, J., Hardle, W., Mori, Y.: Handbook of Computational Statistics. Springer (2004)
21. Wang, G., Wu, J.: Multi-dimensional evidence-based trust management with multi-trusted paths. Future Generation Computer Systems **17**, 529–538 (2011)
22. Liu, G., Wang, Y., Orgun, M.A.: Trust transitivity in complex social networks. In: AAAI 2011, pp. 1222–1229
23. Mccallum, A., Wang, X., Corrada-Emmanuel, A.: Topic and role discovery in social networks with experiments on Enron and academic email. Journal of Artificial Intelligence Research **30**(1), 249–272 (2007)
24. Tang, J., Zhang, J., Yan, L., Li, J., Zhang, L., Su, Z.: Arnetminer: Extraction and mining of academic social networks. In: KDD 2008, pp. 990–998 (2008)
25. Berger, P., Luckmann, T.: The Social Construction of Reality: A Treatise in the Sociology of Knowledge. Anchor Books (1966)
26. Korkmaz, T., Krunz, M.: Multi-constrained optimal path selection. In: INFOCOM 2001, pp. 834–843
27. Morton, D., Popova, E.: Monte-carlo simulation for stochastic optimization. Encyclopedia of Optimization, pp. 2337–2345 (2009)

28. Mislove, A., Marcon, M., Gummadi, K., Druschel, P., Bhattacharjee, B.: Measurement and analysis of online social networks. In: ACM IMC 2007, pp. 29–42 (2007)
29. Chia, P., Pitsilis, G.: Exploring the use of explicit trust link for filtering recommenders: A study on epinions.com. Journal of Information Processing **19**, 332–344 (2011)
30. Chua, F., Lim, E.P.: Trust network inference for online rating data using generative models. In: KDD 2010, pp. 889–898
31. Lo, D., Surian, D., Zhang, K., Lim, E.P.: Mining direct antagonistic communities in explicit trust networks. In: CIKM 2011, pp. 1013–1018

Pricing Strategies for Maximizing Viral Advertising in Social Networks

Bolei Zhang, Zhuzhong Qian$^{(\boxtimes)}$, Wenzhong Li, and Sanglu Lu

State Key Laboratory for Novel Software Technology,
Nanjing University, Nanjing, China
zhangbolei@dislab.nju.edu.cn, {qzz,lwz,sanglu}@nju.edu.cn

Abstract. Viral advertising in social networks is playing an important role for the promotions of new products, ideas and innovations. It usually starts from a set of initial adopters and spreads via social links to become viral. Given a limited budget, one central problem in studying viral advertising is *influence maximization*, in which one needs to target a set of initial adopters such that the number of users accepting the advertising afterwards is maximized. To solve this problem, previous works assume that each user has a fixed cost and will spread the advertising as long as the provider offers a benefit that is equal to the cost. However, the assumption is oversimplified and far from real scenarios. In practice, it is crucial for the provider to understand how to incentivize the initial adopters.

In this paper, we propose the use of concave probability functions to model the user valuation for sharing the advertising. Under the new pricing model, we show that it is NP-hard to find the optimal pricing strategy. Due to the hardness, we then propose a discrete greedy pricing strategy which has a constant approximation performance guarantee. We also discuss how to discretize the budget to provide a good trade-off between the performance and the efficiency. Extensive experiments on different data sets are implemented to validate the effectiveness of our algorithm in practice.

Keywords: Viral advertising · Influence maximization · Social networks

1 Introduction

The emergence and proliferation of online social networks such as Facebook, Twitter and Google+ have greatly boosted the spread of information. People are actively engaged in the social networks and generating contents at an ever-increasing rate. Viral advertising, which utilizes information diffusion for the promotions of new products, ideas and innovations, has attracted enormous attentions from companies and providers. Compared with TVs, newspapers and radios which broadcast advertising, viral advertising in social networks has the effect of "word-of-mouth" which is considered to be more trustworthy. Moreover, the information diffusion between users can spread across multiple links and trigger large cascades of adoption.

© Springer International Publishing Switzerland 2015
M. Renz et al. (Eds.): DASFAA 2015, Part II, LNCS 9050, pp. 418–434, 2015.
DOI: 10.1007/978-3-319-18123-3_25

To start a cascade of viral advertising, one would first incentivize a set of initial adopters and let them spread the advertising further. For example, popular e-commerce platforms like Amazon, e-Bay and JD[1] all encourage people to share information of the products they buy in their social networks for advertising. In JD, users could get vouchers if they comment and share the products information. Suppose there is a limited budget, a question arising naturally is: How should we distribute the budget, so that the spread of viral advertising can be maximized. The process can be described as a two-step **pricing strategy**: we first offer each user a discriminative price as incentive; The users who accept the incentive should then share the advertising in social networks.

JD's strategy is to offer each user uniform price, regardless of their valuations and influence. Alternatively, we may allocate the budget proportionally so that users with high influence can be allocated high price. In general, to determine who should be offered price and how much should be offered to each of them, it is crucial to understand both user valuations for being initial adopters and the information diffusion process. From the perspective of users, their valuations for being initial adopters usually depend on the price offered. For the advertisers, they may expect more users with high influence to share and spread the information.

A similar problem that has been extensively studied is *influence maximization*. The problem aims to select the most influential set of users as initial adopters to maximize the spread of influence. However, the underlying assumption that each user has an inherent *constant value* for being initial adopter may not be reasonable. In practice, users' decisions are often not deterministic: They may decline or ignore the offered price, leading to unpredictability of the information spread. In comparison, in this work, we address that the user decisions for being initial adopters are probabilistic rather than constant. Given the probability distribution, we study optimal pricing strategies to maximize the viral advertising under some well-studied diffusion models.

1.1   Our Results

The main contributions of this paper are:

- We introduce a concave probabilistic model in which users' values are distributed according to some concave functions. The model is practical and can characterize different users' preferences for sharing the advertising in social networks.
- We formalize the optimization problem and show that it can be reduced to NP-hard quadratic programming problem. Due to the hardness, we propose an approximate discrete greedy algorithm with near optimal result. We further analyze how to discretize the budget to provide a good tradeoff between the performance and efficiency.

---

[1] JD(http://www.jd.com) is the largest online direct sales company in China.

– Extensive experiments on different data sets are implemented to validate the effectiveness of our algorithm. Our algorithm significantly outperforms other algorithms in almost all cases. In addition, we evaluate the discrete greedy algorithm with respect to different granularity. The results reveal that our algorithm can converge asymptotically.

1.2 Related Work

Our work has a strong tie with the problem *influence maximization*, which was first proposed by Domingos and Richardson [11,19]. Later, Kempe et al. [14] formulated the problem as a discrete optimization problem, and proposed greedy algorithm with hill-climbing strategies to find the influential nodes. Due to the monotonicity and submodularity of the information diffusion process, the algorithm can be proved to achieve constant approximation ratio. Following their work, extensive researches [8–10,15,21] have studied algorithmic improvement of the spread of influence in social networks. Despite a lot of progress in choosing which nodes to select, the problem of how to incentivize the initial nodes is often neglected.

Another thread of our work is inspired by the problem of *revenue maximization* which was first introduced by Hartline et al. [12]. In order to influence many buyers to buy a product, a seller could first offer some popular buyers discounts. The problem then studies marketing strategies like how large the discounts be and in what sequence should the selling happen. Specifically, the work assumed that the willingness that each user may pay for a product is drawn from some given probability distributions. A lot of following works have studied revenue maximization in social networks. Some of them have considered Nash Equilibrium between users for purchasing one product. Different pricing strategies were designed to maximize the revenue, i.e., uniform pricing [5,7], discriminative pricing [3,22], iterative pricing [1] etc.

In a recent work of Singer [21], the author considered auction based influence maximization in which each user can bid a cost for being an initial adopter. To make sure that each user declares the true cost, they designed incentive compatible mechanisms. However, the mechanism requires extra step for each user to bid a cost and may be cumbersome to implement in practice. Comparatively, we adopt a pricing strategy with a more natural way for incentivizing each user. In [10], Demaine et al. proposed partial incentives in social networks to influence people. The pricing and influence models in our paper generalize the models in [10] and we further analyze the discrete setting of the problem.

2 Preliminaries

We model the social network as a directed (undirected) graph $\mathcal{G} = (\mathcal{N}, \mathcal{E})$, where node set $\mathcal{N}$ represents users and edge set $\mathcal{E} \subseteq \mathcal{N} \times \mathcal{N}$ represents social relationships between them. Given a limited budget B, a pricing strategy is to distribute the budget among users to maximize the spread of the viral advertising. In this section, we first introduce the pricing model and information diffusion models respectively. Then, we will formulate our optimization problem and establish its hardness.

2.1   Pricing Model

After distributing the budget, we use the pricing model to characterize users' valuations for being initial adopters, i.e., how much price is it need to incentivize a user? In *influence maximization*, users are assumed to have some *constant value* for being initial adopters. However in general cases, the users' decisions are often not deterministic. In comparison, we address that users have different valuations and the valuations are dependent on the allocated price.

In this paper, we propose a probabilistic model in which the user values are drawn from some prior known distributions: $\mathcal{F} = \{F_i | i \in \mathcal{N}\}$. Suppose that user i is offered a price p_i as incentive, then $F_i(p_i)$ is the probability that i will accept the price for being an initial adopter. Literatures like [13,17] have observed that the marginal gain of user satisfaction decreases as the price increases, which arises naturally in practical situations as *diminishing returns*. Accordingly, we also regard the cumulative functions F_i as *concave functions*, i.e. for any x and y in the interval and for any $t \in [0, 1]$, $F_i(tx+(1-t)y) \geq tF_i(x)+(1-t)F_i(y)$. Fig. 1 presents some examples of the possible distribution functions. In Fig. 1(a), the user's value is distributed uniformly in the range of $[0, \tau]$, where τ is a constant threshold. In Fig. 1(b), the valuation is drawn from $F_i(p_i) = \sqrt{\frac{p_i}{d_i+1}}$, where d_i is the degree of user i.

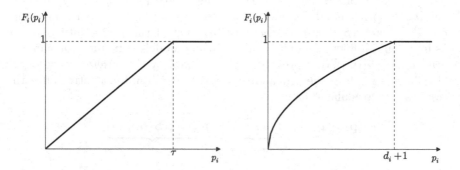

Fig. 1. Examples of user valuation distribution function

2.2   Information Diffusion Models

After distributing the budget, a set of initial adopters S will share the advertising to trigger a cascade. The advertising will spread in the social networks as a piece of information. In an information cascade, we say that a node is *active* if it adopts the information, otherwise it is called *inactive*. Initially, only the users in S are active. The information then spreads via social links to influence more users. The number of active users after the cascade stops is the spread of the viral advertising, or the influence of S, denoted as $\sigma(S)$. Modeling the information diffusion process has been extensively studied [2,4,6,14,16]. We introduce some of the most widely used models here.

Coverage model. In the Coverage model [21], each user i is associated with the set of its neighbors $N(i)$, which is also the influence of user i. The information will not further spread. So the influence of the S will be $\sigma(S) = |\bigcup_{i \in S} N(i)|$.

Independent Cascade model. In IC model [14], the information starts from S as follows: At step t, the newly activated nodes in S_t try to activate their neighbors independently. Each active node u succeeds in activating its neighbor v with probability $\mu_{(u,v)}$. The newly activated nodes are added into set S_{t+1}. The process continues until $S_t = \emptyset$.

Linear Threshold model. In LT model [14], for each neighbor w a node v associates a weight $\omega_{v,w} \geq 0$ where $\sum_{w \in N(v)} \omega_{v,w} \leq 1$, and chooses some threshold $\theta_v \in [0,1]$ uniformly at random. The node v is activated at time step t if $\sum_{w \in N_t(v)} \omega_{v,w} \geq \theta_v$, where $N_t(v)$ denotes the neighbors of v that are active at time step t.

2.3   Problem and Optimization Objective

Given a budget B, a pricing strategy will identify a price vector $\mathbf{p} = (p_1, p_2, ..., p_n), (\sum_i^n p_i = B)$, where $p_i \in \mathbb{R}_{\geq 0}$ is the price offered to user i. We use $\mathbf{p}_{-\mathbf{i}}$ as the price list offered to users except i. There are two ways for a user i to get active: On one hand, i may accept the offered price p_i with probability $F_i(p_i)$; On the other hand, i could be influenced by other users with probability $q(S,i)$, where $q(S,i)$ is a reachability function of the probability that the nodes in S could influence i under some diffusion models. The function can also be written as $q(S(\mathbf{p}_{-\mathbf{i}}), i)$, representing the probability that users accepted $\mathbf{p}_{-\mathbf{i}}$ could influence i. Thus, the probability $w_i(\mathbf{p})$ that i could get active after the cascade, can be formulated as:

$$w_i(\mathbf{p}) = 1 - (\underbrace{1 - F_i(p_i)}_{i \ does \ not \ take \ p_i})(\underbrace{1 - q(S(\mathbf{p}_{-\mathbf{i}}), i)}_{i \ is \ not \ influenced}) \tag{1}$$

Given the probability that each user gets active, the objective can be formulated as an optimization problem in which the goal is to maximize overall expected number of active users after the cascade:

$$\max \ f(\mathbf{p}) = \sum_{i \in \mathcal{N}} w_i(\mathbf{p})$$

$$\text{s.t.} : \sum_{i \in \mathcal{N}} p_i \leq B \tag{2}$$

Hardness. Now we show that the above optimization problem is NP-hard. In particular, it is NP-hard even if the social network forms a line structure.

Theorem 1. *Identifying the optimal pricing strategy is NP-hard even when the social network forms line structure.*

Proof. Consider an instance of the quadratic programming problem:

$$\min \ g(\mathbf{p}) = \frac{1}{2}\mathbf{pQp^T} + \mathbf{cp^T}$$

$$\text{s.t.} : \ \sum_{i=1}^{n} p_i \leq B$$

where $Q = \begin{bmatrix} 0 & 1 & 0 & \cdots & 0 & 0 \\ 1 & 0 & 1 & \cdots & 0 & 0 \\ 0 & 1 & 0 & \cdots & 0 & 0 \\ \vdots & \vdots & \vdots & \ddots & \vdots & \vdots \\ 0 & 0 & 0 & \cdots & 0 & 1 \\ 0 & 0 & 0 & \cdots & 1 & 0 \end{bmatrix}$ and $c = (-2, -2, ..., -2, -1)$. Since $\mathbf{pQp^T} = 2\sum_{i=2}^{n} p_i p_{i-1}$,

Q is indefinite. According to [20], the quadratic programming function $g(\mathbf{p})$ is NP-hard. We show that the programming can be viewed as a special case of the pricing function $f(\mathbf{p})$.

Given the instance of the quadratic programming problem, we define a corresponding instance of the pricing function under the Coverage model where the social network G forms a line structure as presented in Figure 2.

Fig. 2. Line structure of the social network

In this case, the probability that a user gets active with pricing vector $\mathbf{p}$ is

$$w_i(\mathbf{p}) = \begin{cases} p_1 & i = 1 \\ 1 - (1 - p_i)(1 - p_{i-1}) & i \geq 2 \end{cases}$$

In this way, $f(\mathbf{p}) = 2\sum_{i=1}^{n} p_i - \sum_{i=2}^{n} p_i p_{i-1}$ is the negative of $g(\mathbf{p})$. So maximizing function $f(\mathbf{p})$ is equivalent to minimizing the function $g(\mathbf{p})$, which is also NP-hard.

3 A Continuous Greedy Process

Due to the hardness result from Theorem 1, no polynomial algorithm exists for the optimization problem unless $P = NP$. To motivate our approximation algorithm, in this section, we present a continuous greedy process with constant approximation ratio. Even though the process takes infinite steps for implementation, it provides analytic picture behind our main algorithm. Intuitively, the process allocates the budget smoothly to the users with high influence and low valuation. It can be formally regarded as a particle $p(t)$ starting from $p(0) = 0$ and following a certain flow over a unit time interval:

$$\frac{d\mathbf{p}}{dt} = v(p)$$

where $v(p)$ is defined as

$$v(p) = \arg\max_{i \in \mathcal{N}} \left(\frac{f(\mathbf{p} + p_i) - f(\mathbf{p})}{p_i} \right) \quad (p_i = \epsilon \to 0) \tag{3}$$

until $t = 1$.

At each interval, we allocate a small unit of budget ($\epsilon \to 0$) to the user with maximal marginal gain. Since the marginal gain of each user is a function of the allocated budget, we have to run the process continuously to get the local optimum. This process provides good approximation :

Lemma 1. *Let the optimal value be OPT, the continuous greedy process has* $1 - 1/e$ *approximation ratio, i.e.,* $f(p(1)) \geq (1 - 1/e)OPT$.

The proof of Lemma 1 borrows idea from the problem of maximization of *smooth submodular functions*. For the completeness of the proof, we first introduce the concept of *smooth submodular function*.

Definition 1. *A function* $f : [0,1]^X \to \mathbb{R}$ *is* **smooth monotone submodular** *if*

- $f \in C_2([0,1]^X)$, *i.e., it has second partial derivatives everywhere.*
- *For each* $j \in X$, $\frac{\partial f}{\partial y_j} \geq 0$ *everywhere (monotonicity).*
- *For any* $i, j \in X$ *(possibly equal),* $\frac{\partial f}{\partial y_i \partial y_j} \leq 0$ *everywhere (submodularity).*

In Definition 1, $\frac{\partial f}{\partial y_i \partial y_j} \leq 0$ indicates that a function is smooth submodular if it is concave along any non-negative direction. Examining the objective formulation 2, it can be easily proved that f satisfies the above conditions so is smooth submodular. Vondrák et al. [23] showed that the continuous greedy process as presented above can achieve $1 - 1/e$ approximation ratio to maximize a smooth submodular function, which concludes Lemma 1.

4   Pricing Strategies for Viral Advertising

Following the idea of the continuous greedy process, in this section, we consider discretizing the budget so that the process can run in polynomial time. Our main result is a discrete greedy algorithm with constant approximation ratio. We further discuss how the discrete granularity affects the result of the algorithm.

4.1   A Discrete Greedy Strategy

In our discrete pricing strategy, we first divide the budget B into m equal pieces, denoted as $\mathcal{B} = \{b_1, b_2, ..., b_m\}$, where $b_j = B/m$ for all $j \in \{0, 1, 2, ..., m\}$. In this setting, the prices offered to each user are disjoint subsets from $\mathcal{B}$, namely $A_i \subseteq \mathcal{B}, A_i \cap A_j = \varnothing$ for all $i, j \in \mathcal{N}$. Thus, the pricing strategy can be regarded as a mapping problem in which m elements of the budget set are to be allocated to n users. The objective is then a set function where the input is a subset from the ground set: For the user set $\mathcal{N}$ and pricing set $\mathcal{B}$, define the new ground set as $X = \mathcal{N} \times \mathcal{B}$. By associating the variable a_{ij} with user-price pairs, i.e., a_{ij} means assigning price b_j to user i, the ground set $X = \{a_{11}, a_{12}, ..., a_{1m}, a_{21}, a_{22}..., a_{2m}, ..., a_{n1}, a_{n2}, ..., a_{nm}\}$ is the set of all price elements that can be chosen. Note that each piece of price is replicated n times in the set X, e.g. the price b_j is replicated as $a_{1j}, a_{2j}, ..., a_{nj}$, only one of them can be chosen for the solution. Slightly abusing notations, we use $\sigma(A) : 2^X \rightarrow \mathbb{R}_+$ as the expected number of active users after the cascade by choosing price set $A = \bigcup_{i \in \mathcal{N}} A_i, |A| = m$. The problem of finding the optimal pricing strategy in discrete setting can be described as choosing a set A ($|A| = m, A \subseteq X$) of m elements from X to maximize the the expected number of active users $\sigma(A)$, with the constraint that only one of $\{a_{1j}, a_{2j}, ..., a_{nj}\}$ can be chosen for all $j \in \{1, 2, ..., m\}$.

Coverage Model. We begin by maximizing the objective function under Coverage model. There are two reasons we first consider the Coverage model. On one hand, the diffusion process under the Coverage model can be regarded as information exposures to users, i.e, a user is influenced if and only if one of his/her neighbors is an initial adopter; Moreover, the Coverage model is simplistic and exhibits similar properties as the IC and LT model.

To optimize the spread of viral advertising, we first prove that the pricing function $\sigma(\cdot)$ is monotone and submodular in the discrete setting. A function $\sigma(\cdot)$ is submodular if for any element $a \in X$, for all $V \subseteq T$, there is $\sigma(V \cup \{a\}) - \sigma(V) \geq \sigma(T \cup \{a\}) - \sigma(T)$. Submodularity implies that the marginal gain of choosing an element decreases as the number of chosen elements increases.

Theorem 2. *The pricing function $\sigma(\cdot) : 2^X \rightarrow \mathbb{R}_+$ is monotone submodular under Coverage model if the cumulation distribution functions of user valuation functions $F_i(\cdot) : \mathbb{R}_+ \rightarrow [0, 1], i \in \mathcal{N}$ are non-decreasing concave functions.*

Proof. **Monotonicity** Obviously, by adding a new pricing element a_{ij} to the outcome set A, namely assigning the jth piece of price to user i, the probability that user i could get active will not decrease. Accordingly, the probability that user i influences other users $q(\{i\}, \cdot)$ will also not decrease, which concludes that $\sigma(\cdot)$ is non-decreasing monotone.

Submodularity For submodularity, let $\delta(a_{ij})$ denote the marginal gain of $\sigma(\cdot)$ by adding the element a_{ij}. It can be formulated as the sum of increased probability from all users:

$$\delta(a_{ij}) = \sigma(A^j \cup \{a_{ij}\}) - \sigma(A^j) = \delta_i(a_{ij}) + \sum_{k \in \mathcal{N} \setminus \{i\}} \delta_k(a_{ij}) \qquad (4)$$

where A^j is set of first $j-1$ price elements that have been allocated, and $\delta_i(a_{ij})$ is the *marginal gain* of user i. As the class of submodular functions is closed under non-negative linear combinations, we only need to prove that the function for each user is submodular respectively.

For user i, by adding a price element a_{ij}, the increased probability is:

$$\delta_i(a_{ij}) = w_i(A^j \cup \{a_{ij}\}) - w_i(A^j)$$
$$= (F_i(A^j \cup \{a_{ij}\}) - F_i(A^j))(1 - q(S(A^j), i))$$

Since submodularity is the discrete analog of concavity, there is $F_i(V \cup \{a_{ij}\}) - F_i(V) \geq F_i(T \cup \{a_{ij}\}) - F_i(T)$ if $V \subseteq T$. Meanwhile, $1 - q(S(A^j), i)$ does not change with a_{ij}, we can conclude that $w_i(V \cup \{a_{ij}\}) - w_i(V) \geq w_i(T \cup \{a_{ij}\}) - w_i(T)$ if $V \subseteq T$, which indicates that $w_i(\cdot)$ is submodular.

Similarly, for any other user k, the increased probability can be formulated as:

$$\delta_k(a_{ij}) = (q(S(A^j \cup \{a_{ij}\}), k) - q(S(A^j), k))(1 - F_k(A^j))$$

In the Coverage model, $q(S, i) = 1$ means i is a neighbor of S, and 0 otherwise. According to the monotonicity of $S(\cdot)$, $S(V) \subseteq S(T)$ if $V \subseteq T$. So by adding an element a_{ij}, the smaller set $S(V \cup \{a_{ij}\}) - S(V)$ is more likely to influence a node k, i.e., $q(S(V \cup \{a_{ij}\}), k) - q(S(V), k) \geq q(S(T \cup \{a_{ij}\}), k) - q(S(T), k)$. So $w_k(\cdot)$ is also submodular.

Summing up the increased probabilities of all users, we have $\sigma(V \cup \{a_{ij}\}) - \sigma(V) \geq \sigma(T \cup \{a_{ij}\}) - \sigma(T)$ if $V \subseteq T$, showing that $\sigma(\cdot)$ is submodular. □

According to the work of Nemhauser et al. [18], finding a set A of with uniform matroid (m) to maximize the monotone submodular function is NP-hard and a greedy algorithm with hill-climbing strategy approximates the optimal to a factor of $1 - 1/e$. Let the optimal value in the discrete setting be OPT_d, we can conclude our main theorem as:

Theorem 3. *A greedy hill-climbing strategy can achieve $1 - 1/e$ approximation ratio of the optimal pricing strategy in the discrete setting, i.e., $f(A) \geq (1 - 1/e)OPT_d$.*

Following Theorem 3, we now present our greedy pricing strategy in Algorithm 1. In each step of the algorithm, we greedily choose the user that has the largest marginal gain by allocating a piece of price.

IC Model and LT Model. As for the IC model and LT model, we show that the pricing function σ is still monotone and submodular. By adopting a different diffusion model, we have a different diffusion function q. Recall that

Algorithm 1. $DiscreteGreedy(\mathcal{G}, \mathcal{B})$

$A \leftarrow \emptyset;$
for $j \leftarrow 1$ *to* m **do**
    $s \leftarrow 0, max \leftarrow -1;$
    for $i \in \mathcal{N}$ **do**
        $\delta(a_{ij}) \leftarrow \sigma(A^j \cup \{a_{ij}\}) - \sigma(A^j);$
        if $\delta(a_{ij}) > max$ **then**
            $s \leftarrow i, max \leftarrow \delta(a_{ij});$
    $A \leftarrow A \cup \{a_{sj}\};$
return A

---

by adding a new pricing element a_{ij}, the marginal gain of $\sigma(\cdot)$ is the sum of the increased probability from all users. For user i, the function $\sigma_i(\cdot)$ is still submodular since the reachability function remains the same. For any other user k ($k \neq i$), the marginal gain is $\delta_k(a_{ij}) = (q(S(A^j \cup \{a_{ij}\}), k) - q(S(A^j), k))(1 - F_k(A^j))$. Since $S(\cdot)$ is monotone, $\sigma(\cdot)$ is submodular if and only if the reachability function $q(S, k)$ is submodular. In [14], Kempe et al. have already showed that the functions are submodular under the IC and LT model [14]. So the pricing function σ is also submodular.

Lemma 2. *For submodular influence functions like IC and LT model, the pricing function σ is also monotone and submodular, and Algorithm 1 approximates the optimal to a factor of $1 - 1/e$.*

4.2   How to Choose m?

Despite the near optimal results from Algorithm 1, for more general situations, we would like to know how it approximates the optimal value OPT. In this section, we will discuss how to choose the discrete granularity m to achieve a good tradeoff between the performance and efficiency.

Apparently, by increasing m to ∞, the discrete greedy algorithm is close to the continuous greedy process which also takes infinite steps. To the other extreme, if m decreases to 1, the algorithm is simply choosing one user to allocate. To achieve a good trade-off between the performance and the algorithm efficiency, we focus on deriving the gap between the continuous process $f(p(1))$ and the discrete greedy strategy $f(A)$, since the continuous greedy process approximates well to the optimal value OPT.

Lemma 3. *When $m \geq O(n)$, $f(A) \geq (1 - 1/e - o(1))OPT$ with high probability.*

Proof. After allocating first $j - 1$ pieces of the total budget ($A_j = \frac{j-1}{m}B$), the continuous process will get a price vector $\mathbf{p_j}$ and the greedy pricing strategy will get a set of A^{j+1}. Observe the marginal gain by increasing price $a = B/m$. For the continuous process, the marginal gain is $\delta(\mathbf{a}^*) = f(\mathbf{p_{j+1}}) - f(\mathbf{p_j})$, where $\mathbf{a}^* = \mathbf{p_{j+1}} - \mathbf{p_j}$ and $|\mathbf{a}^*| = B/m$. For the discrete greedy strategy, we choose

a_{ij} at the jth step, where $i = arg\max_{i\in\mathcal{N}}(f(A^j \cup \{a_{ij}\}) - f(A^j))$. Consider the marginal gain $\delta(a_{ij}) = f(\mathbf{p_j} + a_{ij}) - f(\mathbf{p_j})$. Taking Taylor series and bounding the lower items, we have:

$$\delta(\mathbf{a}^*) = \delta(a_{ij}) + R(\xi) \tag{5}$$

where $\xi = (1 - c)\mathbf{a}^* + ca_{ij}$, $c \in [0,1]$. For $R(\xi)$, there is:

$$
\begin{aligned}
R(\xi) &= \frac{\partial\delta}{\partial\xi}(\mathbf{a}^* - a_{ij}) = \sum_{k\in\mathcal{N}}\frac{\partial\delta}{\partial\xi_\mathbf{k}}\mathbf{a}_\mathbf{k}^* - \frac{\partial\delta}{\partial\xi_\mathbf{i}}a_{ij} \\
&\leq (\frac{\partial f}{\partial\mathbf{p_j}} - \frac{\partial f}{\partial\mathbf{p_{j+1}}})a - \frac{\partial\delta}{\partial\xi_\mathbf{i}}a = Ca
\end{aligned}
\tag{6}
$$

where C is a constant. So when $a \leq O(1/n)$, namely $m \geq n$, the error $R(\xi)$ is within $O(1/n)$, which concludes that $\delta(\mathbf{a}^*) \approx \delta(a_{ij})$. Accordingly, there is $f(A) \geq (1 - 1/e - o(1))OPT$ with high probability.   □

5   Evaluations

In addition to the provable performance guarantee, in this section, we conduct extensive simulations to evaluate our DiscreteGreedy pricing strategy. We first compare our proposed algorithm with several intuitive heuristics. Then we will observe how the discrete granularity of the budget affects the results of the DiscreteGreedy strategy.

5.1   Experiment Setup

Our experiments are conducted on 3 real social network data sets. The first is CondMat data set which contains $23,133$ nodes and $93,497$ undirected edges. The nodes represent authors and an undirected edge (i,j) represents coauthor relationships between them. The second is Youtube data set which contains $560,123$ nodes and $1,976,329$ undirected edges. The nodes are users and the edges indicates friendship relationships between them. The last data set is Weibo data set which has $877,391$ nodes and $1,419,850$ directed edges. Weibo is a Twitter-like mirco blog in China. The directed edge (i,j) means that user i is following j. All 3 data sets exhibit small world, high clustering complex network structural features.

Since our algorithm is applicable to any forms of concave functions of user valuations, we manually set the distributions in the experiments. We first consider a uniform symmetric setting in which $F_i(p_i) = \frac{p_i}{\tau}$. The user values are distributed uniformly in the range $(0, \tau]$. The second distribution function we consider is $F_i(p_i) = \sqrt{\frac{p_i}{\tau}}$. The function also has a threshold τ and the acceptance probability is proportional to the square root of the offered price. We also consider differential valuation functions where the user valuation is a function of their degree: $F_i(p_i) = \frac{r+d_i+1}{r+p_i}\frac{p_i}{d_i+1}$. The parameter r is used to control the shape of the curve: a larger value of r indicates a steeper curve. The threshold here is $d_i + 1$.

For the information diffusion models, since all of them exhibit monotone submodular properties, we mainly conduct experiments under the Coverage model. In this model, the influence of user i is the set of users that follows i in Weibo data set. In CondMat and Youtube data set, it is the set of friends\coauthors of user i. We also considered the IC and LT model. As the information diffusion process under these two models are stochastic processes, we need to take Monte Carlo methods to generate fixed graphs and take the average influence of different probability results. In the IC model, without loss of generality, we assume uniform diffusion probability μ on each edge and assign μ to be 0.01. In the LT model, the weight of a directed edge (i, j) is $\frac{1}{d_j}$, so the sum of the weights will not exceed 1.

Comparison Methods. In our DiscreteGreedy algorithm, we discretize the budget into $O(n)$ pieces in our DiscreteGreedy pricing strategy to obtain near optimal pricing solutions. We compare our greedy algorithm with the following 3 heuristics:

- *Uniform*: The Uniform pricing strategy adopts a simple idea that all users get the same price $p_i = \frac{B}{n}$, regardless of users' influence and valuation.
- *Proportional*: In the Proportional pricing strategy, the price offered to user i is proportional to d_i, i.e., $p_i = \frac{d_i}{\sum_{j \in \mathcal{N}} d_j} B$.
- *FullGreedy*: We use the greedy algorithm in *influence maximization* to select the initial adopters and offer each of them the threshold price.

5.2   Results

The Spread of Viral Advertising. In Fig. 3 to 5, we separately present the results w.r.t different valuation functions under the Coverage model. The x-axis represents the total budget that is allocated. The y-axis is the active set size of the users after cascade, or the spread of the viral advertising.

We first evaluate the viral advertising spread when user values are distributed as $F_i(p_i) = \frac{p_i}{\tau}$. In this case, since the valuation function grows linearly to the threshold τ, the marginal gain of a user will remain the same in the range $p_i \in (0, \tau]$. Therefore, the FullGreedy algorithm has the same result as the DiscreteGreedy algorithm, which will not be presented here. Obviously from Fig. 3(a) to 3(c), our DiscreteGreedy algorithm outperforms other heuristics significantly. The results of the Proportional strategy grow almost linearly as the allocated budget increases. There are two reasons for this: First, the probability for a user to accept the price is linear as F; Second, there are not much overlap between the influenced users in this case. There is a large gap between the DiscreteGreedy and the Proportional strategies, indicating that allocating the budget more concentratedly may have better result. In the Uniform algorithm, the price offered to each user is quite low. So few users might accept to be initial adopters, leading to poor performance of the algorithm in all 3 data sets. This illustrates that differential pricing strategy is necessary for the viral advertising.

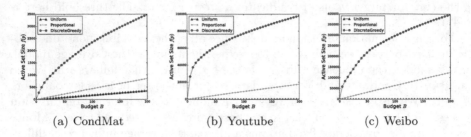

(a) CondMat   (b) Youtube   (c) Weibo

Fig. 3. The spread of viral advertising with valuation function $F_i(p_i) = \frac{p_i}{\tau}$

In Fig. 4, we set the pricing function as $F_i(p_i) = \sqrt{\frac{p_i}{\tau}}$. In this case, the users are more likely to be initial adopters with small amount of price. The user who has the maximal marginal gain at beginning may not hold as the allocated budget increases. The FullGreedy allocates each user full prices τ, so it has a worse result than the DiscreteGreedy strategy. Both the Proportional and the Uniform algorithm perform better than in Fig. 4, due to the reason that users with low allocated price also have a higher probability to be active. However, the performance of the 3 heuristics vary distinctly in different data sets.

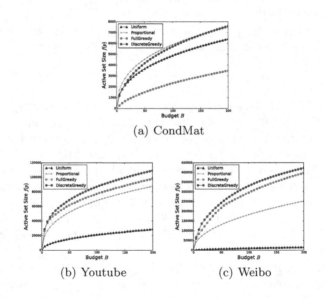

(a) CondMat

(b) Youtube   (c) Weibo

Fig. 4. The spread of viral advertising with valuation function $F_i(p_i) = \sqrt{\frac{p_i}{\tau}}$

Fig. 3 presents the results with the valuation distribution function as $F_i(p_i) = \frac{r+d_i+1}{r+p_i} \frac{p_i}{d_i+1}$. In this function, users have different valuation distribution functions which is related with their degree. For a user with degree d_i, the threshold will be $d_i + 1$. We set $r = 10$ so the valuation function has a steeper curve than the above two functions. In this way, the FullGreedy strategy performs even worse since the marginal gain of each user decreases very quickly when the allocated price increases. Note that the threshold is actually not necessary in practice. The FullGreedy strategy is infeasible if users don't have a threshold. In the Proportional algorithm, as the price offered increases proportionally with users' degree, the probability that different users accept the price are almost the same. Therefore, it still grows linearly in this case. The Uniform strategy is quite inefficient in large data sets such as Youtube and Weibo.

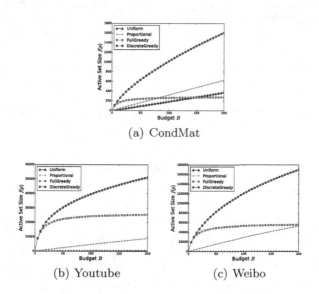

(a) CondMat

(b) Youtube   (c) Weibo

Fig. 5. The spread of viral advertising with valuation function $F_i(p_i) = \frac{r+d_i+1}{r+p_i} \frac{p_i}{d_i+1}$

In Figure 6(a) and Figure 6(b), we fix the user valuation function as $F_i(p_i) = \sqrt{\frac{p_i}{r}}$ and conduct the experiments under the IC model and LT model. As the information diffusion processes are stochastic, we simulate the graph massive times (1000) and conduct the experiments on a smaller sample of the Weibo data set. As shown in Fig 6, our DiscreteGreedy strategy still has the best performance.

The Accuracy. Finally, we run the DiscreteGreedy strategy in CondMat data set w.r.t different granularity. We discretize the budget into m pieces where m ranges from 1 to $10n$. The results are presented in Table 1. We start from $m = 1$ ($n/m = 23,333$) and gradually increase m to n ($n/m \approx 1$). Apparently, the active set size converges when m approaches n, i.e. $n/m \approx 1$. When user valuations are linear functions, i.e., when $F_i(p_i) = \frac{p_i}{r}$, the DiscreteGreedy strategy is likely to pay the current selected user full price. So the results can converge when the granularity is the threshold r. In other cases, though the active set

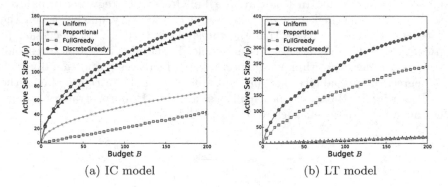

Fig. 6. The spread of advertising in the Weibo data set in the IC and LT model

size converges slower, the results can stabilize when $m \to n$. When $m > n$, the result is not likely to grow much.

Table 1. Active set size in CondMat data set w.r.t. different granularity

$O(n/m)$ $\diagdown$ F_i	$\frac{p_i}{5}$	$\sqrt{\frac{p_i}{\tau}}$	$\frac{r+d_i+1}{r+p_i}\frac{p_i}{d_i+1}$
23,133	281.00	281.00	277.14
1,000	3309.24	3620.32	1541.26
100	3452.58	5296.93	1605.29
10	3485.00	6939.77	1608.26
1	3485.00	7599.42	1609.14
0.1	3485.00	7600.15	1609.21

6   Conclusion

In this work, we studied optimal pricing strategies in a social network to maximize viral advertising with budget constraint. We formalized the pricing strategy as an optimization problem and established its hardness. A novel *DiscreteGreedy* algorithm with near optimal results was proposed, and the tradeoff between the performance and efficiency was discussed. Extensive evaluations showed that our DiscreteGreedy algorithm outperforms other intuitive heuristics significantly in almost all cases. Moreover, the DiscreteGreedy algorithm can converge to a hight accuracy if the budget is discretized properly.

For possible future works, we are interested in the following aspects. First, we would like to study the user valuation distributions empirically, i.e., how much price does it need to incentivize a user. Second, we aim to improve the algorithm efficiency, especially in the IC and LT models. Finally, we may consider other factors that influence users' decisions to implement real applications.

Acknowledgments. This work is partially supported by the National Natural Science Foundation of China under Grant No. 61202113, 61472181, 61321491, 91218302, 61373128; Jiangsu Key Technique Project (industry) under Grant No. BE2013116; EU FP7 IRSES MobileCloud Project under Grant No. 612212; the Fundamental Research Funds for the Central Universities under Grant No. 20620140513; NJU Graduate Innovation Foundation Project (No. 2013CL09). And this work is also partially supported by Collaborative Innovation Center of Novel Software Technology and Industrialization.

References

1. Akhlaghpour, H., Ghodsi, M., Haghpanah, N., Mirrokni, V.S., Mahini, H., Nikzad, A.: Optimal iterative pricing over social networks (extended abstract). In: Saberi, A. (ed.) WINE 2010. LNCS, vol. 6484, pp. 415–423. Springer, Heidelberg (2010)
2. Anagnostopoulos, A., Kumar, R., Mahdian, M.: Influence and correlation in social networks. In: Proceedings of the 14th ACM SIGKDD International Conference on Knowledge Discovery and Data Mining, pp. 7–15. ACM (2008)
3. Arthur, D., Motwani, R., Sharma, A., Xu, Y.: Pricing strategies for viral marketing on social networks. In: Leonardi, S. (ed.) WINE 2009. LNCS, vol. 5929, pp. 101–112. Springer, Heidelberg (2009)
4. Bakshy, E., Karrer, B., Adamic, L.A.: Social influence and the diffusion of user-created content. In: Proceedings of the 10th ACM Conference on Electronic Commerce, pp. 325–334. ACM (2009)
5. Candogan, O., Bimpikis, K., Ozdaglar, A.: Optimal pricing in networks with externalities. Operations Research $60(4)$, 883–905 (2012)
6. Cha, M., Haddadi, H., Benevenuto, F., Gummadi, P.K.: Measuring user influence in twitter: The million follower fallacy. In: ICWSM 2010, pp. 10–17 (2010)
7. Chen, W., Lu, P., Sun, X., Tang, B., Wang, Y., Zhu, Z.A.: Optimal pricing in social networks with incomplete information. In: Chen, N., Elkind, E., Koutsoupias, E. (eds.) WINE 2011. LNCS, vol. 7090, pp. 49–60. Springer, Heidelberg (2011)
8. Chen, W., Wang, Y., Yang, S.: Efficient influence maximization in social networks. In: Proceedings of the 15th ACM SIGKDD International Conference on Knowledge Discovery and Data Mining, pp. 199–208. ACM (2009)
9. Chierichetti, F., Kleinberg, J., Panconesi, A.: How to schedule a cascade in an arbitrary graph. In: Proceedings of the 13th ACM Conference on Electronic Commerce, pp. 355–368. ACM (2012)
10. Demaine, E.D., Hajiaghayi, M., Mahini, H., Malec, D.L., Raghavan, S., Sawant, A., Zadimoghadam, M.: How to influence people with partial incentives. In: Proceedings of the 23rd International Conference on World Wide Web, pp. 937–948. International World Wide Web Conferences Steering Committee (2014)
11. Domingos, P., Richardson, M.: Mining the network value of customers. In: Proceedings of the Seventh ACM SIGKDD International Conference on Knowledge Discovery and Data Mining, pp. 57–66. ACM (2001)
12. Hartline, J., Mirrokni, V., Sundararajan, M.: Optimal marketing strategies over social networks. In: Proceedings of the 17th International Conference on World Wide Web, pp. 189–198. ACM (2008)
13. Ioannidis, S., Chaintreau, A., Massoulié, L.: Optimal and scalable distribution of content updates over a mobile social network. In: INFOCOM 2009, pp. 1422–1430. IEEE (2009)

14. Kempe, D., Kleinberg, J., Tardos, É.: Maximizing the spread of influence through a social network. In: Proceedings of the Ninth ACM SIGKDD International Conference on Knowledge Discovery and Data Mining, pp. 137–146. ACM (2003)
15. Leskovec, J., Krause, A., Guestrin, C., Faloutsos, C., VanBriesen, J., Glance, N.: Cost-effective outbreak detection in networks. In: Proceedings of the 13th ACM SIGKDD International Conference on Knowledge Discovery and Data Mining, pp. 420–429. ACM (2007)
16. Lin, S., Wang, F., Hu, Q., Yu, P.S.: Extracting social events for learning better information diffusion models. In: Proceedings of the 19th ACM SIGKDD International Conference on Knowledge Discovery and Data Mining, pp. 365–373. ACM (2013)
17. Marshall, A.: Principles of economics. Digireads. com (2004)
18. Nemhauser, G.L., Wolsey, L.A., Fisher, M.L.: An analysis of approximations for maximizing submodular set functions. Mathematical Programming **14**(1), 265–294 (1978)
19. Richardson, M., Domingos, P.: Mining knowledge-sharing sites for viral marketing. In: Proceedings of the Eighth ACM SIGKDD International Conference on Knowledge Discovery and Data Mining, pp. 61–70. ACM (2002)
20. Sahni, S.: Computationally related problems. SIAM Journal on Computing **3**(4), 262–279 (1974)
21. Singer, Y.: How to win friends and influence people, truthfully: influence maximization mechanisms for social networks. In: Proceedings of the Fifth ACM International Conference on Web Search and Data Mining, pp. 733–742. ACM (2012)
22. Singer, Y., Mittal, M.: Pricing mechanisms for crowdsourcing markets. In: Proceedings of the 22nd International Conference on World Wide Web, pp. 1157–1166. International World Wide Web Conferences Steering Committee (2013)
23. Vondrak, J.: Optimal approximation for the submodular welfare problem in the value oracle model. In: Proceedings of the 40th Annual ACM Symposium on Theory of Computing, pp. 67–74. ACM (2008)

Boosting Financial Trend Prediction
with Twitter Mood
Based on Selective Hidden Markov Models

Yifu Huang[1], Shuigeng Zhou[1]([✉]), Kai Huang[1], and Jihong Guan[2]

[1] Shanghai Key Lab of Intelligent Information Processing,
School of Computer Science, Fudan University, Shanghai 200433, China
{huangyifu,sgzhou,kaihuang14}@fudan.edu.cn
[2] Department of Computer Science and Technology,
Tongji University, Shanghai 201804, China
jhguan@tongji.edu.cn

Abstract. Financial trend prediction has been a hot topic in both academia and industry. This paper proposes to exploit Twitter mood to boost financial trend prediction based on *selective hidden Markov models* (sHMM). First, we expand the *profile of mood states* (POMS) Bipolar lexicon to extract rich society moods from massive tweets. Then, we determine which mood has the most predictive power on the financial index based on *Granger causality analysis* (GCA). Finally, we extend sHMM to combine financial index and the selected Twitter mood to predict next-day trend. Extensive experiments show that our method not only outperforms the state-of-the-art methods, but also provides controllability to financial trend prediction.

1 Introduction

Financial trend prediction has been a hot research and practice topic in both academia and industry. The purpose of financial trend prediction is to predict the ups and downs of financial trends by building models based on historical financial data [11,15,23,36]. Besides historical financial data, more and more additional indictors such as news reports [14,26], Twitter mood [2,20,27,28] and trading relationship [30] have been used to improve financial trend prediction.

According to behavior finance [21], society mood is correlated with and even has predictive power on public financial index. Si *et al.* [27] modeled society mood of Twitter to support financial trend prediction. They utilized topic-based model to extract sentiments from Twitter posts, and then regressed the stock index and the sentiment series in an autoregressive framework. They achieved improved prediction performance when taking advantage of Twitter mood. However, in addition to prediction accuracy, *controllability* is also an important issue in financial trend prediction. Some works [4,7] modeled controllability on *selective prediction*—a prediction framework that can qualify its own prediction results and reject the outputs when they are not confident enough. Selective prediction can provide a trade-off between *coverage* (indicating how many predictions

© Springer International Publishing Switzerland 2015
M. Renz et al. (Eds.): DASFAA 2015, Part II, LNCS 9050, pp. 435–451, 2015.
DOI: 10.1007/978-3-319-18123-3_26

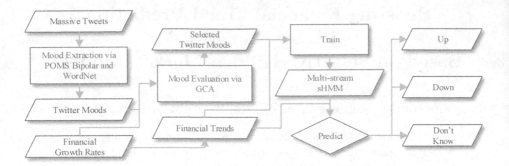

Fig. 1. The workflow of our approach

are made) and *accuracy* (indicating how many predictions are correct). This mechanism allows the users to interfere the prediction process, which is a desirable feature from the practice point of view. Three years ago, *hidden Markov model* (HMM) was introduced to the selective prediction framework, which leads to the *selective hidden Markov models* (sHMM) [23].

In this paper, to further boost financial trend prediction, we propose to combine Twitter mood and sHMM so that we can not only achieve high prediction performance but also obtain good controllability to financial trend prediction. To the best of our knowledge, this is the first attempt to exploit Twitter mood to predict financial trend with controllability. Concretely, we first use the *profile of mood states* (POMS) [18] Bipolar lexicon expanded by WordNet [19] to efficiently extract six-dimensional society moods from massive tweets, including *composed-anxious, agreeable-hostile, elated-depressed, confident-unsure, energetic-tired* and *clearheaded-confused*. Second, we perform *Granger causality analysis* (GCA) [9] between financial index and each Twitter mood to effectively determine which Twitter mood has the most predictive power on the index.

Then, we extend sHMM to multiple data streams so that historical financial data and selected Twitter mood can be combined to train the sHMM. Finally, we identify low-quality states of the trained sHMM according to given coverage and prevent predictions that are made from the low-quality states. Extensive experiments over real datasets show that our method not only outperforms the state-of-the-art methods, but also provides controllability to financial trend prediction. Fig. 1 illustrates the workflow of our approach.

The rest of this paper is organized as follows: Section 2 reviews the related work and highlights the differences between our work and major existing methods. Section 3 introduces the details of Twitter mood extraction and evaluation. Section 4 extends sHMM to combine index data and Twitter mood to predict financial trend. Section 5 presents performance evaluation and comparison. Finally, Section 6 concludes the paper.

2   Related Work

The purpose of financial trend prediction is to build models based on historical financial data and then employ the built models to predict the trend of future financial data. Up to now, various models have been proposed and different data were exploited. Existing works focus mainly on two aspects: data selection and model selection. For data selection, Idvall and Jonsson [11], Lin et al. [15], Pidan and El-Yaniv [23] and Zhang [36] used historical financial data only to predict financial trend, while Bollen et al. [2], Mittal et al. [20], Si et al. [27] and Sprenger et al. [28] combined historical financial data with some additional indicators such as Twitter mood to predict. As for model selection, Bollen et al. [2] exploited non-linear Self-Organizing Fusion Neural Networks (SOFNN) model, Si et al. [27] employed linear Vector Autoregression (VAR) model, while Idvall and Jonsson [11], Pidan and El-Yaniv [23], and Zhang [36] proposed HMM based methods. Since we do prediction by considering historical financial trends and Twitter moods together, some works [3,33,34] about how to model multiple time series that may correlate with each other and how to use multiple time series together to make prediction, are also highly related. The closest works to ours are [2], [23] and [27]. In what follows, we give a detailed description of the three methods and highlight the differences between them and our method.

Bollen et al. [2] investigated whether society moods have predictive on stock index. The authors extracted society moods from tweets through a lexicon called Google-POMS. Then, they performed a non-linear model called SOFNN to predict future trend. Major technical differences between their work and ours are: They used a lexicon called Google-POMS for Twitter mood extraction, which is not publicly available. Instead, we build a new lexicon by expanding POMS Bipolar with WordNet. Considering the importance of controllability, we utilize the sHMM rather than their SOFNN model to provide controllability to the prediction process.

Pidan and El-Yaniv [23] introduced the selective prediction framework into HMM, and addressed the importance of compromising coverage to gain better accuracy in a controlled manner. This method can identify low-quality HMM states and prevents predictions at those states. However, the training sequence of this model is only historical financial data. In this paper, we extend the sHMM to adopt both historical financial data and Twitter mood as input to boost the prediction performance.

Si et al. [27] proposed to leverage topic based sentiments from Twitter to help to predict the stock market. The authors utilized continuous Dirichlet process mixture (DPM) model to extract sentiments from Twitter posts, and then regressed the stock index and the sentiment time series in an autoregressive framework. They focused mainly on topic based sentiment analysis of tweets and used a simple prediction model, while we combine lexicon-based Twitter mood extraction and sHMM — an advanced model with controllability.

3   Mood Extraction and Evaluation

In this section, we extract and evaluate Twitter moods. Our aim is to predict public financial index, so we focus on analyzing sentiments of global tweets. First, we build a sentiment lexicon based on POMS Bipolar and WordNet. Then, we leverage the MapReduce framework to retrieve Twitter moods from massive tweets. Finally, we evaluate the predictive power of different Twitter moods via GCA, and determine the most predictive Twitter mood.

3.1   Basics of Sentiment Analysis

Behavior finance [21] shows that society mood has powerful influence on society decision. Furthermore, society mood is correlated with and even has predictive power on public financial index. To acquire society moods, we can analyze the sentiments of social media such as tweets that are fresh opinions shared by citizens. Sentiment analysis, *a.k.a.* opinion mining [17,22], is an application of *Natural Language Processing* (NLP) that aims at extracting subjective information such as author attitudes from texts [16,25]. Author attitudes may reflect the judgements or opinions of the authors, mood states or sentiments that the authors want to disseminate to the public. A major task of sentiment analysis is to extract multi-dimensional polarities from texts. Generally, there should be a pre-defined sentiment lexicon [10,32] for each polarity. A polarity is a time series obtained by first counting the sentiment word frequencies and then aggregating the frequencies in terms of a certain time granularity.

3.2   Expanding POMS Bipolar Lexicon by WordNet

According to the research of multi-dimensional sentiment analysis, human mood is very rich in social media, and a piece of text may contain multiple sentiments such as calm and agreement. POMS [18] is a questionnaire designed by psychologists to assess human mood states, and it already has three versions, namely, POMS Standard, POMS Brief and POMS Bipolar. POMS Bipolar consists of 6 polarities called *composed-anxious* (**Com.**), *agreeable-hostile* (**Agr.**), *elated-depressed* (**Ela.**), *confident-unsure* (**Con.**), *energetic-tired* (**Ene.**) and *clearheaded-confused* (**Cle.**), respectively. Each polarity contains 12 adjectives, and each of the 12 adjectives can increase either positive or negative polarity.

Due to the small size of the POMS Bipolar lexicon, it cannot capture all sentiments from texts in practice. So there should be some method to expand the POMS Bipolar lexicon. This paper employs WordNet synsets to expand the POMS Bipolr lexicon. WordNet [19] is an English language lexicon that subsumes English words into groups of synonyms called synsets. By mapping 72 words in the POMS Bipolar lexicon to their WordNet synsets, we get an expanded lexicon consisting of 638 words.

Algorithm 1. Twitter mood extraction

```
1: def MAP(date d, tweet t)                    1: def REDUCE(date d, vectors [v₁, v₂, ...])
2:     v ← ANALYZE(STEM(FILTER(t)))            2:     v_avg ← AVERAGE([v₁, v₂, ...])
3:     EMIT(date d, vector v)                   3:     EMIT(date d, vector v_avg)
4: end def                                      4: end def
```

3.3 Mood Extraction from Massive Tweets

Because of the massive amount of tweets, we leverage the MapReduce [6] framework to efficiently extract Twitter moods. The algorithm is outlined in Algorithm 1.

In the Map stage, the FILTER method discards tweets containing spam keywords such as "http:" and "www.", and keeps tweets containing subjective phrases such as "i feel" and "makes me". The STEM method normalizes terms in a tweet by eliminating prefixes and suffixes. The ANALYZE method computes the six-dimensional sentiment vector of a tweet using the expanded POMS Bipolar lexicon. In the Reduce stage, we compute the average sentiment vector of each trading day.

3.4 Mood Evaluation via Granger Causality Analysis

After sentiment analysis of tweets, we get six-dimensional sentiment series. Following that, we need to effectively evaluate them to determine which mood can help mostly predict market trend. GCA was first proposed by Clive Granger [9], it is a statistical hypothesis test for determining whether one time series is useful in forecasting another. Clive Granger argued that causality in economics could be reflected by measuring the ability of predicting the future values of a time series using past values of another time series. Formally, a time series X is said to Granger-cause another time series Y if it can be shown, usually through a series of t-test and F-test on some lagged values of X and Y, that those X values have statistically significant influence on the future values of Y. Formally, the following two equations hold:

$$Y_t = y_0 + \sum_{i=1}^{lag} y_i Y_{t-i} + \varepsilon_t, \tag{1}$$

$$Y_t = y_0 + \sum_{i=1}^{lag} y_i Y_{t-i} + \sum_{i=1}^{lag} x_i X_{t-i} + \varepsilon_t. \tag{2}$$

Above, t and i are time variables (in days), and lag is the upper bound of lagged days. We perform GCA in the same way as [8] between the growth rate of financial index (Y) and each Twitter mood (X). We determine the Twitter mood that has the most predictive power and its corresponding lagged value according to the following two rules: 1) find which p_{value} is at statistically significant level ($p_{value} \leq 0.1$); 2) find which p_{value} decreases significantly comparing to its precursor ($difference$ <-0.25). The first rule is for selecting the Twitter mood with the corresponding lagged days that has significant predictive power to the growth rate of financial index. The second rule is for guaranteeing that the selected Twitter mood of a certain day can provide significant improvement on predictive power comparing to that of the preceding day. The parameter lag of the selected Twitter mood is further used as the encoding length of the observation in our prediction model.

4   The Multi-stream sHMM

Here we extend sHMM to handle multi-streams. We call the extended model *multi-stream sHMM*, or *msHMM* in short. First, we introduce the basic concepts of HMM. Then, we briefly introduce sHMM. And finally, we present multi-stream sHMM, including the training and prediction processes as well as model evaluation with a large number of random starts based on the MapReduce framework.

4.1   HMM

HMM is a generative probabilistic model with latent states, where hidden state transitions and visible observation emissions are assumed to be Markov processes. Given an observation sequence $O=\{o_1, o_2, ..., o_T\}$ that is generated by a HMM λ, we associate O with a latent state sequence $S=\{s_1, s_2, ..., s_T\}$ that most likely produces O. λ can be formally defined as a quintuple $\{N, M, \boldsymbol{\pi}, \boldsymbol{A}, \boldsymbol{B}\}$. Here, N is the number of states in the state set $Q=\{q_1, q_2, ..., q_N\}$; M is the number of observations in the observation set $U=\{u_1, u_2, ..., u_M\}$; $\boldsymbol{\pi}$ is the initial probability vector of states and $\pi_i=P(s_1=q_i)$ is the initial probability of state q_i; $\boldsymbol{A}$ is the transition probability matrix of states and $a_{ij}=P(s_{t+1}=q_j|s_t=q_i)$ is the transition probability from state q_i to state q_j; $\boldsymbol{B}$ is the observation emission probability matrix of states and $b_{ij}=P(o_t=u_j|s_t=q_i)$ is the emission probability of observation u_j at state q_i.

4.2   sHMM

Selective prediction [4,7] is a prediction framework that can qualify its own prediction results and reject outputs if they are not confident enough. Pidan and El-Yaniv [23] introduced the selective prediction framework [7] to HMM, and thus developed sHMM. As in [23], we add state label p_i, empirical visit rate v_i and empirical state risk r_i to each state q_i, and add reject subset RS and heavy state q_h to HMM λ. For better understanding sHMM and the following multi-stream sHMM, we recall the major definitions of sHMM as follows in the context of financial trend prediction.

Definition 1. *Given an observation sequence $O=\{o_1, o_2, ..., o_T\}$ (indicating historical financial trend), a relative label sequence $L=\{l_1, l_2, ..., l_T\}$ (indicating* *next-day financial trend) and a HMM λ, the state label p_i denotes the most probable label that state q_i should have. Formally,*

$$p_i = \arg \max_{l=up,down} \sum_{t=1,l_t=l}^{T} \gamma_{ti}. \tag{3}$$

Above, $\gamma_{ti}=P(s_t=q_i|O, \lambda)$ denotes the probability that the HMM λ stays at state q_i at time t, which can be computed by the forward-backward procedure [24].

Definition 2. *Given an observation sequence $O=\{o_1, o_2, ..., o_T\}$ and a HMM λ, the empirical visit rate v_i denotes the fraction of time that the HMM λ spends at state q_i, i.e.,*

$$v_i = \frac{1}{T} \sum_{t=1}^{T} \gamma_{ti}. \tag{4}$$

Definition 3. *Given an observation sequence $O=\{o_1, o_2, ..., o_T\}$, a relative label sequence $L=\{l_1, l_2, ..., l_T\}$ and a HMM λ, the empirical state risk r_i denotes the rate of erroneous visits to state q_i. Formally,*

$$r_i = \frac{\frac{1}{T} \sum_{t=1,l_t \neq p_i}^{T} \gamma_{ti}}{v_i}. \tag{5}$$

Furthermore, we sort all HMM states by their empirical state risks in descending order and record them as $Q_d=\{q_{d_1}, q_{d_2}, ..., q_{d_N}\}$ (for each $j < k$, $r_{d_j} \geq r_{d_k}$). The low-quality HMM states, also called *reject states*, constitute the *reject subset* RS. Predictions at those states are prevented.

Definition 4. *Given a coverage bound C_B, we label the reject states sequentially until their cumulative empirical visit rate $\sum_{j=1}^{K} v_{d_j}$ exceeds 1-C_B. Formally, the reject subset RS is defined as*

$$RS = \{q_{d_1}, q_{d_2}, ..., q_{d_K} | \sum_{j=1}^{K} v_{d_j} \leq 1 - C_B, \sum_{j=1}^{K+1} v_{d_j} > 1 - C_B\}. \tag{6}$$

Definition 5. *Given a visit bound V_B, state $q_{d_{K+1}}$ is identified as a heavy state q_h if its visit rate $v_{d_{K+1}} > V_B$.*

The heavy state q_h is the cause of coarseness problem as described in [23], and it should be recursively refined in the training stage. Another issue should be taken into consideration in practice is scaling, because floating point underflow can easily happen in the forward-backward procedure that is the fundamental of HMM. We handle it with the solution provided by [24].

4.3   Multi-stream sHMM

To combine financial index and Twitter moods to sHMM, we extend sHMM to process multiple data streams. We treat historical financial trend and Twitter mood trends as multiple observation sequences generated by sHMM, and formulate multiple observation sequences as $O_K=\{O^{(1)}, O^{(2)}, ..., O^{(K)}\}$ where $O^{(k)}=\{o_1^{(k)}, o_2^{(k)}, ..., o_{T_k}^{(k)}\}$. The observation is gained by encoding the trend with lag length, which is determined in "Mood Evaluation via Granger Causality Analysis" subsection. In the training stage, likelihood function $P(O_K|\lambda)$ (indicating the probability that multiple observation sequences are produced by the model) is maximized via a variation of the Baum-Welch algorithm [1]:

$$P(O_K|\lambda) = \prod_{k=1}^{K} P(O^{(k)}|\lambda) = \prod_{k=1}^{K} P_k. \tag{7}$$

For example, denote 1 as *up trend* and 0 as *down trend*. Say we have the financial trend sequence $A=\{0,0,0,1,1,0,1\}$, the selected Twitter mood trend sequence $B=\{1,0,1,0,1,0,1\}$ and the encoding length $lag = 3$. Based on A, B and lag, we can get the encoding financial trend sequence $A_e=\{0,1,3,6,5\}$ and the encoding selected Twitter mood trend sequence $B_e=\{5,2,5,2,5\}$. Our prediction model multi-stream sHMM use A_e and B_e as training sequences to get model parameters.

Training. Given a coverage bound C_B, multiple observation sequences O_K and a label sequence L, we train a multi-stream sHMM and recursively refine the heavy state q_h until there is no heavy state remaining. The training process consists of the following steps:

1. Initialize the root HMM λ_0 with random parameters, and train it with the set O_K of historical financial trend and Twitter mood trends. The train algorithm is Baum-Welch [1] variation adjusted to multiple observation sequences.
2. Compute state label p_i, empirical visit rate v_i and empirical state risk r_i for each state q_i in the root HMM λ_0. Under the coverage bound C_B, compute the reject subset RS of the root HMM λ_0 to identify which state is the heavy state q_h. If there is no heavy state, then the training is done.
3. Initialize a random HMM λ_{random} to replace the heavy state q_h, so a refined HMM λ_{refine} is obtained. Train the refined HMM λ_{refine} with the previous set O_K of multiple observation sequences until it converges. Details of this step are described in Algorithm 2.

$$\pi_j = \frac{\sum_{i=1}^{K} \frac{1}{P_i}(\gamma_{1j}^{(i)} + \sum_{t=1}^{T_i-1} \sum_{k=1,k\neq h}^{N} \xi_{t,k,j}^{(i)})}{Z}, \tag{8}$$

$$a_{jk} = \frac{\sum_{i=1}^{K} \frac{1}{P_i} \sum_{t=1}^{T_i-1} \xi_{t,j,k}^{(i)}}{\sum_{l=N+1}^{N+n} \sum_{i=1}^{K} \frac{1}{P_i} \sum_{t=1}^{T_i-1} \xi_{t,j,l}^{(i)}}, \tag{9}$$

$$b_{jm} = \frac{\sum_{i=1}^{K} \frac{1}{P_i} \sum_{t=1,o_t^{(i)}=u_m}^{T_i} \gamma_{tj}^{(i)}}{\sum_{i=1}^{K} \frac{1}{P_i} \sum_{t=1}^{T_i} \gamma_{tj}^{(i)}}. \tag{10}$$

Algorithm 2. Training the refined model

Input: HMM λ with N states, heavy state q_h, multiple sequences $O_K=\{O^{(1)}, O^{(2)}, ..., O^{(K)}\}$
1: Initialize a random HMM λ_{random} with n states
2: For each $j=1, 2, ..., N$, $j\neq h$, replace transition q_jq_h to $q_jq_{N+1}, q_jq_{N+2}, ..., q_jq_{N+n}$ and transition q_hq_j to $q_{N+1}q_j, q_{N+2}q_j, ..., q_{N+n}q_j$
3: Record the heavy state q_h as a refined state q_{refine} and remove the observation emission probability vector from it. For each $j=N+1, N+2, ..., N+n$, set state label $p_j=p_h$
4: **while** not converged **do**
5:     For each $j=1, 2, ..., N$, $j\neq h$, $k=N+1, N+2, ..., N+n$, update $a_{jk}=a_{jh}\pi_k$, $a_{kj}=a_{hj}$
6:     For each $j=N+1, N+2, ..., N+n$, update $\pi_j=\pi_h\pi_j$
7:     For each j, $k=N+1, N+2, ..., N+n$, update $a_{jk}=a_{hh}a_{jk}$
8:     For each $j=N+1, N+2, ..., N+n$, re-estimate π_j by Eq. (8)
9:     For each j, $k=N+1, N+2, ..., N+n$, re-estimate a_{jk} by Eq. (9)
10:     For each $j=N+1, N+2, ..., N+n$, $m=1, 2, ... M$, re-estimate b_{jm} by Eq. (10)
11: **end while**
12: Perform the operations of Lines 5-7 once again
Output: HMM λ with N-$1+n$ states

Above, Z is the normalization factor of π_j, and $\xi_{t,j,k}=P(s_t=q_j, s_{t+1}=q_k|O, \lambda)$ is the transition probability of HMM λ from state q_j to state q_k at time t, which can be efficiently computed by the forward-backward procedure [24].

4. Compute empirical visit rate v_i and empirical state risk r_i for each state q_i in the refined HMM λ_{refine}. Under the coverage bound C_B, compute the reject subset RS of the refined HMM λ_{refine} to identify which state is the heavy state q_h. If there is a heavy state, go to Step 3.

Prediction. Given a trained multi-stream sHMM and a new observation sequence $O=\{o_1, o_2, ..., o_T\}$, we predict the last label l_T in the relative label sequence L according to O through recursively finding the most probable state q_{most}. The prediction process consists of the following steps:

1. Find the most probable state q_{most} at the last time T by computing γ_{Ti} of all states in the root HMM λ_0.
2. If q_{most} is a refined state $q_{refined}$, reset the most probable state q_{most} by computing γ_{Ti} of new states added to the next level refined HMM λ_{refine}, and go to Step 2.
3. If q_{most} is in the reject subset RS, no prediction is made; otherwise, the label p_{most} of state q_{most} is output as the prediction result of the last label l_T in the relative L according to O.

Large-Scale Evaluation. As the parameters of HMM are randomly initialized, and the train algorithm such as Baum-Welch [1] is sensitive to the initial parameters, it may converge to different local maxima for different initializations. So we may get different predictions with the same training and test sequences. To reduce the random effect caused by parameter initialization, we run the algorithm a number N_{rs} of times, and evaluate the averaged empirical error rate as the performance measure. To make the evaluation efficient, we adopt the MapReduce [6] framework. The procedure is outlined in Algorithm 3. In the Map stage, given a coverage bound C_B, we train a multi-stream sHMM (by the

---

Algorithm 3. Large-scale performance evaluation

1: **def** MAP(id i, coverage C_B)
2:     $e \leftarrow$ PREDICT(TRAIN(C_B))
3:     EMIT(coverage C_B, error e)
4: **end def**

1: **def** REDUCE(coverage C_B, errors $[e_1, e_2, \ldots]$)
2:     $e_{avg} \leftarrow$ AVERAGE($[e_1, e_2, \ldots]$)
3:     EMIT(coverage C_B, error e_{avg})
4: **end def**

TRAIN method) using different parameter values, then do prediction (by the PREDICT method) based on the trained model, and get the error rate for different parameter values. In the Reduce stage, we compute the averaged error rate for the given C_B.

5   Experimental Evaluation

In this section, we present experimental evaluation results. First, we introduce experimental datasets and computing environment. Then, we present and analyze the results of GCA.

Finally, we compare our method with seven existing approaches to demonstrate the advantage of our method.

5.1   Experimental Setup

There are two Twitter datasets used in our experiments. The first one is from [35]. It contains 467 million Twitter posts that were published on Twitter in a seven-month period from Jun. 2009 to Dec. 2009. We call it *Twitter2009*. The second is from [13]. It contains 50 million tweets that cover a 20 month period from Jan. 2010 to Aug. 2011. We call it *Twitter2011*.

The financial data used are the S&P500 Index and NYSE Composite Index from Yahoo! Finance. For *Twitter2009* that covers the time period from 06/12/2009 to 12/21/2009, we train the model with data of the first 100 days (before 11/09/2009), and test the model using data of the next 30 days. As for *Twitter2011* that covers the time period from 01/04/2010 to 07/26/2011, we train the model with data of the first 300 days (before 03/13/2011), and test the model using data of the next 90 days. Here, we focus on predicting the trend of daily close price data.

We implemented our method in the MapReduce framework at a Hadoop platform, which was built on a Hadoop cluster that consists of 1 namenode/jobtracker and 24 datanodes/tasktrackers. Each node is equipped with an Intel(R) Core(TM)2 Duo CPU E7500 @ 2.93GHz and 4GB RAM.

5.2   Statistics of Extracted Twitter Moods

We utilize Algorithm 1 to extract Twitter moods from massive tweets, and keep the extracted daily Twitter moods during the period of trading days. For these

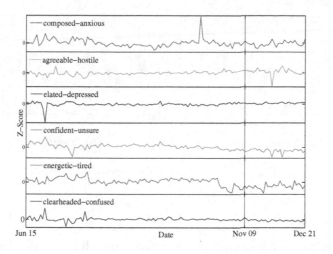

(a) *Twitter 2009*

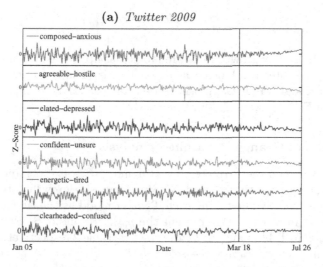

(b) *Twitter 2011*

Fig. 2. Extracted Twitter Moods

days that have not enough tweets, we get the Twitter moods by linear interpolation. We compute the z-scores of all Twitter moods, including *composed-anxious* (**Com.**), *agreeable-hostile* (**Agr.**), *elated-depressed* (**Ela.**), *confident-unsure* (**Con.**), *energetic-tired* (**Ene.**) and *clearheaded-confused* (**Cle.**). The results of *Twitter 2009* and *Twitter 2011* are plotted in Fig. 2a and Fig. 2b respectively. We also discretize Tweet moods in daily trends, which will be fed to our multi-stream sHMM later.

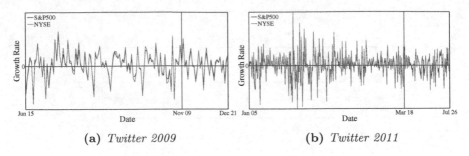

(a) *Twitter 2009*   (b) *Twitter 2011*

Fig. 3. S&P500 and NYSE growth rates

5.3   Growth Rates of Financial Indexes

Given the daily close price $close_t$ at day t, daily growth rate of day t is evaluated as follows:

$$growth_t = \frac{close_t - close_{t-1}}{close_{t-1}}. \tag{11}$$

The days with positive growth rates are labeled as up trends, while the days with negative growth rates are labeled as down trends. The computation results of S&P500 Index and NYSE Composite Index for *Twitter 2009* and *Twitter 2011* are shown in Fig. 3a and Fig. 3b respectively.

5.4   Results of Granger Causality Analysis

We perform GCA on the extracted Twitter moods via Eq. (1) and Eq. (2). We treat the S&P500 growth rate and NYSE Composite growth rate as the time series Y respectively, while taking each Twitter mood as the time series X. The *lag* range is set from 1 to 7. We present the p_{value} results of S&P500 and NYSE for *twitter2009* in Table 1 and for *twitter2011* in Table 2 respectively. By checking the results in the two tables, we select 3-day (*i.e.*, *lag*=3) lagged *agreeable-hostile* (**Agr.**) Twitter mood as the predictive indicator of public financial index due to following two reasons: 1) Most p-values under 3-day lagged **Agr.** Twitter mood achieve significant level (≤ 0.1), and their values are 0.159, **0.096**, **0.071** and **0.054**, respectively. 2) All p-values under 3-day lagged **Agr.** Twitter mood decrease significantly (*difference* $<$-0.25) comparing to p-values under 2-day lagged **Agr.** Twitter mood. The differences are -0.278, -0.276, -0.284 and -0.409, respectively.

5.5   Prediction Performance Comparison

We compare our method with seven existing methods, in which six methods exploit Twitter mood. These six methods and our method all incorporate 3-day lagged **Agr.** Twitter mood to predict public financial trend. The six methods are:

Table 1. p_{value} results of S&P500 and NYSE for *Twitter2009* (all $p_{value}^{*} \leq 0.1$)

Lag	S&P500						NYSE					
	Com.	Agr.	Ela.	Con.	Ene.	Cle.	Com.	Agr.	Ela.	Con.	Ene.	Cle.
1	0.704	0.226	0.681	0.696	0.535	0.270	0.739	0.179	0.756	0.625	0.529	0.385
2	0.764	0.437	0.648	0.588	0.722	0.305	0.851	0.372	0.664	0.444	0.746	0.417
3	0.228	0.159	0.856	0.276	0.741	0.338	0.231	**0.096***	0.876	0.238	0.772	0.489
4	0.234	0.233	0.516	0.386	0.886	0.127	0.214	0.134	0.615	0.349	0.900	0.232
5	0.379	0.389	0.515	0.315	0.966	0.159	0.348	0.258	0.569	0.275	0.974	0.241
6	0.301	0.145	0.186	0.439	0.949	0.180	0.277	**0.061***	0.228	0.405	0.948	0.277
7	0.428	0.148	0.331	0.262	0.955	0.218	0.364	**0.094***	0.418	0.231	0.941	0.296

Table 2. p_{value} results of S&P500 and NYSE for *Twitter2011* (all $p_{value}^{*} \leq 0.1$)

Lag	S&P500						NYSE					
	Com.	Agr.	Ela.	Con.	Ene.	Cle.	Com.	Agr.	Ela.	Con.	Ene.	Cle.
1	0.352	0.153	0.991	0.565	0.223	0.596	0.401	0.209	0.811	0.584	0.137	0.542
2	0.690	0.355	0.924	0.450	**0.082***	0.747	0.707	0.463	0.885	0.463	**0.060***	0.772
3	0.876	**0.071***	0.897	0.415	0.172	0.842	0.855	**0.054***	0.950	0.409	0.132	0.821
4	0.886	0.131	0.963	0.524	0.241	0.525	0.864	**0.099***	0.986	0.490	0.216	0.490
5	0.929	0.215	0.981	0.647	0.328	0.498	0.913	0.174	0.993	0.629	0.270	0.475
6	0.872	0.309	0.994	0.705	0.266	0.559	0.837	0.261	0.999	0.646	0.156	0.523
7	0.885	0.476	0.999	0.524	0.109	0.621	0.840	0.413	1.000	0.472	**0.071***	0.587

1. **VAR.** The Vector Autoregressive (**VAR**) framework [29] treats historical financial data and Twitter moods as an integrated vector to make prediction based linear regression.
2. **HMM.** Hidden Markov model (**HMM**) [24] considers two states of "up" and "down", and treats both historical financial data and Twitter moods as observation sequences to make prediction.
3. **CRF.** Conditional Random Field (**CRF**) [12] model also considers "up" and "down" states, and uses historical financial data and Twitter moods as observation sequences to make prediction.
4. **SVM.** Support Vector Machine (**SVM**) [5] combines both historical financial data and Twitter moods as a feature vector to make prediction.
5. **NN.** Neural Network (**NN**) [31] uses two nodes of "up" and "down" as output layer, and puts both historical financial data and Twitter moods to input layer to make prediction.
6. **cDPM.** It was proposed in [27], which uses a topic-model based approach to extract Twitter sentiments and then combines historical financial data and Twitter sentiments into an autoregressive framework.

The only compared method that does not use Twitter mood is sHMM. It was developed in [23], which uses sHMM for financial trend prediction without using any social media information. This is the best existing model for financial trend prediction with controllability. For discrimination, we call our method **msHMM** as it uses multiple sequences for training.

Comparison with existing methods using Twitter mood. As **msHMM**'s performance is adjustable by the coverage bound C_B, we set four values between 1.0 and 0.1 for C_B to evaluate **msHMM**. A smaller C_B value means that **msHMM**

Table 3. Comparison with six methods using Twitter mood

Model	Error Rate (%)			
	Twitter2009		*Twitter2011*	
	S&P500	**NYSE**	**S&P500**	**NYSE**
VAR	*26.667*	*33.333*	46.667	44.444
HMM	36.667	46.667	44.444	54.444
CRF	40.000	40.000	45.556	44.444
SVM	40.000	46.667	50.000	44.444
NN	36.667	46.667	*37.778*	*32.222*
cDPM	40.000	43.333	48.889	38.889
msHMM(C_B)	39.715(1.0)	45.802(1.0)	45.796(1.0)	45.802(1.0)
	30.055(0.5)	35.972(0.5)	42.408(0.5)	35.972(0.5)
	22.209(0.3)*	33.227(0.3)*	36.622(0.2)*	31.869(0.1)*
	8.033(0.1)	**32.694**(0.1)	**35.380**(0.1)	**31.869**(0.1)

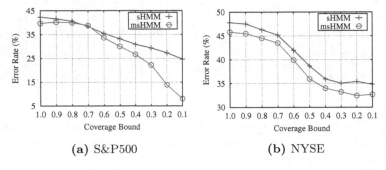

(a) S&P500   (b) NYSE

Fig. 4. Risk Coverage Curves for *Twitter2009*

puts more restriction on prediction output, which leads to a smaller error rate. For the six existing methods, we tune their parameters to get the best results. All experimental results are presented in Table 3. We can see that among the six existing methods, **VAR** has the smallest error rates (26.667% for S&P500, 33.333% for NYSE) for *Twitter2009* and **NN** obtains the smallest error rates (37.778% for S&P500, 32.222% for NYSE) for *Twitter2011*. For the four cases, by reducing C_B's value to 0.3, 0.3, 0.2 and 0.1 respectively, **msHMM** can get smaller error rates (22.209%, 33.227%, 36.622% and 31.869% respectively) than the six existing methods. And with C_B=0.1 **msHMM** achieves the lowest error rate on all two datasets. More importantly, the error rate of **msHMM** is controllable, while the six existing methods do not have such a feature.

Comparison with sHMM. For different coverage bound C_B values from 1.0 to 0.1, we first run **sHMM** on the S&P500 and NYSE index data, and then combine historical financial data and 3-days lagged agreeable-hostile Twitter mood to run our method **msHMM**. The results are plotted in Fig. 4 and Fig. 5, which show the *Risk Coverage* (RC) curves for both **sHMM** and **msHMM**. When taking a smaller coverage, **msHMM** rejects to make predictions if not confident enough, so a smaller error rate is obtained. For example, as shown in Fig. 4a, when C_B=0.1, **sHMM** gets a 24.67% error rate, while **msHMM** achieves a 8.03% error rate. From Fig. 4 and 5, we can see that our method

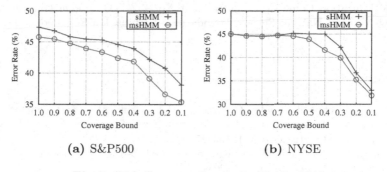

(a) S&P500   (b) NYSE

Fig. 5. Risk Coverage Curves for *Twitter2011*

msHMM obviously outperforms **sHMM**. On the one hand, given a certain error rate, **msHMM** can achieve a larger coverage than **sHMM**; On the other hand, given a certain coverage bound, **msHMM** can obtain a smaller error rate than **sHMM**. Another advantage of **msHMM** over **sHMM** is that **msHMM** lays down a way to utilize more additional indicators to boost financial trend prediction performance.

6 Conclusion

We proposed to utilize Twitter moods to boost financial trend prediction based on sHMM. First, we used the POMS Bipolar lexicon expanded by WordNet to extract six-dimensional society moods from large scale tweets, then we performed GCA between financial index and each Twitter mood to determine which Twitter mood has the most predictive power on financial index. Finally, we extended sHMM to combine financial index and Twitter moods to predict next-day trend. Experiments on the S&P500 and NYSE Composite index show that our method with 3-days lagged agreeable-hostile Twitter mood not only performs better than the state-of-the-art methods, but also provides a controllability mechanism to financial trend prediction.

Note that the major contribution of our work is combining financial data and Twitter moods into sHMM. We treat financial trend and Twitter mood trends as multiple observation sequences generated by sHMM. In this work, we use bivariate GCA to determine which Twitter mood has the most predictive power. For future work, we will explore multivariate GCA to select the optimal combination of multiple Twitter moods to improve prediction performance. Furthermore, we will investigate more advanced data combination methods for sHMM, and try other prediction models with controllability.

Acknowledgments. This work was partially supported by the Key Projects of Fundamental Research Program of Shanghai Municipal Commission of Science and Technology under grant No. 14JC1400300, and the Innovation Research Program of Shanghai

Municipal Education Commission under grant No. 13ZZ003. Jihong Guan was supported by National Natural Science Foundation of China (NSFC) under grant No. 61373036.

References

1. Baum, L.E., Petrie, T., Soules, G., Weiss, N.: A maximization technique occurring in the statistical analysis of probabilistic functions of markov chains. The annals of mathematical statistics **41**(1), 164–171 (1970)
2. Bollen, J., Mao, H., Zeng, X.: Twitter mood predicts the stock market. Journal of Computational Science **2**(1), 1–8 (2011)
3. Brand, M.: Coupled hidden markov models for modeling interacting processes. Tech. rep., MIT (1997)
4. Chow, C.K.: On optimum recognition error and reject tradeoff. IEEE Transactions on Information Theory **16**(1), 41–46 (1970)
5. Cortes, C., Vapnik, V.: Support-vector networks. Machine Learning **20**(3), 273–297 (1995). http://dx.doi.org/10.1007/BF00994018
6. Dean, J., Ghemawat, S.: Mapreduce: simplified data processing on large clusters. Communications of the ACM **51**(1), 107–113 (2008)
7. El-Yaniv, R., Wiener, Y.: On the foundations of noise-free selective classification. The Journal of Machine Learning Research **11**, 1605–1641 (2010)
8. Gilbert, E., Karahalios, K.: Widespread worry and the stock market. In: Proceedings of the Fourth International Conference on Weblogs and Social Media, pp. 59–65 (2010)
9. Granger, C.W.J.: Investigating causal relations by econometric models and cross-spectral methods. Econometrica: Journal of the Econometric Society **37**(3), 424–438 (1969)
10. Hu, M., Liu, B.: Mining and summarizing customer reviews. In: Proceedings of the Tenth ACM SIGKDD International Conference on Knowledge Discovery and Data Mining, pp. 168–177 (2004)
11. Idvall, P., Jonsson, C.: Algorithmic trading: hidden markov models on foreign exchange data. Master's thesis, Södertörn University (2008)
12. Lafferty, J.D., McCallum, A., Pereira, F.C.N.: Conditional random fields: probabilistic models for segmenting and labeling sequence data. In: Proceedings of the Eighteenth International Conference on Machine Learning, pp. 282–289 (2001). http://dl.acm.org/citation.cfm?id=645530.655813
13. Li, R., Wang, S., Deng, H., Wang, R., Chang, K.C.C.: Towards social user profiling: unified and discriminative influence model for inferring home locations. In: Proceedings of the 18th ACM SIGKDD International Conference on Knowledge Discovery and Data Mining, pp. 1023–1031 (2012). http://doi.acm.org/10.1145/2339530.2339692
14. Li, X., Wang, C., Dong, J., Wang, F., Deng, X., Zhu, S.: Improving stock market prediction by integrating both market news and stock prices. In: Hameurlain, A., Liddle, S.W., Schewe, K.-D., Zhou, X. (eds.) DEXA 2011, Part II. LNCS, vol. 6861, pp. 279–293. Springer, Heidelberg (2011)
15. Lin, Y., Guo, H., Hu, J.: An svm-based approach for stock market trend prediction. In: The 2013 International Joint Conference on Neural Networks, pp. 1–7 (2013)
16. Liu, B.: Sentiment analysis and subjectivity. In: Handbook of Natural Language Processing, 2nd edn (2010)

17. Liu, B.: Sentiment Analysis and Opinion Mining. Morgan & Claypool Publishers (2012)
18. McNair, D.M., Lorr, M., Droppleman, L.F.: Profile of mood states. Educational and Industrial Testing Service (1971)
19. Miller, G.A.: Wordnet: a lexical database for english. Communications of the ACM **38**(11), 39–41 (1995)
20. Mittal, A., Goel, A.: Stock prediction using twitter sentiment analysis. Tech. rep., Stanford University
21. Nofsinger, J.R.: Social mood and financial economics. The Journal of Behavioral Finance **6**(3), 144–160 (2005)
22. Pang, B., Lee, L.: Opinion mining and sentiment analysis. Found. Trends Inf. Retr. **2**(1–2), 1–135 (2008). http://dx.doi.org/10.1561/1500000011
23. Pidan, D., El-Yaniv, R.: Selective prediction of financial trends with hidden markov models. In: Advances in Neural Information Processing Systems, pp. 855–863 (2011)
24. Rabiner, L.R.: A tutorial on hidden markov models and selected applications in speech recognition. Proceedings of the IEEE **77**(2), 257–286 (1989)
25. Riloff, E., Wiebe, J.: Learning extraction patterns for subjective expressions. In: Proceedings of the 2003 Conference on Empirical Methods in Natural Language Processing, pp. 105–112 (2003)
26. Schumaker, R.P., Chen, H.: A discrete stock price prediction engine based on financial news. Computer **43**(1), 51–56 (2010)
27. Si, J., Mukherjee, A., Liu, B., Li, Q., Li, H., Deng, X.: Exploiting topic based twitter sentiment for stock prediction. In: Proceedings of the 51st Annual Meeting of the Association for Computational Linguistics (vol. 2: Short Papers), pp. 24–29 (2013)
28. Sprenger, T.O., Tumasjan, A., Sandner, P.G., Welpe, I.M.: Tweets and trades: the information content of stock microblogs. European Financial Management (2013). http://onlinelibrary.wiley.com/doi/10.1111/j.1468-036X.2013.12007.x/abstract
29. Stock, J.H., Watson, M.W.: Vector autoregressions. The Journal of Economic Perspectives **15**(4), 101–115 (2001). http://www.jstor.org/stable/2696519
30. Sun, X.Q., Shen, H.W., Cheng, X.Q.: Trading network predicts stock price. Scientific Reports **4**(3711), 1–6 (2014)
31. Trippi, R.R., Turban, E.: Neural Networks in Finance and Investing: Using Artificial Intelligence to Improve Real World Performance. McGraw-Hill, Inc (1992)
32. Wilson, T., Hoffmann, P., Somasundaran, S., Kessler, J., Wiebe, J., Choi, Y., Cardie, C., Riloff, E., Patwardhan, S.: Opinionfinder: a system for subjectivity analysis. In: Proceedings of HLT/EMNLP on Interactive Demonstrations, pp. 34–35 (2005)
33. Wu, D., Ke, Y., Yu, J.X., Yu, P.S., Chen, L.: Detecting leaders from correlated time series. In: Kitagawa, H., Ishikawa, Y., Li, Q., Watanabe, C. (eds.) DASFAA 2010. LNCS, vol. 5981, pp. 352–367. Springer, Heidelberg (2010)
34. Yang, B., Guo, C., Jensen, C.S.: Travel cost inference from sparse, spatio temporally correlated time series using markov models. Proc. VLDB Endow. **6**(9), 769–780 (2013). http://dx.doi.org/10.14778/2536360.2536375
35. Yang, J., Leskovec, J.: Patterns of temporal variation in online media. In: Proceedings of the fourth ACM international conference on Web search and data mining, pp. 177–186 (2011)
36. Zhang, Y.: Prediction of financial time series with Hidden Markov Models. Master's thesis, Simon Fraser University (2004)

k-Consistent Influencers in Network Data

Enliang Xu[1]([✉]), Wynne Hsu[1], Mong Li Lee[1], and Dhaval Patel[2]

[1] School of Computing, National University of Singapore, Singapore, Singapore
{xuenliang,whsu,leeml}@comp.nus.edu.sg
[2] Department of Electronics and Computer Engineering, IIT Roorkee, Roorkee, India
patelfec@iitr.ernet.in

Abstract. With the prevalence of online social media such as Facebook, Twitter and YouTube, social influence analysis has attracted considerable research interests recently. Existing works on top-k influential nodes discovery find influential users at single time point only and do not capture whether the users are consistently influential over a period of time. Finding top-k consistent influencers has many interesting applications, such as targeted marketing, recommendation, experts finding, and stock market. Identifying top-k consistent influencers is a challenging task. First, we need to dynamically compute the total influence of each user at each time point from an action log. However, to find the consistent top-scorers, we need to sort and rank them at each time point. This is computationally expensive and not scalable. In this paper, we define the consistency of a node based on its influence and volatility over time. With the help of grid index, we develop an efficient algorithm called TCI to obtain the top-k consistent influencers given a time period. We conduct extensive experiments on three real world datasets to evaluate the proposed methods. We also demonstrate the usefulness of top-k consistent influencers in identifying information sources and finding experts. The experimental results demonstrate the efficiency and effectiveness of our methods.

1 Introduction

Social networking sites such as Facebook, Twitter, Delicious and YouTube have provided a platform where user can express their ideas and share information. With the prevalence of these sites, social networks now play a significant role in the spread of information. Recognizing this, researchers have focused on influence analysis to discover influential nodes (users, entities) and influence relationships (who influences whom) among nodes in the network. Existing works on influential nodes discovery define influential user as one who posts/tweets frequently and/or with a large number of followers/friends. However, from a psychological perspective, frequency and popularity are not sufficient to develop influence and loyalty. Instead, it is consistency that builds trusts and thereby resulting in the greatest influence.

© Springer International Publishing Switzerland 2015
M. Renz et al. (Eds.): DASFAA 2015, Part II, LNCS 9050, pp. 452–468, 2015.
DOI: 10.1007/978-3-319-18123-3_27

We observe that consistency comes in two forms. The first form of consistency is known as personal consistency. This refers to one who is consistent in his behavior, for example, a user could tweet regularly on the same topic over a period of time. This user tends to gain greater authority as other users' trusts in him grow, and thereby increases his influence.

The second form refers to our preference for consistent behavior. We have a tendency to remain consistent with our previous actions. In the case of social networking, if a user u_2 has retweeted a post from another user u_1, there is a much higher probability that u_2 will retweet other posts from the same user u_1. In other words, the u_2 has a strong preference for u_1.

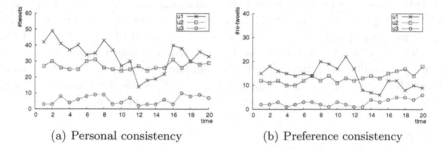

(a) Personal consistency   (b) Preference consistency

Fig. 1. Example of two forms of consistency

Figure 1(a) shows an example of 3 users' tweeting frequency over 20 time points and the number of followers they have. We observe that both u_1 and u_2 have a large number of followers. However, u_1's tweeting frequency appears random whereas u_2 consistently tweets at regular interval. On the other hand, u_3 has a small number of followers but he tweets regularly.

Figure 1(b) shows the number of followers retweeting their tweets over the 20 time points. In the beginning, u_1 appears to have the most number of followers retweeting his tweets. However, over time, the number of followers retweeting his tweets declines. In contrast, both u_2 and u_3 maintain the same number of followers retweeting their tweets. However, since u_3's base of followers is small, his influence is not as great as u_2.

Clearly, an accurate measure of degree of influence must take into account these two forms of consistency. A user is highly influential if he has high personal consistency and he has established consistent preferences to his tweets/posts in a large number of users.

We can depict the 3 users in a 2D personal-preference consistency space over 5 time points as shown in Figure 2. We observe that users near the top right corner are high in both personal and preference consistency. For example, u_1 has the highest personal and preference consistency at $t = 2$ and $t = 8$, but its personal consistency drops at $t = 13$ and $t = 14$. On the other hand, u_2 has the second highest personal and preference consistency at $t = 2$ and $t = 8$, and leads

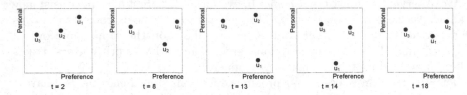

Fig. 2. Personal-preference 2D space

at time points $t = 13$, $t = 14$ and $t = 18$. Clearly, u_2 is more consistent and hence, can exert a greater influence over time compared to u_1 who seems to be more volatile.

In this paper, we define the notion of k-consistent influential users and devise an efficient algorithm to identify these users. Our algorithm linearizes the 2D personal-preference consistency space to construct a GridIndex. Based on the GridIndex, we can quickly obtain the k-consistent influencers for a given time interval. We conduct extensive experiments on three real world datasets to evaluate the efficiency of the proposed approach, as well as the effectiveness of using k-consistent influencers to identify information sources and experts.

2   Problem Formulation

In this section, we first introduce some terminologies, and then give the formal problem definition.

Definition 1 (Action Log). *An* action log *is a relation D where a tuple $< t, u, a > \in D$ indicates that node u has performed action a at time t.*

Figure 3 shows an action log and the corresponding user relation graph. Note that the relation graph is given as input. For example, node u_1 performs action a at time point 0 and u_4 performs the same action a following u_1 at time point 1.

Definition 2 (Degree of Influence). *Let $G = (V, E)$ denote a social network where V and E are the sets of nodes and edges respectively. An edge $(u, v) \in E$ represents a relationship between node u and v. We say a node u influences node v on action a if we have $(u, v) \in E$, $< t_u, u, a >$, $< t_v, v, a > \in D$, and $t_v - t_u \leq \tau$, where τ is the time threshold. The* degree of influence *that node u has on v for action a, denoted as $p(u, v, a)$, is defined by:*

$$p(u, v, a) = \begin{cases} 0 & if \ t_v - t_u > \tau \\ e^{(t_u - t_v)} & otherwise \end{cases} \tag{1}$$

This implies that if node u performs an action, and shortly thereafter node v repeats the same action, then it is highly likely that u has an influence on v. On the other hand, if v repeats the action only after a long lapse, then we may conclude that it is an independent action and that u has little influence on v. Let the time threshold $\tau = 1$. The degree of influence that node u_1 has on u_4 for action a is $p(u_1, u_4, a) = e^{(-1)} = 0.37$.

time	user	action
0	u1, u2, u3	a
0	u1, u3	b
0	u3	c
1	u1, u3, u4, u5, u7	a
1	u1, u2, u4, u7	b
1	u1, u2, u7	c
2	u1, u2, u4, u7	a
2	u2, u3, u4, u5	b
2	u2, u3, u4, u6	c
3	u2, u3, u4, u5	a
3	u1, u2, u5, u7	b
3	u2, u6, u7	c
4	u1, u2, u5, u7	a
4	u1, u2, u3, u4, u5	b
4	u1, u2, u6	c
5	u2, u3, u4, u5	a
5	u2, u3, u4, u5, u7	b
5	u2, u3, u4, u6	c

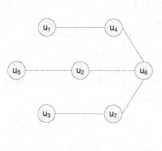

Fig. 3. Action log and user relation graph

Definition 3 (Preference Consistency). *Let A_t denote the set of actions taken by nodes u and v between the start time t_s and a given time point t. The preference of a node u for the node v is given by:*

$$Preference(u, v, t) = \sum_{a \in A_t} p(u, v, a) \qquad (2)$$

The preference consistency of node u at time point t is defined by:

$$PrefCon(u, t) = \sum_{v \in V} Preference(u, v, t) \qquad (3)$$

In the time interval [3,4], node u_5 performs action a and b following u_2, and node u_6 performs action c following u_2, so the preference of node u_2 for u_5 at time point 4 is $Preference(u_2, u_5, 4) = p(u_2, u_5, a) + p(u_2, u_5, b) = 0.74$, and the preference of node u_2 for u_6 is $Preference(u_2, u_6, 4) = p(u_2, u_6, c) = 0.37$. The preference consistency of node u_2 at time point 4 is $PrefCon(u_2, 4) = Preference(u_2, u_5, 4) + Preference(u_2, u_6, 4) = 1.11$. On the other hand, node u_4 performs action b following u_1, so the preference consistency of node u_1 at time point 4 is $PrefCon(u_1, 4) = Preference(u_1, u_4, 4) = 0.37$.

Definition 4 (Personal Consistency). *Let $M = \{m_{t_s}, \cdots, m_t\}$ be the number of actions taken by user u from the start time t_s to time point t. Let μ be the mean of M. Then the personal consistency of u at time point t is given by:*

$$PersonCon(u, t) = \frac{t - t_s + 1}{\sum_{j \in [t_s, t]} (m_j - \mu)^2} \qquad (4)$$

This is equivalent to the inverse of the standard deviation of the number of posts made by u. A higher value in $PersonCon(u,t)$ implies a smaller deviation in the number of postings over time, implying that user u is more consistent. For example, the personal consistency of node u_1 and u_2 at time point 4 is as follows.

$$PersonCon(u_1, 4) = \frac{5}{(2-2)^2+(3-2)^2+(1-2)^2+(1-2)^2+(3-2)^2} = 1.25.$$

$$PersonCon(u_2, 4) = \frac{5}{(1-2.4)^2+(2-2.4)^2+(3-2.4)^2+(3-2.4)^2+(3-2.4)^2} = 1.56.$$

Definition 5 (Overall Consistency). *The consistency of node u at time point t is defined as:*

$$Consistency(u,t) = \Theta(PrefCon(u,t), PersonCon(u,t)) \tag{5}$$

where Θ can be any function that maps the pair $(PrefCon(u,t), PersonCon(u,t))$ to a real number. In our experiment, we set Θ as the sum of the two terms.

Given the preference and personal consistency of node u_1 at time point 4, the overall consistency of u_1 is $Consistency(u_1, 4) = PrefCon(u_1, 4) + PersonCon(u_1, 4) = 0.37 + 1.25 = 1.62$.

We rank the users based on their overall consistency values at each time point.

Definition 6 (Rank). *Given a node u at time point t, let $S = \{v \in V \mid Consistency(v,t) > Consistency(u,t)\}$. Then, the rank of u at t is given by:*

$$rank(u,t) = |S|$$

Similarly, for node u_2 and u_3 we have: $Consistency(u_2, 4) = 2.67$, $Consistency(u_3, 4) = 2.3$. So the rank of node u_1, u_2 and u_3 at time point 4 is 3, 1 and 2 respectively.

Definition 7 (Volatility). *Let $\mu_{rank}(u)$ denote the mean rank of u in the query interval $[q_s, q_e]$. The volatility of node u in the interval $[q_s, q_e]$ is given by:*

$$Volatility(u) = \frac{\sum_{t \in [q_s, q_e]} (rank(u,t) - \mu_{rank}(u))^2}{q_e - q_s + 1} \tag{6}$$

For node u_1, we can get its rank at each time point in the time interval $[1,5]$. The volatility of u_1 is $Volatility(u_1) = \frac{(1-2)^2+(1-2)^2+(3-2)^2+(3-2)^2+(2-2)^2}{5} = 0.8$. Similarly, for node u_2 we have $Volatility(u_2) = \frac{(2-1.4)^2+(2-1.4)^2+(1-1.4)^2+(1-1.4)^2+(1-1.4)^2}{5} = 0.24$.

Definition 8 (Score). *The score of node u in the query interval $[q_s, q_e]$ is the weighted sum of consistency and volatility:*

$$Score(u) = w_1 * \sum_{t \in [q_s, q_e]} Consistency(u,t) - w_2 * Volatility(u), \tag{7}$$

where $w_1 + w_2 = 1$, $w_1 > 0$ and $w_2 > 0$.

Let $w_1 = w_2 = 0.5$. The score of node u_1 in the time interval $[1,5]$ is $Score(u_1)$ $= 0.5 \times 12.71 - 0.5 \times 0.8 = 5.96$. Similarly, $Score(u_2) = 0.5 \times 14.66 - 0.5 \times 0.24 = 7.21$. We can see that node u_2 is more consistent than u_1.

Problem Statement: Given an action log D, a social network graph G, a query time interval $[q_s, q_e]$, and time threshold τ, we want to identify a subset of users $U \subset V$ such that $|U| = k$ and $\forall u \in U$, $\nexists v \in V \backslash U$ such that $Score(v) > Score(u)$. We call the users in U the k-consistent influencers in G.

3   Proposed Method

Given an action log and a user relationship graph, we compute the personal and preference consistency of each user u at time point t. We compute $PrefCon(u, t)$ by examining all users who have performed the same action following u's action. If v has previously followed u and the time lapse between v's and u's actions is smaller than time threshold τ, we conclude that u's post has influenced v to some degree and this influence will be included in computing the preference consistency of node u for v according to Equation 3. Otherwise, the influence of u's post on v is said to be negligible and will be ignored.

For $PersonCon(u, t)$, we keep track of the number of posts made by user u from the start time point t_s till current time point t and obtain the variance of these numbers.

Each pair of $(PrefCon(u, t), PersonCon(u, t))$ values is a point in the personal-preference 2D space. To find users with top-k overall consistency values $Consistency(u, t)$, the naive way of sorting users by their consistency values is computationally expensive as there may be millions of users at each time point. Given that k is typically a small fraction compared to the total number of users, this is certainly not efficient.

Instead, we partition this 2D space into cells of size $\delta \times \delta$ and assign a user u to the cell $\left(\left\lfloor \frac{PrefCon(u,t)}{\delta} \right\rfloor, \left\lfloor \frac{PersonCon(u,t)}{\delta} \right\rfloor \right)$. We observe that the top-right grid has the highest overall consistency value. As we slide the black line from this top-right cell towards the bottom-left cell, the consistency values of the users in the cells will decrease. In other words, if we wish to find the top-k influencers, we only need to process the cells in the zig-zag order as shown by the arrows in Figure 4. In this manner, only the likely candidates for k-consistent users in the shaded cells are processed, resulting in great savings of computational time. The zig-zag traversal applies to any scoring function, which is monotone on all dimensions.

We map the users to the cells in a grid based on their personal and preference consistency values at time point t. Figure 5 shows the grids at the various time points.

Next, we design a function Φ to linearize the grids so that the cells can be processed in the desired zig-zag order as follows:

$$\Phi(i, j) = (N + M) - (\lfloor i \rfloor + \lfloor j \rfloor)$$

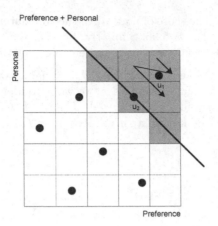

Fig. 4. Illustration of zig-zag traversal

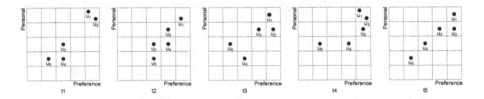

Fig. 5. Grids at different time points

where N is the maximum $\lfloor \frac{PrefCon(u,t)}{\delta} \rfloor$ value and M is the maximum $\lfloor \frac{PersonCon(u,t)}{\delta} \rfloor$ value.

Note that $\Phi(N, M) = 0$, $\Phi(N-1, M) = \Phi(N, M-1) = 1$, and $\Phi(N-2, M) = \Phi(N-1, M-1) = \Phi(N, M-2) = 2$, etc. We call this set of linearized grids the GridIndex. Figure 6 shows the GridIndex obtained from Figure 5.

Based on the GridIndex, we design an algorithm called TCI to find the top-k consistent influencers. We obtain the initial lists of the top-k candidate users at each time point from the set of linearized grids. If a candidate user u does not appear in the lists for any of the time point, say t, then we traverse the grid at t till we find u, and compute its score $Score(u)$. To handle big data, we can deploy our TCI algorithm to the MapReduce framework.

Algorithm 1 shows the details of TCI. The algorithm first scans each action log D_a backwards with a sliding window of size τ (Lines 1-6). For each tuple $< t, u, a > \in D_a$, we increment the number of posts made by user u at time point t and utilize the user relationship graph G to compute the preference of u at t.

After scanning all action logs, we compute the preference consistency $PrefCon(u,t)$ and personal consistency $PersonCon(u,t)$ for each user at each

---

Algorithm 1. TCI

Require: action log D, graph G, query interval $[q_s, q_e]$, time threshold τ, and integer k
Ensure: set of k-consistent influencers *Result*
 1: **for each** $D_a \subset D$ where D_a is a projection of D on action a **do**
 2: initialize $numPost_{u,t}$ to 0 for all u and t
 3: **for each** tuple $< t, u, a > \in D_a$ **do**
 4:   increment $numPost_{u,t}$
 5:   $V = \{v \mid <t', v, a > \in D_a, (u,v) \in G, t' \in [t+1, t+\tau]\}$
 6:   $Preference(u, v, t') + = p(u, v, a)$
 7: let G_t be the linearized grid at time t
 8: **for each** user u and time point t **do**
 9: $PrefCon(u, t) + = Preference(u, v, t)$
10: compute $PersonCon(u, t)$ from $avg(numPost_{u,t})$ and $sum(numPost_{u,t})$ using Equation 4
11: insert u to $G_t[\Phi(\lfloor \frac{PrefCon(u,t)}{\delta} \rfloor, \lfloor \frac{PersonCon(u,t)}{\delta} \rfloor)]$
12: $Result \leftarrow \emptyset$
13: initialize $threshold$, $Score_{min}$ to 0 and position p to 1
14: **for** $t = q_s$ to q_e **do**
15: $ptr_t \leftarrow 0$; $candSet_t \leftarrow G_t[ptr_t]$
16: **while** $(|Result| < k$ or $threshold > Score_{min})$ **do**
17: **for each** $t \in [q_s, q_e]$ **do**
18:   **while** $|candSet_t| < p$ **do**
19:   increment ptr_t; $candSet_t = candSet_t \bigcup G_t[ptr_t]$
20: let θ_t be the consistency value of the user at p in $candSet_t$
21: $threshold = \sum_{t \in [q_s, q_e]} \theta_t$
22: let $C = \bigcup candSet_t - \bigcap candSet_t$
23: **for each** user $u \in C$ **do**
24:   let T be the set of time points that u has not appeared
25:   **for each** $t \in T$ **do**
26:   **while** $u \notin candSet_t$ **do**
27:    increment ptr_t; $candSet_t = candSet_t \bigcup G_t[ptr_t]$
28: **for each** user $u \in \bigcup candSet_t$ **do**
29:   $rank_u = 0$
30:   **for each** $t \in [q_s, q_e]$ **do**
31:   $rank_u \mathrel{+}=$ position of u in $candSet_t$
32:   $ave\_rank_u = \frac{rank_u}{q_e - q_s + 1}$
33:   compute $Volatility(u)$ using Equation 6
34:   compute $Score(u)$ using Equation 7
35: increment position p
36: **if** $|Result| < k$ **then**
37:   $Result = Result \cup \{u\}$
38: **else**
39:   let u' be the user with lowest score in $Result$ and $Score_{min} = Score(u')$
40:   **if** $Score(u) > Score(u')$ **then**
41:   $R = R - \{u'\}$; $R = R \cup \{u\}$
42: **return** $Result$

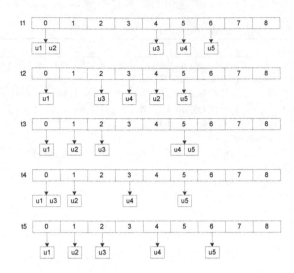

Fig. 6. GridIndex obtained from Figure 5

time point (Lines 7-10). Then we insert the users into the various linearized grids in the GridIndex (Line 11).

Once we have constructed the GridIndex, the algorithm tries to identify the set of top-k consistent influencers, *Result*. For each time point in the given query interval, we first obtain the initial list of candidate influencers (Lines 17-19). For early termination, we compute a threshold value by summing the consistency values of the candidate users at position p in each grid G_t (Lines 20-21).

For each candidate user u who does not appear in all the lists, we expand the candidate sets corresponding to the time points that u is missing from until u is included in the candidate set (Lines 22-27). When this is completed, we obtain the rank of u at all the time points and compute the volatility of u (Lines 28-33). Finally, we compute the score of u (Line 34).

If the size of the result set is less than k, we add u to R (Lines 36-37). Otherwise, we check whether the score of u is larger than that of the kth user in *Result*. If yes, we replace the kth user with u (Lines 38-41). The algorithm terminates when the size of *Result* is k and *threshold* is smaller than $Score_{min}$.

Let us illustrate how the constructed GridIndex in Figure 6 is used to find the 2-consistent influencers. The initial lists obtained for the 5 time points are shown in Figure 7(a). We observe that u_2 has not appeared in the time points t_2 to t_5, so we proceed to traverse the GridIndex at time points t_2 to t_5 to retrieve additional users till u_2 is found. Similarly, we traverse the GridIndex where u_3 has not appeared to retrieve additional users till u_3 is found. Figure 7(b) shows the updated lists.

Initially, the *threshold* and $Score_{min}$ are set to 0. We first compute the score of u_1 and get $Score(u_1) = 2.25$. At this time, *threshold* is 2.25. We continue to compute the score of users until the score of the 2nd user is larger than

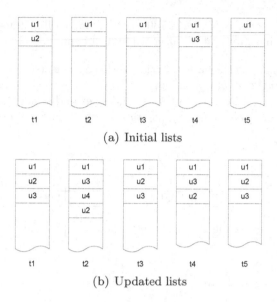

(a) Initial lists

(b) Updated lists

Fig. 7. Rank lists

the threshold. For u_2 and u_3, we have $Score(u_2) = 1.0$ and $Score(u_3) = 1.6$. We update *threshold* to 1.2. Since the current 2nd user is u_3 and $Score(u_3) > threshold$, we can be sure that the 2-consistent influencers are u_1 and u_3.

4   Experimental Evaluation

In this section, we present the results of experiments conducted to evaluate the effectiveness and efficiency of our methods. We implement all the algorithms in Java. The experiments are performed using an Intel Core 2 Quad CPU 2.83 GHz system with 3GB of main memory and running Windows XP operating system.
    We use the following real world datasets in our experiments:

1. Citation dataset [11,12]. This dataset is part of the DBLP computer science bibliography. It contains 1,397,240 papers and 3,021,489 citation relationships between these papers. Each paper is associated with attributes such as abstract, authors, year, venue, and title, etc.
2. Flixster dataset [13]. This is a social network for movies in which users to share their opinion on movies with friends by rating and reviewing movies. The Flixster dataset has 1M users, 26.7M friendship relations among users, and 8.2M ratings that range from half a star (rating 0.5) to five stars (rating 5). On average each user has 27 friends and each user has rated 8.2 movies.
3. Twitter dataset [9,10]. This dataset consists of 476 million tweets published by 20 million users over a 7 month period from June 1 2009 to December 31 2009. To make our experiments manageable, we use a subset of the Twitter

dataset, which consists of 17,214,780 tweets from 1,746,259 users. Each tweet has the following information: user, time and content.

Table 1 summarizes the characteristics of these datasets. In our experiments, we set the query interval $[q_s, q_e]$ to be the whole period of the datasets. The grid size is set to 10×10. We can try different grid sizes and select the one that gives the best performance. The default value for weight w_1 and w_2 is 0.5 respectively.

Table 1. Dataset statistics

Datasets	# Nodes	# Edges	Avg Edges	Max Edges
Citation	1,397,240	3,021,489	2.16	4,090
Flixster	1M	26.7M	26.70	1,045
Twitter	1,746,259	92,286,461	52.85	241,428

4.1   Efficiency Experiments

We first evaluate the efficiency of TCI. For comparison, we also implement TCI-NoGrid, a variant of TCI that does not utilize the GridIndex structure. TCI-NoGrid sorts all users by their consistency values at each time point to obtain their ranks. Then it retrieves candidate users from the rank lists at each time point and computes their scores. If a retrieved user u does not appear in the lists for all time points, TCI-NoGrid will retrieve the rank lists where u does not appear to find u.

We vary the size of the action logs from 100k to 900k, and set $k = 5$. For the Citation dataset, we set τ to 10 years. For the Flixster dataset, $\tau = 10$ days. For the Twitter dataset, τ is set to 10 hours.

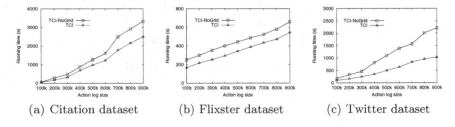

(a) Citation dataset   (b) Flixster dataset   (c) Twitter dataset

Fig. 8. Runtime of TCI for varying action log size

Figure 8 shows the runtime for TCI and TCI-NoGrid for the three datasets. We observe that TCI outperforms TCI-NoGrid, and the gap widens as the action log size increases. This demonstrates that the grid index is effective in reducing the runtime. For the Flixster dataset, the grid index is not very beneficial. This is because the ranks of users in Flixster dataset vary greatly.

4.2   Sensitivity Experiments

We also examine the effect of the parameters k and τ on the performance of TCI and TCI-NoGrid. We fix the size of the action log at 100k, and vary k from 5 to 25. Figure 9 shows the runtime for both methods. We observe that the runtime does not change much as k increases. This is because both algorithms have to scan the action log, the time of which dominates the total running time.

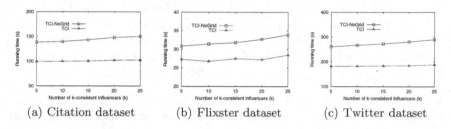

(a) Citation dataset   (b) Flixster dataset   (c) Twitter dataset

Fig. 9. Effect of varying k

Next, we set the number of consistent influencers k to 5, action log size to 100k and vary the time threshold τ from 10 to 50. Increasing τ is equivalent to increasing the search space, i.e. the number of potential consistent influencers. Figure 10(a) shows that the runtime for both algorithms slightly increases as τ increases on the Citation dataset. Similar trend is observed for the Flixster dataset (see Figure 10(b)). However, both algorithms are sensitive on the Twitter dataset, as can be seen in Figure 10(c). This is because the Twitter dataset is "dense", which means in a very short time interval hundreds or thousands of tweets are posted.

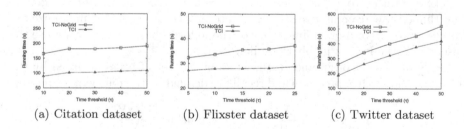

(a) Citation dataset   (b) Flixster dataset   (c) Twitter dataset

Fig. 10. Effect of varying τ

4.3   Effectiveness Experiments

In this section, we demonstrate how the proposed k-consistent influencers is useful for two tasks:

1. Identifying information sources [14,15]. Identifying information sources is useful for user recommendation. A social network user who is interested in receiving information about a particular topic would subscribe to the information sources for the same topic in order to receive up-to-date and relevant information.
2. Finding experts [20,21]. Expert finding aims to find persons who are knowledgeable on a given topic. It has many applications in expertise search, social networks, recommendation and collaboration.

We use the Twitter dataset for the first task and the Citation dataset for the second task. We use the public dissemination accounts in Twitter (e.g. @Yahoo) as the ground truth, provided that these accounts performed actions in the Twitter dataset. For Citation dataset, we use the ground truth given in [16].

We compare the TCI algorithm with the following methods:

1. TES [19]. TES is designed to answer durable top-k queries. By exploiting the fact that the changes in the top-k set at adjacent time points are usually small, TES indexes these changes and incrementally computes the snapshot top-k sets at each time point of the query window.
2. Greedy [4]. The greedy algorithm finds k influential nodes such that the expected number of nodes influenced by these k nodes is maximized [2,3,5]. At each iteration, the greedy algorithm selects a node that leads to the largest increase in the number of nodes influenced. The algorithm stops when k nodes are selected.
3. Follower-based. Given the following relationships between users, the follower-based method returns the k users with the largest number of followers.

For each method, we determine the top-k users. We first apply the TCI algorithm on the action log to obtain the top-k consistent influencers. For the TES algorithm, we first compute the consistency value of each user at each time point and construct the rank lists based on their consistency values. Then we run the TES algorithm to get the top-k users from the rank lists. We run the greedy algorithm on the given user relation graph to find the k users that maximize the expected number of users influenced. For the follower-based method, we use the k users with the largest number of followers.

Let X be the set of ground truth, let Y be the set of users returned by the various methods, then precision and recall are defined by the following equations:

$$precision = \frac{|X \cap Y|}{|Y|} \qquad recall = \frac{|X \cap Y|}{|X|} \tag{8}$$

Figure 11 shows the precision and recall for finding information sources on Twitter dataset as we vary k from 5 to 25. We observe that the precision of TCI outperforms that of TES algorithm, the greedy algorithm and the follower-based method for all values of k. The recall for all four methods increases as k increases. Further, the gaps in recall widen as k increases. This is because all the methods will predict more information sources with the increase of k, leading to better recall.

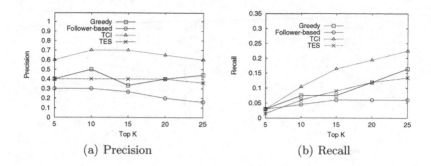

(a) Precision   (b) Recall

Fig. 11. Effectiveness of finding information sources on Twitter dataset

Figure 12 shows the precision and recall of the various methods for finding data mining experts in the Citation dataset. Again, the precision of TCI algorithm outperforms the other three methods, especially when k is large. The recall for all four methods increases as k increases, because all the methods will find more experts with larger k. Further, the gaps in recall widen as k increases. Similar results and trends are observed for information retrieval experts as shown in Figure 13.

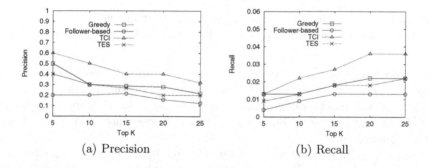

(a) Precision   (b) Recall

Fig. 12. Effectiveness of finding data mining experts in Citation dataset

Table 2 shows the top-5 experts on data mining and information retrieval returned by our TCI method. Among the results, some well-known authors, such as Jiawei Han and Christos Faloutsos (Data Mining), Bruce Croft and Ricardo Baeza-Yates (Information Retrieval), are all ranked among the top-5 experts. This is because these commonly ranked authors are not only highly cited, but also in the top at each time point. In our setting, high citation counts means high consistency, and high rank at each time point means little volatility. Hence, the score values of these authors are likely to be high, making them among the top-5 results.

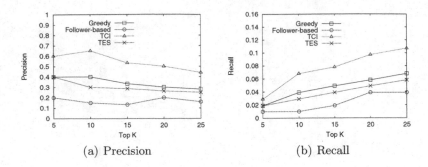

(a) Precision   (b) Recall

Fig. 13. Effectiveness of finding information retrieval experts in Citation dataset

Table 2. Top-5 experts on data mining and information retrieval

Data Mining		Information Retrieval	
consistency + volatility	consistency	consistency + volatility	consistency
Jiawei Han	Jiawei Han	Bruce Croft	Bruce Croft
Christos Faloutsos	Philip S. Yu	Ricardo Baeza-Yates	Gerard Salton
Philip S. Yu	Christos Faloutsos	Chengxiang Zhai	Oded Goldreich
Vipin Kumar	Mohammed J. Zaki	Anil K. Jain	Michael I. Jordan
Mohammed J. Zaki	Rakesh Agrawal	H. Garcia	Christopher D. Manning

5   Related Work

In this section, we review and summarize works that are most relevant to our research. These include works in social influence analysis and top-k query processing.

Social Influence Analysis. Research on influence analysis has focused on validating the existence of influence [1], studying the maximization of influence spread in the whole network [4], modeling direct influence in homogeneous networks [7], mining topic-level influence on heterogeneous networks [6], and conformity influence [8].

Tang et al. [7] introduce the problem of topic-based influence analysis and present a method to quantify the influential strength in social networks. In [6], Liu et al. introduce a probabilistic model for mining direct and indirect influence between the nodes of heterogeneous networks. They measure influence based on the clearly observable "following" behaviors and study how the influence varies with number of hops in the network. Thus far, the influence analysis is based on the observed behavior at a given snapshot. Yet, consistent behavior is an important factor that has not been taken into account by these works.

Top-k Query Processing. Fagin et al. [17] introduce the TA algorithm for computing the top-k queries over multiple sources, where each source provides a ranking of a subset of attributes only.

Lee et al. [18] were the first to study consistent top-k query. They construct a RankList for each time series to store the rank information. During query processing, they traverse the list of each time series and search for entries with rank values greater than k. The process terminates whenever an entry in the list with rank value greater than k is encountered. Wang et al. [19] proposed an efficient method called TES for durable top-k queries. TES exploits the fact that the changes in the top-k set at adjacent time stamps are usually small. TES indexes these changes and incrementally computes the snapshot top-k sets at each time stamp of the query window.

These works assume the scores are precomputed at each time point and do not consider the influence between users. However, we need to compute the score of each user at each time point from an action log. In our setting, the ranked lists correspond to users who are high in consistency values. Yet, these users may not have high scores if their rank positions differ vastly at different time points. To account for this, our proposed algorithm dynamically computes the total score that combines consistency and volatility, and outputs the k-consistent users.

6   Conclusion

Social influence plays a key role in many social networks, e.g., Facebook, Twitter and YouTube. In this paper, we introduce the notion of k-consistent influencers, and propose an efficient approach to identify them in a social network. We define the consistency of a node based on its influence and volatility over time. With the help of grid index, we develop an algorithm called TCI to obtain the k-consistent influencers given a time period. We conduct extensive experiments on three real world datasets to evaluate the proposed method. We also demonstrate the applicability of k-consistent influencers in identifying information sources and finding experts. The experimental results demonstrate the efficiency and effectiveness of our approach.

References

1. Agarwal, N., Liu, H., Tang, L., Yu, P.S.: Identifying the influential bloggers in a community. In: WSDM, pp. 207–218 (2008)
2. Chen, W., Wang, C., Wang, Y.: Scalable influence maximization for prevalent viral marketing in large-scale social networks. In: KDD, pp. 1029–1038 (2010)
3. Chen, W., Wang, Y., Yang, S.: Efficient influence maximization in social networks. In: KDD, pp. 199–208 (2009)
4. Kempe, D., Kleinberg, J., Tardos, É.: Maximizing the spread of influence through a social network. In: KDD, pp. 137–146 (2003)
5. Leskovec, J., Krause, A., Guestrin, C., Faloutsos, C., VanBriesen, J., Glance, N.: Cost-effective outbreak detection in networks. In: KDD, pp. 420–429 (2007)
6. Liu, L., Tang, J., Han, J., Jiang, M., Yang, S.: Mining topic-level influence in heterogeneous networks. In: CIKM, pp. 199–208 (2010)
7. Tang, J., Sun, J., Wang, C., Yang, Z.: Social influence analysis in large-scale networks. In: KDD, pp. 807–816 (2009)
8. Tang, J., Wu, S., Sun, J.: Confluence: conformity influence in large social networks. In: KDD, pp. 347–355 (2013)

9. Yang, J., Leskovec, J.: Patterns of temporal variation in online media. In: WSDM, pp. 177–186 (2011)
10. Kwak, H., Lee, C., Park, H., Moon, S.: What is twitter, a social network or a news media?. In: WWW, pp. 591–600 (2010)
11. Tang, J., Zhang, J., Jin, R., Yang, Z., Cai, K., Zhang, L., Su, Z.: Topic Level Expertise Search over Heterogeneous Networks. Machine Learning Journal **82**(2), 211–237 (2011)
12. Tang, J., Zhang, J., Yao, L., Li, J., Zhang, L., Su, Z.: Arnetminer: extraction and mining of academic social networks. In: KDD, pp. 990–998 (2008)
13. Jamali, M., Ester, M.: A matrix factorization technique with trust propagation for recommendation in social networks. In: RecSys, pp. 135–142 (2010)
14. Canini, K.R., Suh, B., Pirolli, P.L.: Finding credible information sources in social networks based on content and social structure. In: SocialCom, pp. 1–8 (2011)
15. Mahata, D., Agarwal, N.: What does everybody know? identifying event-specific sources from social media. In: CASoN, pp. 63–68 (2012)
16. Zhang, J., Tang, J., Li, J.: Expert finding in a social network. In: Kotagiri, R., Radha Krishna, P., Mohania, M., Nantajeewarawat, E. (eds.) DASFAA 2007. LNCS, vol. 4443, pp. 1066–1069. Springer, Heidelberg (2007)
17. Fagin, R., Lotem, A., Naor, M.: Optimal aggregation algorithms for middleware. In: PODS, pp. 102–113 (2001)
18. Lee, M.L., Hsu, W., Li, L., Tok, W.H.: Consistent top-k queries over time. In: Zhou, X., Yokota, H., Deng, K., Liu, Q. (eds.) DASFAA 2009. LNCS, vol. 5463, pp. 51–65. Springer, Heidelberg (2009)
19. Wang, H., Cai, Y., Yang, Y., Zhang, S., Mamoulis, N.: Durable Queries over Historical Time Series. IEEE TKDE **26**(3), 595–607 (2014)
20. Balog, K., Azzopardi, L., de Rijke, M.: Formal models for expert finding in enterprise corpora. In: SIGIR, pp. 43–50 (2006)
21. Zhu, J., Song, D., Rger, S., Huang, X.: Modeling document features for expert finding. In: CIKM, pp. 1421–1422 (2008)

Industrial Papers

Analyzing Electric Vehicle Energy Consumption Using Very Large Data Sets

Benjamin Krogh[✉], Ove Andersen, and Kristian Torp

Department of Computer Science, Aalborg University,
Selma Lagerlöfs vej 300, Aalborg, Denmark
{bkrogh,xcalibur,torp}@cs.aau.dk

Abstract. An electric vehicle (EV) is an interesting vehicle type because it has the potential of reducing the dependence on fossil fuels by using electricity from, e.g., wind turbines. A significant disadvantage of EVs is a very limited range, typically less than 200 km. This paper compares EVs to conventional vehicles (CVs) for private transportation using two very large data sets. The EV data set is collected from 164 vehicles (126 million rows) and the CV data set from 447 vehicles (206 million rows). Both data sets are collected in Denmark throughout 2012, with a logging frequency of 1 Hz. GPS data is collected from both vehicle types. In addition, EVs also log the actual energy consumption every second using the vehicle's CAN bus. By comparing the two data sets, we observe that EVs are significantly slower on motorways, faster in cities, and drive shorter distances compared to CVs. Further, we study the effects of temperature, wind direction, wind speed, and road inclination. We conclude that the energy consumption (and range) of an EV is very sensitive to head wind, low temperatures, and steep road inclinations.

1 Introduction

The electric vehicle (EV) type is gaining traction as an alternative to the conventional vehicle (CV) with an internal combustion engine. The EV has the potential of lowering the greenhouse gas emissions and reducing the dependence on fossil fuels. Furthermore, the EV is an interesting vehicle type because it has a set of new features, such as energy recuperation, close to ideal speed-torque profile, and zero tail-pipe emissions [11].

EU has a goal of reducing the CO_2 and other greenhouse gas emissions by 80% by 2050, compared to 1990 levels [9]. Because the transportation sector is responsible for approximately 30% of all CO_2 emissions [4], this sector has to adapt new energy sources. Transportation by EV is seen as one of the technological solutions that will help reach this goal. Especially, as more of the electricity is generated from renewable energy sources such as solar panels and wind turbines.

Although the EV has a number of advantages over the CV, it also has some significant drawbacks. The major drawback is the limited range of EVs compared to CVs. A CV usually has a range of 500-600 km [27], but the range of an EV is usually between 150 and 200 km [7,15,20,23,24]. Furthermore, a CV can be

© Springer International Publishing Switzerland 2015
M. Renz et al. (Eds.): DASFAA 2015, Part II, LNCS 9050, pp. 471–487, 2015.
DOI: 10.1007/978-3-319-18123-3_28

refilled with gasoline in a few minutes whereas it may take hours to recharge the battery in an EV [18]. However, only limited research has compared the usage of the two vehicle types for private transport, and not using very large data sets. Such a comparison is required in order to examine how well an EV satisfy the transportation needs of families and how restrictive the range limitations are.

The main contributions of this paper are a detailed comparison of the driving patterns for EVs and CVs, and a thorough analysis of the energy consumption of EV driving patterns. The work is based on two high-resolution GPS data sets (1 Hz) from a fleet of 164 EVs and a fleet of 447 CVs, recorded throughout 2012. We study the effects of the limited range of EVs by analyzing and comparing the average length of trajectories and daily driven distance of both EVs and CVs.

The EVs log the actual energy consumption using the vehicle Controller Area Network Bus (CAN Bus) [1]. The combination of the GPS and CAN Bus data is used to analyze the energy consumption of EVs with respect to road inclination, temperature, wind direction, wind speed, season, and vehicle speed. These analyses are possible due to state-of-the-art data processing of moving object data, and sophisticated data integration with fine-grained meteorological data from weather stations. This paper is an extended version of [13].

The rest of the paper is organized as follows. Sect. 2 describes the two data sets used, Sect. 3 presents our method for comparing and evaluating the energy consumption, Sect. 4 presents the results, Sect. 5 reviews the related work, and Sect. 6 concludes the paper.

2   Data Foundation

The EV data set used is from the project "Test en Elbil" ("Try an Electric Vehicle") [8] and collected from January to December 2012. In this project, families in Denmark could try an EV as the main household vehicle for a period of three to six months. A fleet of 164 vehicles were used in the project, consisting of 33 Citroën C-Zero [7], 56 Mitsubishi i-MiEV [15], and 75 Peugeot iOn [23]. The three vehicle types are produced by the same manufacturer and are practically identical, i.e., the same body, weight, 16 kWh battery, and 47 kW (63 hp) engine. The vehicles are therefore treated identically in the following analyses.

126.5 million records in total were collected from the 164 EVs, with a total driven distance of 1.4 million km, and 159 862 trajectories (or trips). An EV record consists of GPS information and other data from the EV. The GPS data include the location, altitude, direction, speed, and time-stamp. The EV data include State of Charge (SoC), charging status, and odometer speed. The SoC is collected from the EVs CAN Bus [1]. All EVs log these parameters with a frequency of 1 Hz.

We compare the EV data set to a large data set from CVs. 205.6 million records were collected from 447 vehicles, with a total driven distance of 3.4 million km, and 187 303 trajectories. This data set is collected in the "ITS Platform" project [2]. Records are logged with a 1 Hz frequency, during the period January to December in 2012 (only limited data in January and February). The EV

Table 1. Overall Statistics for Data Sets (left) and Road Network Coverage (right)

	EV	CV		Edges	EV Covered (%)	CV Covered (%)
Vehicles	164	447	Motorway	2226	2111 (95)	2187 (98)
GPS records	126.5M	205.6M	Primary	22 175	14 798 (67)	19 985 (90)
Trajectories	159 862	187 303	Secondary	53 271	38 274 (72)	39 020 (73)
Total km	1.4M	3.4M	Residential	568 307	82 799 (15)	59 383 (10)

and CV data sets are thus very similar, with the exception that the CV data set contains only GPS information and no fuel consumption data. Tab. 1 (left) summarizes the overall statistics for the EV and CV data sets.

Tab. 1 (right) shows the coverage of the most important road categories. The *Edges* column shows the total number of edges in each category as defined in the road network [3]. The *EV Covered* and *CV Covered* columns show the number of edges that the EV and CV data sets cover, respectively. Tab. 1 shows that the data sets covers most of the important road-network infrastructure in Denmark. That is, the EV and CV data sets cover 95% and 98% of the motorways, respectively, and 67% and 90% of the primary roads, respectively. Note that, motorway edges are generally 10 times longer than other edges. For instance, the total length of the motorway network in Denmark is 2428 km, whereas the total length of the primary network edges is 2841 km.

3   Method

In this section, we describe the notations used, the approach for performing a comparative analysis of the two data sets, and the processing of the data in order to conduct our study.

The road network is modeled as a directed graph $\mathbf{G} = (\mathbf{V}, \mathbf{E})$, where $\mathbf{V}$ is a set of vertices and $\mathbf{E}$ is a set of directed edges $\mathbf{E} \subseteq \mathbf{V} \times \mathbf{V}$. The road network is from the OpenStreetMap project [3], and consists of 1.5 million directed edges.

Before the EV and CV data sets can be compared, it is necessary to map-match the location updates to the road network, and convert the location updates into a network-constrained representation. Further, the set of location updates from each vehicle needs to be divided logically into network constrained trajectories.

A trajectory t is a sequence of network constrained location updates $t = [m_1, m_2, \ldots, m_n]$. Each m_i is a tuple $(e_{id}, time_{enter}, time_{leave}, SoC)$, where e_{id} is the id of the network edge, $time_{enter}$ and $time_{leave}$ are the times at which the moving object entered and left edge e_{id}, respectively. Both $time_{enter}$ and $time_{leave}$ are linearly interpolated between two location updates. $time_{enter}$ is linearly interpolated between the location update prior to entering e_{id} and the first location update on e_{id}. $time_{leave}$ is interpolated between the last location update on e_{id} and the immediately following location update. Finally, SoC shows the current state of charge on the battery in percent. A $SoC = 100$ indicates that the battery is fully charged, whereas $SoC = 0$ indicates that the battery is

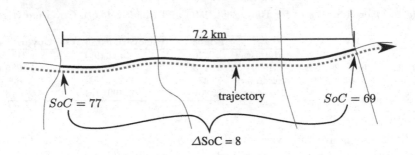

Fig. 1. Path-based Measurements of SoC

completely discharged. The *SoC* value is collected from the EV batteries through the CAN Bus. A trajectory describes the movement of a vehicle during the course of one trip, e.g., driving from home to work. A new trajectory is created when there is a temporal gap of more than three minutes between consecutive location updates from a vehicle. This gap allows for small outages in the recording of location updates, e.g., when going through a tunnel.

The algorithm in [19] is used to map-match both data sets. A very high map-matching accuracy is reported in [19] for moving object data with a logging frequency higher than 0.2 Hz. Since both data sets used in this paper are recorded with a logging frequency of 1 Hz, this map-matching algorithm is very well-suited. 91% and 93% of all records are map-matched for EVs and CVs, respectively. The reason that a lower percentage of records are map-matched for EVs is that EVs sometimes continue to log when parked, and that these records are ignored by the algorithm.

To study energy consumption, we convert the change in SoC into the corresponding energy consumption. According to the company CLEVER[1] that collected the EV data, a change in SoC of 1 percentage point (the smallest observable in the EV data set), corresponds to an energy consumption of 154 Wh.

3.1 Path Based Analysis

The change in SoC is a relatively coarse measure of energy consumption. For this reason, we only use SoC to compute the energy consumption of trajectories on paths that are 3 km or longer, see the details in Sect. 4.4. In many cases, it is useful to select a specific path through the road network, and study vehicles on this path throughout the year. In the rest of the paper, we refer to this approach as *path-based analysis*. In a path-based analysis, we select a path through the road network and retrieve the trajectories that strictly follow this path, i.e., have no detours or stops. To find the trajectories that strictly follow a path the approach described in [14] is used.

The strategy for deriving energy consumption data using path-based analysis is visualized in Fig. 1. The solid gray lines show a road network, and the thick

---

[1] Private email communications with data provider.

solid line is the selected path. One trajectory (the dotted line) strictly follows this selected path, i.e., enters the solid path at its beginning and follows the path until its end without any detours. If we assume that the SoC of this trajectory is 77% and 69% when entering and leaving the path, respectively, the energy consumed is 77% - 69% = 8 percentage points, or 1.2 kWh (8 x 0.154 kWh). Because the path is 7.2 km long, the vehicle has an average energy consumption of 166 Wh per km throughout the path.

Note that, on Fig. 1, the trajectories can contain location updates on edges both before and after the path selected for analysis, i.e., only the sub trajectory on the path being analyzed is considered. In addition, note that the selected path may consist of any number of edges, turns and intersections, and that the trajectory may start and end anywhere in the road network, as long as it at some point enters the selected path and strictly follows this path to its end.

By applying the strategy shown in Fig. 1 we can study the energy consumption, because we can choose paths long enough to overcome the coarse granularity of the SoC.

3.2   Weather Measurements

In Sect. 4.4 and Sect. 4.5, the effects of temperature and wind on the energy consumption of EVs are analyzed. To evaluate these effects, we have integrated detailed weather data from NOAA [21] and annotated all location updates with this weather data. 73 official weather stations covered Denmark in 2012 with one measurement every hour for each station, seven days a week.

Each weather record includes information on temperature, wind direction, and wind speed. Each location update from the EVs is matched to the nearest station using the Euclidean distance. If the nearest station does not have any records within the same hour (weather records are sometimes missing), the second nearest station is selected, and so on. With this approach, 90% of the location updates are matched to a station within 26 km, and 99% to a station within 36 km. The average distance from a location update to the nearest weather station with a temporally matching weather record is 13.3 km. Denmark is relative flat and weather conditions do therefore not vary significantly over these short distances. The weather measurements from the nearest weather station can therefore be used, with reasonable accuracy, to examine the effects of temperature and wind speed on the energy consumption when driving an EV.

3.3   Wind Speed and Direction Analysis

To examine the effects of the wind direction, we use the angular difference between the wind direction and the GPS direction of the EV. If the difference is below a given threshold, β, the location update is classified as either head-wind or tail-wind. If the angle is larger than β, the location update is classified as cross-wind. Fig. 2 illustrates this classification. The blue dotted line represents a trajectory and the blue arrow the direction of a location update being examined. Using the direction of the location update, we create two fans with the angle

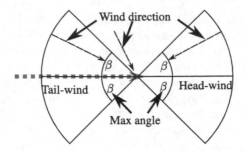

Fig. 2. Head, Tail, or Cross Wind Classification

2β: one backward fan (left) and a forward fan (right). If the angle between the wind direction and the direction of the location update (the vehicle) is less than β, the location update is marked as either tail-wind or head-wind, depending on which direction the wind comes from. As an example, in Fig. 2 we show three wind directions using dashed arrows. The wind direction represented by the leftmost arrow is classified as tail-wind and the wind direction represented by the rightmost arrow is classified as head-wind. The wind direction represented by the middle arrow is classified as cross-wind.

To analyze the effects of wind speed, we further classify location updates into the classes shown in Tab. 2. A location update is classified as H1, for a given angle β, if and only if the wind has an angle of attack less than β, a wind speed between 1 and 5 m/s, and the wind is head-wind. Most location updates have a wind speed within either H1, H2, H3, T1, T2, or T3. However, 2.3% of all location updates have a wind speed of 0 m/s, and 0.08% has a wind speed above 15 m/s. Because the EV data set has only limited data with a wind speed of zero or above 15 m/s, H0, T0, H4 and T4 are not used in any analysis, but included for completeness. Classifications for cross-wind are not included in Tab. 2, because we are only interested in studying head-wind and tail-wind.

Table 2. Wind Classifications

	0 m/s	1-5 m/s	6-10 m/s	11-15 m/s	16- m/s
Head-wind	H0	H1	H2	H3	H4
Tail-wind	T0	T1	T2	T3	T4

Recall that the SoC value is a relatively coarse measure that needs to be observed over longer distances. As such, the classification of individual location updates is not immediately useful for studying the energy consumption. To overcome this challenge, we classify entire trajectories based on the location updates in the trajectory. If more than a fraction, α, of the location updates of a trajectory have a specific classification the trajectory adopts this classification.

We can then study how a varying angle β, $0 \leq \beta \leq 90$ degrees, affects the average energy consumption of trajectories, how a larger fraction of location updates, α, with head-wind or tail-wind affects the energy consumption, and how the wind-speed affects the energy consumption, see Sect. 4.5.

4 Results

We first study and compare overall statistics of both EV and CV trajectories. We then compare the average speed of both EVs and CVs on different types of road network infrastructure throughout the year. Finally, we analyze the energy consumption with respect to environmental factors such as season, temperature, road inclination, wind speed, and wind direction.

4.1 Trajectory Comparisons

A significant challenge for EVs is the limited range compared to CVs. For instance, the vehicles in the EV data set have a specified maximum range of 160 km [15]. CVs typically have a range of 500-600 km [27]. Additionally, an EV may require several hours to recharge, whereas CVs can be refueled within a few minutes. To see whether the lower range of EVs affects the individual trajectory, we compare the relative frequency of trajectories of a certain length.

Fig. 3a shows the relative frequency of trajectories of a specific extent. From this figure it can be observed that most trajectories from both the EV and CV data sets are short. In fact, 99% and 90% of trajectories are shorter than 50 km for EVs and CVs, respectively. CVs have relatively fewer trajectories shorter than 20 km, and relatively more trajectories longer than 30 km. Overall however, the two data sets appear to have comparable trajectory lengths, which suggests that the limited range of these EVs only to a limited extend affects the individual trajectory.

Fig. 3c shows the specific energy consumption per trajectory for EVs. Note that the y-axis is logarithmic. Most trajectories (93%) use less than 4 kWh, which is less than 25% of the battery capacity. Thus, in most cases the battery has more than enough capacity to complete the individual trajectory.

Although the EV trajectories appear to be only slightly affected by the limited range, this does not mean that the EVs satisfy the transportation requirements of users. Indeed, one may argue that an EV does not satisfy the transportation requirements, if the EV can get you to work but not back again. To analyze this, we compare the total driven distance per day of both EVs and CVs. Fig. 3b shows the relative frequency of days with a certain total travelled distance for both EVs and CVs. From Fig. 3b we observe that EVs drive significantly shorter daily distances than CVs. More than 99% of all days for EVs have a total travel distance of less than 160 km (the specified range of the EV). For CVs, only 86% of all days travel less than 160 km. As such, on 14% of all days the limited range of EVs cannot satisfy the transportation needs of families without recharging during the day.

Fig. 3d shows the number of days with a specific total energy consumption per vehicle (y-axis logarithmic). Note that it is possible to exceed the 16 kWh capacity of the battery by recharging the battery during the day. From Fig. 3d

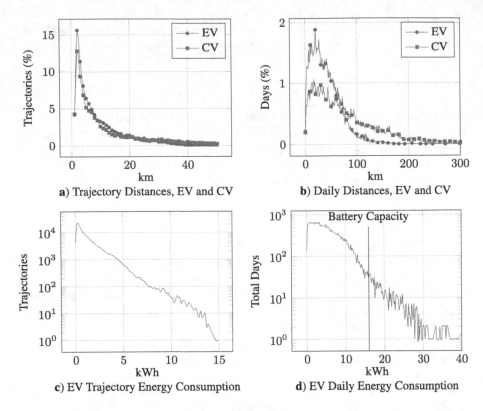

Fig. 3. Statistics per Trajectory and per Day

we observe that the battery capacity is usually sufficient for the transportation needs for a single day, without recharging. For 90% of all days, less than 73% of the battery capacity is used. The usage patterns of these EVs therefore only rarely require recharging during the day.

4.2 Speed Comparisons

In Fig. 4a and Fig. 4b, we analyze the speed of EVs and CVs throughout the year on motorways and residential roads, respectively. In both figures, the x-axis shows the month of the year, and the y-axis shows the average speed. The overall speed distributions of EVs and CVs on motorways with 130 km/h speed limits (maximum and default for motorways in Denmark) and on residential roads are shown on Fig. 4c and Fig. 4d, respectively. To identify the road network edges in each of these categories, we used the road categories motorway and residential from OpenStreetMap [3].

To ensure that the results are comparable, we include only location updates on edges in the road network that are touched by both the EV and CV data in each month. This is also the reason that only 10 months are shown on both

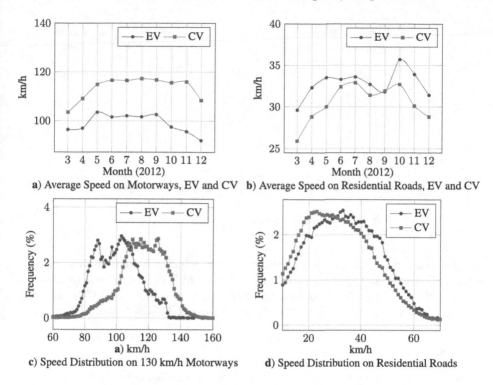

a) Average Speed on Motorways, EV and CV b) Average Speed on Residential Roads, EV and CV

c) Speed Distribution on 130 km/h Motorways   d) Speed Distribution on Residential Roads

Fig. 4. Highway and Residential Road Speeds

figures, since only limited data from CVs is available for the first two months of 2012.

Fig. 4a shows the average speed of EVs and CVs on motorways with 130 km/h speed limits. We observe that the EVs drive significantly slower than CVs on motorways. On edges with 130 km/h speed limits, EVs are between 7 and 20 km/h slower than EVs. We note that the top speed of the EVs is limited to 130 km/h [17], thus the EVs should be able to maintain the average speed of CVs. We assume this difference between EVs and CVs is because EVs lower their speed to increase range. The EVs continuously report the expected range to the driver based on the energy consumption of the last 25 km driven [18]. Since a high speed significantly increases the energy consumption, see Fig. 5b, the vehicles will report a significantly lower expected range when driving with a high speed. The speed distribution diagram on Fig. 4c confirms this theory. On motorway edges with a 130 km/h speed limit, the speed distribution of EVs corresponds to the speed distribution of CVs reduced by 20 km/h.

Fig. 4b shows the speeds of EVs and CVs on residential roads. Surprisingly, the EVs are slightly faster than the CVs. We assume this is because EVs accelerate more quickly at low speeds than CVs, partly due to differences in transmission and engine torque. Most CVs in Denmark have a manual transmission whereas the EVs have a single speed transmission. Further, the speed-torque

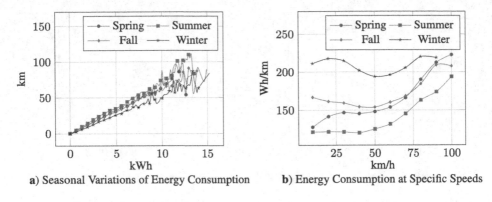

a) Seasonal Variations of Energy Consumption   b) Energy Consumption at Specific Speeds

Fig. 5. Seasonal Variations in Energy Consumption

profile of electric engines is close to the ideal [11], which results in higher acceleration at low speeds. The overall speed distribution of EVs and CVs on residential roads is shown in Fig. 4d. Note that the distributions for EVs and CVs are very similar. However, in contrast to Fig. 4c, the speed of EVs is higher than CVs.

4.3   Seasonal Variations

Denmark has significant seasonal variations in the weather over the four seasons. In the winter season, it is generally necessary to heat the cabin, whereas in the summer it is necessary to cool the cabin. The EVs use the battery for both heating/cooling the cabin and driving [16]. The range of EVs is therefore affected by the outside temperature. The magnitude of this effect is examined in the following.

Fig. 5a shows the relation between the energy consumed by a trajectory, and the length of the trajectory. We divide the set of trajectories into four groups, one for each season, and compute the average length of trajectories that used a specific amount of energy. The fluctuations above 8 kWh are due to a limited number of trajectories that consume more than 8 kWh.

From Fig. 5a we observe that the energy consumption varies significantly over the seasons. For trajectories that use less than 8 kWh, the driven distance per kWh is always less in winter than in the summer. The energy consumption in winter is approximately 20% higher than in the summer. The difference in energy consumption between summer and spring/fall is between 5% and 10%.

Based on Fig. 5a we compute the average energy consumption per kilometer and the resulting range of the EVs for each season. Tab. 3 shows the results.

Fig. 5b shows the average energy consumption as a function of the average speed of trajectories. Four series are shown, one for each season. Each trajectory included has a minimum length of 5 km.

Fig. 5b shows a significant increase in energy consumption as the average speed approaches 100 km/h. For instance, there is a 47% increase in energy consumption per km between having an average speed of 60 km/h and 100 km/h

Table 3. Energy Consumption/km and Range

Season	Consumption (Wh/km)	Range (km)
Winter	203	76
Spring	151	102
Summer	130	118
Fall	159	97

in summer. The winter series appears to decrease Wh/km until reaching average speeds of 50 km/h. We assume that this is because the heating system in the EVs uses a significant amount of energy (up to 5 kW [16]). The difference in average energy consumption between a speed of 60 km/h in winter and summer is 3.9 kW, and is reasonably explained by the heating system's energy consumption. The difference in average energy consumption between a speed of 60 km/h in winter and summer is 3.9 kW. This suggests that the least energy consuming path varies significantly throughout the seasons, i.e., an average speed of 40 km/h is most energy efficient in summer, whereas an average speed of 50 km/h is energy efficient in winter.

Based on Fig. 5b we compute the energy consumption per km for EVs as a function of season and speed. Tab. 4 shows the results. All units are in Wh/km. There are no trajectories with an average speed of 100 km/h in the winter season. The energy consumption is therefore omitted for this entry.

Table 4. Wh/km Seasons and Speed

	20 km/h	40 km/h	60 km/h	80 km/h	100 km/h
Winter	217	201	196	220	-
Spring	141	145	154	190	222
Summer	121	120	131	163	193
Fall	161	154	160	184	207

4.4 Path-Based Energy Comparisons for EVs

We now select two frequently used paths in the road network, and compare the energy usage for each of these paths for each month. The purpose is to study the effects of the outside temperature and going uphill/downhill on the energy consumption. As described in Sect. 3, we only include trajectories that strictly follow the selected paths, and compute the energy consumption per trajectory as the difference in SoC between entering and leaving the path, see Fig. 1. The two paths are the Esbjerg-Varde path (a path between two cities), and the Universitetsboulevarden (Uni. Blvd) path in Aalborg. Both paths are described in detail in [5].

The Esbjerg-Varde path is relatively flat with a difference in altitude of 10 meters between start and end. The difference in altitude between the highest and

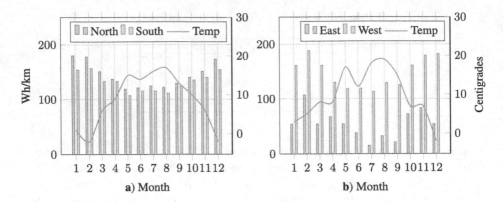

Fig. 6. Wh/km per Trajectory on Esbjerg-Varde (a), and Uni. Blvd (b)

lowest peak is 26 meters. The road inclination is between -4.7% and 4.9%, with an average inclination of -0.1% in south direction. The Uni. Blvd path has either descending (going east) or ascending (going west) altitude, with a difference in altitude of 42 meters between start and end. The average inclination of Uni. Blvd is -1.5% in east direction, and at most -5.9%.

Fig. 6a and Fig. 6b show the monthly average energy consumption per trajectory on the Esbjerg-Varde path and the Uni. Blvd path, respectively. On both figures, a green line shows the average temperature along the path. This temperature is based on the weather data annotations as described in Sect. 3. First, the average temperature of each trajectory is found, by computing the average temperature of the location updates on the selected path for the given trajectory. Then the average temperature for a month is computed by averaging the temperatures of all trajectories in the month, on the specific path.

On the Esbjerg-Varde path, Fig. 6a, we observe a clear correlation between temperature and energy consumption. There is a difference between the coldest and warmest months of up to 25% in average energy consumption. The two directions have similar energy consumptions, but the north direction always has a slightly higher energy consumption than the south direction. We believe this is due to the 10 m difference in altitude. The weight of an empty vehicle is 1070 kg [17]. Assuming a driver with a weight of 60 kg, a 10 m increase (decrease) in altitude therefore has a difference in potential energy of 31 Wh (-31 Wh) [25]. The predicted difference in energy consumption between going uphill and downhill is therefore 62 Wh. The actual difference in energy consumption between uphill and downhill in the summer months on this path is 75 Wh. We assume the remaining differences in energy consumption are due to other factors such as differences in speed when entering and leaving the path.

On the Uni. Blvd path, Fig. 6b, we observe again that energy consumption varies significantly with the temperature. Further, we observe that the inclination of the path has a significant effect on the energy consumption. The difference between going uphill and downhill on this path is up to a factor of nine in energy consumption. The 42 m increase (decrease) in altitude has a difference in potential energy of 129 Wh (-129 Wh). The predicted difference in energy consumption on this path is therefore 258 Wh. The actual energy difference between going uphill and downhill in summer is 266 Wh. These results show that there are huge variations in energy consumption due to increasing/decreasing altitude and variations in temperature. As such, the accuracy of a range estimation system can be improved by incorporating the path to be followed and the temperature forecast. In [5], we study other paths, with similar results.

4.5   Effects of Wind

We now analyze how the wind direction and wind speed affect the energy consumption. A challenge in this study is that the angle of attack, i.e., the angle between the wind direction and the vehicle direction, is an inherently *local* phenomenon, whereas measuring the energy consumption requires a distance of several kilometers.

We overcome this challenge by first classifying each location update as head-wind, tail-wind, or cross-wind as described in Sect. 3. Second, if a trajectory has more than 70% of its location updates in a specific class, e.g., H1 the entire trajectory is classified as H1. Ideally, all location updates in each trajectory would be in just one class. This is rarely the case, however, as the path of a vehicle usually includes turns, hills, changing wind directions and wind speed. As such, the 70% percentage requirement is a trade-off between isolating the effects of a specific wind class, and filtering out most trajectories. We study how varying this percentage affects the results in Fig. 8a and Fig. 8b. To cancel out effects of temperature and geography, each data-point in the following figures is an average of at least 100 trajectories.

First, we study the effects of tail-wind with varying angles, i.e., the β parameter in Fig. 2. We include only trajectories with a length of at least 5 km and measure the energy consumption per trajectory using the difference in SoC between start and end.

Fig. 7a shows the effect of tail-wind on the energy consumption as the angle β varies from 30 to 90 degrees. A low angle implies that the wind has a direction similar to the direction of the vehicle. None of the series has enough data for $\beta < 30$, i.e., there are less than 100 trajectories. For the 11-15 m/s (T3) series there is not enough data when $\beta < 50$.

As can be seen from Fig. 7a, tail-wind with a narrow angle of attack reduces the energy consumption of vehicles with approximately 5 Wh/km compared to a wider angle. There is not a significant difference between 1-5 m/s (T1) and 6-10 m/s (T2) trajectories. However, the 11-15 m/s (T3) trajectories use significantly less energy. Independently of wind speed, the energy consumption increases as the angle of attack is increased. That is, as more cross-wind is included, the benefit of tail-wind is reduced.

Fig. 7b shows the corresponding results for head-wind. We observe that for trajectories with head-wind, the wind speed has a significant effect on the energy consumption. In fact, 11-15 m/s (T3) trajectories always use more energy than

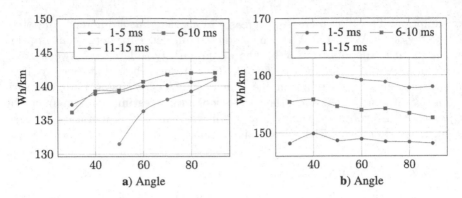

Fig. 7. Tail-wind (a) and Head-wind (b) Effects on Energy Consumption

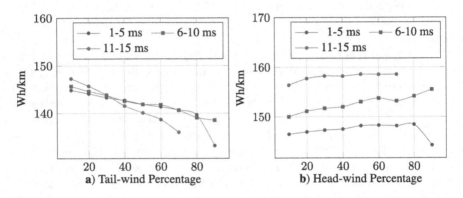

Fig. 8. Percentage of Trajectory with Tail-wind (a) and Head-wind (b)

6-10 m/s (T2) trajectories, which always use more than the 1-5 m/s (T1) trajectories. The difference in energy consumption between lowest and highest wind speed is approximately 10 Wh/km. As the angle of attack is increased, i.e., more cross-wind is included, less energy is generally consumed. Finally, by comparing Fig. 7a and Fig. 7b, we note that trajectories with head-wind always consume more energy than trajectories with tail-wind, independently of the angle and the wind-speed. The difference between H3 and T3 trajectories is up to 28 Wh/km, which corresponds to a difference in energy consumption of more than 20%.

We now study the effects of the α parameter, i.e., the percentage of location updates with tail-wind or head-wind. To do this, we use a fixed angle of attack, β, of 60 degrees. Fig. 8a shows the average energy consumption as a function of the percentage of location updates with tail-wind. When a low percentage of location updates have tail-wind, higher wind speeds result in a higher energy consumption. When a high percentage of location updates have tail-wind, higher wind speeds reduce energy consumption. Intuitively, a low percentage of location updates with tail-wind allows a high percentage of head-wind location updates.

This explains why the line for 11-15 m/s wind speed starts with the highest energy consumption and ends with the lowest energy consumption.

Fig. 8b shows the average energy consumption of trajectories as a function of the percentage of location updates with head-wind. We observe that the wind-speed classes are clearly separated for head-wind, in contrast to tail-wind. Each line shows an overall increasing trend with respect to the percentage of location updates with head-wind. The series, 1-5 m/s, has an outlier point towards the end, which we assume is due to limited data, i.e., only few trajectories have more than 90% of all location updates with head-wind. When more than 20% of the location updates of trajectories are with head-wind, the average energy consumption is always higher than the corresponding energy consumption for tail-wind in Fig. 8a.

5 Related Work

In [12], the psychological implications of the limited range of EVs are studied. 40 EVs were used in a 6-month period, after which data were collected using questionnaires and interviews. The authors then evaluated, among other things, the fraction of battery capacity most persons are comfortable utilizing. They conclude that most drivers are comfortable using between 75% and 80% of the total battery capacity.

In [22], 484 CVs, instrumented with GPS loggers, were monitored for a period of up to three years. The authors used the collected data to estimate the percentage of vehicles that could be substituted with EVs, the required range of EVs, and how many days the vehicle owner would be required to adapt to the lower vehicle range. They conclude that limited range EVs can be satisfactory for a significant fraction of the population, as long as they are willing to adapt, e.g., by recharging during the day.

Socio-technical barriers to EV adoption are studied in [10]. The data foundation is an online survey, completed by 481 persons. The authors find that the dominating concern in the sampled population is the limited range and vehicle cost.

Routing for electric vehicles [6,10,26], has proven to be challenging because the vehicle consumption model is more advanced than for CVs due to, e.g., recuperation. Many EVs generate power when going downhill, which gives some edges a negative energy consumption. This makes it harder to find the energy optimal routes, because the algorithm needs to ensure that the predicted battery charge is always between 0% and 100%. Our results suggests, that this line of research should account for forecasted environmental parameters, such as temperature and wind, and the predicted speed profile of the vehicle in order to return better results.

6 Conclusion

To the best of our knowledge, this paper presents the first large scale study of how electric vehicles (EVs) are used and which factors affect the energy consumption of the EVs. We compared the EV data set with a data set from conventional vehicles (CVs). Both data sets are collected in Denmark in 2012. Further, we integrated fine-grained weather measurements, which we used to study the effects of temperature and wind on the energy consumption.

Compared to CVs, we have found that EVs generally drive shorter distances, both in terms of the individual trajectory and the total travelled distance per day. We have shown that EVs are significantly slower on motorway, most likely because the drivers want to preserve energy. Surprisingly, the EVs are slightly faster than CVs in cities, which we assume is due to a close to ideal speed-torque profile [11] and single-speed transmission.

We have shown that the average range of the EV is much lower than the specified range of 160 km. In summer, the average range is 118 km (25% less), and in winter the average range is 76 km (53% less). The large difference between winter and summer is most likely due to heating of the cabin (which consumes up to 5 kW [16]).

Acknowledgments. The authors are financially supported by the Danish Energy Agency, www.ens.dk and by the EU REDUCTION project, www.reduction-project. eu. We thank the ITS platform project [2], for providing the CV data set.

References

1. CAN Bus. http://en.wikipedia.org/wiki/CAN_bus
2. ITS Platform. http://www.itsplatform.eu
3. OpenStreetMap. www.openstreetmap.org
4. Agency, E.E.: Final energy consumption by sector. http://www.eea.europa.eu/data-and-maps/indicators/final-energy-consumption-by-sector-5/assessment
5. Andersen, O., Krogh, B., Torp, K.: TR-34: Analyse af elbilers forbrug (in Danish). Tech. rep., Aalborg University (2014)
6. Artmeier, A., Haselmayr, J., Leucker, M., Sachenbacher, M.: The shortest path problem revisited: optimal routing for electric vehicles. In: Dillmann, R., Beyerer, J., Hanebeck, U.D., Schultz, T. (eds.) KI 2010. LNCS, vol. 6359, pp. 309–316. Springer, Heidelberg (2010)
7. Citroën: Citroën C-Zero. http://info.citroen.co.uk/Assets/pdf/new-cars/c-zero/brochure.pdf
8. Clever: Projekt Test-en-elbil. testenelbil.dk
9. Commission, E.: Roadmap for moving to a low-carbon economy in 2050. http://ec.europa.eu/clima/policies/roadmap/index_en.html
10. Egbue, O., Long, S.: Barriers to widespread adoption of electric vehicles: An analysis of consumer attitudes and perceptions. Energy Policy **48**, 717–729 (2012)
11. Ehsani, M., Gao, Y., Emadi, A.: Modern electric, hybrid electric, and fuel cell vehicles: fundamentals, theory, and design. CRC Press (2009)
12. Franke, T., Krems, J.F.: Interacting with limited mobility resources: Psychological range levels in electric vehicle use. Transportation Research Part A: Policy and Practice **48**, 109–122 (2013)
13. Krogh, B., Andersen, O., Torp, K.: Electric and conventional vehicle driving patterns. In: ACM SIGSPATIAL. ACM (2014)

14. Krogh, B., Pelekis, N., Theodoridis, Y., Torp, K.: Path-based queries on trajectory data. In: ACM SIGSPATIAL (2014)
15. Mitsubishi Motors: About i MiEV. http://www.mitsubishi-motors.com/special/ev/whatis/index.html
16. Mitsubishi Motors: Air-conditioning system for electric vehicles (i-miev). http://www.sae.org/events/aars/presentations/2010/W2.pdf
17. Mitsubishi Motors: Mitsubishi i-MiEV, Full Specifications. http://www.mitsubishi-cars.co.uk/imiev/specifications.aspx
18. Mitsubishi Motors: Technology, Mitsubishi i-MiEV. http://www.mitsubishi-cars.co.uk/imiev/technology.aspx
19. Newson, P., Krumm, J.: Hidden markov map matching through noise and sparseness. In: ACM SIGSPATIAL, pp. 336–343. ACM (2009)
20. Nissan USA: Nissan Leaf, Price and Specs. http://www.nissanusa.com/electric-cars/leaf/versions-specs/version.s.html
21. NOAA: National oceanic and atmospheric administration. http://www.noaa.gov
22. Pearre, N.S., Kempton, W., Guensler, R.L., Elango, V.V.: Electric vehicles: How much range is required for a days driving? Transportation Research Part C: Emerging Technologies 19(6), 1171–1184 (2011)
23. Peugeot: Prices, Equipment, and Technical Specifications. http://www.peugeot.co.uk/media/peugeot-ion-prices-and-specifications-brochure.pdf
24. Renault: Renault Fluence Z.E. http://www.renault.com/en/vehicules/aujourd-hui/renault-vehicules-electriques/pages/fluence-ze.aspx
25. Serway, R., Vuille, C.: College Physics. Cengage Learning (2011)
26. Storandt, S.: Quick and energy-efficient routes: computing constrained shortest paths for electric vehicles. In: ACM SIGSPATIAL IWCTS, pp. 20–25. ACM (2012)
27. Today, D.: Hybrid electric vehicles. www.drivingtoday.com/features/archive/hybrid_electrics/index.html

Interactive, Flexible, and Generic What-If Analyses Using In-Memory Column Stores

Stefan Klauck[1]([✉]), Lars Butzmann[1], Stephan Müller[1], Martin Faust[1], David Schwalb[1], Matthias Uflacker[1], Werner Sinzig[2], and Hasso Plattner[1]

[1] Hasso Plattner Institute, University of Potsdam, Potsdam, Germany
{stefan.klauck,lars.butzmann,stephan.mueller,martin.faust,
david.schwalb,matthias.uflacker,hasso.plattner}@hpi.de
[2] SAP SE, Walldorf, Germany
werner.sinzig@sap.com

Abstract. One well established method of measuring the success of companies are key performance indicators, whose inter-dependencies can be represented by mathematical models, such as value driver trees. While such models have commonly agreed semantics, they lack the right tool support for business simulations, because a flexible implementation that supports multi-dimensional and hierarchical structures on large data sets is complex and computationally challenging. However, in-memory column stores as the backbone of enterprise applications provide incredible performance that enables to calculate flexible simulation scenarios interactively even on large sets of enterprise data.

In this paper, we present the HPI Business Simulator as a tool to model and run generic what-if analyses in an interactive mode that allows the exploration of scenarios backed by the full enterprise database on the finest level of granularity. The tool comprises a meta-model to describe the dependencies of key performance indicators as a graph, a method to define data bindings for nodes, and a framework to specify rules that describe how to calculate simulation scenarios.

Keywords: Business simulation · Column store · What-If analysis

1 Introduction

Today's reporting tools offer an unprecedented flexibility. Companies can dive into their data, i.e filter for arbitrary criteria and drill down into hierarchies to explore the data at the finest level of granularity. Companies wish to exploit this flexibility not only for reporting but also for forecasting and simulating. They want to define potential future scenarios and calculate how these influence their businesses. With the help of what-if analyses, they can evaluate simulation scenarios in terms of their goal fulfillment and support decisions in day-to-day operations.

Mathematical models, hereinafter also called calculation models, are one way to define the dependencies between measures, i.e. the logic how changes of one

© Springer International Publishing Switzerland 2015
M. Renz et al. (Eds.): DASFAA 2015, Part II, LNCS 9050, pp. 488–497, 2015.
DOI: 10.1007/978-3-319-18123-3_29

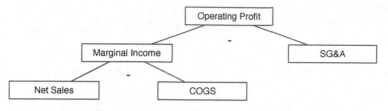

Fig. 1. Value driver tree for the operation profit

key performance indicator (KPI) influence other KPIs. Value driver trees, such as the DuPont model [1], are a well-known method to model KPIs with linear dependencies among each other [2]. Figure 1 shows a driver tree for the operating profit, which can be calculated by subtracting selling, general and administrative expenses (SG&A) from the marginal income. The marginal income depends on net sales and cost of goods sold (COGS). The values of the nodes base on enterprise data, e.g. on G/L account transactions as *actuals* or a combination of the actuals and planned sales, production costs, and expenses as forecasted KPIs. On the basis of enterprise data, companies want to define scenarios, i.e. changes of the data, which reflect changing KPIs, and calculate the effects on other KPIs. The challenge to define and run simulation scenarios is not the mathematical complexity of the calculation, but the combination of a generic calculation model and the large amount of data the model builds on, which enables the users to define flexible scenarios and calculate them interactively.

For many years, the biggest obstacle has been the speed to access enterprise data with all the relevant criteria to allow flexible and interactive simulations. To run such simulations for net sales requires scanning sales documents with all their associated line items. The corresponding tables can comprise billions of records, specifying relevant attributes like sales date, sales volume, price, product, and customer, but also hundreds of other attributes. The advent of columnar in-memory databases has increased the performance of queries accessing few attributes of large data sets, which enables the development of new enterprise applications on top of it [3,4].

This paper presents the HPI Business Simulator, a tool to create, modify and run what-if analyses interactively. This comprises two things: First, a way to define what-if simulations, which consist of a calculation model, the specification of data bindings between KPIs of the model and data of database tables, and the support to specify simulation scenarios. Second, the concept of a simulation tool to define and edit what-if simulations as well as to calculate scenarios.

After this short introduction into the problem domain, Section 2 presents the theoretical concepts of the HPI Business Simulator. Its implementation is shown in Section 3. Section 4 presents related work. The paper closes with a presentation of the conclusions and offering an outlook for future work.

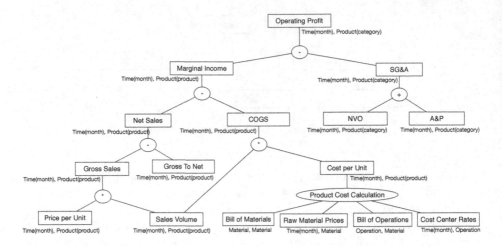

Fig. 2. Calculation model for the operating profit as extension of the value driver tree in Figure 1

2   Simulation Model

The HPI Business Simulator is a proof of concept for implementing generic what-if analyses. This section presents its theoretical concepts called *simulation model*. The simulation model consists of three parts: the generic calculation model to describe the drivers and the dependencies between them, the data binding to connect drivers with data, and a way to specify scenarios and to calculate their results.

2.1   Calculation Model

Calculation models are hypergraphs and extend value driver trees by supporting complex operations and the loosening of the tree structure for dependencies. Each node has a language dependent name, a measure with a unit, and freely definable dimensions for specifying further criteria of a measure. Nodes can be connected with other nodes by *operations*, which are hyperedges. The simulation model specifies available dimensions including their hierarchy levels. The time, for example, can be structured hierarchically into years and months. Operations define the dependencies between nodes, i.e. the way one can calculate the values for the result node based on the data of input nodes. Operations can be one of the four elementary operations addition, subtraction, multiplication, and division. Besides, users can define own, more complex operations, e.g. a product cost calculation based on raw material prices, the bill of materials, cost center rates, and the bill of operations. Figure 2 shows a calculation model for the operating profit.

Data of a node can be seen as a data cube. Combining data cubes with elementary operations works as for one-dimensional data. However, one has to define rules for dimension handling. For additions and subtractions, the result data cube has the intersection of dimensions from all input data cubes. The records of the input data cubes with the same dimension values are combined to a single output record. Result data cubes for multiplications receive the union of dimensions from all input cubes. When combining data cubes with divisions, the calculation model has to define the dimensions for the result data cube so that combining these dimensions with the dimensions of the divisor results in the dimensions of the dividend. Table 1 shows a multiplication and subtraction of data cubes.

Table 1. Data cube calculations

Product	Time	Quantity
Product 1	01/15	5
Product 1	02/15	10
Product 2	02/15	5

(a) Sales Volume

Product	Price
Product 1	20
Product 2	5

(b) Product Prices

Time	Expenses
01/15	50
02/15	100

(c) Expenses

Product	Time	Sales
Product 1	01/15	100
Product 1	02/15	200
Product 2	02/15	25

(d) Sales (= Sales Volume * Product Prices)

Time	Profit
01/15	50
02/15	125

(e) Sales - Expenses

User defined operations work in a similar way as elementary operations with the distinction that arbitrary algorithms can define the calculation logic for the result data cube. A simplified version of a product cost calculation with resolving a hierarchical bill of materials (BOM) is described in the following. We assume that product costs are only affected by raw material prices and thus ignore labor, energy, and machine costs, which would occur in a real world scenario. A calculation of these costs bases on the bill of operations (BOO) and follows the same calculation logic as the BOM resolution. Figure 3 and Table 2 present an exemplary BOM and its database representation.

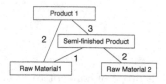

Fig. 3. BOM as graph

Table 2. BOM as table

Child Material	Parent Material	Quantity
Raw Material 1	Product 1	2
Semi-finished Product	Product 1	3
Raw Material 1	Semi-finished Product	1
Raw Material 2	Semi-finished Product	2

Thereby, the table stores *parent-child* relationships, i.e. how much of a *child material* is needed to produce a *parent material*. Child materials can be raw materials and semi-finished products. Parent materials are semi-finished or end products. The second input of our cost calculation scenario is a table with raw material prices. To calculate product costs, we have to resolve the BOM so that products are represented as costs of raw materials, but no semi-finished products. Therefore, we traverse the BOM graph recursively or iteratively. The following equations present the resolution for the exemplary BOM.

$$costs_{Product1} = 2 * costs_{RawMaterial1} + 3 * costs_{Semi-finishedProduct}$$
$$= 5 * costs_{RawMaterial1} + 6 * costs_{RawMaterial2}$$
(1)

2.2 Data Binding

Nodes of the calculation model can obtain their data cubes in two ways. First, they can query their data directly from data sources. Second, they can calculate their values by solving the equation defined by the operation between connected nodes and themselves. For the first case, data bindings are required. Our work focuses on relational databases as data sources. Data bindings define the database connection and query to calculate the data cube with all dimension values. Additionally, data bindings have to specify how to filter the cube and reduce the level of detail to calculate aggregates for higher levels of hierarchies. When specifying the data binding, we have to ensure that the values of all nodes can be calculated unambiguously, meaning that the data has to be sufficient and consistent.

2.3 Simulation Scenarios

Based on the calculation model and data binding, the data cubes of all nodes can be calculated. These data cubes are the basis of what-if analyses. In addition, it is required to specify in which direction changes propagate through the model. Thereby the direction of propagation is not allowed to contain cycles. A *simulation scenario* is a set of changes to the data of nodes. A single change specifies a node, optionally filter conditions to limit the change to a subset of records, and how the specified values are changed. The HPI Business Simulator implements three types of simulation changes: an overwrite for records of the data cube, an absolute adjustment by a delta value, and a relative adjustment by a linear factor.

3 HPI Business Simulator

This section describes the HPI Business Simulator, a proof of concept to implement generic what-if analyses. For the implementation, we have engaged with a Fortune 500 company in the consumer goods industry and discussed their needs

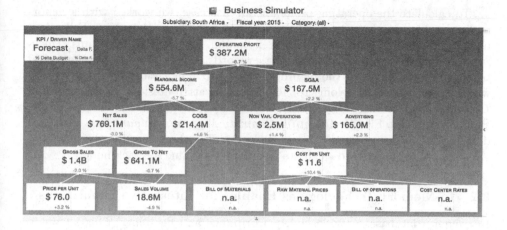

Fig. 4. Screenshot of the HPI Business Simulator with anonymized data

in the area of what-if analyses. Based on their input and the data set they provided to us, we implemented the HPI Business Simulator. In this section, we explain features and implementation details of the HPI Business Simulator, why in-memory column stores enable flexible and interactive simulations, and the benefits of the HPI Business Simulator compared to existing tools.

3.1 Features and Implementation

The HPI Business Simulator is a browser-based graphical tool to define simulation model instances and calculate simulation scenarios using SAP HANA. Figure 4 shows a screenshot of the business simulator with the calculation model from Figure 2. The legend in the top left corner shows the visualized metrics of the calculation model. Each node contains the driver name, a *forecast* value, and the delta of the forecast to the *budget*. Thereby, the forecast is calculated as a combination of the actuals up to today and the budget until the end of the planning horizon. After specifying simulation changes, the effects are included in the forecast and two additional values are displayed: the difference between the old and new forecast and the change in percentage.

The calculation model can be edited by adding, deleting, and dragging the graph components, i.e. nodes and edges. To change node properties as the name, unit of measure, available dimensions, and data binding, one can open a detailed view on the right side. The information to store simulation model instances consists of three parts, describing the available dimensions with their hierarchies, the nodes, and the operations. The top dropdown menu offers a way to drill down into the data, e.g. by selecting a specific product category. In this case, the HPI Business Simulator recalculates the drivers with the filter condition and displays the new values. To define simulation scenarios, the user can select a node and open the simulation interface in the bottom.

To calculate the operating profit for our use case, we worked with a denormalized table, which contained transactional as well as plan data. Following, important attributes of the table are explained. The first two columns, *Document ID* and *is Plan Data*, identify single records and declare whether the record belongs to the actuals or the plan data. *G/L Account Description* indicates to which driver the record belongs to. The following attributes, i.e. *Product, Brand, Category, and Time*, specify criteria, which are mapped to dimensions of the simulation model. Finally, *Quantity, Quantity Measure, Amount, and Currency* describe the measures for the record. In particular, the values for *Quantity* and *Amount* are aggregated to calculate the displayed value for each node.

3.2 In-Memory Technology as Enabler for Interactive Simulations

In-memory databases provide high performance and flexibility. This allows to access the complete business data dynamically even of large companies at the finest level of granularity, opening completely new opportunities. However, the new dynamic nature and flexibility require adapting the way data is accessed and consumed. Recently, self-service tools for business analytics fulfill this need by providing intuitive interfaces to explore and analyze data. Typically, those tools are focused on historic data and do not consider dependencies of the underlying value drivers and metrics. With the HPI Business Simulator, we want to extend self-service tools for business analytics and provide an intuitive approach to model value dependencies in enterprises with an interactive simulation environment, leveraging the full computational power of in-memory databases.

In-memory column stores, such as SAP HANA [5], are the key enabler for interactive simulations on large enterprise data. First of all, analytical queries are accelerated compared to traditional disk-based systems by keeping all data in main memory. Analytical queries are the basis to retrieve the data for business simulations. Based on the specified parameters, the HPI Business Simulator calculates aggregates of disjoint data sets: the ones which are affected by simulation parameters and the ones which are not. The values of the nodes can then be calculated by applying changes and combining theses aggregates. The second benefit is the columnar data layout. Since the data bindings of the simulation model describe only columns that are required for a calculation, the amount of processed data is kept to a minimum. This functionality is especially beneficial in the context of enterprise systems where data is typically very wide and sparse, with up to 400 columns per table [6]. A third benefit is an aggregate cache, which is a transparent caching engine inside the database [7]. Other than classic materialized views, the aggregate cache does not create a hard copy of the data and as a result does not return stale data. The aggregate cache leverages the internal table representation in certain column stores like SAP HANA or Hyrise [8] and always works on the latest data. During a typical simulation session, different scenarios are analyzed and compared. The differences between the scenarios can vary, but are usually small. Consequently, the executed queries are similar and can therefore benefit from the aggregate cache.

The performance to calculate a simulation scenario depends on many factors like the number of records stored in the underlying tables, the number of nodes with a data source, the number of simulation changes, the number of columns used to specify the filter criteria, the granularity of filter criteria, and the data distribution. For a preliminary performance test, our HPI Business Simulator ran on a data set comprising 300 million records of customer data. The initial on-the-fly calculation of an aggregate on a single enterprise class server with 128 cores and 256GB of RAM running SAP HANA was calculated within a second. In that way, simulations can be defined and run interactively.

3.3   Benefits Compared to Existing Tools

Generic model. Existing simulation tools are targeted for specific processes and are difficult to modify or extend to capture new use cases. Users may want to extend the calculation model for the operating profit (see Figure 2) to distinguish between advertising channels, which enables to simulate changes for a specific kind of advertisement. The HPI Business Simulator allows the modification of existing calculation models and definition of new ones without changing the source code of the simulation tool or rewriting the application.

Support for complex calculations. Calculation models should not be limited to tree structures and elementary operations as in the DuPont model. Instead, graph structures and custom calculations should be supported.

Using in-memory column stores. Exploiting information at the finest level of granularity requires the capability to operate on terabytes of data. Simulation tools without an enterprise database as backbone have to load pre-aggregated measures, which come along with a loss of information. The speed of current in-memory databases supports aggregating large amounts of data on the fly within seconds, which enables us to define and calculate flexible what-if scenarios interactively. In addition, simulations are always calculated on consistent and up-to-date data.

Collaborations. The HPI Business Simulator supports iterative what-if analyses of users in different roles: The management defines KPIs and adapts the definitions in case the calculation model does not support a desired simulation scenario. More technical staff with extensive knowledge about the data schema is responsible for providing the model with data and the integration of new data sources. Concrete simulation scenarios are discussed and worked out by potentially multiple controllers, which are responsible for different business areas.

4   Related Work

Golfarelli et al. introduce a methodology and process to design what-if analyses [9]. Compared to our approach, they describe what-if simulations in a more general way. In particular, they divide the process to design simulations into seven phases: goal analysis, business modeling, data source analysis, multi-dimensional modeling, simulation modeling, data design and implementation, and validation. Our approach uses a multi-dimensional data model, which Golfarelli et al. see as most suitable to design what-if analyses. To describe the actual simulation model, they propose an extension of UML 2 activity diagrams [10]. In comparison to our work, they do not implement the simulation model instances as applications, specify data bindings, nor define how to calculate simulation scenarios. Furthermore, we see the dependencies described by a calculation model instance as subject to changes and extensions.

In the field of data cubes, which are the basis to calculate our simulation scenarios, most research focuses on materialized data cubes [11–15]. Gray et al. introduce a data cube operator as generalization and unification of following database concepts: aggregates, group by, histograms roll-ups and drill-downs, and cross tabs [11]. Further papers cover efficient implementation [12,13] and maintenance techniques [14,15] for materialized data cubes. In contrast to previous work, we calculate the required aggregates of the data cubes on the fly.

A further research area in the field of what-if analyses is the combination of spreadsheets and SQL. Spreadsheets have an easy to understand interface, but do not build on consolidated enterprise data, which is usually stored in a RDBMS. On the other side, SQL lacks the support for array-like calculations as Witkowski et al. claim in [16]. Their idea is to combine both and offer a spreadsheet-like computation in RDBMS through SQL extensions. In [17], they continue that research and introduce a way to translate MS Excel computations in SQL. Using their approaches for what-if analysis comes with two drawbacks. First, it does not encapsulate the definition of the simulation model so that it is not tangible, but only expressed by multiple formulas spread over many table cells. Second, MS Excel is limited to the two-dimensional representation and cannot visualize graphical dependencies between nodes appropriately.

5   Conclusion

In this paper, we presented our vision of generic, flexible, and interactive business simulations, enabled by the performance capabilities of columnar in-memory databases. In particular, we presented the HPI Business Simulator and its theoretical concepts to specify and run enterprise simulations. We proposed a meta-model to describe what-if analyses comprising a calculation model, the data binding and simulation parameters. Implementing this meta-model, simulation model instances can be created and edited, such as one for the operating profit. Based on a simulation model instance, scenarios are specified and calculated interactively. With the performance of in-memory column stores, such as SAP HANA, what-if analyses can include millions of records and work on the finest level of granularity to enable interactive and fully flexible simulations.

This paper names performance factors that influence the calculation of simulation scenarios. In this field a deep analysis can be conducted. Furthermore, future work can investigate how to optimize the calculation of scenarios, for example by exploiting cases in which nodes query the same table or by optimizing queried data cubes so that their granularity is sufficient to calculate connected nodes. The visualization of complex calculation models can be another field for future investigations.

References

1. Chandler, A., Salsbury, S.: Pierre S. Du Pont and the Making of the Modern Corporation. BeardBooks (2000)
2. Zwicker, E.: Prozeßkostenrechnung und ihr Einsatz im System der integrierten Zielverpflichtungsplanung. Techn. Univ, Berlin (2003)
3. Plattner, H.: A common database approach for OLTP and OLAP using an in-memory column database. In: SIGMOD (2009)
4. Plattner, H.: The impact of columnar in-memory databases on enterprise systems. In: VLDB (2014)
5. Färber, F., Cha, S.K., Primsch, J., Bornhövd, C., Sigg, S., Lehner, W.: SAP HANA database - data management for modern business applications. In: SIGMOD (2011)
6. Krüger, J., Kim, C., Grund, M., Satish, N., Schwalb, D., Chhugani, J., Dubey, P., Plattner, H., Zeier, A.: Fast updates on read-optimized databases using multi-core CPUs. In: VLDB (2011)
7. Müller, S., Plattner, H.: Aggregates caching in columnar in-memory databases. In: VLDB (2013)
8. Grund, M., Krüger, J., Plattner, H., Zeier, A., Cudre-Mauroux, P., Madden, S.: HYRISE: a main memory hybrid storage engine. In: VLDB (2010)
9. Golfarelli, M., Rizzi, S., Proli, A.: Designing what-if analysis: towards a methodology. In: DOLAP (2006)
10. Golfarelli, M., Rizzi, S.: UML-based modeling for what-if analysis. In: DaWak (2008)
11. Gray, J., Bosworth, A., Layman, A., Pirahesh, H.: Data Cube: A Relational Aggregation Operator Generalizing Group-By, Cross-Tab, and Sub-Totals. Data Min. Knowl, Discov (1997)
12. Harinarayan, V., Rajaraman, A., Ullman, J.D.: Implementing data cubes efficiently. In: SIGMOD (1996)
13. Sismanis, Y., Deligiannakis, A., Roussopoulos, N., Kotidis, Y.: Dwarf: shrinking the petacube. In: SIGMOD (2002)
14. Mumick, I.S., Quass, D., Mumick, B.S.: Maintenance of data cubes and summary tables in a warehouse. In: SIGMOD (1997)
15. Roussopoulos, N., Kotidis, Y., Roussopoulos, M.: Cubetree: organization of and bulk incremental updates on the data cube. In: SIGMOD (1997)
16. Witkowski, A., Bellamkonda, S., Bozkaya, T., Dorman, G., Folkert, N., Gupta, A., Shen, L., Subramanian, S.: Spreadsheets in RDBMS for OLAP. In: SIGMOD (2003)
17. Witkowski, A., Bellamkonda, S., Bozkaya, T., Naimat, A., Sheng, L., Subramanian, S., Waingold, A.: Query by Excel. In: VLDB (2005)

Energy Efficient Scheduling
of Fine-Granularity Tasks in a Sensor Cloud

Rashmi Dalvi and Sanjay Kumar Madria[✉]

Department of Computer Science,
Misouri University of Science and Technology, Rolla, Missouri65401, USA
{rgd8t6,madrias}@mst.edu

Abstract. Wireless Sensor Networks (WSNs) are frequently used in number of applications like unattended environmental monitoring. WSNs have low battery power hence schemes have been proposed to reduce the energy consumption during sensor task processing. Consider a Sensor Cloud where owners of heterogeneous WSNs come together to offer sensing as a service to the users of multiple applications. In a Sensor Cloud environment, it is important to manage different types of tasks requests from multiple applications efficiently. In our work, we have proposed a scheduling scheme suitable for the multiple applications in a Sensor Cloud. The scheduling scheme proposed is based on TDMA which considers the fine granularity of tasks. In our performance evaluation, we show that the proposed scheme saves energy of sensors and provides better throughput and response time in comparison to a most recent work.

1 Introduction

Wireless sensor networks (WSNs) are popular because they can be used in a wide range of applications, easy to deploy, withstand adverse conditions, work in unmonitored networks, and provide dynamic data access. Unfortunately, WSNs have limited battery power and thus, they have shorter lifespan. Therefore, sensor tasks must be efficiently scheduled so that their lifespan increases. We consider a Sensor Cloud of heterogeneous WSNs as described in our work [7]. The WSN owners within a Sensor Cloud collaborate with one another to provide sensing as a service at the same time and thereby gain profit from underutilized WSNs. The users of the sensing as a service can select sensors either from a single or multiple WSNs within the Sensor Cloud at the same time. Therefore, we need a sensor task scheduling scheme for WSNs within a Sensor Cloud.

Pantazis et al. [2] proposed a TDMA based scheduling scheme that balances power saving and end-to-end delay in WSNs. The scheme schedules the wakeup intervals such that data packets are delayed by only one sleep interval from sensor to gateway. However, while scheduling the sensors, they did not consider multi-application environment. In a Sensor Cloud, users of multiple applications want to consume the data differently which brings forth the need of categorizing user tasks. The scheme proposed in [2] fails to account for various types of tasks

© Springer International Publishing Switzerland 2015
M. Renz et al. (Eds.): DASFAA 2015, Part II, LNCS 9050, pp. 498–513, 2015.
DOI: 10.1007/978-3-319-18123-3_30

and thus, it cannot achieve the goal of energy conservation for different types of requests. Similar to push requests [1], a number of users may like to consume data from WSNs at a specific frequency. Other users prefer to do so once and in ad-hoc fashion like push requests in [1]. Some users may like to be notified when a specific event occurs. In all such needs, in-spite of being energy efficient [2], it does not conserve enough energy because it did not consider fine granularity of the tasks received from user applications.

Kapoor et al. [4], addressed the issue of allocating and scheduling sensors in a multi-application environment. This work eventually led to energy conservation in WSNs. Each sensor in the scheme can be utilized by a single application. A scheduling algorithm was used to reduce the sensors response time. However, this [4] proposed approach is not suitable for a Sensor Cloud where many users of different applications might be interested in obtaining data from the same geographical region (i.e. from a wireless sensor).

In this paper, we have extended the scheduling scheme proposed in [2] by considering fine granularity of the tasks. In multi-application environment, tasks are treated differently depending on their types. This increases the user satisfaction by reducing the response time of the tasks and improving throughput greatly. It also provides the optimal energy conservation. Unlike scheduling scheme proposed in [4], our scheme enables multiple application users to retrieve data from same set of sensors by using TDMA. Also, we compared our experimental results with the results in [4].

The major contributions of this work include showing experimentally that our scheduling scheme provides better throughput and response time than Least Number Distance Product (LNDP) scheduling scheme proposed in [4]. Our scheme treats each task differently which leads to optimal energy conservation.

The rest of the paper is organized as follows: Section 2 presents the related work. Section 3 introduces the architecture of Sensor Cloud briefly. Section 4 describes the scheduling scheme in detail. Section 5 contains the implementation details and experimental results. Finally, Section 6 concludes the paper.

2   Related Work

Andrei et al. [5] proposed the approach of scheduling tasks. They suggested that the Minimum cut theorem be used to partition a WSN into zones, reducing the number of transmission and receptions. Thus, each convergent transmission (from any node to the BS) will have a minimum number of hops. Such a strategy may, however, lead to communication overhead in the process of partitioning and gossiping between the sensors in WSNs. Additionally, the minimum cut algorithm runs only once at the network initialization. Therefore, this scheme is not suitable for dynamic networks.

Cao et al. [6] proposed scheduling in low duty cycled WSNs. Here, multiple paths are used to transmit the data from the sensor to the BS when data transmission fails to increase the reliability. Although this scheme provides a greater fault tolerance, it requires an extensive amount of energy.

Xiong et al. [3] proposed a multi task scheduling technique that uses load balancing for low-duty-cycled WSNs. They concentrated on load balancing problem for multiple tasks among sensor nodes in both spatial and temporal dimensions. Although this scheme is dynamic, did not account for fine granularity of the tasks. They also failed to address how the scheme would work if the length of each task was unequal to the next.

3   Sensor Cloud Architecture

A Sensor Clouds architecture [7] is illustrated in Figure 1. This architecture is broadly divided into three layers: Client centric, Middleware, and Sensor centric. A Client centric layer connects end users to the Sensor Cloud; it manages a users membership, session, and GUI. Middleware is the heart of the Sensor Cloud. It uses virtualization, with help from various components (eg. provision management, image life-cycle management, and billing management) to connect all participating WSNs to the end user. Sensor centric layer connects the Middleware to the physical WSNs. It also registers participating WSNs, maintains participating WSNs, and collects data.

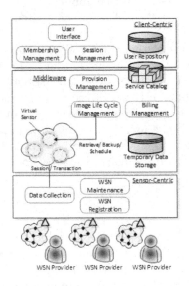

Fig. 1. Sensor Cloud Architecture

3.1   Challenges in Multi-Application Evironment

A Sensor Cloud is comprised of multiple WSNs. Thus, various types of multiple application users may want to consume the data from Sensor Cloud at the same instance of time. Hence, we need a scheduling scheme that schedules WSNs such that it caters maximum number of applications and their users.

Conventional schemes that are typically used to schedule WSNs will not work efficiently in multi-application environment, because those schemes are designed for single application environment. The scheme proposed here focuses on scheduling sensors in multi-application environment to increase user satisfaction.

Assumptions. In WSNs, tree topology and cluster topology are widely used over other network topologies. Because of its structure, a tree topology provides a wider area coverage than a cluster topology. Thus, in this work a tree topology is assumed for all WSNs that participate in a Sensor Cloud.

We assume that the proposed scheduling scheme will have a sensor allocation scheme. In a WSN, according to the user's request, allocation scheme aimed to select some number of sensors from available sensors. Allocation scheme is then followed by the scheduling scheme which then serves maximum number of application users.

4   Scheduling Tasks

The sensing requests were broadly classified into three categories. In multi-applications environment, applications that consume Sensor Cloud services are expected to receive requests from one or more of the categories given below. These requests are also known as Tasks.

4.1   Types of Tasks

The types of tasks are classified as Task T1, Task T2, and Task T3.

Task T1. Task Type 1 (referred to as T1) is a task that is requested by users when they need sensor data at a specific frequency. These tasks may include a request for information on a specific topic (eg. weather broadcast). This information is sampled periodically and then broadcast back to the user. These requests have also been referred to as Push requests [1]. When this type of request is made, the BS sends data to the sensors only once at the beginning of the request. The sensors sense and then send the data to the BS at a given sampling frequency and time duration. Task T1 becomes the most expensive of all tasks in a WSN that is running at a maximum sampling frequency because this task preempts other tasks for the same WSN. The cost of this task is high at a higher frequency. It decreases as the sampling frequency decreases.

In a WSN, consider a scenario when n users have requested for n tasks of type T1. These tasks are requested for same set of sensors and for same sensing phenomenon, but for different data frequencies (different time intervals between requests). In this case, the physical data frequency of WSN will be the minimum data frequency of all n tasks. Remaining n-1 frequencies will be virtual data frequencies. For n-1 tasks, selection of minimum frequency will lead to some delay in data received from sensors. However, this approach will conserve the sensors energy because *minimum frequency = min(f1, f2, ... fn)*. In another approach, LCM (Least Common Multiple) of all n frequencies can be set as the minimum frequency. LCM approach will reduce delay in response, but will significantly increase sensors energy because *minimum frequency* $<=$ *LCM(f1, f2, ... fn)*.

We prioritized saving sensors energy over latency in receiving data. Therefore, the minimum frequency of all requests was used in this study. Users in need of data (based on type T1) need to send information (e.g., location, sensing phenomenon, sampling frequency, and sampling duration) to the sensor cloud.

Task T2: Type 2 tasks (referred as T2) are requested by users who need one time data on the fly. This type of task also known as a Pull Task [1], is designed to serve ad-hoc requests. BS will send request details (e.g. location, sensing phenomenon) to the sensors, which will respond with the latest data.

Task T3: Task Type T3 (referred as T3) is used for event-based requests. During this type of request, the BS sets the event on sensors according to the requested condition on the sensor data. These applications will need several inputs, including location, sensing phenomena, event condition, monitoring frequency, and monitoring duration. Sensors continue sensing the data at a requested frequency and respond when the event condition occurs. Task T3 is useful for a number of applications, including fire detection, intrusion detection, and so forth. The algorithm used for T3 is a trade-off between the event occurrence frequency and the cost of event detection. If the event occurs very frequently, then a large number of duty cycles must be assigned to T3, increasing the effective cost.

T3 task can be further classified into 'Notify once' and 'Notify until the condition is false'.

The behavior of 'Notify once' would be similar to Task T2. In case of 'T3 with Notify Once' sensors send data to the BS only once when event condition is met. On the contrary, for T2 tasks, on request, sensors respond with data.

Behavior of 'Notify until the condition is false' would be similar to Task T1. When event occurs, sensors executing task 'T3 with Notify until condition is false' send data to BS until condition turns false. However, for task T1, sensors send data to the BS at a given frequency. As T3 does not require data to be always sent from sensors to the BS, cost of T3 is lesser than T1.

4.2   Handling Redundant Requests

The handling of redundant requests benefits application users, WSN owners, and cloud service providers within a sensor cloud. The BS handles the redundant requests without affecting the physical frequency. For an instance, assume that a node 'A' is already serving request 'R1' for task T1 at frequency 'f'. The following redundant requests to node 'A' can be served by the BS from the data received for request 'R1'.

1. Other T1 requests for node 'A', with a frequency in multiples of 'f', until the duration of R1 ends
2. T2 requests for node 'A'. The maximum delay is $=$ 'f'

3. T3 requests for node 'A'. The maximum delay is = 'f' until the duration of R1 ends

Similarly, BS will serve requests for other nodes according to requests already made on the WSN.

4.3 Scheduling Scheme

Our proposed scheme is divided into Sensor Scheduling and Task Scheduling.

Sensor Scheduling. The scheduling scheme proposed in [2] is extended here to avoid the problem of packet collision during transmission and reception. Both wake up and sleep modes are used for all types of tasks. For tasks T1 and T2, the BS determines in what order the sensors must transmit and receive data. Thus path wakeup is not required for tasks T1 and T2. In task T3, however the event triggers the transmission of data from node(s) to the BS. The schedule for T3 is not defined beforehand. On the occurrence of event, data is pushed into the vacant duty cycle. Hence, T3 needs the path wakeup strategy proposed in [2] to transmit the data from the node(s) to BS.

Consider the tree topology given in Fig. 2. Messages between sensors are divided into following categories according to the direction of packet transmission. They are further classified based on the level of granularity:

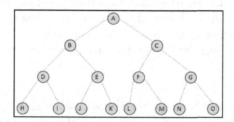

Fig. 2. WSN in a tree topology

BS to Node(s) Command messages: This type of message has both wakeup and command packets [2]. Time is divided (see Fig. 3) to avoid a collision between packets. Wakeup messages have a short duration; Command messages have a longer duration. The command packets time period is determined by the commands maximum length of the command. These messages are transmitted from the BS to the node(s) for every new request. They are not, however, transmitted for redundant requests.

Node 'A' sends data to nodes 'B' and 'C'. Transmitter of 'A' is turned on for the first two slots. Receivers of 'B' and 'C' are turned on in the first and second timeslots, respectively. Next, nodes 'B' and 'C', in parallel, send data to 'D' and 'F', followed by 'E' and 'G'. This procedure continues until the algorithm reaches the leaf nodes. This sequential transmission of packets helps in avoiding collision between packets.

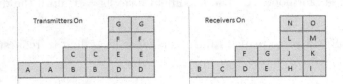

Fig. 3. BS to Node(s) messages

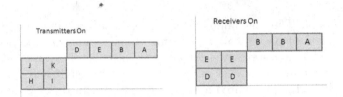

Fig. 4. Node(s) to BS messages

Node(s) to BS Data messages: This type of messages has both wakeup and data packets [2]. The data packets in task T1 are pushed to the BS at a scheduled frequency. The data for tasks T2 and T3 is pushed at the next available duty cycle. Data transmission occurs as shown in Fig. 4. Nodes 'J', 'K', 'H', and 'I' are leaf level nodes that need to send data to the BS. The 'J' and 'H' leaf nodes send data to the 'E' and 'D' leaf nodes, respectively. Later, nodes 'K', 'I' send data to 'E' and 'D', in order. Accordingly, the transmitters that are sending data are turned on, and receivers receiving data are turned on for a specified amount of time.

Task Scheduling. The scheme for task scheduling is an extension of the sensor scheduling explained in Section 3.1. Two types of cycles are given in Fig. 5.

Duty Cycle: The duty cycle includes time slots for the transmission of data from BS to all nodes and from the node to the BS. For a particular WSN, the time duration of a duty cycle is equal to the time duration of the duty cycle that contains the longest task. Any task (T1/ T2/T3) can be assigned to a duty cycle according to whether or not it can serve the task. The duty cycles length is selected as the length of the largest task in the requests currently served. Each slot in Fig. 5 separated by a brown color line is a duty cycle. Duty cycle T1 is further divided into two messages: BS to Node(s) and Node(s) to BS. Duty cycles are used for tasks T2 and T3 as per allocated tasks.

Task Cycle: A task cycle is a combination of multiple consecutive duty cycles that may contain tasks T1, T2, and T3. The task cycles length changes when the physical operating frequency of the WSN changes. Essentially, task cycle depends on the T1 frequency.

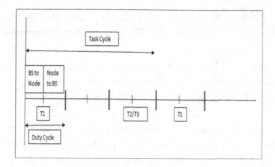

Fig. 5. Duty cycle and Task Cycle

4.4 Duty Cycles in Detail

A duty cycle has a variable length. The minimum length of all requested tasks is considered to be the length of a duty cycle for a given WSN. When tasks are short in length, this strategy reduces the delay in response. Correlation between tasks and messages in duty cycle is explained further.

Duty Cycle for Task T1. Duty cycles for T1 are divided into two types of messages, BS to Node(s) messages and Node(s) to BS messages.

BS to Node(s) message: The BS to Node(s) message is transmitted on a T1 request to the WSN. For the first T1 duty cycle, time is allocated on BS to Node(s) message. Transmission will start from BS and propagate to the leaf level nodes.

Node(s) to BS message: A Node(s) to BS message is transmitted at each occurrence of a sampling frequency. Time slots are assigned by beginning at the leaf level node and moving to the BS. These messages are driven by both, as the BS and the Node(s) are each aware of the transmission time slots.

Duty Cycle for task T2. Duty cycles for T2 are divided into two types of messages, BS to Node(s) messages and Node(s) to BS messages.

BS to Node(s) message: Task T2 are designed to serve ad-hoc requests. Thus, these messages are transmitted when nodes find an empty duty cycle.

Node(s) to BS message: If a request is already received by a node, then the response is sent in next available duty cycle. Node(s) to BS slot of the duty cycle is selected to transfer the message. Unlike T1 Node(s) to BS messages, these messages are driven completely by the BS because the BS tells the node(s) when to transmit the data.

Duty Cycle for Task T3. Duty cycles for T3 are divided into two types of messages, BS to Node(s) messages and Node(s) to BS messages.

BS to Node(s) message: These messages can be transmitted in next available duty cycle. The event is set to the given node(s) once a request is sent to the node(s) with monitoring frequency and condition.

Nodes(s) to BS message This message transmission uses path wakeups from node to the BS. Wake up messages are sent at the beginning of this message to all nodes in the path. Data messages are sent once path to the BS is established. These messages are completely driven by the monitoring nodes because the BS is unaware of the data received.

4.5   Pre-emption Condition

The pre-emption of nodes in a WSN depends on two factors: the availability of duty cycles at the root level node and the frequency of requests in T1 mode. The minimum frequency of T1 limits a network by restricting other nodes from transmitting the data.

$$\frac{1}{\sum_{p=1}^{exisingT1Freq}(1/Freq_p)} >= D \tag{1}$$

Equation (1) defines the pre-emption condition by rejecting requests that do not satisfy the above condition. Here, existingT1Freq is the list of T1 frequencies operating on a WSN, Freqp is the pth operating frequency (of non-redundant requests), and D is the length of the duty cycle.

Algorithm 1 ensures that at least one duty cycle is available in the task cycle for T2 and T3 requests. This algorithm, however accepts T1 and T3 requests only when the value at L.H.S. in (1) is greater than the value of R.H.S. The algorithm rejects the request when the condition is not satisfied. Rejected requests inform users that the WSN does not currently have the capacity to serve.

Algorithm 1 takes request R as an input. It returns *true* if the request is served and *false* if the request is not served. This algorithm evaluates the pre-emption condition (in Equation (1)) first. If its value is *true*, the request can be scheduled else the request will be rejected by returning *false*. If the value of preemption condition is *true* and *R.task* is either *T1* or *T3* (e.g. a task with a frequency), then it evaluates whether or not the desired frequency is in multiples of a duty cycle or vice versa by finding either of *mod(R.freq, Task Cycle)* and *(mod(TaskCycle,R.freq)=0))* condition is *true*. If the frequency is in multiples i.e., one of these conditions is *true*, then the algorithm assigns values to *minslot* and *majslot* by using methods *GetMinorSlot(R)* and *GetMajorSlot(R)* respectively. These methods get *minslot* and *majslot* by finding the next vacant timeslot that can be scheduled. Further *ScheduleWSN(R, minslot, majslot)* method schedules the request on appropriate duty cycle according to the values of *minslot*, *majslot* and returns *true*. Returned value true means that the request is scheduled for *T1* or *T3*. In other case when preemption condition is met and *R.task* is *T2*, steps same as *T1* and *T3* will be executed except a difference that, as task *T2* doesnt have any frequency; condition *mod(R.freq, Task Cycle) and (mod(TaskCycle,R.freq)=0))* will not be checked. Rest of the steps for *T2* will be same as *T1* and *T3*.

Algorithm 1. BS Scheduling

Objective: Scheduling the input request

Input: *R* {*task, phenom, sens, fwd, freq, cond*}

Output: *true* if request is scheduled, *false* otherwise

---

if

$$\frac{1}{\sum_{p=1}^{exisingT1Freq}(1/Freq_p)} <= D$$

then
    if *R.task* = *T1* **or** *R.task* = *T3* **then**
        if *R.freq* mod *D* = 0 **or** *D* mod *R.freq* = 0 **then**
            minslot = GetMinorSlot(*R*)
            majslot = GetMajorSlot(*R*)
            ScheduleWSN(*R, minslot, majslot*)
            Return *true*
        end if
    end if
    if *R.task* = *T2* **then**
        minslot = GetMinorSlot(*R*)
        majslot = GetMajorSlot(*R*)
        ScheduleWSN(*R, minslot, majslot*)
        Return *true*
    end if
else
    Return *false*
end if

The DutyCycleTimer fires every *D* seconds. The currentRequests refers to the collection of all requests running on a given sensor, *mincurrentslot* is the minor slot id for the current Duty Cycle, *majcurrentslot* is the major slot id for the current Duty Cycle, *currentnode.id* is the id for the sensor on which code is executing, and *currentnode.parent* is the id for parent sensor of the sensor on which the code is executing. The *notified* is a flag; its value is true if the child of the current node has detected an event.

After Algorithm 1 sends requests to sensors, a new request is added into the *currentRequests* in Algorithm 2. The *DutyCycleTimer* is set to fire (*DutyCycleTimer fired event*) on each duty cycle frequency tick. On this event, for each request *R* in *currentRequests*, Algorithm 2 checks whether a *mincurrentslot* is equal to *R.minslot* and *majcurrentslot* is equal to *R.majslot*. If both slots match for any request *R*, the Algorithm 2 continues else no action is taken.

Further if $R.task$ is $T1$ or $T2$, and current sensor is selected for sensing (i.e. $R.sens.contains(currentnode.id)$) then, data is sensed for the phenomenon $R.phenom$ using method $SenseData(R.phenom)$. Sensed data is aggregated with the data received from immediate children by method $AggregateData()$ and sent to the parent using $SendData(currentnode.parent)$. If the task is either $T1$ or $T2$ and the current sensor is selected to forward the data (i.e. $R.fwd.contains(currentnode.id)$), then sensor just forwards the data to its parent using method $Send-Data(currentnode.parent)$. Now if the request is $T1$ i.e. $R.task$ is $T1$, algorithm completes. In case if $R.task$ is $T2$, it continues further. The $T2$ requests are one-time requests, hence after processing $T2$ request $R.minslot$ and $R.majslot$ are set to 0, so that this request will not be served further.

Algorithm 2. Sensor Scheduling
Objective: Serving the scheduled requests

DutyCycleTimer fired Event
for all R in $currentRequests$ **do**
  if $R.minslot = mincurrrentslot$ **and** $R.majslot = majcurrentslot$ **then**
    if $R.task = T1$ **or** $R.task = T2$ **then**
      if $R.sens.contains(currentnode.id)$ **then**
        SenseData($R.phenom$)
        AggregateData()
        SendData($currentnode.parent$)
      end if
      if $R.fwd.contains(currentnode.id)$ **then**
        SendData($currentnode.parent$)
      end if
      if $R.task = T2$ **then**
        $R.minslot = 0$
        $R.majslot = 0$
      end if
    end if
    if $R.task = T3$ **then**
      if $R.sens.contains(currentnode.id)$ **then**
        if CheckEvent(SenseData($R.phenom$), $R.cond$) = true **then**
          SendNotification($currentnode.parent$)
        end if
      end if
      if $R.fwd.contains(currentnode.id)$ **and** $notified = true$ **then**
        SendNotification($currentnode.parent$)
      end if
    end if
  end if
end for

In another case, if the timeslots match, i.e. ($mincurrentslot$ is equal to $R.minslot$ and $majcurrentslot$ is equal to $R.majslot$), and $R.task$ is $T3$, it checks whether the current sensor is selected to sense the phenomena by using a check, $R.sens.contains(currentnode.id)$. Then a phenomenon is sensed and condition is tested by checking *CheckEvent(SenseData(R.phenom), R.cond)*. If the condition check has met the event criteria, i.e. if this check returns *true* then the sensor sends a notification to its parent using method *SendNotification (currentnode.parent)*. However, if the current sensor is selected as the forwarding node (i.e. value of *R.fwd.contains(currentnode.id)* is *true*), then sensor assesses whether or not it has received any notifications from its children by checking whether *notified* flag is *true*. If *notified* flag is *true*, it notifies its parent using method *SendNotification(currentnode.parent)*. If it has not received a notification (*notified* flag is *false*), it remains idle.

5   System Implementation, Experiments and Performance Evaluation

For scheduling scheme, we have developed the software to run on Telosb motes using TinyOS 2.0 and java. The Telosb motes were powered by two AA rechargeable batteries (1.2V - 2600mA). A Sensor Cloud web application programmed using java communicates with the BSs of different WSNs using RMI and socket communication. BS developed in java then transfers the messages to the motes which are programmed using TinyOS.

In order to show the efficiency of our design, we compared scheduling scheme with the LNDP scheme in [4]. For LNDP scheme, we programmed motes using TinyOS. It had just one type of task, which can be executed once at a time because in [4], authors did not consider granularity of the tasks. Also, this task had a frequency like task T1 in our scheduling scheme. Similarly, we have developed our TDMA scheduling scheme having tasks T1, T2, and T3, using TinyOS. For the experiment, we deployed a WSN with 5 nodes for LNDP and for our scheduling algorithm. The measures we used for performance comparison are, Response Time, Throughput, Network Lifetime and Power Consumption.

Response Time. To find the Response Time, we considered the tasks of type T1 in our experiment because they are most expensive tasks. For increasing number of tasks we found the values of response time for the best case, the average case and the worst case. Fig. 6 shows the tasks table used in performing the experiment. Also, Fig. 7 shows the graph of Number of Tasks v/s Response Time (in sec) for scheduling algorithm. In the best-case scenario, the new request for the task arrives when all other previous requests are saved and next duty cycle is vacant to be served. On the other hand we consider a worst case when a vacant duty cycle was just got over and a new request arrives. When the number of tasks are more in worst case, we have taken the tasks with a higher frequency rate which causes the response time to increase rapidly. However, in average case, the new request arrives between the two vacant duty cycles. The response time for average case increases linearly with the number of tasks. The response time for the scheme in [4] is shown in Fig. 8, where the overall response time is significantly greater than the response time of scheduling algorithm. The response time of scheduling algorithm is shorter than [4] because in our scheduling algorithm, we allow tasks to execute in parallel. On the contrary, for scheme in [4], tasks are executed serially and hence, the response time is longer.

Number of tasks	Scenarios - Task Frequency (sec)		
	Worst case	Average case	Best case
1	1. T1=5/ T2/ T3 = 5	1. T1=5/ T2/ T3 = 5	1. T1=5/ T2/ T3=5
	1. T1=10	1. T1=10	
2	2. T1=10/ T2/ T3 = 10	2. T1=10/ T2/ T3 = 10	1. T1=5/ T2/ T3=5
	1. T1=10	1. T1=15	
	2. T1=20	2. T1=15	
3	3. T1=20/ T2/ T3=20	3. T1=15/ T2/ T3=15	1. T1=5/ T2/ T3=5
	1. T1=10	1. T1=20	
	2. T1=20	2. T1=20	
	3. T1=40	3. T1=20	
4	4. T1=40/ T2/ T3=40	4. T1=20/ T2/ T3=20	1. T1=5/ T2/ T3=5
	1. T1=10	1. T1=25	
	2. T1=20	2. T1=25	
	3. T1=40	3. T1=25	
	4. T1=80	4. T1=25	
5	5. T1=80/ T2/ T3=80	5. T1=25/ T2/ T3=25	1. T1=5/ T2/ T3=5
	1. T1=10	1. T1=30	
	2. T1=20	2. T1=30	
	3. T1=40	3. T1=30	
	4. T1=80	4. T1=30	
	5. T1=160	5. T1=30	
6	6. T1=160/ T2/ T3=160	6. T1=30/ T2/ T3=30	1. T1=5/ T2/ T3=5

Fig. 6. Tasks for the experiment

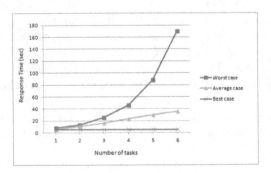

Fig. 7. Number of tasks v/s Response time (*sec*) for scheduling algorithm

Throughput. For the results of throughput, refer to the table in Fig. 6. The number of tasks completed with elapsed time for scheduling algorithm is shown in Fig. 9. Throughput of best and worst case increase linearly, however in the worst case, with increase in time, initially the number of tasks executed go on increasing, but later they go down. As the elapsed time goes on increasing we select the higher frequencies which cause WSNs to accommodate less number of tasks. Hence in worst case throughput reflects downward slope at higher values of time elapsed. On the other hand, the throughput for the LNDP scheme in [4] (in Fig. 10) was very low as compared to throughput of scheduling algorithm because [4] allows serial execution of tasks only.

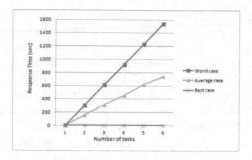

Fig. 8. Number of tasks v/s Response time (*sec*) in LNDP scheduling [4]

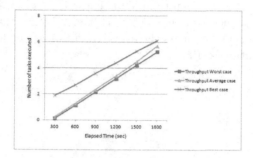

Fig. 9. Elapsed time v/s Number of tasks executed for scheduling algorithm

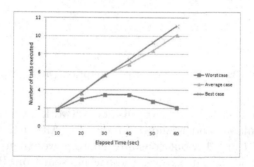

Fig. 10. Elapsed time v/s Number of tasks executed in LNDP scheduling [4]

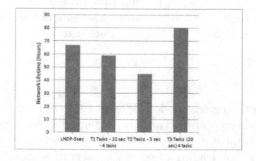

Fig. 11. Tasks v/s Network Lifetime (*hours*)

Network Lifetime. In Fig. 11, we have compared the network lifetime of scheduling algorithm with the network lifetime of the LNDP schem in [4]. The length of the duty cycle was set as 5 sec. When a task with frequency 5 seconds having least number distance product was executed for infinite duration for the scheme [4], the batteries were discharged in 67 hours. Similarly with scheduling algorithm, we deployed three WSNs each having duty cycle as 5 seconds and were assigned task T1, T2 and T3 respectively. The networks for tasks T1, T2 and T3 lasted for 59, 45 and 80 hours respectively. Four tasks of T1 and T3 are set for WSNs, with frequency of each task as 20 seconds. However, for task T2, we sent a new request every 5 seconds. It is observed that in a WSN with task T2 has shortest life because it involves the packets sent from BS to node(s) and node(s) to BS both. T1 has more lifespan than T2 because it was a push request, which requires data being pushed from node(s) to BS only. T3 has the highest lifespan as it mostly involves sensing unless the event condition is not met when node(s) have to send notification to BS.

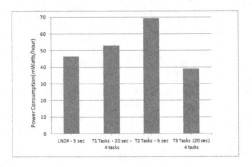

Fig. 12. Tasks v/s Power consumption ($mW/hour$)

Power Consumption. Based on the results of Network lifetime the power consumption was calculated and is shown in Fig. 12. Although the [4] consumes lesser power than T1 and T2, but consumes more power than T3. Task T2 consumes more energy because it involves the packets sent from BS to node(s) and node(s) to BS both. T1 consumed lesser energy than T2 because it was a push request, which requires data being pushed from node(s) to BS only. T3 was the least expensive task because it mostly involves sensing unless the event condition is not met when node(s) have to send notification to BS.

6   Conclusions

In this paper, a scheduling scheme for different types of tasks initiated by multiple applications has been proposed and implemented under a Sensor Cloud environment. In the implementation, we showed that our scheduling scheme provides energy efficient operation leading to energy conservation in WSNs. The scheduling scheme improves response time and throughput, and only consuming some additional energy for tasks T1 (when data is requested at specific frequency) and T2 (when data is requested only one time) but not for tasks T3 (event based data collection) in comparison to a recent scheme handling multiple applications.

References

1. Xu, Y., Helal, S., Scmalz, M.: Optimizing push/pull envelopes for energy-efficient cloud-sensor systems. In: Proceedings of the 14th ACM International Conference on Modeling, Analysis and Simulation of Wireless and Mobile Systems., ACM (2011)
2. Pantazis, N.A., Vergados, D.J., Vergados, D.D., Douligeris, C.: Energy efficiency in wireless sensor networks using sleep mode TDMA scheduling. Ad. Hoc. Networks **7**(2), 322–343 (2009)
3. Xiong, S., Li, J., Li, Z., Wang, J., Liu, Y.: Multiple task scheduling for low-duty-cycled wireless sensor networks. In: 2011 Proceedings IEEE INFOCOM, IEEE (2011)
4. Kapoor, N.K., Majumdar, S., Nandy, B.: Scheduling on wireless sensor networks hosting multiple applications. In: 2011 IEEE International Conference on Communications (ICC), IEEE (2011)
5. Voinescu, A., Tudose, D.S., Tapus, N.: Task scheduling in wireless sensor networks, 2010. In: 2010 Sixth International Conference on Networking and Services (ICNS), IEEE (2010)
6. Yongle, C., Guo, S., He, T.: Robust multi-pipeline scheduling in low-duty-cycle wireless sensor networks. In: 2012 Proceedings IEEE INFOCOM, IEEE (2012)
7. Madria, S., Kumar, V., Dalvi, R.: Sensor cloud: a cloud of virtual sensors. IEEE Software **31**(2), 70–77 (2014). IEEE

Demo

Invariant Event Tracking on Social Networks

Sayan Unankard[1,2]([✉]), Xue Li[2], and Guodong Long[3]

[1] Information Technology Division, Maejo University, Chiang Mai, Thailand
sayan@mju.ac.th
[2] School of ITEE, The University of Queensland, Brisbane, Australia
xueli@itee.uq.edu.au
[3] Centre for Quantum Computation and Intelligent Systems,
University of Technology Sydney, Sydney, Australia
guodong.long@uts.edu.au

Abstract. When an event is emerging and actively discussed on social networks, its related issues may change from time to time. People may focus on different issues of an event at different times. An invariant event is an event with changing subsequent issues that last for a period of time. Examples of invariant events include government elections, natural disasters, and breaking news. This paper describes our demonstration system for tracking invariant events over social networks. Our system is able to summarize continuous invariant events and track their developments along a timeline. We propose invariant event detection by utilizing an approach of Clique Percolation Method (CPM) community mining. We also present an approach to event tracking based on the relationships between communities. The *Twitter* messages related to the 2013 Australian Federal Election are used to demonstrate the effectiveness of our approach. As the first of this kind, our system provides a benchmark for further development of monitoring tools for social events.

1 Introduction

Micro-blog services like *Twitter* have become useful sources for watching real-world events. The monitoring of events over social networks has many applications such as decision making and situation awareness. As a particular event develops, people may be interested in seeing an overview of the situation. An event may have several related topics that develop over time. In this paper, we introduce a new concept called invariant event tracking. An event is a social activity or a phenomena that occurs in a certain place during a certain time period. Event tracking is to monitor streams of topic-discussions in order to understand the event. A series of changing topics derived from an event over time is called an invariant event. In general, a topic is associated with a set of keywords. At any point in time, there are multiple topics discussed on social networks. Invariant event tracking is important for analyzing the overall situation of a particular event on social networks. For example, during a natural disaster, government may need to analyze the development of situations in order to

M. Renz et al. (Eds.): DASFAA 2015, Part II, LNCS 9050, pp. 517–521, 2015.
DOI: 10.1007/978-3-319-18123-3_31

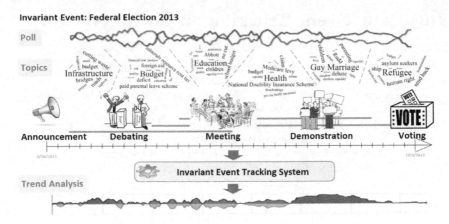

Fig. 1. Invariant event tracking conceptual diagram

make the right responses at the right times. For a longer-running event like a government election, people may wish to track the event with-respect-to multiple issues such as campaign-launch speeches and a number of open TV-debates under different topics, in order to cast their votes.

However, general micro-blog searches for given keywords return large amounts of messages that are not grouped or organized in any meaningful way. It is difficult for people to comprehend a large number of messages in a chronological order and to monitor an event as it unfolds. Although traditional techniques such as clustering are able to capture major events in social networks [1,4], it is difficult to capture the incidental events that may or may not be relevant to the current focused event. For example, when a natural disaster has just occurred, people may initially talk about the natural phenomenon that they have just witnessed. Then, damages, casualties, or the consequences of the disaster might be reported. Topics related to volunteer organizations and rescue activities might also be discussed later. All these topics are related to the same event, yet a general clustering approach is not able to correlate them into a single event.

To the best of our knowledge, every event develops in a unique way and is transitionally coherent to a set of keywords at different stages of its life time. So the only restriction should be on its time of emerging and time of finishing. In other words, we should monitor an event not by any pre-specified patterns, but by the subsequently developed topics related to the event and follow their changes within the given time frame. In this paper, we propose an approach of invariant event tracking on social networks. We use our system to track an event based on micro-blog messages and monitor the topic changes over time for an event that is rendered to the system as a set of keywords. The research challenges are: (1) effectively summarizing the given event-search query (termed as an invariant event), and (2) tracking the evolution of an event within a given time period.

2   System Architecture

In order to show a comprehensive understanding of our invariant event tracking framework, a conceptual diagram is presented in Fig. 1. The architecture of our system consists of two components, including micro-blog loader and pre-processing, and invariant event tracking.

2.1   Micro-blog Loader and Pre-processing

A micro-blog loader is developed to collect the *Twitter* messages from public users via the Java library API service. The user's initial query (i.e., a set of keywords) is used for specifying an event. The pre-processing was designed to ignore common words that carry less important meaning than keywords and to remove irrelevant data such as *re-tweet* keyword and web address. The stop words are removed and all words are stemmed.

2.2   Invariant Event Tracking

Event detection is to identify hierarchically nested event topics that break down an event into more refined parts. Then, event tracking will be performed to discover an invariant event.

Event Detection: we aim to group the co-occurring keywords for topic discovery. Note that the concepts of event and topic are different; an event may have several topics at different stages in its life cycle. We adopt the idea of community detection in graphs for locating and analyzing overlapping dense user groups in social networks [2]. In our approach, a so-called community that represents a set of users is termed as an episode that includes the topics related to an event at a certain time frame. Therefore, in our approach, the migration of members amongst communities is treated as the evolution of the topics amongst different episodes in an invariant event. We partition the messages into time frames. For each time frame, co-occurring keywords that appear together in at least *min_occur* are extracted. To compute co-occurring words, we exclude *re-tweet* messages. Networks of keywords are then constructed as graphs. Finally, the keywords in an episode are grouped along with the topics of the episode. Each episode represents one or more event topics in a particular time frame.

Event Tracking: at this stage, we aim to identify an invariant event by tracking all the event topics detected at each time frame. The event evolution is represented by a series of episodes from different time frames. In order to capture the changes of episodes, we consider five types of transitions (i.e., form, dissolve, survive, split, and merge) [3]. Topic evolution is a sequence of changes succeeding each other in the consecutive time frame. All event topics, which are linked together over time frames, are represented as an invariant event. For each time frame, node is the keyword and edge between the nodes is formed when those keywords co-occur in at least *min_occur* times.

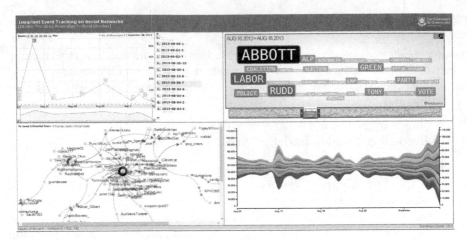

Fig. 2. Dashboard of our system

3   Demonstration Scenario

For demonstration, a collection of messages posted by Australian locals (given latitude, longitude and radius) via the *Twitter Search API* service from 4 August 2013 to 8 September 2013 with 808,661 messages with the user's initial event-query, is used. We define an event by specifying the keyword query (i.e., *"#ausvotes13"*, *"#election2013"*, *"#AusVotes"*, *"#auspol"*, *"Kevin Rudd"* and *"Tony Abbott"*). We decided to choose this period because the election date was announced on 4 August 2013 and people started discussions on this event. Also, we chose the keywords related to the two candidates because in Australian politics the candidates will be from the two major parties.

We designed a dashboard to display an invariant event and topic evolution over time. Events are presented via *Annotated Time Line Chart* as shown in Fig. 2 (top-left) for each day (represented by letters A to Z). For a given invariant event, the size of *Timecloud* indicates the frequency of words over the selected time period as shown in Fig. 2 (top-right). A re-tweet network is constructed and influential users are shown in Fig. 2 (bottom-left). *Stream graph*, which is a visualization for displaying multiple time series, is used to show the number of people using the words over time as shown in Fig. 2 (bottom-right).

4   Conclusions

In this paper, we proposed an approach to tracking invariant events and topic evolution within a given time period. The main contributions of this paper are twofold. (1) An effective approach of tracking invariant events is proposed by incorporating CPM community mining and community evolution discovery techniques. (2) We have implemented an invariant event tracking system which provides user with an overview of the development of an event. Due to the limited space of this paper, we will give the detailed algorithms and the further performance evaluation in our future articles.

References

1. Li, C., Sun, A., Datta, A.: Twevent: segment-based event detection from tweets. In: CIKM, pp. 155–164 (2012)
2. Palla, G., Derényi, I., Farkas, I., Vicsek, T.: Uncovering the overlapping community structure of complex networks in nature and society. Nature **435**, 814–818 (2005)
3. Takaffoli, M., Sangi, F., Fagnan, J., Zaiane, O.R.: Modec - modeling and detecting evolutions of communities. In: ICWSM, pp. 626–629 (2011)
4. Unankard, S., Li, X., Sharaf, M.A.: Emerging event detection in social networks with location sensitivity. World Wide Web Journal, 1–25 (2014)

EmoTrend: Emotion Trends for Events

Yi-Shin Chen, Carlos Argueta, and Chun-Hao Chang[✉]

Intelligent Data Engineering and Applications Laboratory, Institute of Information
Systems and Applications, National Tsing Hua University , No. 101, Section 2,
Kuang-Fu Road, Hsinchu 30013 , Taiwan, Republic of China
ccha97u@gmail.com

Abstract. In this demo paper we present EmoTrend, a web-based
system that supports event-centric temporal analytics of the global mood,
as expressed in Twitter. Given a time range, and optionally a set of keywords, the system relies on peak frequencies, and the social graph, to identify relevant events. Subsequently, by performing sentiment analysis on
related tweets, the global impact and reception of the events are presented
by a visualization of the overall mood trend in the time range.

1 Introduction

The widespread use of connected portable devices has allowed users to spread
news as they happen, making microblog services the most rapid news outlets
in the world. In this demo paper, we present EmoTrend, a system that automatically identifies events from Twitter streams and summarizes the impact
on society in a meaningful way. The system first clusters keywords with frequency peaks as candidate event descriptors. It then uses evolving social graphs
to identify meaningful events. Batches of tweets related to the detected events
are sentiment-analyzed based on emotion-bearing patterns. Finally, the user is
presented with a timeline of events. By selecting an event, the evolution through
time of the global mood towards the event is summarized based on six basic
emotions—anger, fear, hope, joy, sadness, surprise. Such tool can help a user to
better understand how an event has evolved and how it has impacted society.

2 Methodology

The system presented in this demo paper has two major components: the Event
Detection component and the Mood Summarization component.

2.1 Event Detection

Good keywords in tweets must satisfy two special criteria: meaningfulness and
burstiness. To detect meaningful keywords, profanity is first filtered out. Then,
an adaptation of the m-function used by the Porter Stemmer [1] is used to
remove words without meaning. To determine whether or not a keyword is bursty

© Springer International Publishing Switzerland 2015
M. Renz et al. (Eds.): DASFAA 2015, Part II, LNCS 9050, pp. 522–525, 2015.
DOI: 10.1007/978-3-319-18123-3_32

(i.e. it has been frequently used during a specific period of time), we calculate its frequency inside a sliding window. Using z-scores, we obtain a probability statistic determining how likely the word frequency in the past window is less than or equal to the frequency in the current timeframe. If that value is big enough, then the word is considered bursty and with high temporal use.

The filtered keywords are then grouped into event candidates. To achieve this, an event graph is constructed for each time frame. The vertices correspond to keywords and each edge connects co-occurrent keywords within a same tweet. A weight for each directed edge is computed based on frequency and co-occurrence statistics. PageRank [2] is then used to rank the keywords. The top keywords and their strongest neighbors are grouped as event candidates.

Concept-Based Evolving Graph Sequences (cEGS) [3] is a sequence of directed graphs representing information propagation within social streams. In this system, one such sequence of graphs is built for each event candidate that is monitored. Given a cEGS for a specific event, a directed graph is built for every day, with its vertices being the users that mention one or more event keywords on that day, and its edges representing a "following" relationship between two users.

2.2   Mood Summarization

After the events in the time frame have been identified, related tweets are fed, in order of publication, to the sentiment analyzer. Using emotion-bearing patterns, the analyzer summarizes the overall mood of the global community towards the event, and how it evolves in time. The steps performed are described in the following subsections.

By monitoring word frequencies in tweets, the system separates words into high-frequency (HW) and low-frequency words. Infrequent words that appear in a dictionary obtained from LIWC's [4] [5] psychological categories are deemed as psychological-words (PW). Subsequences of words pertaining to a combination of HW (e.g. "this") and PW (e.g. "hate", "love", "beach"), and appearing frequently in tweets, are grouped together based on matching HWs (e.g. "hate this weather" and "love this beach"). Subsequently, by having their PWs replaced by a wildcard (e.g. ".+ this .+"), the subsequences form emotion-bearing patterns. Patterns that fall below a frequency threshold are discarded.

An adaptation of the term frequency-inverse document frequency (tf-idf) statistic is used. By viewing patterns as terms, and collections of tweets per emotion as documents (6 documents total), the tf-idf is adapted to include a third score based on how many PWs in a collection can be captured by a pattern with its wildcard. The result is one ranking per emotion class, where top patterns are both more relevant to the class and bear a high level of emotion.

For each tweet fed to the sentiment analyzer, two different classifiers are used, a bag-of-words style classifier, and a Neural Networks classifier. The top two emotions obtained from the classifiers are used for each tweet. Finally, the overall mood state towards an event is computed for every day based on the individual emotions expressed in tweets related to that event in specific days.

3   Demonstration Overview

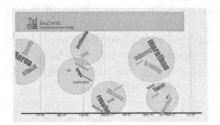
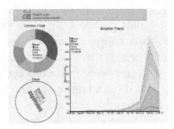

Fig. 1. Main page of the system The Timeline

Fig. 2. Detailed view for event Boston Marathon Bombing

In this section the functionality of the proposed system is at http://www. cs.ccu.edu.tw/~ccha97u/emotrend/. When opening the system's main page, the user will be presented with the screen in Figure 1. The screen presents a timeline showing the different detected events. Scrolling horizontally on the timeline the user can move forward or backward in time. As the user moves, the timeline will update to show the corresponding events. Each event is represented by a circle which extends across the timeline according to its duration. Inside the circle the user is presented with the most important keywords representing the event. As can be seen in Figure 1, the event for the Marathon bombing contains relevant keywords, such as *marathon*. The size of the font for each keyword is proportional to the frequency with which it was used in related tweets. Clicking a circle brings the detailed view for the specific event.

The detailed view for each event has different components. In this demo scenario let's assume a user clicks on the circle representing the Boston Marathon bombing event. The detailed view, presented in Figure 2, shows three main components. The Emotion Pizza, like a regular pie chart, presents the proportion of tweets expressing each of the six emotions detected by the emotion detection algorithm. The Event circle is the same circle representing the event in the timeline and contains the most important keywords for the event. The Emotion Trend is an area chart used to represent the emotion trend over time.

The detailed view allows a user to inspect the impact of the event in society by observing how the global emotion evolves. Considering the Boston bombing event, from this view it can be seen at a glance that the predominant emotions are fear and sadness as direct consequences of such a tragedy, but also hope as expressed in tweets wishing the survivors to recover soon.

Additionally, the interface allows the user to restrict the summary to a specific emotion or subinterval of time. For instance, by clicking on an emotion name, or corresponding portion in the Emotion Pizza, the Emotion Trend will show only the area over time for that specific emotion. The resulting peaks give the user an idea of when the event caused specific emotional reactions on the society. Clicking on other emotions disable or enable them, updating the Emotion Trend accordingly. Additionally, if the user clicks and drags the mouse along the Emotion Trend graph, the Emotion Pizza gets updated to show the proportions of tweets with corresponding emotions within the selected time interval only. This is particularly useful to better understand how the emotions evolved by observing the predominant reactions from people at specific times.

4 Conclusions

In this paper we presented EmoTrend, a system that provides temporal summarization of the global mood towards interesting events mined from Twitter. For the regular user, EmoTrend constitutes an interesting tool to understand how a society is affected, and its mood evolves, during and after events happen (e.g. natural disasters, royal weddings). For corporations, politicians, and stars, EmoTrend represents a clear chance to grasp the overall reception of their events as they happen (product releases, campaign speeches), giving the possibility of better decision-making in order to improve image and products. We have demonstrated the ability of our system to extract events, and summarize their impact in a comprehensive way, an ability that traditional news articles and encyclopedia entries lack.

References

1. Porter, M.F.: An algorithm for suffix stripping. Program: Electronic Library and Information Systems 14(3), 130–137 (1980)
2. Page, L., et al.: The PageRank citation ranking: Bringing order to the web (1999)
3. Kwan, E., et al.: Event identification for social streams using keyword-based evolving graph sequences. In: Proceedings of the 2013 IEEE/ACM International Conference on Advances in Social Networks Analysis and Mining. ACM (2013); APA
4. Pennebaker, J.W., et al.: The development and psychometric properties of LIWC 2007, Austin, TX, LIWC. Net (2007)
5. Tausczik, Y.R., Pennebaker, J.W.: The psychological meaning of words: LIWC and computerized text analysis methods. Journal of Language and Social Psychology 29(1), 24–54 (2010)

A Restaurant Recommendation System
by Analyzing Ratings and Aspects in Reviews

Yifan Gao[1], Wenzhe Yu[1], Pingfu Chao[1], Rong Zhang[1]([⊠]),
Aoying Zhou[1], and Xiaoyan Yang[2]

[1] Shanghai Key Laboratory of Trustworthy Computing, Institute for Data Science
and Engineering, East China Normal University, Shanghai, China
{yfgao,wyu,pfchao}@ecnu.cn, {rzhang,ayzhou}@sei.ecnu.edu.cn
[2] Advanced Digital Sciences Center, Illinois at Singapore Pte. Ltd.,
Singapore, Singapore
xiaoyan.yang@adsc.com.sg

Abstract. Recommender systems are widely deployed to predict the
preferences of users to items. They are popular in helping users find
movies, books and products in general. In this work, we design a restaurant recommender system based on a novel model that captures correlations between hidden aspects in reviews and numeric ratings. It is
motivated by the observation that a user's preference against an item
is affected by different aspects discussed in reviews. Our method first
explores topic modeling to discover hidden aspects from review text.
Profiles are then created for users and restaurants separately based on
aspects discovered in their reviews. Finally, we utilize regression models
to detect the user-restaurant relationship. Experiments demonstrate the
advantages.

Keywords: Recommender systems · Review analysis · Hidden aspect ·
Regression model

1 Introduction

Recommendation systems have been widely deployed in Web applications to
predict the preferences of users to items. There are two major approaches to
produce recommendations [3]: *content-based filtering* (CBF) and *collaborative
filtering* (CF). CBF explores properties of items to recommend additional items
with similar properties. However, CBF systems tend to make overspecialized
recommendations. CF models users' past behavior and preferences from similar
users. CF systems concentrate on modeling user-item relationship to make recommendations [3]. In this demo, we explore the second approach and examine
the relationship between users and restaurants by analyzing ratings and aspects
in reviews to build a restaurant recommendation system.

Customer reviews, which are usually associated with numeric ratings, have
become commonly available on various review websites. Web portals such as
Yelp and *Dianping*[1] accommodate extensive reviews covering large numbers of

---

[1] yelp: http://www.yelp.com; Dianping: http://www.dianping.com

© Springer International Publishing Switzerland 2015
M. Renz et al. (Eds.): DASFAA 2015, Part II, LNCS 9050, pp. 526–530, 2015.
DOI: 10.1007/978-3-319-18123-3_33

restaurants from different countries. These reviews contain abundant information about users' opinions and preferences, which are valuable to any recommendation system.

2   System

The flowchart of our restaurant recommender system is presented in Figure 1. The input to our system is a review corpus and a target user u. The output is a list of top-k restaurants recommended for u. Our system consists of two main components: *profile generator* and *rating predictor*.

2.1   Profile Generator

We map users and restaurants to a common latent space $\mathcal{S}$ discovered from the review text. The intuition is that reviews, though in the format of unstructured free text, contain information about user preferences and opinions on different aspects of items. These aspects 'hidden' in the review text may well reflect the latent factors that affect the user ratings. To find these hidden aspects and construct the latent space, we apply standard Latent Dirichlet Allocation (LDA)[1] to reviews, which is an effective tool for extracting topics from pure texts.

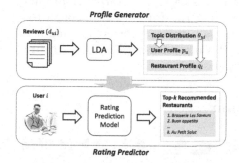

Fig. 1. System Flowchart

Let d_{ui} denote the review of restaurant i by user u and we treat each d_{ui} as one document. We apply LDA on the review corpus $\{d_{ui}\}$ and discover K topics. Let θ_{ui} denote the topic distribution of d_{ui} generated by LDA. Define D_u as the set of reviews written by user u, and D_i as the set of reviews written for restaurant i. Each user u (or restaurant i) is associated with a profile p_u (or q_i), which is a vector from $\mathcal{S}$. In our system, $\mathcal{S} = [0, 1]^K$. For a given user u, we define her profile p_u as: $p'_{uj} = \frac{\sum_i \theta_{uij}}{|D_u|}$, and $p_{uj} = \frac{p'_{uj}}{\sum_j p'_{uj}}, j \in [1, K]$. $p_u = (p_{u1}, p_{u2}, \cdots, p_{uK})$, p_{uj} is the distribution on the jth topic for user u, and θ_{uij} is the distribution on the jth topic for review d_{ui}. Similarly, we define profile q_i for restaurant i as: $q'_{ij} = \frac{\sum_u \theta_{uij}}{|D_i|}$, and $q_{ij} = \frac{q'_{ij}}{\sum_j q'_{ij}}, j \in [1, K]$. In summary, profile p_u/q_i is the normalized average topic distribution over all reviews of a given user u/restaurant i.

2.2 Rating Predictor

Given a user u, we want to predict the rating $\hat{r}_{ui}$ for restaurant i. Recommendations are then made based on $\hat{r}_{ui}$ of restaurants that u have not rated/visited. To predict ratings, we rely on the intuition that hidden topics discovered from review text define the latent factors that affect the ratings. Our rating prediction model is build on linear/logistic regression to model the relationship between ratings $\hat{r}_{ui}$ and topic distributions θ_{ui} of d_{ui}.

Linear Regression A rating is predicted by the following function: $\hat{r}_{ui} = W^T \theta_{ui} + \varepsilon_{ui}$. $W = (W_1, ..., W_K)$, W_j is the weight of the jth topic, and ε_{ui} is an error variable.

Logistic Regression Assume ratings $r_{ui} \in \{1, 2, ..., N\}$. We build a multinormal logistic regression model as : $\Pr(\hat{r}_{ui} = n) = \frac{e^{\beta_n^T \theta_{ui}}}{1 + \sum_{n'=1}^{N-1} e^{\beta_{n'}^T \theta_{ui}}}$, and $\Pr(\hat{r}_{ui} = N) = \frac{1}{1 + \sum_{n'=1}^{N-1} e^{\beta_{n'}^T \theta_{ui}}}$. $n = 1, 2, ..., N-1$, and $\beta_n = (\beta_{n1}, \beta_{n2}, ..., \beta_{nK})$ are the weights.

Rating Prediction Given a user u and a restaurant i that u has not rated, we estimate the topic distribution $\hat{\theta}_{ui}$ based on p_u and q_i as: $\theta'_{uij} = p_{uj} q_{ij}$, and $\hat{\theta}_{uij} = \frac{\theta'_{uij}}{\sum_j \theta'_{uij}}, j \in [1, K]$. $\hat{\theta}_{ui}$ is then fed into one of the learned regression model to predict $\hat{r}_{ui}$. Restaurants with top-k $\hat{r}_{ui}$ are returned as recommendations to user u.

Representative Review Selection Our system provides a new browsing tool that enables efficient access to representative restaurant reviews. A review is considered to be representative if it is 'close' enough to the restaurant profile. We measure the closeness between a review r_{ui} and a restaurant i as follows: $d(r_{ui}, i) = ||\theta_{ui} - q_i||_2^2$. Reviews with smallest $d(r_{ui}, i)$ are presented to users as representatives of restaurant i.

3 Demonstration

(a) User Selection   (b) User Profile   (c)   Recommended (d) Restaurant Profile
                                                    Restaurants   file

Fig. 2. Interface

For demonstration like Figure 2, we collect review data and restaurant details from *Dianping*, the largest restaurant review website in China. To make a recommendation, we first display several *Dianping* users randomly to choose from in Figure 2a. As a target user u is selected, u's preference and review history would be created (Figure 2b). Then in the next interface (Figure 2c), our system recommends top-k restaurantlist to u. These restaurants are selected from those that u has not rated/visited, based on their ratings produced by our rating predictor. To gain further information on recommendations, our system leads to the restaurant profile interface (Figure 2d) including recommended food according to its frequency of occurrence in review text and a list of representative reviews for efficient browsing and access to mainstream viewpoints on the restaurant.

4 Experiments

Let R-Linear and R-Logistic denote our rating predictor using linear and logistic regression respectively. We test on reviews crawled from *Dianping*. It consists of 1,168,420 reviews written by 316,702 users for 42,274 restaurants in Shanghai. Each review contains a user id, a numeric rating, a timestamp and a piece of review text. The ratings r_{ui} take value from $\{1, 2, ..., 5\}$. For evaluation purpose, we split the dataset into two subsets for training and test purpose respectively with the ratio 9 to 1 according to two different ways: "*random*" with test data selected randomly and "*by time*" with the latest 1/9 data as test data. We use MSE (mean squared error) and ACC (Accuracy) as the evaluation metrics following [2]. $MSE = \frac{1}{M} \sum_{u,i} (\hat{r}_{ui} - r_{ui})^2$, where M is the total number of predicted ratings. We define $acc = \frac{m}{M}$. Let m denote the number of ratings correctly predicted. For R-Linear (linear regression), the predicted ratings $\hat{r}_{ui}$ is first rounded to the nearest integer and then compared with r_{ui}.

Baseline We compare our rating predictor with HFT [2], the work most related to ours.

Parameters of LDA We set the total number of topics $K = 10$ and the hyperparameters $\alpha = 0.2$, $\beta = 0.1$. The number of iterations for LDA is set to 100.

Table 1. MSE of Dianping Data

	Random	by Time
HFT	0.66	1.31
R-Linear	0.69	**0.66**
R-Logistic	0.79	0.74

Table 2. Accuracy of Rating Prediction on Dianping Data

	Random	by Time
HFT	50.0%	36.4%
R-Linear	**52.3%**	48.7%
R-Logistic	52.0%	**50.0%**

Results The results are shown in Table 1 and Table 2. In general, our model performs better than HFT on test data with different partitions on both MSE and Acc.

Acknowledgments. Rong Zhang and Aoying Zhou are partially supported by National 863 Program under grant No. 2015AA011508, National Science Foundation of China under grant No.61232002 and No.61332006, and the Natural Science Foundation of Yunnan Province (2014FA023). Xiaoyan Yang is supported by Human-Centered Cyber-physical Systems (HCCS) programme by A*STAR in Singapore.

References

1. Blei, D.M., Ng, A.Y., Jordan, M.I.: Latent dirichlet allocation. The Journal of Machine Learning Research **3**, 993–1022 (2003)
2. McAuley, J., Leskovec, J.: Hidden factors and hidden topics: understanding rating dimensions with review text. In: Proceedings of the 7th ACM conference on Recommender systems, pp. 165–172. ACM (2013)
3. Rajaraman, A., Ullman, J.D.: Mining of Massive Datasets, Cha. 9. University Press, Cambridge (2012)

ENRS: An Effective Recommender System Using Bayesian Model

Yingyuan Xiao[1,2(✉)], Pengqiang Ai[1,2], Hongya Wang[3],
Ching-Hsien Hsu[2], and Yukun Li[1]

[1] Tianjin University of Technology, Tianjin 300384, China
yyxiao@tjut.edu.cn
[2] Tianjin Key Lab of Intelligence Computing and Novel Software Technology,
Tianjin 300384, China
[3] Donghua University, Shanghai 201620, China

Abstract. Traditional content-based news recommender systems strive to use a bag of words or a topic distribution to capture readers' reading preference. However, they didn't take advantage of the named entities extracted from news articles and the relations among different named entities to model readers' reading preference. Named entities contain much more semantic information and relations than a bag of words or a topic distribution. In this paper, we design and implement a prototype system named ENRS, which combines the named entity with the naïve Bayesian algorithm, to recommend readers news articles. The key technical merit of our work is that we built a probabilistic entity graph to capture the relations among different named entities, based on which ENRS can increase the diversity of recommendation significantly. The architecture of ENRS and the recommendation algorithm are discussed and a demonstration of ENRS is also presented.

1 Introduction

Recommender system has become one of the most popular and profitable applications using state-of-art knowledge over the past decades. Recently, recommending news articles has attracted more research attention. Many companies or websites have developed news recommender systems, such as Google News and Yahoo! News, provide personalized news recommendation services for substantial amount of online users [1]. In general, existing news recommender systems are usually classified into three different categories: content-based system, collaborative filtering and hybrid recommendation which combine content-based technology and collaborative filtering.

Currently, content-based news recommender systems are mainly focusing on a bag of words [2] or a topic distribution [3] extracted from news content. However, few news recommender systems take advantage of named entities extracted from news articles and even the relations among the named entities. Typically, in news articles, named entities can provide readers information that describing what happened, when the event happened, where it happened, who were involved, and so on. Named entities contain much more semantic information than a bag of words or a topic distribution.

© Springer International Publishing Switzerland 2015
M. Renz et al. (Eds.): DASFAA 2015, Part II, LNCS 9050, pp. 531–535, 2015.
DOI: 10.1007/978-3-319-18123-3_34

ENRS is a novel news recommender system, which takes advantage of the named entities extracted from news articles and the relations among them. Based on readers' reading
history, ENRS can model readers' named entities preference and recommend readers news
articles that they might be interested in. Besides, we build a probabilistic entity graph in
our demo to help readers to explore fresh preference. The probabilistic entity graph increases the diversity of recommendation as well. In this paper, we give an overview of the
architecture of ENRS and its major modules including a briefing on the algorithms in the
Section 2. We then demonstrate a case in Section 3.

2   Overview and Architecture

ENRS is a news recommender system, which takes advantage of named entities extracted
from news articles and the relations among these named entities. In ENRS, all news articles are crawled from news websites on the Internet. For each news article, ENRS extracts three types of named entities – Person, Place, and Organization - from news article.
Based on reader's reading history, ENRS can model reader's named entity preference. We
also build a probabilistic entity graph to capture relations among named entities and increase the diversity of recommended news articles.

2.1   Probabilistic Entity Graph

Named entities can provide readers important information, which is important and useful
to capture readers' preference. However, there are some relations and semantic information among different named entities. For example, there are some relations between
Google and Apple. Both of them are technology companies. At the same time, Apple is a
competitor of Google. Therefore, a reader who is interested in news about Google might
also prefer to read news about Apple.

Fig. 1. Part of the Probabilistic Entity Graph

In ENRS, by using the conditional probability, we create a probabilistic entity graph to
capture and represent the relations among different named entities. Figure 1 illustrates part of
our probabilistic entity graph. Based on the probabilistic entity graph, ENRS recommends

readers various news articles that he/she may be interested in, which can also help readers to explore fresh preference according to their existing preference. We build a corpus of technology news from news websites to capture the relations among different named entities extracted from news articles.

We assume that, if two different named entities occur in the same news article, the two named entities have a certain relation. In particular, we define that,

- e is a named entity extracted from news articles;
- $E = \{e_i \mid i = 1,2, \cdots, N\}$ denotes the set of all N named entities extracted from the corpus;
- $W = \{\omega_{i,j} \mid$ the weight of e_i to e_j, $e_i, e_j \in E\}$ represents the set of weight.

Therefore, our probabilistic entity graph is $G = \{E, W\}$, where

$$\omega_{i,j} = P\big(e_j \mid e_i\big) = \frac{P(e_i, e_j)}{P(e_i)}, \; \omega_{j,i} = P\big(e_i \mid e_j\big) = \frac{P(e_i, e_j)}{P(e_j)} .$$

In our demo, $\omega_{i,j}$ represents the probability that a reader who is interested in e_i may also has a preference for e_j . Based on the corpus, it is easy to calculate $P(e_i), P(e_j)$ and $P(e_i, e_j)$. Then, $\omega_{i,j}$ can be calculated. Therefore, our system can recommend readers news articles about e_j, if $\omega_{i,j}$ is big enough.

2.2   Reader Profile

In our system, two kinds of data will be recorded into readers' history data: (1) news articles that are clicked and read by readers; (2) news articles that are showed to readers, but the readers don't click and read. After named entity detection, each news article can be represented as a named entity vector that describes what named entities are involved in the article. In particular, we denote $V = (e_1, e_2, \cdots, e_k)$ k named entities extracted from a news article. Then, based on reader' reading history, the reader's profile can be parameterized with a two-attribute tuple $\mathcal{U} =< L, T >$, where

- $\mathcal{L}$ represents the named entity distribution of news articles that the reader clicked and read, in the format of a named entity vector $\{< e_1, \omega_1 >, < e_2, \omega_2 >, \cdots\}$, where each entry consists of an named entity and the corresponding weight;
- $\mathcal{T}$ denotes the named entity distribution of news articles that the reader didn't click and read when the news articles were showed to the reader, in the format of a named entity vector $\{< e_1, \omega_1 >, < e_2, \omega_2 >, \cdots\}$, where each entry consists of an named entity and the corresponding weight.

2.3   Naïve Bayesian Recommendation

Formally, a news article is represented as a named entity vector $V = (e_1, e_2, \cdots, e_k)$. Given a news article $V = (e_1, e_2, \cdots, e_k)$, $P(read \mid V)$ denotes the probability of the reader read this news article, and $P(unread \mid V)$ represents the probability of the reader doesn't read this news article. We assume that the reader will read a news article if $\theta > 1$, where

$$\theta = \frac{P(read \mid V)}{P(unread \mid V)} .$$

According to the Bayesian' theorem, given a reader's profile $\mathcal{U} =<L, T>$ and a news article $\mathcal{V} = (e_1, e_2, \cdots, e_k)$,

$$P(read \mid \mathcal{V}) = \frac{P(\mathcal{V} \mid read) \cdot P(read)}{P(\mathcal{V})}, \quad P(unread \mid \mathcal{V}) = \frac{P(\mathcal{V} \mid unread) \cdot P(unread)}{P(\mathcal{V})}.$$

According to the naïve Bayesian theorem assume that each named entity is conditionally independent of every other named entity for $i \neq j$, given the news article $\mathcal{V} = (e_1, e_2, \cdots, e_k)$. This means that

$$P(\mathcal{V} \mid read) = P(e_1, e_2, \cdots, e_k \mid \mathcal{V}) = \prod_{i=1}^{k} P(e_i \mid \mathcal{V}).$$

However, we found that $P(e_i \mid \mathcal{V})$ sometimes is very small and even is 0. This will make $P(\mathcal{V} \mid read)$ very small or even equal to 0. Therefore, we define

$$\varepsilon = \ln \theta = \ln \frac{P(\mathcal{V} \mid read) \cdot P(read)}{P(\mathcal{V} \mid unread) \cdot P(unread)}.$$

If $\varepsilon > 0$, which means that the reader is more likely to read this news article, our system will recommend the news article to readers.

Fig. 2. Recommendation interface of ENRS

3   Demonstration

In this section, we demonstrate our ENRS by registering a reader. After registering as a reader in ENRS, we clicked some news articles about Google and Microsoft. When we log onto ENRS again, we can find that news articles about Google and Microsoft are recommended to reader, and news articles about Yahoo and Amazon are also recommended. Figure 2 is the recommendation interface of our system. Therefore, we can conclude that both probabilistic entity graph and naïve Bayesian recommendation in ENRS are useful and effective.

Acknowledgment. This work is supported by the NSF of China (No. 61170174, 61370205, 61170027) and Tianjin Training plan of University Innovation Team (No.TD12-5016).

References

1. Liu, J., Dolan, P., Pedersen, E.R.: Personalized news recommendation based on click behavior. In: Proc. of IUI, Hong Kong, China (2010)
2. Capelle, M., Frasincar, F., Moerland, M., Hogenboom, F.: Semantics-based news recommendation. In: Proc. of WIMS, New York, USA (2012)
3. Li, L., Wang, D., Li, T., Knox, D., Padmanabhan, B.: Scene: a scalable two-stage personalized news recommendation system. In: Proc. of SIGIR, New York, USA (2011)

EPSCS: Simulating and Measuring Energy Proportionality of Server Clusters

Jiazhuang Xie[1], Peiquan Jin[1,2(✉)], Shouhong Wan[1,2], and Lihua Yue[1,2]

[1] School of Computer Science and Technology, University of Science
and Technology of China, Hefei 230027, China
[2] Key Laboratory of Electromagnetic Space Information, Chinese Academy
of Sciences, Hefei 230027, China
jpq@ustc.edu.cn

Abstract. Energy proportionality for a server cluster means energy consumption is proportional to the workloads running on the cluster. One problem is that it is too costly, time-consuming, and complex to build a real cluster to evaluate energy-proportional algorithms. Aiming to solve this problem, we propose to build a prototype system that is able to simulate and test energy proportionality of a server cluster quickly and easily. Our system, named *Energy-Proportional Server Cluster Simulator (EPSCS)*, allows users to configure a virtual server cluster and test energy proportionality on real traces continuously. We implement three energy-proportional algorithms in *EPSCS* and visualize the real-time results to evaluate time performance and energy consumption. New algorithms can be integrated into *EPSCS* and be compared with existing ones. We first describe the architecture and key designs of *EPSCS*. And finally, a case study of *EPSCS's* demonstration is presented.

Keywords: Energy proportionality · Server cluster · Measurement · Simulation

1   Introduction

Energy consumption has been recognized as a critical problem in both mobile applications [1] and database systems [2]. Previous studies showed that a Google server cluster only used about 20% of their CPU capabilities [3]. In other words, we need not run all the servers to provide services in most cases; thus we can adjust the size of a server cluster with workload changes by dynamically turn on/off some nodes so that we can realize better *energy proportionality*.

Energy proportionality means that the energy consumption of a system is proportional to the workloads running on it [3]. Constructing an energy-proportional server cluster is an effective way to reduce the energy consumption. However, one problem is that it is difficult to perform studies towards energy proportionality over server clusters, because it is costly and time-consuming to build a real large-scale server cluster and integrate energy-proportional algorithms into the cluster. Although there are some previous works focusing on simulating cloud computing environments [4] or distributed systems [5], they do not consider the energy consumption of each server and how to change the power states of servers.

© Springer International Publishing Switzerland 2015
M. Renz et al. (Eds.): DASFAA 2015, Part II, LNCS 9050, pp. 536–540, 2015.
DOI: 10.1007/978-3-319-18123-3_35

In this paper, aiming at providing a platform for quickly testing the effectiveness of energy-proportional algorithms, we present a flexible simulation tool, called *Energy-Proportional Server Cluster Simulator (EPSCS)*. The unique features of *EPSCS* are as follows:

(1) *EPSCS* allows users to configure different server clusters with different settings w.r.t. number of nodes, power parameters, and boot/shutdown times and powers. In addition, it is able to simulate the behaviors of a server cluster, such as job distribution and execution.

(2) *EPSCS* can visualize the real-time time performance and energy consumption of a server cluster continuously. In addition, it can output the results of execution time, energy consumption and other useful information for further studies.

(3) *EPSCS* implement several energy-proportional algorithms and provides real traces, which can be used as baseline algorithms and workloads for future research. New energy-proportional algorithms can be integrated into *EPSCS* easily.

2   Overview

2.1   Architecture of *EPSCS*

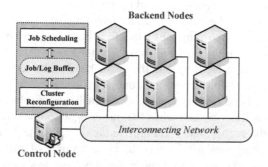

Fig. 1. Architecture of the Server Cluster Simulated   **Fig. 2.** Architecture of *EPSCS*

The architecture of the server cluster simulated is shown in Fig. 1, and Fig. 2 shows the software architecture of *EPSCS*, which consists of the following modules.

(1) **Cluster Setup**. The *Cluster Setup* module provides an interface for configuring the environment of the server cluster, such as the number of servers as well as the computing capacity and power metrics of each server.

(2) **Job Scheduling**. This module reads job information from the trace file and then distributed jobs among servers. We use a job buffer to maintain jobs. In addition, this module simulates the executing process of the server cluster by reading the jobs from the job buffer and modifying the log data in the log buffer shown in Fig. 2.

(3) **System Monitoring**. This module monitors the execution of the workloads and records necessary log data about the status of the cluster in a log buffer, which is then used by the *Cluster Reconfiguration* module to check whether a reconfiguration on the cluster is required.

(4) Cluster Reconfiguration. This module periodically checks the log-buffer data and employs a specific energy-proportional algorithm to determine whether the cluster needs to be reconfigured, i.e., some active servers need to be turned off and some power-off servers should be turned on. We suppose that each server can be in one of three states: *active, idle, standby*, where *standby* refers to power-off and *active* is power-on. We have to determine the number of active servers according to workload changes and determine which servers should be turned on/off. These tasks are performed by the energy-proportional algorithm.

(5) Visualization. This module visualizes the execution process of the trace. It continuously shows the current figures about time performance and energy consumption.

2.2 Key Designs in *EPSCS*

2.2.1 Job Quantification and Buffering

Previous studies showed that CPU is the most energy-consuming module in a server [2]. Thus, we use CPU time-slices to quantify the power metric of a job, and use I/O bandwidth and memory utilization to measure its time performance. We sum up the CPU time-slices of all the jobs in the server to indicate the current load on this server and aggregate the loads of all servers to represent the load of the cluster.

Further, we introduce a *Job Buffer* in the control node for realizing job scheduling. The job buffer is used to cache the new incoming jobs when the cluster is overloaded. We use the *FIFO* mechanism to maintain the buffered jobs. When a server is ready to receive new jobs, the control node will move the job in the queue head to the server.

The benefits of *Job Buffer* are three-fold. First, the control node can cache extra jobs to avoid the overloading of backend nodes. Second, as we cache extra jobs temporarily, powerful backend nodes will not be *idle* when the buffer is not empty. Thus, we can let the powerful nodes in the *active* state and improve the overall time performance of the cluster. Third, by caching new jobs during the wake-up time of *standby* nodes, we can avoid the overloading of current backend nodes, because new jobs are distributed to newly wake-up nodes.

2.2.2 Server Load Estimation

Server load estimation refers to predicting the future load on a server according to historical load information recorded in the log buffer shown in Fig. 2. We adopt a slide-window mechanism to figure out this issue. First, we record the load information of the cluster in the log buffer during a time window, forming a series of sampling data storing the load information at different time points. Next, when the time window is ended, we choose a set of critical points from all the sampling time instants in order to remove the noisy time points in the sample. Then, we use a linear fitting approach on the time series composed of critical points to estimate the future load of each server in the next time window.

2.2.3 Energy-Proportional Algorithms for Cluster Reconfiguration

We use an energy-proportional algorithm in the *Cluster Reconfiguration* module to determine whether a cluster reconfiguration is necessary as well as the policy of adjusting node states. We implement three algorithms in *EPSCS* as the baseline ones: (1) *AlwaysOn*. This policy always leaves the backend nodes active regardless of the workload change and thus behaves the best performance but consumes the most energy. (2) *Reactive*. This policy reacts to the current number of jobs, attempting to keep the right number of active nodes at each time instant.   (3) *Linear Regression*. This policy attempts to predict the future workload by using the linear regression method. It adjusts the number of active nodes according to the prediction result and the current capacity.

Users can integrate their new energy-proportional algorithms into *EPSCS* by extending internal classes and functions. While a new algorithm is added, it will automatically be included in the list shown at left-bottom in Fig. 3 for testing.

3   Demonstration

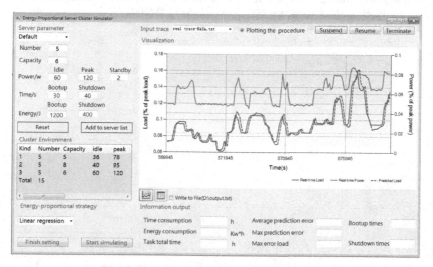

Fig. 3. User Interface of *EPSCS* Demonstration

EPSCS is implemented in C# under Windows 7. Our demonstration will use *EPSCS* to simulate different configurations of server clusters and visualize different energy-proportional algorithms. The user interface of the *EPSCS* is shown in Fig. 3. We will first show the configuring procedure of the cluster, and then show the process of simulation on real traces. We will also show how different energy-proportional algorithms differ in time performance and energy savings.

Acknowledgements. This work is supported by the National Science Foundation of China (61379037, 61472376, & 61272317) and the OATF project funded by University of Science and Technology of China.

References

1. Yang, P., Jin, P., Yue, L.: Exploiting the Performance-Energy Tradeoffs for Mobile Database Applications. Journal of Universal Computer Science **20**(10), 1488–1498 (2014)
2. Jin, Y., Xing, B., Jin, P.: Towards a Benchmark Platform for Measuring the Energy Consumption of Database Systems. Advanced Science and Technology Letters **29**, 385–389 (2013)
3. Barroso, L., et al.: The Case for Energy-Proportional Computing. IEEE Computer **40**(12), 33–37 (2007)
4. Calheiros, R.N., Ranjan, R., et al.: CloudSim: A Toolkit for Modeling and Simulation of Cloud Computing Environments and Evaluation of Resource Provisioning Algorithms. Software: Practice and Experience **41**(1), 23–50 (2011)
5. Buyya, R., Murshed, M.: Gridsim: A Toolkit for the Modeling and Simulation of Distributed Resource Management and Scheduling for Grid Computing. Concurrency and Computation: Practice and Experience **14**(13–15), 1175–1220 (2002)

MAVis: A Multiple Microblogs Analysis and Visualization Tool

Changping Wang, Chaokun Wang(✉), Jingchao Hao,
Hao Wang, and Xiaojun Ye

School of Software, Tsinghua University, Beijing 100084, China
wang-cp12@mails.tsinghua.edu.cn, {chaokun,yexj}@tsinghua.edu.cn

Abstract. An increasing number of people obtain and share information on social networks through short text messages, a.k.a. microblogs. These microblogs propagate widely online based on the followship between users as well as the retweeting mechanism. The regular pattern of retweeting behaviors can be discovered by mining the historical retweet data, and the key users in the information diffusion process can also be found in this way. This paper gives the novel definition of information diffusion network and three categories of nodes in the network. A tool designed to mine the information diffusion network and visualize the result is implemented. This paper introduces related definitions, the architecture, mining algorithms and the visualization interface.

Keywords: Online social network · Information diffusion · Visualization

1 Introduction

Recently, online social networks play an important role in information diffusion. A lot of works have been made to study the diffusion process, including detecting popular topics [1][2] and analyzing the network structure [3]. Several tools that analyze and visualize information diffusion process have emerged, such as SONDY [4] which can identify influential spreaders on a user-specified topic. However, data analysts sometimes hope to analyze the diffusion of multiple microblogs, which belong to different topics, posted by one particular user to discover diffusion patterns of this user's microblogs. There exist no tools that satisfy the above demand exactly, so we design and implement a Multiple Microblogs Analysis and Visualization Tool (**MAVis**).

Assuming the diffusion process of multiple microblogs from user u is known, we define the **information diffusion network** (**IDN**) of u as the collection of users and their corresponding retweeting behaviors to microblogs of u. User u is represented by the source node. Edges represent retweeting behaviors. To

This work was supported in part by the National Natural Science Foundation of China (No. 61373023, No. 61170064) and the National High Technology Research and Development Program of China (No. 2013AA013204)

M. Renz et al. (Eds.): DASFAA 2015, Part II, LNCS 9050, pp. 541–545, 2015.
DOI: 10.1007/978-3-319-18123-3_36

distinguish different patterns of users, nodes in the network are classified into three categories: nodes that can significantly increase the scope of microblogs are **explosive nodes**; nodes on the shortest paths from the source node to explosive nodes are **bridge nodes**; the remaining nodes are **normal nodes**.

MAVis processes the historical retweet data to discover the IDN of specific users and shows the network. The major contribution of this paper can be summarized as follows:

1. **Key User Discovery:** Through analyzing the historical retweet data of a specific user, MAVis discovers key users who significantly contribute to the diffusion process.
2. **Optimized Visualization:** The visualization module uses an optimization strategy named *incremental loading*, that a reduced network without normal nodes is shown at first, to improve the legibility of the visualization result.
3. **Extensibility and Portability:** The analysis module outputs the result in XML format and the visualization module processes them independently. Hence, the tool can be extended easily.

2   Architecture

As shown in Fig. 1, **MAVis** is composed of three parts: input and preprocessing module, analysis module and visualization module.

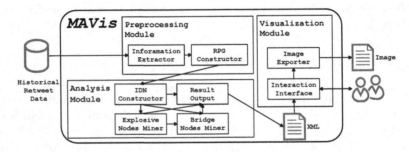

Fig. 1. The architecture of MAVis

2.1   Preprocessing Module

The preprocessing module extracts information from historical retweet data. It constructs the **retweeting process graph (RPG)** for each microblog. Different from IDN, RPG describes the retweeting behaviors of one single microblog. Each node represents a user and each edge represents a retweeting behavior in RPG. Besides the source node indicating the author, other nodes have at least one parent. RPGs are stored in the form of graph structure.

2.2   Analysis Module

Through mining the RPGs, the analysis module constructs the IDN and discovers explosive nodes and bridge nodes in the network. The analysis result is stored in XML files, rather than being showed directly. The independence between the two modules makes the tool more user-friendly and extendable.

2.3   Visualization Module

The visualization module implemented on Prefuse[1] shows the analysis result to end users via a user-friendly interface. Due to the large volume of the result, it uses *incremental loading* strategy to show the results efficiently. It provides rich user interaction features and is able to export the result as images.

3   Key Techniques

To implement MAVis, some key techniques have been used, such as mining explosive nodes and incremental loading. For the limitation of space, only the technique of mining explosive nodes is presented in this section.

In the IDN, a node u_e is an explosive node if it satisfies the following condition:

$$(\sum_{i=1}^{N} \frac{|\omega(u_e, m_i)|}{|\psi(m_i)|})/N > \lambda$$

where N is the cardinality of microblog set $M(u_r) = \{m_1, m_2, ..., m_N\}$, $\omega(u_e, m_i)$ represents the retweet number of microblog m_i caused by u_e's retweeting behaviour, $\psi(m_i)$ is the number of users who retweet m_i, and λ is an user-specified threshold.

The **explosion value** $\theta(u_e)$ which measures the importance of an explosive node u_e is defined as:

$$\theta(u_e) = \left(\sum_{i=1}^{N} \frac{\kappa_G(u_r, m_i) - \eta_G(u_r, u_e)}{\kappa_G(u_r, m_i)} * \frac{\omega(u_e, m_i)}{(\sum_{u_k \in U_G} \omega(u_k, m_i))/|U_G|}\right)/N$$

where $G(m)$ represents the RPG of microblog m, $\kappa_G(u_r, m_i)$ is the maximum of the shortest distance from the source node u_r to any other node in $G(m_i)$, $\eta_G(u_r, u_e)$ is the shortest distance from u_r to u_e and U_G is the node set of G.

4   Demonstration

This section will introduce the usage and features of MAVis, and the demonstration is based on a real data set of Sina Weibo (Chinese microblogging website).

---

[1] http://www.prefuse.org/

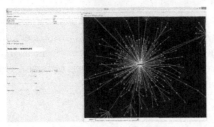 

(a) The reduced IDN

(b) The reduced IDN together with a detailed local network around a clicked node

Fig. 2. Screenshots of MAVis

4.1   Preprocessing and Analyzing

First, we load the historical retweet data stored in text form through the preprocessing module. Then, we construct the IDN and discover the key users after specifying the value of parameter λ. Finally, the mining result is output in XML format.

4.2   Visualization

The visualization module can show the mining result which is in XML format. The source node is shown in blue, while bridge nodes and explosive nodes are green and red respectively. Besides, the higher explosion value one explosive node has, the darker red it is. Due to the large size of the IDN, MAVis uses the incremental loading strategy to show the network. The visualization module shows the reduced network with only explosive nodes and bridge nodes (Fig. 2(a)). When one node in the reduced network is clicked, a new window will open and show the detail of a part of the network, in which this node is the source node (Fig. 2(b)).

There exist several parameters to adjust the display effect, one of which is proportion parameter α. Once α is specified (e.g. 20%), only the explosive nodes which are ranked top α on explosion value will be shown (Fig. 3).

Fig. 3. Display effects with different values of proportion parameter α (20% and 90%)

References

1. Rong, Q.: Trends analysis of news topics on twitter. International Journal of Machine Learning and Computing **2**(3), 327–332 (2012)
2. Takahashi, T.: Yamanishi, K.: Discovering emerging topics in social streams via link anomaly detection. In: ICDM 2011, pp. 1230–1235. IEEE (2011)
3. Gomez-Rodriguez, L.: Schölkopf: Structure and dynamics of information pathways in online media. In: WSDM 2013, pp. 23–32. ACM (2013)
4. Guille, F., Hacid, Z.: Sondy: An open source platform for social dynamics mining and analysis. In: SIGMOD 2013, pp. 1005–1008. ACM (2013)

Author Index

Printed in the United States
By Bookmasters